Analog Circuits and Signal Processing

The *Analog Circuits and Signal Processing* book series, formerly known as the *Kluwer International Series in Engineering and Computer Science*, is a high level academic and professional series publishing research on the design and applications of analog integrated circuits and signal processing circuits and systems. Typically per year we publish between 5-15 research monographs, professional books, handbooks, and edited volumes with worldwide distribution to engineers, researchers, educators, and libraries. The book series promotes and expedites the dissemination of new research results and tutorial views in the analog field. There is an exciting and large volume of research activity in the field worldwide. Researchers are striving to bridge the gap between classical analog work and recent advances in very large scale integration (VLSI) technologies with improved analog capabilities. Analog VLSI has been recognized as a major technology for future information processing. Analog work is showing signs of dramatic changes with emphasis on interdisciplinary research efforts combining device/circuit/technology issues. Consequently, new design concepts, strategies and design tools are being unveiled. Topics of interest include: Analog Interface Circuits and Systems; Data converters; Active-RC, switched-capacitor and continuous-time integrated filters; Mixed analog/digital VLSI; Simulation and modeling, mixed-mode simulation; Analog nonlinear and computational circuits and signal processing; Analog Artificial Neural Networks/Artificial Intelligence; Current-mode Signal Processing; Computer-Aided Design (CAD) tools; Analog Design in emerging technologies (Scalable CMOS, BiCMOS, GaAs, heterojunction and floating gate technologies, etc.); Analog Design for Test; Integrated sensors and actuators; Analog Design Automation/Knowledge-based Systems; Analog VLSI cell libraries; Analog product development; RF Front ends, Wireless communications and Microwave Circuits; Analog behavioral modeling, Analog HDL.

Cosmin Radu Popa

Analog Current-Mode Computational Circuits for Artificial Neural Networks

Springer

Cosmin Radu Popa
National University of Science
and Technology Politehnica Bucharest
Bucharest, Romania

ISSN 1872-082X ISSN 2197-1854 (electronic)
Analog Circuits and Signal Processing
ISBN 978-3-032-03988-0 ISBN 978-3-032-03989-7 (eBook)
https://doi.org/10.1007/978-3-032-03989-7

This Springer imprint is published by the registered company Springer Nature Switzerland AG
The registered company address is: Gewerbestrasse 11, 6330 Cham, Switzerland

This book is dedicated to my beloved wife Natalia.

Preface

Artificial neural networks (ANNs) are important entities because they enable computers to learn from data, recognize complex patterns and make intelligent decisions—often outperforming traditional algorithms in many tasks. They power modern AI, being the core technology behind computer vision (facial recognition, medical imaging), natural language processing (ChatGPT, machine translation), speech recognition (Siri, Alexa) or autonomous systems (self-driving cars, drones). ANNs have the ability to learn from data and, unlike rule-based systems, they are capable to learn patterns and rules automatically from large datasets, making them highly adaptable to real-world complexity. Additionally, artificial neural networks handle nonlinear and complex relationships, comparing with traditional algorithms, that often struggle with nonlinear or high-dimensional data. ANNs excel at capturing subtle, non-obvious relationships and modeling intricate dependencies across variables. Artificial neural networks are scalable with data and compute; as data and computational power grow, ANN performance scales up. Deep learning (deep ANNs) has shown superhuman performance in games, vision and more and, also, they can be trained end-to-end with millions of parameters. The areas of utilization of artificial neural networks are extremely diverse: healthcare (cancer detection from images), finance (fraud detection, algorithmic trading), robotics (sensor data fusion, control systems), entertainment (music and art generation) or industry (predictive maintenance, quality control).

Activation functions in neural networks are mathematical functions applied to a neuron's output, in order to introduce nonlinearity into the model. This nonlinearity allows the network to learn complex patterns and perform tasks like image recognition, language translation and more. Without an activation function, a neural network would behave like a linear regression model, no matter how many layers it has. With activation functions, the network can approximate any complex function (universal approximation theorem).

The nonlinearity of activation functions is extremely important because real-world data is rarely linear. Without activation functions, a neural network (even with many layers) is just a linear transformation, which can't solve complex tasks. Activation functions break this linearity, allowing networks to learn curved, multi-dimensional decision boundaries. Activation functions help networks learn intricate features, such as: edges and textures in images (CNNs), dependencies in sequences (recurrent neural networks, transformers) or latent representations (autoencoders, generative adversarial networks). Because, during training, neural networks use gradient descent, activation functions must be differentiable, so gradients can be calculated. They directly affect the flow of gradients, influencing how well and how fast a network learns.

The method proposed in this book for generating with improved accuracy a multitude of activation functions for neural network applications is based on specific superior-order approximation functions. In order to implement in CMOS technology all the previous approximation functions, the most convenient analog CMOS computational circuits must be designed. Analog current-mode computational circuits for artificial neural networks (ANNs) are hardware circuits that perform neural computations using currents rather than voltages. These circuits are designed to emulate neural operations directly in hardware, offering ultra-fast, low-power and compact implementations compared to digital processors. The advantages of using the current-mode operation of CMOS computational circuits are multiple. First, the speed of circuits can be strongly increased, because no charging/discharging of large capacitive loads (as in voltage-mode), leading to faster operation. Second, the power efficiency is better than the same parameter of voltage-mode computational circuits, as a consequence of low dynamic power consumption, making them ideal for edge computing or IoT applications. Third, current-mode circuits are more compact than voltage-mode circuits, because current mirrors and translinear circuits are often simpler than their voltage-mode counterparts.

Chapter 1 analyzes sigmoidal activation functions from the point of view of their mathematical particularities and proposes a multitude of superior-order approximation functions for generating them. Having a sigmoidal shape (S-shape), sigmoidal activation function has the important advantage of squishing input values into a limited range (usually between 0 and 1 or between –1 and 1). The sigmoidal activation function that are described and generated in this chapter are: unipolar sigmoidal activation function, bipolar sigmoidal activation function, Einstein activation functions, hyperbolic tangent activation function or continuous-log sigmoidal activation functions.

The neural networks based on radial basis functions (RBF) represent a class of neural networks that are especially used for approximating functions, interpolations, regressions and classifications. The RBF neural networks have a multitude of applications: models recognition, medical classifications (very useful for deciding the diagnostic), fuzzy systems or function interpolation. These functions are detailed analyzed in Chap. 2 and superior-order approximation functions are developed, in order to accurately generate them.

In addition to the functions from previous two chapters (sigmoidal activation functions (presented in Chap. 1) and radial basis activation functions (described in Chap. 2), it exists other important mathematical functions, having a multitude of applications in analog signal processing and artificial neural networks. These functions will be detailed and analyzed in Chap. 3 and the most accurate approximation functions will be proposed, in order to generate them. Two important objectives will be considered: as better as possible accuracy of the approximation and the hardware resources required for their implementation in CMOS technology using fundamental CMOS computational circuits (further described in Chap. 4).

In order to implement in CMOS technology all the previous approximation functions, the most convenient analog CMOS computational circuits must be designed. A multitude of very important objectives must be taken into account for a proper choice of these fundamental computational circuits: the accuracy of implementing the required mathematical function, the immunity of the computation on technological errors and, also, on temperature variations, the low-voltage low-power operation of the computational circuits, as well as the hardware resources required for their implementing in CMOS technology. Considering all these very important criteria, the current-mode squaring and the current-mode multiplier/divider circuits are designed and simulated in Chap. 4.

Chapter 5 presents a multitude of analog function synthesizers, developed for implementing in CMOS technology sigmoidal activation functions, analyzed in Chap. 1 (unipolar and bipolar sigmoidal activation functions, Einstein activation functions, hyperbolic tangent activation function and continuous-log sigmoid activation functions). The superior-order approximation functions described in Chap. 1 represent the basis for elaborating the block diagrams from this chapter, while the concrete implementations in CMOS technology based on reasonable hardware resources use the fundamental CMOS computational circuits (current-mode squaring and multiplier/divider circuits), analyzed in Chap. 4.

Chapter 6 analyzes many analog function synthesizers, developed for implementing in CMOS technology the radial basis activation functions, detailed studied in Chap. 2 (cubic activation function, Gaussian activation function, quadratic, inverse quadratic activation functions, generalized multiquadratic and generalized inverse multiquadratic activation functions). The superior-order approximation functions described in Chap. 2 represent the basis for elaborating the block diagrams from this chapter, while the concrete implementations in CMOS technology consuming reasonable hardware resources uses the fundamental CMOS computational circuits (current-mode squaring and multiplier/divider circuits), analyzed in Chap. 4.

Chapter 7 presents analog function synthesizers, designed for implementing in CMOS technology important functions, having a multitude of applications in analog signal processing and artificial neural networks, analyzed in Chap. 3 (exponential function, exponential ramp function, hyperbolic sinusoidal function, hyperbolic cosinusoidal function and, also, any mathematical function with continuous behaviour, useful for previous mentioned applications). The superior-order approximation functions described in Chap. 3 represent the basis for elaborating the block diagrams from this chapter, while the concrete implementation in CMOS technology based on reasonable hardware resources use the fundamental CMOS computational circuits (current-mode squaring and multiplier/divider circuits), analyzed in Chap. 4.

Bucharest, Romania — Cosmin Radu Popa

Acknowledgements This book was funded from the project "National Platform for Semiconductor Technologies," contract no. G 2024-85828/390008/27.11.2024, SMIS code 304244, co-funded by the European Regional Development Fund under the Program for Intelligent Growth, Digitization, and Financial Instruments.

Competing Interests The author has no competing interests to declare that are relevant to the content of this manuscript.

Contents

1 Superior-Order Approximation Functions for Generating Sigmoidal Activation Functions 1
1.1 Introduction 1
1.2 Approximation Functions for Unipolar Sigmoidal Activation Function 1
1.2.1 General Form of the Unipolar Sigmoidal Activation Function 2
1.2.2 First Approximation Function for Generating the Unipolar Sigmoidal Activation Function 2
1.2.3 Second Approximation Function for Generating the Unipolar Sigmoidal Activation Function 4
1.2.4 Third Approximation Function for Generating the Unipolar Sigmoidal Activation Function 8
1.3 Approximation Functions for Bipolar Sigmoidal Activation Function 9
1.3.1 General Form of the Bipolar Sigmoidal Activation Function 11
1.3.2 First Approximation Function for Generating the Bipolar Sigmoidal Activation Function 11
1.3.3 Second Approximation Function for Generating the Bipolar Sigmoidal Activation Function 13
1.4 Approximation Functions for Einstein Activation Functions 14
1.4.1 General Form of the First Einstein Activation Function 17
1.4.2 First Approximation Function for Generating the First Einstein Activation Function 18
1.4.3 Second Approximation Function for Generating the First Einstein Activation Function 18
1.4.4 Third Approximation Function for Generating the First Einstein Activation Function 21

1.4.5 General Form of the Second Einstein Activation Function ... 24
1.4.6 First Approximation Function for Generating the Second Einstein Activation Function ... 24
1.4.7 Second Approximation Function for Generating the Second Einstein Activation Function ... 27
1.5 Approximation Functions for Hyperbolic Tangent (Tan-Sigmoid) Activation Function ... 28
1.5.1 General Form of the Hyperbolic Tangent Activation Function ... 30
1.5.2 First Approximation Function for Generating the Hyperbolic Tangent Activation Function ... 31
1.5.3 Second Approximation Function for Generating the Hyperbolic Tangent Activation Function ... 34
1.5.4 Third Approximation Function for Generating the Hyperbolic Tangent Activation Function ... 35
1.5.5 Fourth Approximation Function for Generating the Hyperbolic Tangent Activation Function ... 38
1.6 Approximation Functions for Continuous-Log Sigmoid Activation Function ... 39
1.6.1 General Form of the Continuous-Log Sigmoid Activation Function ... 41
1.6.2 First Approximation Function for Generating the Continuous-Log Sigmoid Activation Function (for b = 2) ... 42
1.6.3 Second Approximation Function for Generating the Continuous-Log Sigmoid Activation Function (for b = 2) ... 45
1.6.4 First Approximation Function for Generating the Continuous-Log Sigmoid Activation Function (for b = 3) ... 45
1.6.5 Second Approximation Function for Generating the Continuous-Log Sigmoid Activation Function (for b = 3) ... 48
Reference ... 52

2 Superior-Order Approximation Functions for Generating Radial Basis Activation Functions ... 53
2.1 Introduction ... 53
2.2 Approximation Functions for Cubic Activation Function ... 53
2.3 Approximation Functions for Gaussian Activation Function ... 54
2.3.1 General Form of the Gaussian Activation Function ... 56
2.3.2 First Approximation Function for Generating the Gaussian Activation Function ... 60

2.3.3 Second Approximation Function for Generating the Gaussian Activation Function 62
2.3.4 Third Approximation Function for Generating the Gaussian Activation Function 65
2.3.5 Fourth Approximation Function for Generating the Gaussian Activation Function 67
2.3.6 Fiveth Approximation Function for Generating the Gaussian Activation Function 70
2.3.7 Sixth Approximation Function for Generating the Gaussian Activation Function 72
2.3.8 Seventh Approximation Function for Generating the Gaussian Activation Function 74
2.3.9 Eighth Approximation Function for Generating the Gaussian Activation Function 77
2.3.10 Nineth Approximation Function for Generating the Gaussian Activation Function 78
2.3.11 Tenth Approximation Function for Generating the Gaussian Activation Function 82
2.3.12 11th Approximation Function for Generating the Gaussian Activation Function 83
2.3.13 12th Approximation Function for Generating the Gaussian Activation Function 87
2.3.14 13th Approximation Function for Generating the Gaussian Activation Function 89
2.3.15 14th Approximation Function for Generating the Gaussian Activation Function 91
2.3.16 15th Approximation Function for Generating the Gaussian Activation Function 93
2.3.17 16th Approximation Function for Generating the Gaussian Activation Function 95
2.4 Quadratic and Inverse Quadratic Activation Functions 98
2.4.1 General Form of the Quadratic Function 99
2.4.2 General Form of the Inverse Quadratic Function 100
2.5 Approximation Functions for Multiquadratic Activation Functions 101
2.5.1 General Form of the Multiquadratic Function 102
2.5.2 First Approximation Function for Generating the Multiquadratic Function for a = 1 102
2.5.3 Second Approximation Function for Generating the Multiquadratic Function for a = 1 105
2.5.4 Third Approximation Function for Generating the Multiquadratic Function for a = 1 105
2.5.5 Fourth Approximation Function for Generating the Multiquadratic Function for a = 1 109

2.5.6 Fifth Approximation Function for Generating the Multiquadratic Function for a = 1 ... 110
2.5.7 Sixth Approximation Function for Generating the Multiquadratic Function for a = 1 ... 114
2.5.8 First Approximation Function for Generating the Multiquadratic Function for a = 2 ... 114
2.5.9 Second Approximation Function for Generating the Multiquadratic Function for a = 2 ... 118
2.5.10 Third Approximation Function for Generating the Multiquadratic Function for a = 2 ... 119
2.5.11 Fourth Approximation Function for Generating the Multiquadratic Function for a = 2 ... 123
2.5.12 First Approximation Function for Generating the Multiquadratic Function for a = 3 ... 124
2.5.13 Second Approximation Function for Generating the Multiquadratic Function for a = 3 ... 128
2.5.14 Third Approximation Function for Generating the Multiquadratic Function for a = 3 ... 128
2.5.15 Fourth Approximation Function for Generating the Multiquadratic Function for a = 3 ... 132
2.6 Approximation Functions for Inverse Multiquadratic Functions ... 133
2.6.1 General Form of the Inverse Multiquadratic Function ... 135
2.6.2 First Approximation Function for Generating the Inverse Multiquadratic Function for a = 1 ... 135
2.6.3 Second Approximation Function for Generating the Inverse Multiquadratic Function for a = 1 ... 138
2.6.4 First Approximation Function for Generating the Inverse Multiquadratic Function for a = 2 ... 138
2.6.5 Second Approximation Function for Generating the Inverse Multiquadratic Function for a = 2 ... 142
2.6.6 Third Approximation Function for Generating the Inverse Multiquadratic Function for a = 2 ... 143
2.6.7 First Approximation Function for Generating the Inverse Multiquadratic Function for a = 3 ... 147
2.6.8 Second Approximation Function for Generating the Inverse Multiquadratic Function for a = 3 ... 147
References ... 151

3 Superior-Order Approximation Functions for Artificial Neural Networks Applications ... 153
3.1 Introduction ... 153
3.2 Approximation Functions for Exponential Function ... 153
3.2.1 General Form of the Exponential Function ... 154

3.2.2 First Approximation Function for Generating the Exponential Function . . . 155
3.2.3 Second Approximation Function for Generating the Exponential Function . . . 156
3.2.4 Third Approximation Function for Generating the Exponential Function . . . 160
3.2.5 Fourth Approximation Function for Generating the Exponential Function . . . 160
3.2.6 Fifth Approximation Function for Generating the Exponential Function . . . 163
3.2.7 Sixth Approximation Function for Generating the Exponential Function . . . 166
3.2.8 Seventh Approximation Function for Generating the Exponential Function . . . 166
3.2.9 Eighth Approximation Function for Generating the Exponential Function . . . 169
3.2.10 Nineth Approximation Function for Generating the Exponential Function . . . 172
3.2.11 Tenth Approximation Function for Generating the Exponential Function . . . 174
3.2.12 Eleventh Approximation Function for Generating the Exponential Function . . . 175
3.2.13 Twelfth Approximation Function for Generating the Exponential Function . . . 179
3.2.14 Thirteenth Approximation Function for Generating the Exponential Function . . . 179
3.2.15 Fourteenth Approximation Function for Generating the Exponential Function . . . 182
3.3 Approximation Functions for Exponential Ramp Function . . . 185
3.3.1 General Form of the Exponential Ramp Function . . . 187
3.3.2 First Approximation Function for Generating the Exponential Ramp Function (for a = 1) . . . 187
3.3.3 Second Approximation Function for Generating the Exponential Ramp Function (for a = 1) . . . 188
3.3.4 Third Approximation Function for Generating the Exponential Ramp Function (for a = 1) . . . 192
3.3.5 Fourth Approximation Function for Generating the Exponential Ramp Function (for a = 1) . . . 193
3.3.6 First Approximation Function for Generating the Exponential Ramp Function (for a = 2) . . . 197
3.3.7 First Approximation Function for Generating the Exponential Ramp Function (for a = 1/2) . . . 199
3.3.8 Second Approximation Function for Generating the Exponential Ramp Function (for a = 1/2) . . . 200

3.3.9 Third Approximation Function for Generating the Exponential Ramp Function (for a = 1/2) 204
3.4 Approximation Functions for Hyperbolic Sinusoidal Function 204
3.4.1 General Form of the Hyperbolic Sinusoidal Function 207
3.4.2 First Approximation Function for Generating the Hyperbolic Sinusoidal Function 207
3.4.3 Second Approximation Function for Generating the Hyperbolic Sinusoidal Function 210
3.4.4 Third Approximation Function for Generating the Hyperbolic Sinusoidal Function 210
3.4.5 Fourth Approximation Function for Generating the Hyperbolic Sinusoidal Function 214
3.5 Approximation Functions for Hyperbolic Cosinusoidal Function 215
3.5.1 General Form of the Hyperbolic Cosinusoidal Function 217
3.5.2 First Approximation Function for Generating the Hyperbolic Cosinusoidal Function 217
3.5.3 Second Approximation Function for Generating the Hyperbolic Cosinusoidal Function 220
3.6 Approximation Functions for General Analog Function Synthesizers 221
3.6.1 First Approximation Function for Generating Continuous Mathematical Functions 223
3.6.2 Second Approximation Function for Generating Continuous Mathematical Functions 223
3.6.3 Third Approximation Function for Generating Continuous Mathematical Functions 224
References 224

4 Low-Voltage Low-Power Current-Mode CMOS Computational Circuits for Implementing Activation Functions 225
4.1 Introduction 225
4.2 CMOS Increased Accuracy Current-Mode Computational Circuits 225
4.3 Low-Voltage Low-Power Techniques for Current-Mode Computational Circuits 226
4.4 Improved Accuracy CMOS Current-Mode Squaring Circuits 226
4.4.1 First CMOS Implementation of the Current-Mode Squaring Circuit 226
4.4.2 Second CMOS Implementation of the Current-Mode Squaring Circuit 229
4.4.3 Third CMOS Implementation of the Current-Mode Squaring Circuit 231

4.4.4 Fourth CMOS Implementation of the Current-Mode Squaring Circuit ... 233
4.4.5 Graphical Representation of the Current-Mode Squaring Circuits (SQ) ... 236
4.5 Improved Accuracy CMOS Current-Mode Multiplier/Divider Circuits ... 237
4.5.1 First CMOS Implementation of the Current-Mode Multiplier/Divider Circuit ... 237
4.5.2 Second CMOS Implementation of the Current-Mode Multiplier/Divider Circuit ... 239
4.5.3 Third CMOS Implementation of the Current-Mode Multiplier/Divider Circuit ... 242
4.5.4 Fourth CMOS Implementation of the Current-Mode Multiplier/Divider Circuit ... 245
4.5.5 Graphical Representation of the Current-Mode Multiplier/Divider Circuits (MD) ... 248
References ... 249

5 Analysis and Design of Analog Function Synthesizers for Generating Sigmoidal Activation Functions ... 251
5.1 Introduction ... 251
5.2 Analog Function Synthesizers for Unipolar Sigmoidal Activation Function ... 251
5.2.1 First Analog Function Synthesizer for Generating the Unipolar Sigmoidal Activation Function ... 251
5.2.2 Second Analog Function Synthesizer for Generating the Unipolar Sigmoidal Activation Function ... 253
5.2.3 Third Analog Function Synthesizer for Generating the Unipolar Sigmoidal Activation Function ... 255
5.3 Analog Function Synthesizers for Bipolar Sigmoidal Activation Function ... 256
5.3.1 First Analog Function Synthesizer for Generating the Bipolar Sigmoidal Activation Function ... 257
5.3.2 Second Analog Function Synthesizer for Generating the Bipolar Sigmoidal Activation Function ... 258
5.4 Analog Function Synthesizers for Einstein Activation Functions ... 260
5.4.1 First Analog Function Synthesizer for Generating the First Einstein Activation Function ... 260
5.4.2 Second Analog Function Synthesizer for Generating the First Einstein Activation Function ... 261
5.4.3 Third Analog Function Synthesizer for Generating the First Einstein Activation Function ... 263
5.4.4 First Analog Function Synthesizer for Generating the Second Einstein Activation Function ... 264

5.4.5 Second Analog Function Synthesizer for Generating the Second Einstein Activation Function 266
5.5 Analog Function Synthesizers for Hyperbolic Tangent (Tan-Sigmoid) Activation Function 268
5.5.1 First Analog Function Synthesizer for Generating the Hyperbolic Tangent Activation Function 268
5.5.2 Second Analog Function Synthesizer for Generating the Hyperbolic Tangent Activation Function 269
5.5.3 Third Analog Function Synthesizer for Generating the Hyperbolic Tangent Activation Function 271
5.5.4 Fourth Analog Function Synthesizer for Generating the Hyperbolic Tangent Activation Function 272
5.6 Analog Function Synthesizers for Continuous-Log Sigmoid Activation Function 274
5.6.1 First Analog Function Synthesizer for Generating the Continuous-Log Sigmoid Activation Function (for b = 2) 274
5.6.2 Second Analog Function Synthesizer for Generating the Continuous-Log Sigmoid Activation Function (for b = 2) 276
5.6.3 First Analog Function Synthesizer for Generating the Continuous-Log Sigmoid Activation Function (for b = 3) 278
5.6.4 Second Analog Function Synthesizer for Generating the Continuous-Log Sigmoid Activation Function (for b = 3) 279
References 281

6 Analysis and Design of Analog Function Synthesizers for Generating Radial Basis Activation Functions 283
6.1 Introduction 283
6.2 Analog Function Synthesizers for Cubic Activation Function 283
6.3 Analog Function Synthesizers for Gaussian Activation Function 284
6.3.1 First Analog Function Synthesizer for Generating the Gaussian Activation Function 284
6.3.2 Second Analog Function Synthesizer for Generating the Gaussian Activation Function 286
6.3.3 Third Analog Function Synthesizer for Generating the Gaussian Activation Function 287
6.3.4 Fourth Analog Function Synthesizer for Generating the Gaussian Activation Function 289
6.3.5 Fifth Analog Function Synthesizer for Generating the Gaussian Activation Function 290

6.3.6 Sixth Analog Function Synthesizer for Generating the Gaussian Activation Function 292
6.3.7 Seventh Analog Function Synthesizer for Generating the Gaussian Activation Function 294
6.3.8 Eighth Analog Function Synthesizer for Generating the Gaussian Activation Function 295
6.3.9 Ninth Analog Function Synthesizer for Generating the Gaussian Activation Function 297
6.3.10 Tenth Analog Function Synthesizer for Generating the Gaussian Activation Function 299
6.3.11 Eleventh Analog Function Synthesizer for Generating the Gaussian Activation Function 301
6.3.12 12th Analog Function Synthesizer for Generating the Gaussian Activation Function 303
6.3.13 13th Analog Function Synthesizer for Generating the Gaussian Activation Function 305
6.3.14 14th Analog Function Synthesizer for Generating the Gaussian Activation Function 306
6.3.15 15th Analog Function Synthesizer for Generating the Gaussian Activation Function 308
6.3.16 16th Analog Function Synthesizer for Generating the Gaussian Activation Function 310
6.4 Analog Function Synthesizers for Quadratic and Inverse Quadratic Activation Function 312
6.5 Analog Function Synthesizers for Multiquadratic Activation Function 313
6.5.1 First Analog Function Synthesizer for Generating the Multiquadratic Activation Function for a = 1 314
6.5.2 Second Analog Function Synthesizer for Generating the Multiquadratic Activation Function for a = 1 315
6.5.3 Third Analog Function Synthesizer for Generating the Multiquadratic Activation Function for a = 1 317
6.5.4 Fourth Analog Function Synthesizer for Generating the Multiquadratic Activation Function for a = 1 318
6.5.5 Fifth Analog Function Synthesizer for Generating the Multiquadratic Activation Function for a = 1 320
6.5.6 Sixth Analog Function Synthesizer for Generating the Multiquadratic Activation Function for a = 1 322
6.5.7 First Analog Function Synthesizer for Generating the Multiquadratic Activation Function for a = 2 324
6.5.8 Second Analog Function Synthesizer for Generating the Multiquadratic Activation Function for a = 2 325
6.5.9 Third Analog Function Synthesizer for Generating the Multiquadratic Activation Function for a = 2 327

6.5.10 Fourth Analog Function Synthesizer for Generating the Multiquadratic Activation Function for a = 2 329
6.5.11 First Analog Function Synthesizer for Generating the Multiquadratic Activation Function for a = 3 331
6.5.12 Second Analog Function Synthesizer for Generating the Multiquadratic Activation Function for a = 3 332
6.5.13 Third Analog Function Synthesizer for Generating the Multiquadratic Activation Function for a = 3 334
6.5.14 Fourth Analog Function Synthesizer for Generating the Multiquadratic Activation Function for a = 3 335
6.6 Analog Function Synthesizers for Inverse Multiquadratic Activation Function 337
6.6.1 First Analog Function Synthesizer for Generating the Inverse Multiquadratic Activation Function for a = 1 337
6.6.2 Second Analog Function Synthesizer for Generating the Inverse Multiquadratic Activation Function for a = 1 339
6.6.3 First Analog Function Synthesizer for Generating the Inverse Multiquadratic Activation Function for a = 2 341
6.6.4 Second Analog Function Synthesizer for Generating the Inverse Multiquadratic Activation Function for a = 2 343
6.6.5 Third Analog Function Synthesizer for Generating the Inverse Multiquadratic Activation Function for a = 2 344
6.6.6 First Analog Function Synthesizer for Generating the Inverse Multiquadratic Activation Function for a = 3 346
6.6.7 Second Analog Function Synthesizer for Generating the Inverse Multiquadratic Activation Function for a = 3 348
References 350

7 Analysis and Design of Analog Function Synthesizers for Artificial Neural Networks Applications 351
7.1 Introduction 351
7.2 Analog Function Synthesizers for the Exponential Function 351
7.2.1 First Analog Function Synthesizer for Generating the Exponential Function 352
7.2.2 Second Analog Function Synthesizer for Generating the Exponential Function 353
7.2.3 Third Analog Function Synthesizer for Generating the Exponential Function 354

7.2.4 Fourth Analog Function Synthesizer for Generating the Exponential Function ... 355
7.2.5 Fifth Analog Function Synthesizer for Generating the Exponential Function ... 356
7.2.6 Sixth Analog Function Synthesizer for Generating the Exponential Function ... 358
7.2.7 Seventh Analog Function Synthesizer for Generating the Exponential Function ... 359
7.2.8 Eight Analog Function Synthesizer for Generating the Exponential Function ... 361
7.2.9 Ninth Analog Function Synthesizer for Generating the Exponential Function ... 363
7.2.10 Tenth Analog Function Synthesizer for Generating the Exponential Function ... 365
7.2.11 Eleventh Analog Function Synthesizer for Generating the Exponential Function ... 366
7.2.12 Twelfth Analog Function Synthesizer for Generating the Exponential Function ... 368
7.2.13 Thirteenth Analog Function Synthesizer for Generating the Exponential Function ... 370
7.2.14 Fourteenth Analog Function Synthesizer for Generating the Exponential Function ... 371
7.3 Analog Function Synthesizers for the Exponential Ramp Function ... 372
7.3.1 First Analog Function Synthesizer for Generating the Exponential Ramp Function (for a = 1) ... 372
7.3.2 Second Analog Function Synthesizer for Generating the Exponential Ramp Function (for a = 1) ... 374
7.3.3 Third Analog Function Synthesizer for Generating the Exponential Ramp Function (for a = 1) ... 375
7.3.4 Fourth Analog Function Synthesizer for Generating the Exponential Ramp Function (for a = 1) ... 377
7.3.5 First Analog Function Synthesizer for Generating the Exponential Ramp Function (for a = 2) ... 378
7.3.6 First Analog Function Synthesizer for Generating the Exponential Ramp Function (for a = 1/2) ... 380
7.3.7 Second Analog Function Synthesizer for Generating the Exponential Ramp Function (for a = 1/2) ... 381
7.3.8 Third Analog Function Synthesizer for Generating the Exponential Ramp Function (for a = 1/2) ... 383
7.4 Analog Function Synthesizers for the Hyperbolic Sinusoidal Function ... 385
7.4.1 First Analog Function Synthesizer for Generating the Hyperbolic Sinusoidal Function ... 385

7.4.2 Second Analog Function Synthesizer for Generating the Hyperbolic Sinusoidal Function 386
7.4.3 Third Analog Function Synthesizer for Generating the Hyperbolic Sinusoidal Function 388
7.4.4 Fourth Analog Function Synthesizer for Generating the Hyperbolic Sinusoidal Function 389
7.5 Analog Function Synthesizers for the Hyperbolic Cosinusoidal Function 390
7.5.1 First Analog Function Synthesizer for Generating the Hyperbolic Cosinusoidal Function 391
7.5.2 Second Analog Function Synthesizer for Generating the Hyperbolic Cosinusoidal Function 392
7.6 Analog Function Synthesizers for Continuous Mathematical Functions 394
7.6.1 First Analog Function Synthesizer for Generating the Continuous Mathematical Functions 394
7.6.2 Second Analog Function Synthesizer for Generating the Continuous Mathematical Functions 395
7.6.3 Third Analog Function Synthesizer for Generating the Continuous Mathematical Functions 396
References 397

Abbreviations

ANN	Artificial Neural Network
CT	Computed Tomography
ELU	Exponential Linear Unit
FNN	Fuzzy Neural Network
GIS	Geographic Information System
GMM	Gaussian Mixture Model
MD	Multiplier/Divider
MLP	Multilayer Perceptron
MRI	Magnetic Resonance Imaging
ODE	Ordinary Differential Equation
PDE	Partial Differential Equation
PINN	Physics-Informed Neural Network
QP	Quadratic Programming
RBF	Radial Basis Function
RBF-FD	Radial Basis Function—Finite Differences
ReLU	Rectified Linear Unit
RNN	Recurrent Neural Networks
SELU	Scaled ELU
SGD	Stochastic Gradient Descent
SQ	Squaring Circuit
SVM	Support Vector Machine
VaR	Value at Risk
XAI	Explainable Artificial Intelligence

Chapter 1
Superior-Order Approximation Functions for Generating Sigmoidal Activation Functions

1.1 Introduction

Sigmoidal activation functions are frequently used in neural networks for introducing nonlinearity (Popa 2023). Having a sigmoidal shape (S-shape), they have the important advantage of squishing input values into a limited range (usually between 0 and 1 or between − 1 and 1).

It exists a multitude of sigmoidal activation functions, such as unipolar sigmoidal activation function, bipolar sigmoidal activation function, Einstein activation functions, hyperbolic tangent activation function or continuous-log sigmoidal activation functions. They are detailed and analyzed in this chapter and many possibilities of generating them are presented, based on superior-order approximation functions (Popa 2023).

1.2 Approximation Functions for Unipolar Sigmoidal Activation Function

The unipolar sigmoidal activation function is a type of sigmoid function commonly used in neural networks. This function can map any real-valued input into a domain comprise between 0 and 1, hence the term *unipolar* (as opposed to *bipolar*, which ranges from − 1 to 1). Its areas of utilization include:

a. ***Binary classification***

- output range between 0 and 1 makes it ideal for representing probabilities.
- it is used in the output layer of classifiers where the goal is to predict a binary outcome (for example, spam vs. not spam).
- common in models like logistic regression, binary perceptrons and feedforward neural networks.

C. R. Popa, *Analog Current-Mode Computational Circuits for Artificial Neural Networks*, Analog Circuits and Signal Processing,
https://doi.org/10.1007/978-3-032-03989-7_1

b. ***Neural networks***

- frequently used in multilayer perceptrons (MLPs) during the early stages of deep learning development.
- particularly in hidden layers of shallow networks before the rise of ReLU and its variants.

c. ***Backpropagation and gradient-based optimization***

- the function is smooth and differentiable, essential for backpropagation.
- its simple derivative made it computationally convenient in early neural network training.

d. ***Probabilistic models***

- because its output can be interpreted as a probability, it is used in models where probabilistic. interpretation of outputs is needed.

e. ***Control systems and cognitive modeling***

- used in fuzzy logic, cognitive systems and biological modeling due to its soft transition from 0 to 1, mimicking gradual activation.

f. ***Embedded systems and hardware neural networks***

- when low-complexity nonlinear functions are needed for approximation in hardware implementations, the unipolar sigmoidal activation function is sometimes used because of its mathematical simplicity.

1.2.1 General Form of the Unipolar Sigmoidal Activation Function

The mathematical form of this function can be expressed as follows:

$$f_1(x) = \frac{1}{1 + \exp(-x)}. \tag{1.1}$$

The graphical representation of the unipolar sigmoidal function $f_1(x)$ is presented in Fig. 1.1.

1.2.2 First Approximation Function for Generating the Unipolar Sigmoidal Activation Function

In order to approximate the unipolar sigmoidal activation function, three new approximation functions $a_1(x)$, $a_2(x)$ and $a_3(x)$ are presented, able to accurately approximate

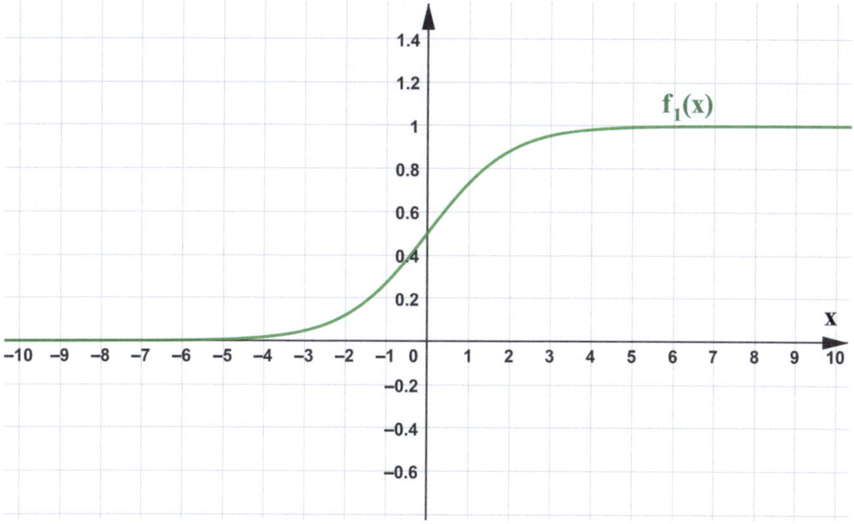

Fig. 1.1 Graphical representation of the unipolar sigmoidal function

the previous activation function. Their particular mathematical forms are developed having two very important objectives. The first objective is represented by the accuracy of the approximation that could be obtained using these approximation functions. The second objective is related to the possibility of their facile analog implementation in CMOS technology using as reasonable as possible hardware resources (see Chaps. 4 and 5).

The general form of the first proposed approximation function $a_1(x)$ is the following:

$$a_{1GEN}(x) = \frac{c_2x^2 + c_1x + c_0}{d_2x^2 + d_0}, \tag{1.2}$$

c_0, c_1, c_2, d_0 and d_2 being constant coefficients having particular values that can be determined by the condition that the superior-order Taylor series expansion of $f_1(x)$ unipolar sigmoidal function, expressed by (1.1), to be identical with the superior-order Taylor series expansion of the general form of the approximation function $a_{1GEN}(x)$, expressed by (1.2). The values of these coefficients, obtained considering the previous conditions, can be written as: $c_0 = \frac{1}{2}$, $c_1 = \frac{1}{4}$, $c_2 = \frac{1}{24}$, $d_0 = 1$ and $d_2 = \frac{1}{12}$. Replacing these particular values in relation (1.2), the first approximation function $a_1(x)$ has the following mathematical expression:

$$a_1(x) = \frac{\frac{x^2}{24} + \frac{x}{4} + \frac{1}{2}}{\frac{x^2}{12} + 1}. \tag{1.3}$$

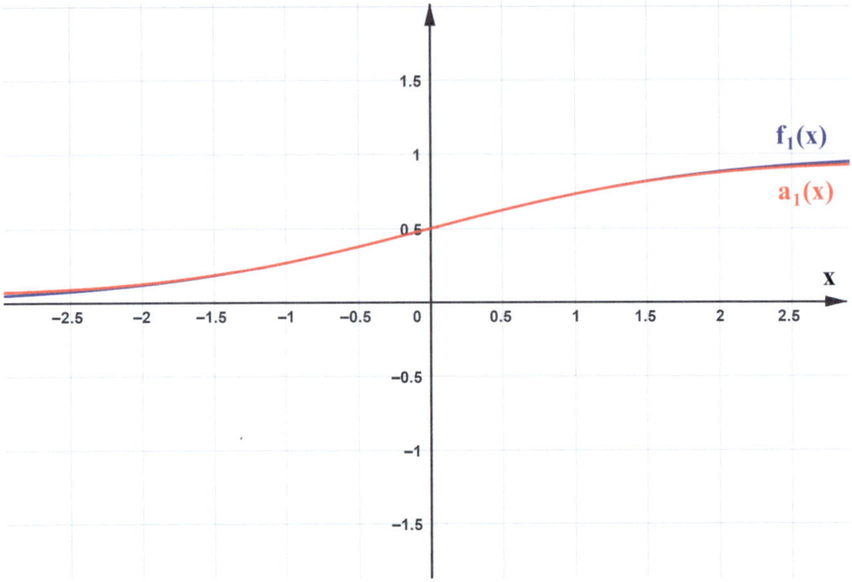

Fig. 1.2 Unipolar sigmoidal function $f_1(x)$ and the first approximation function $a_1(x)$

The graphical representation of the unipolar sigmoidal function $f_1(x)$ and of the first approximation function $a_1(x)$ is presented in Fig. 1.2.

The absolute error of approximation of the unipolar sigmoidal activation function $f_1(x)$ using the approximation function $a_1(x)$ is illustrated in Fig. 1.3. It is defined as the difference between the previous functions, error1 $= f_1(x) - a_1(x)$.

For more accurately evaluating the error of approximation, the relative error of approximation is shown in Fig. 1.4. This error is defined as error2 $= 100\frac{f_1(x)-a_1(x)}{f_1(x)}$ (in percents). Considering as acceptable a maximal approximation error of $\pm 1\%$, the maximal range of the input variable x is $[-1.5; 2.25]$.

1.2.3 *Second Approximation Function for Generating the Unipolar Sigmoidal Activation Function*

The general form of the second proposed approximation function $a_2(x)$ is the following:

$$a_{2\mathrm{GEN}}(x) = \frac{c_3x^3 + c_2x^2 + c_1x + c_0}{d_2x^2 + d_0}, \quad (1.4)$$

c_0, c_1, c_2, c_3, d_0 and d_2 being constant coefficients having particular values that can be determined by the condition that the superior-order Taylor series expansion of $f_1(x)$

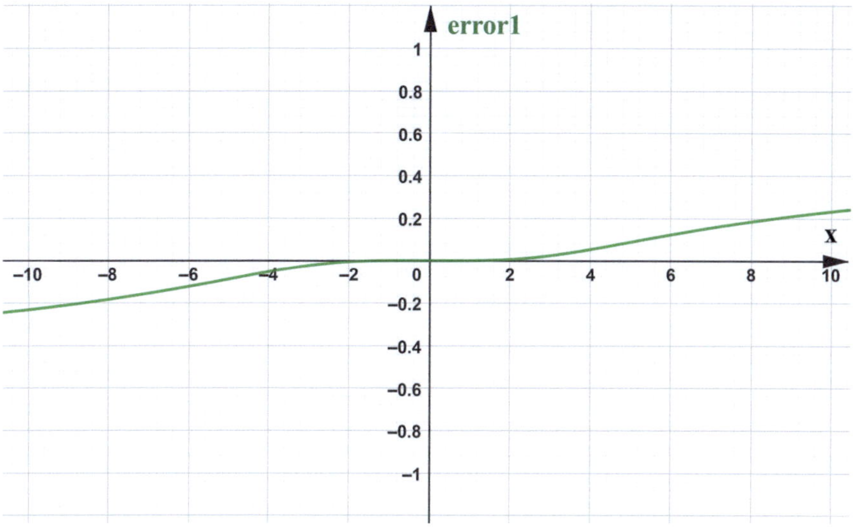

Fig. 1.3 Absolute error of approximation of the unipolar sigmoidal function using approximation function $a_1(x)$

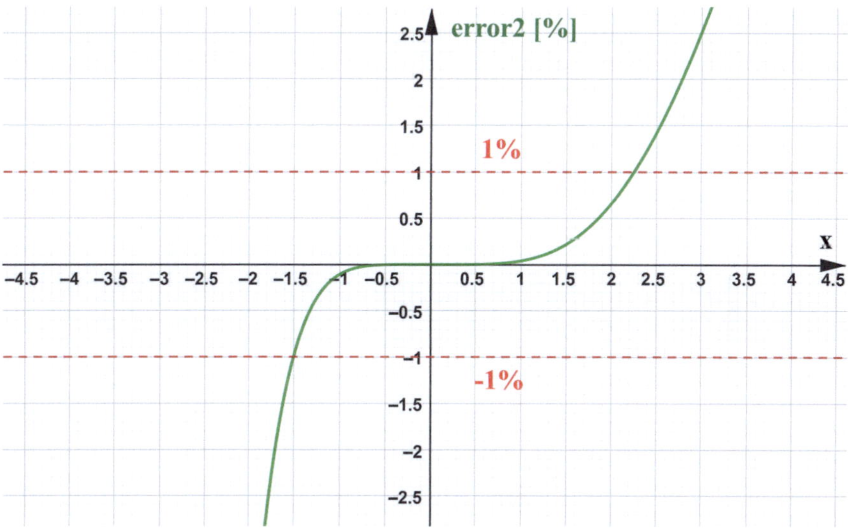

Fig. 1.4 Relative error of approximation of the unipolar sigmoidal activation function $f_1(x)$ using approximation function $a_1(x)$

unipolar sigmoidal function, expressed by (1.1), to be identical with the superior-order Taylor series expansion of the general form of the approximation function $a_{2GEN}(x)$, expressed by (1.4). The values of these coefficients, obtained considering the previous conditions, can be written as: $c_0 = \frac{1}{2}$, $c_1 = \frac{1}{4}$, $c_2 = \frac{1}{20}$, $c_3 = \frac{1}{240}$, $d_0 = 1$ and $d_2 = \frac{1}{10}$. Replacing these particular values in relation (1.4), the second approximation function $a_2(x)$ has the following mathematical expression:

$$a_2(x) = \frac{\frac{x^3}{240} + \frac{x^2}{20} + \frac{x}{4} + \frac{1}{2}}{\frac{x^2}{10} + 1}. \tag{1.5}$$

The graphical representation of the unipolar sigmoidal function $f_1(x)$ and of the second approximation function $a_2(x)$ is presented in Fig. 1.5.

The absolute error of approximation of the unipolar sigmoidal activation function $f_1(x)$ using the approximation function $a_2(x)$ is illustrated in Fig. 1.6. It is defined as the difference between the previous functions, $\text{error1} = f_1(x) - a_2(x)$.

For more accurately evaluating the error of approximation, the relative error of approximation is shown in Fig. 1.7. This error is defined as $\text{error2} = 100\frac{f_1(x)-a_2(x)}{f_1(x)}$ (in percents). Considering as acceptable a maximal approximation error of $\pm 1\%$, the maximal range of the input variable x is $[-2.6\,;\ 4.7]$.

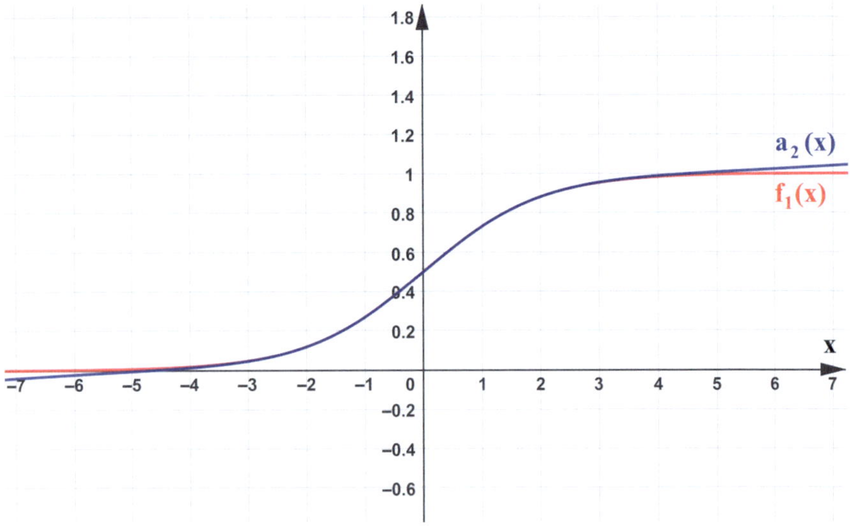

Fig. 1.5 Unipolar sigmoidal function $f_1(x)$ and the second approximation function $a_2(x)$

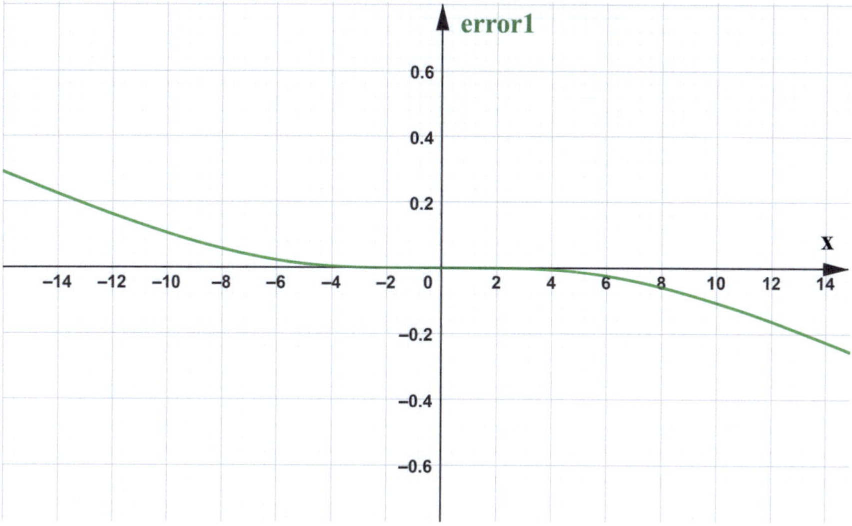

Fig. 1.6 Absolute error of approximation of the unipolar sigmoidal function using approximation function $a_2(x)$

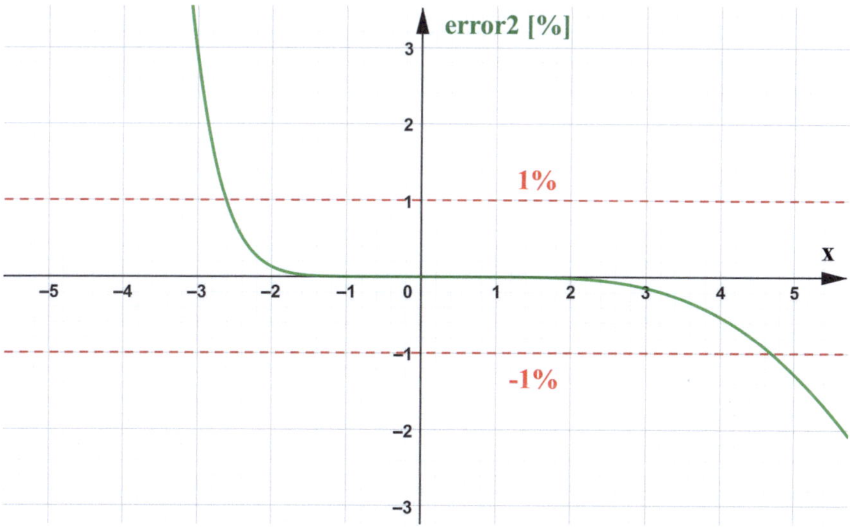

Fig. 1.7 Relative error of approximation of the unipolar sigmoidal activation function $f_1(x)$ using approximation function $a_2(x)$

1.2.4 Third Approximation Function for Generating the Unipolar Sigmoidal Activation Function

The general form of the third proposed approximation function $a_3(x)$ is the following:

$$a_{3\mathrm{GEN}}(x) = \frac{c_4x^4 + c_3x^3 + c_2x^2 + c_1x + c_0}{d_4x^4 + d_2x^2 + d_0}, \tag{1.6}$$

$c_0, c_1, c_2, c_3, c_4, d_0, d_2$ and d_4 being constant coefficients having particular values that can be determined by the condition that the superior-order Taylor series expansion of $f_1(x)$ unipolar sigmoidal function, expressed by (1.1), to be identical with the superior-order Taylor series expansion of the general form of the approximation function $a_{3\mathrm{GEN}}(x)$, expressed by (1.6). The values of these coefficients, obtained considering the previous conditions, can be written as: $c_0 = \frac{1}{2}$, $c_1 = \frac{1}{4}$, $c_2 = \frac{3}{56}$, $c_3 = \frac{1}{168}$, $c_4 = \frac{1}{3360}$, $d_0 = 1$, $d_2 = \frac{3}{28}$ and $d_4 = \frac{1}{1680}$. Replacing these particular values in relation (1.6), the third approximation function $a_3(x)$ has the following mathematical expression:

$$a_3(x) = \frac{\frac{x^4}{3360} + \frac{x^3}{168} + \frac{3x^2}{56} + \frac{x}{4} + \frac{1}{2}}{\frac{x^4}{1680} + \frac{3x^2}{28} + 1}. \tag{1.7}$$

The graphical representation of the unipolar sigmoidal function $f_1(x)$ and of the third approximation function $a_3(x)$ is presented in Fig. 1.8.

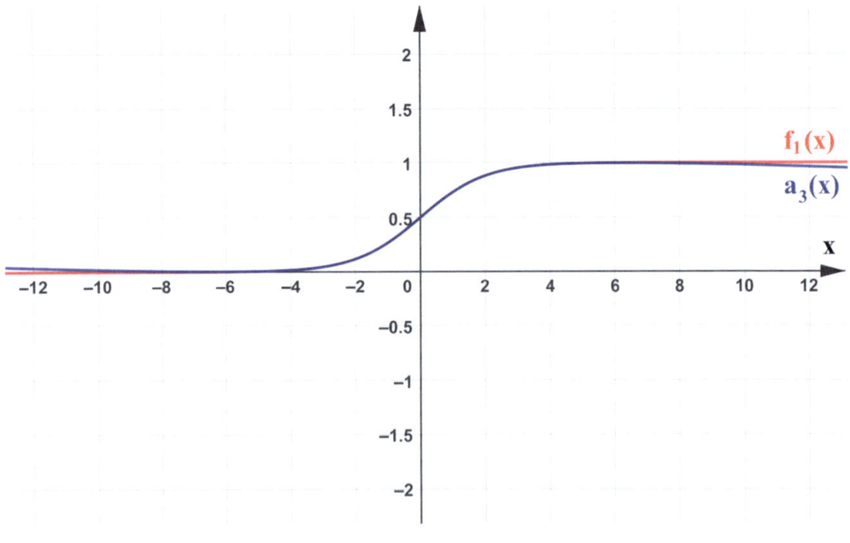

Fig. 1.8 Unipolar sigmoidal function $f_1(x)$ and the third approximation function $a_3(x)$

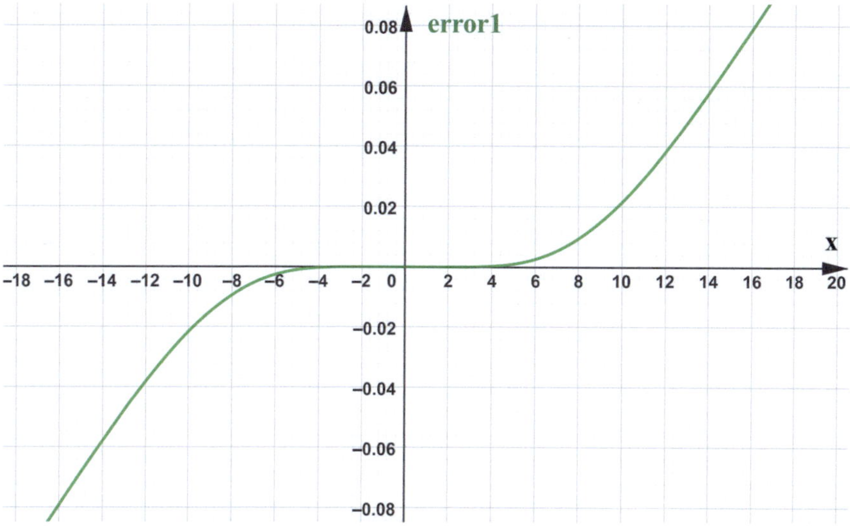

Fig. 1.9 Absolute error of approximation of the unipolar sigmoidal function using approximation function $a_3(x)$

The absolute error of approximation of the unipolar sigmoidal activation function $f_1(x)$ using the approximation function $a_3(x)$ is illustrated in Fig. 1.9. It is defined as the difference between the previous functions, error1 $= f_1(x) - a_3(x)$.

For more accurately evaluating the error of approximation, the relative error of approximation is shown in Fig. 1.10. This error is defined as error2 $= 100\frac{f_1(x)-a_3(x)}{f_1(x)}$ (in percents). Considering as acceptable a maximal approximation error of $\pm 1\%$, the maximal range of the input variable x is $[-3.8\,;\ 8.1]$.

1.3 Approximation Functions for Bipolar Sigmoidal Activation Function

The bipolar sigmoidal activation function is a sigmoid-shaped function whose output range is between − 1 and 1, unlike the unipolar version, whose ranges from 0 to 1. Its areas of utilization include:

a. ***Artificial neural networks***

- used extensively in early multilayer perceptrons (MLPs).
- especially in hidden layers, as the zero-centered output improves convergence compared to the unipolar sigmoid.

b. ***Backpropagation training***

- because it is differentiable and smooth, it supports gradient-based training.

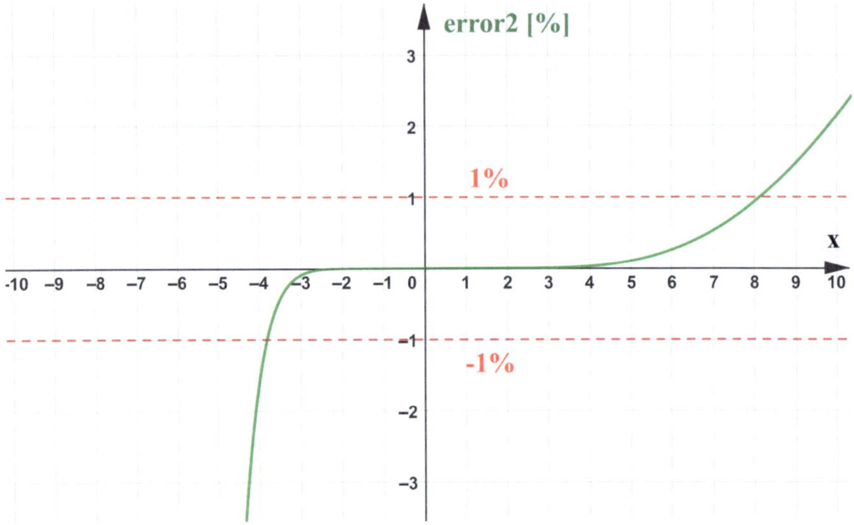

Fig. 1.10 Relative error of approximation of the unipolar sigmoidal activation function $f_1(x)$ using approximation function $a_3(x)$

- being zero-centered, it helps avoid bias shifts and leads to faster convergence during optimization compared to unipolar sigmoidal activation function.

c. ***Control systems and signal processing***

- useful in systems where positive and negative feedback must be represented.
- common in adaptive control, robotics and sensor fusion systems, where a response in both directions is needed.

d. ***Cognitive and biological modeling***

- models neuronal activation more realistically in systems where inhibition (negative response) is as important as excitation (positive response).
- used in neuro-fuzzy systems and psychological behavior modeling.

e. ***Hardware implementations and analog circuits***

- simple to approximate using electronic components.
- used in analog VLSI designs and early neuromorphic computing efforts.

The advantages comparing with the unipolar sigmoid activation functions are: the range of variation between − 1 and 1 (equivalent with a zero-centered behavior), the gradient symmetry that makes it better for training and the biological realism, that is higher than the realism of the unipolar sigmoid activation function.

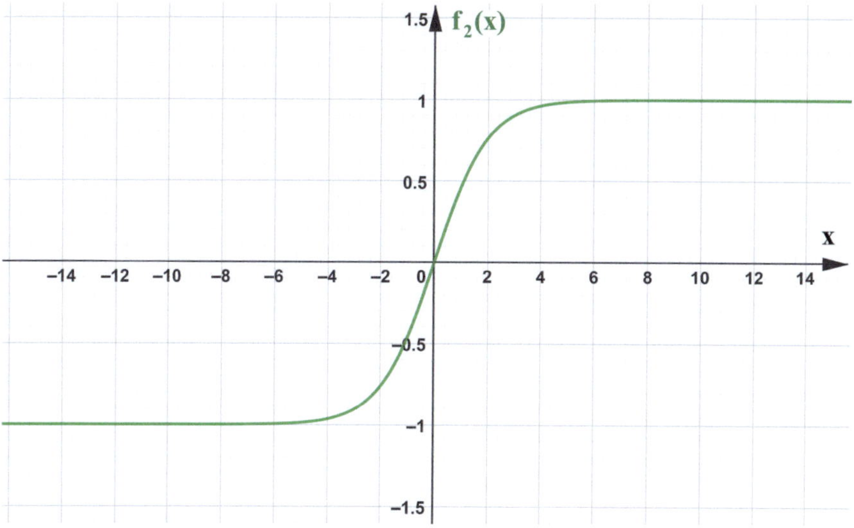

Fig. 1.11 Graphical representation of the bipolar sigmoidal function

1.3.1 General Form of the Bipolar Sigmoidal Activation Function

The mathematical form of this function can be expressed as follows:

$$f_2(x) = \frac{1 - \exp(-x)}{1 + \exp(-x)}. \tag{1.8}$$

The graphical representation of the bipolar sigmoidal function $f_2(x)$ is presented in Fig. 1.11.

1.3.2 First Approximation Function for Generating the Bipolar Sigmoidal Activation Function

In order to approximate the bipolar sigmoidal activation function, two new approximation functions $a_4(x)$ and $a_5(x)$ are presented, able to accurately approximate the previous activation function. Their particular mathematical forms are developed for increasing the accuracy of the approximation and, also, for requiring reasonable hardware resources for their implementations in CMOS technology (see Chaps. 4 and 5).

The general form of the first proposed approximation function $a_4(x)$ is the following:

$$a_{4\mathrm{GEN}}(x) = \frac{c_3 x^3 + c_1 x}{d_2 x^2 + d_0}, \tag{1.9}$$

c_1, c_3, d_0 and d_2 being constant coefficients having particular values that can be determined by the condition that the superior-order Taylor series expansion of $f_2(x)$ bipolar sigmoidal function, expressed by (1.8), to be identical with the superior-order Taylor series expansion of the general form of the approximation function $a_{4\mathrm{GEN}}(x)$, expressed by (1.9). The values of these coefficients, obtained considering the previous conditions, can be written as: $c_1 = \frac{1}{2}$, $c_3 = \frac{1}{120}$, $d_0 = 1$ and $d_2 = \frac{1}{10}$. Replacing these particular values in relation (1.9), the first approximation function $a_4(x)$ has the following mathematical expression:

$$a_4(x) = \frac{\frac{x^3}{120} + \frac{x}{2}}{\frac{x^2}{10} + 1}. \tag{1.10}$$

The graphical representation of the bipolar sigmoidal function $f_2(x)$ and of the first approximation function $a_4(x)$ is presented in Fig. 1.12.

The absolute error of approximation of the bipolar sigmoidal activation function $f_2(x)$ using the approximation function $a_4(x)$ is illustrated in Fig. 1.13. It is defined as the difference between the previous functions, error1 $= f_2(x) - a_4(x)$.

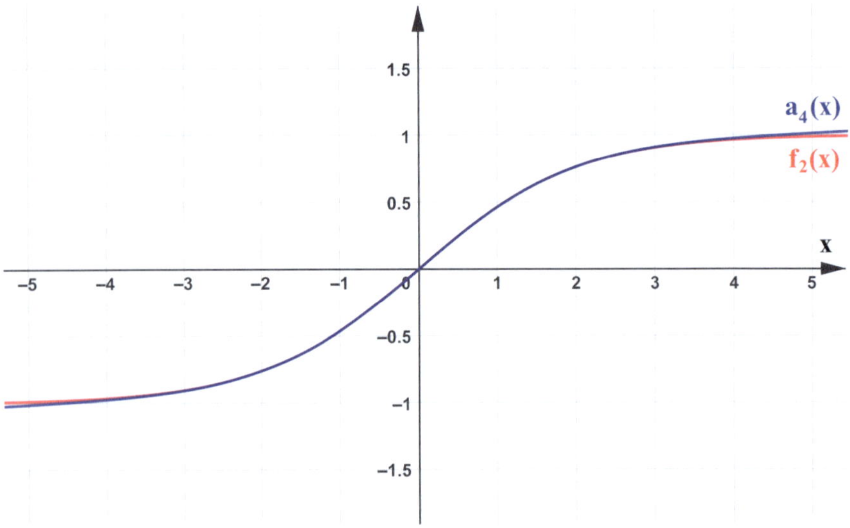

Fig. 1.12 Bipolar sigmoidal function $f_2(x)$ and the first approximation function $a_4(x)$

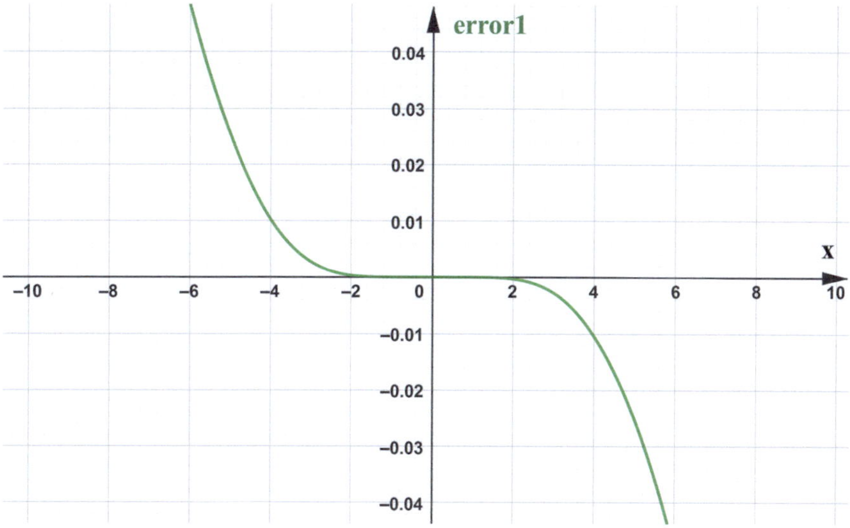

Fig. 1.13 Absolute error of approximation of the bipolar sigmoidal function using first approximation function $a_4(x)$

For more accurately evaluating the error of approximation, the relative error of approximation is shown in Fig. 1.14. This error is defined as error2 $= 100\frac{f_2(x)-a_4(x)}{f_2(x)}$ (in percents). Considering as acceptable a maximal approximation error of $\pm 1\%$, the maximal range of the input variable x is $[-3.9\,;\,3.9]$.

1.3.3 Second Approximation Function for Generating the Bipolar Sigmoidal Activation Function

The general form of the second proposed approximation function $a_5(x)$ is the following:

$$a_{5\text{GEN}}(x) = \frac{c_3x^3 + c_1x}{d_4x^4 + d_2x^2 + d_0}, \tag{1.11}$$

c_1, c_3, d_0, d_2 and d_4 being constant coefficients having particular values that can be determined by the condition that the superior-order Taylor series expansion of $f_2(x)$ bipolar sigmoidal function, expressed by (1.8), to be identical with the superior-order Taylor series expansion of the general form of the approximation function $a_{5\text{GEN}}(x)$, expressed by (1.11). The values of these coefficients, obtained considering the previous conditions, can be written as: $c_1 = \frac{1}{2}, c_3 = \frac{1}{84}, d_0 = 1, d_2 = \frac{3}{28}$ and $d_4 = \frac{1}{1680}$. Replacing these particular values in relation (1.11), the second approximation function $a_5(x)$ has the following mathematical expression:

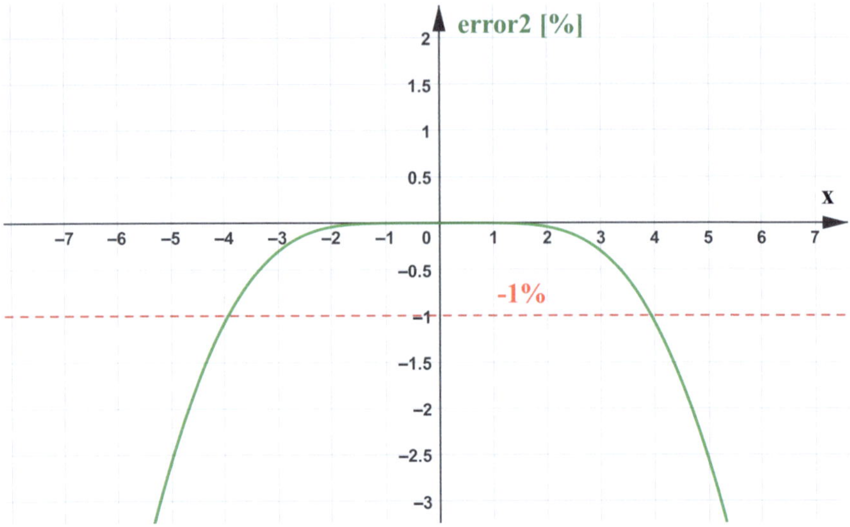

Fig. 1.14 Relative error of approximation of the bipolar sigmoidal activation function $f_2(x)$ using approximation function $a_4(x)$

$$a_5(x) = \frac{\frac{x^3}{84} + \frac{x}{2}}{\frac{x^4}{1680} + \frac{3x^2}{28} + 1}. \tag{1.12}$$

The graphical representation of the bipolar sigmoidal function $f_2(x)$ and of the second approximation function $a_5(x)$ is presented in Fig. 1.15.

The absolute error of approximation of the bipolar sigmoidal activation function $f_2(x)$ using the approximation function $a_5(x)$ is illustrated in Fig. 1.16. It is defined as the difference between the previous functions, error1 $= f_2(x) - a_5(x)$.

For more accurately evaluating the error of approximation, the relative error of approximation is shown in Fig. 1.17. This error is defined as error2 $= 100\frac{f_2(x)-a_5(x)}{f_2(x)}$ (in percents). Considering as acceptable a maximal approximation error of $\pm 1\%$, the maximal range of the input variable x is $[-6.9; 6.9]$.

1.4 Approximation Functions for Einstein Activation Functions

The Einstein activation functions are a family of nonlinear functions inspired by Einstein's addition formulas from relativistic physics, adapted for use in neural networks and fuzzy logic systems. They are part of the Einstein algebra, primarily

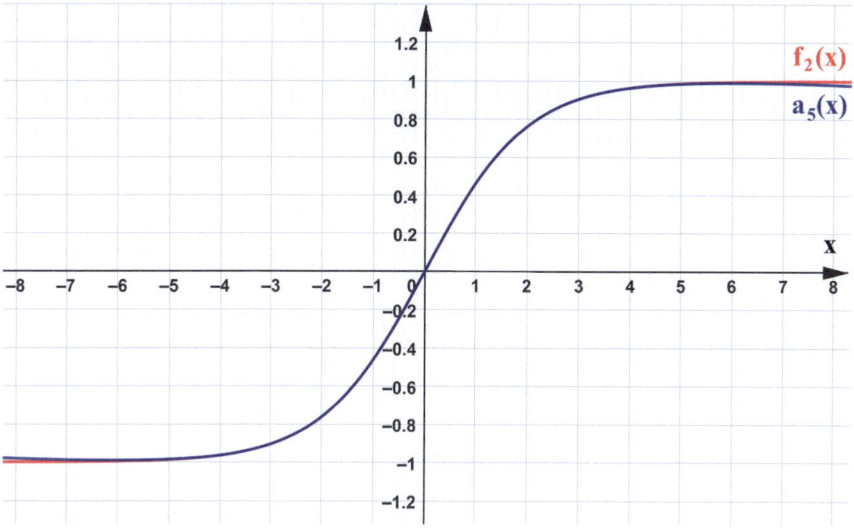

Fig. 1.15 Bipolar sigmoidal function $f_2(x)$ and the second approximation function $a_5(x)$

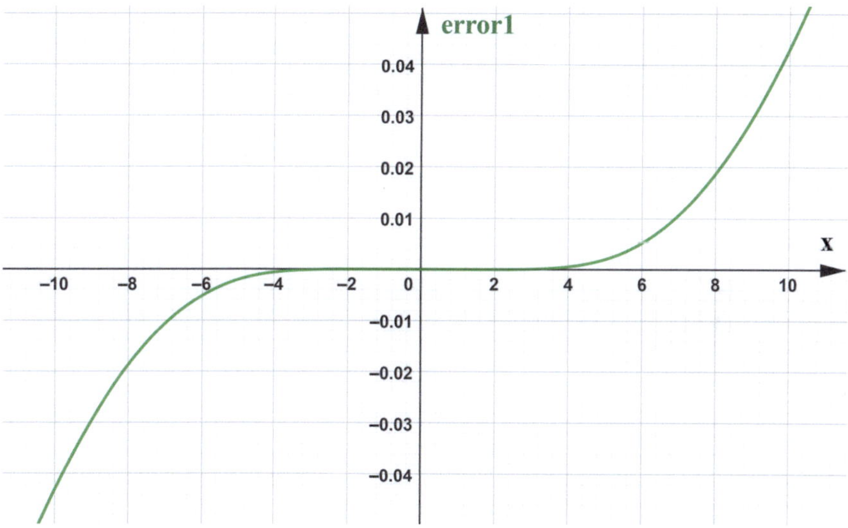

Fig. 1.16 Absolute error of approximation of the bipolar sigmoidal function using second approximation function $a_5(x)$

used in generalized fuzzy logic operations (like Einstein sum and Einstein product). These forms can be used to define activation functions that mimic physical-like nonlinearities or model fuzzy interactions more precisely than classical sigmoids or ReLU. The areas of utilization of Einstein activation functions are the following:

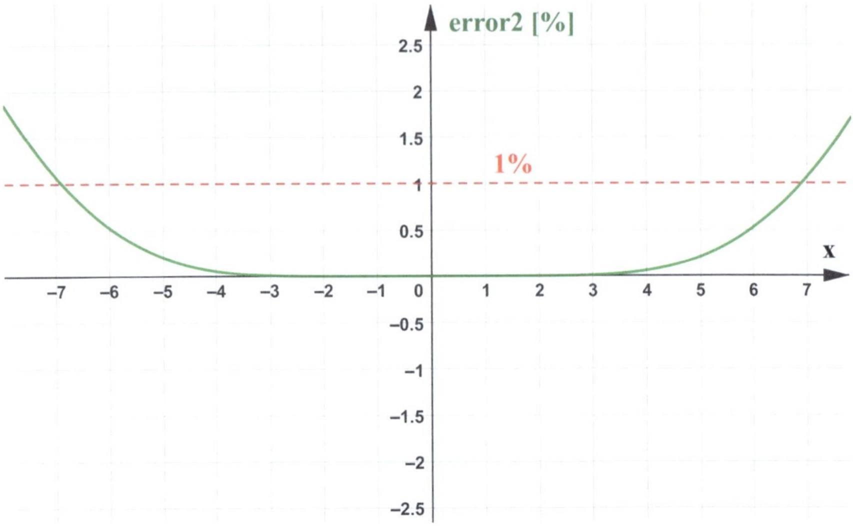

Fig. 1.17 Relative error of approximation of the bipolar sigmoidal activation function $f_2(x)$ using approximation function $a_5(x)$

a. ***Fuzzy neural networks (FNNs)***
   - Einstein activation functions are used in neuro-fuzzy systems where operations like fuzzy conjunction (AND) and disjunction (OR) are modeled using Einstein algebra.
   - particularly useful in approximate reasoning and rule-based learning systems.

b. ***Explainable AI (XAI)***
   - because Einstein operators have interpretable mathematical properties, they are suitable for models requiring interpretability and logical consistency.

c. ***Computational intelligence/soft computing***
   - applied in systems where nonlinear interaction of features (not captured well by ReLU or hyperbolic tangent functions) is crucial.
   - often used alongside other t-norms and s-norms in soft logic frameworks.

d. ***Control systems and decision-making systems***
   - useful in fuzzy control and multi-criteria decision-making, especially where inputs must be combined using more nuanced nonlinearity than traditional activation functions.

e. ***Relational and graph-based models***

- Einstein functions can help model relativistic combinations of features, which is conceptually appealing in graph neural networks or relation-aware models.

The advantages of using Einstein activation functions are: their quality for gradient-based training (because they are nonlinear, but smooth), their fuzzy logic compatibility (because they are comparable with t-norms and s-norms), the capability of relational modeling, making them useful in tasks with complex input interactions.

1.4.1 General Form of the First Einstein Activation Function

The mathematical form of the first Einstein function can be expressed as follows:

$$f_3(x) = \frac{x^2 \exp(x)}{\left[\exp(x) - 1\right]^2}. \tag{1.13}$$

The graphical representation of the first Einstein function $f_3(x)$ is presented in Fig. 1.18.

In order to approximate the first Einstein function, three new approximation functions $a_6(x)$, $a_7(x)$ and $a_8(x)$ are presented, able to accurately approximate $f_3(x)$ function.

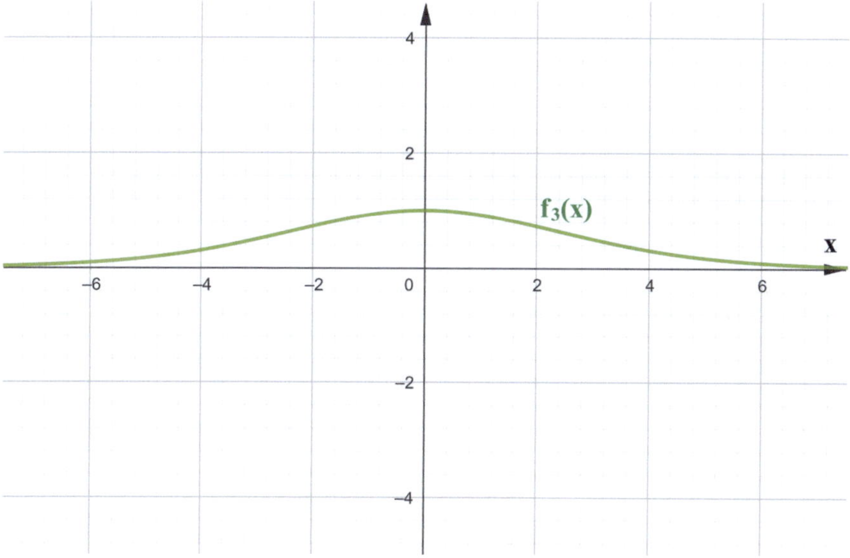

Fig. 1.18 Graphical representation of the first Einstein function

1.4.2 First Approximation Function for Generating the First Einstein Activation Function

The general form of the first proposed approximation function $a_6(x)$ is the following:

$$a_{6\mathrm{GEN}}(x) = \frac{c_2x^2 + c_0}{d_2x^2 + d_0} \tag{1.14}$$

c_0, c_2, d_0 and d_2 being constant coefficients having particular values that can be determined by the condition that the superior-order Taylor series expansion of $f_3(x)$ Einstein function, expressed by (1.13), to be identical with the superior-order Taylor series expansion of the general form of the approximation function $a_{6\mathrm{GEN}}(x)$, expressed by (1.14). The values of these coefficients, obtained considering the previous conditions, can be written as: $c_0 = 1$, $c_2 = -\frac{1}{30}$, $d_0 = 1$ and $d_2 = \frac{1}{20}$. Replacing these particular values in relation (1.14), the first approximation function $a_6(x)$ has the following mathematical expression:

$$a_6(x) = \frac{1 - \frac{x^2}{30}}{\frac{x^2}{20} + 1}. \tag{1.15}$$

The graphical representation of the first Einstein function $f_3(x)$ and of the first approximation function $a_6(x)$ is presented in Fig. 1.19.

The absolute error of approximation of the first Einstein activation function $f_3(x)$ using the approximation function $a_6(x)$ is illustrated in Fig. 1.20. It is defined as the difference between the previous functions, error1 $= f_3(x) - a_6(x)$.

For more accurately evaluating the error of approximation, the relative error of approximation is shown in Fig. 1.21. This error is defined as error2 $= 100\frac{f_3(x)-a_6(x)}{f_3(x)}$ (in percents). Considering as acceptable a maximal approximation error of $\pm 1\%$, the maximal range of the input variable x is $[-2\,;\ \infty]$.

1.4.3 Second Approximation Function for Generating the First Einstein Activation Function

The general form of the second proposed approximation function $a_7(x)$ is the following:

$$a_{7\mathrm{GEN}}(x) = \frac{c_4x^4 + c_2x^2 + c_0}{d_4x^4 + d_2x^2 + d_0}, \tag{1.16}$$

c_0, c_2, c_4, d_0, d_2 and d_4 being constant coefficients having particular values that can be determined by the condition that the superior-order Taylor series expansion

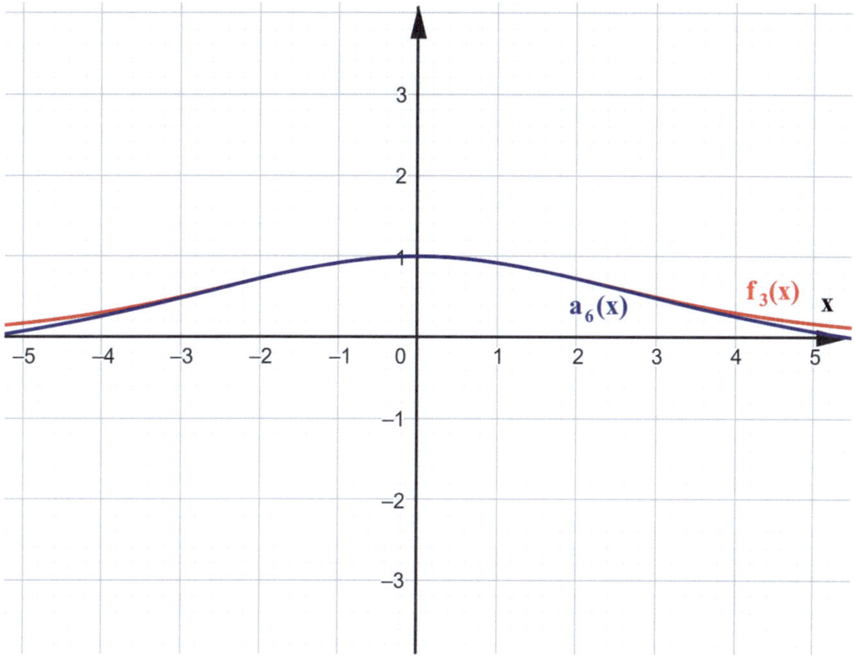

Fig. 1.19 First Einstein function $f_3(x)$ and the first approximation function $a_6(x)$

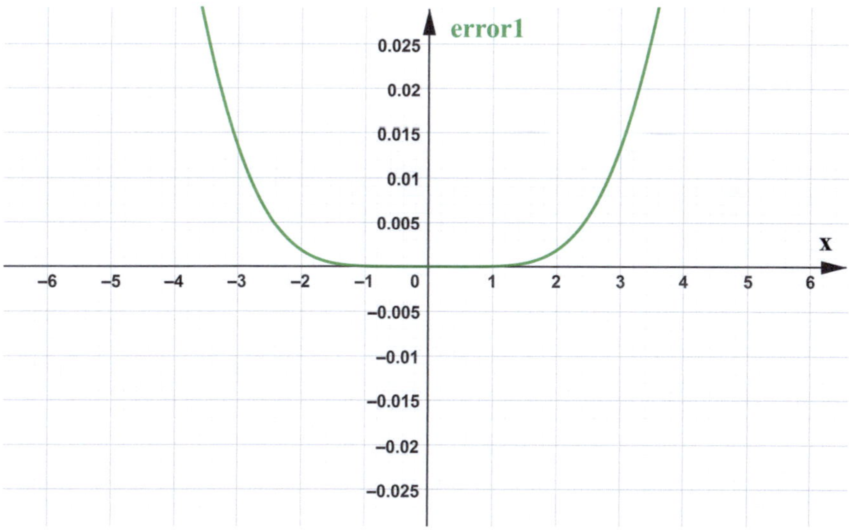

Fig. 1.20 Absolute error of approximation of the first Einstein function using first approximation function $a_6(x)$

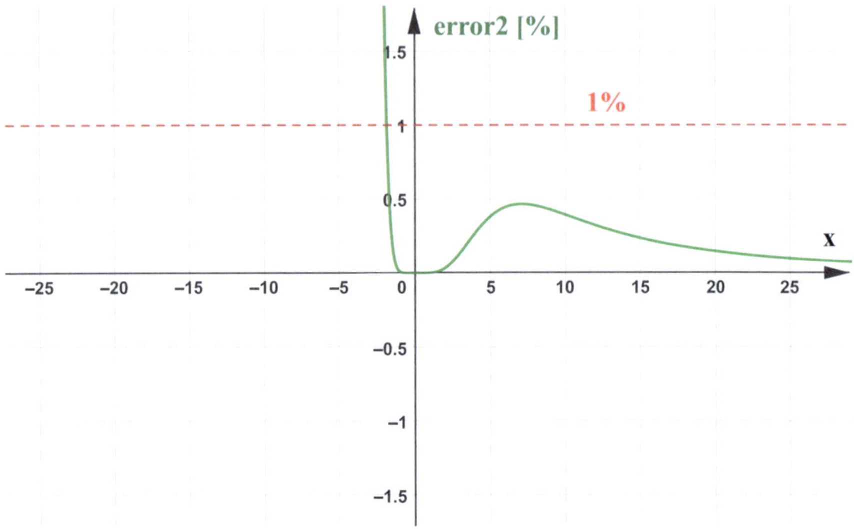

Fig. 1.21 Relative error of approximation of the first Einstein activation function $f_3(x)$ using approximation function $a_6(x)$

of $f_3(x)$ Einstein function, expressed by (1.13), to be identical with the superior-order Taylor series expansion of the general form of the approximation function $a_{7\mathrm{GEN}}(x)$, expressed by (1.16). The values of these coefficients, obtained considering the previous conditions, can be written as: $c_0 = 1$, $c_2 = -\frac{1}{39}$, $c_4 = \frac{17}{65{,}520}$ $d_0 = 1$, $d_2 = \frac{3}{52}$ and $d_4 = \frac{59}{65{,}520}$. Replacing these particular values in relation (1.16), the second approximation function $a_7(x)$ has the following mathematical expression:

$$a_7(x) = \frac{\frac{17x^4}{65{,}520} - \frac{x^2}{39} + 1}{\frac{59x^4}{65{,}520} + \frac{3x^2}{52} + 1}. \tag{1.17}$$

The graphical representation of the first Einstein function $f_3(x)$ and of the second approximation function $a_7(x)$ is presented in Fig. 1.22.

The absolute error of approximation of the first Einstein activation function $f_3(x)$ using the approximation function $a_7(x)$ is illustrated in Fig. 1.23. It is defined as the difference between the previous functions, error1 $= f_3(x) - a_7(x)$.

For more accurately evaluating the error of approximation, the relative error of approximation is shown in Fig. 1.24. This error is defined as error2 $= 100\frac{f_3(x)-a_7(x)}{f_3(x)}$ (in percents). Considering as acceptable a maximal approximation error of $\pm 1\%$, the maximal range of the input variable x is $[-3.2; \infty]$.

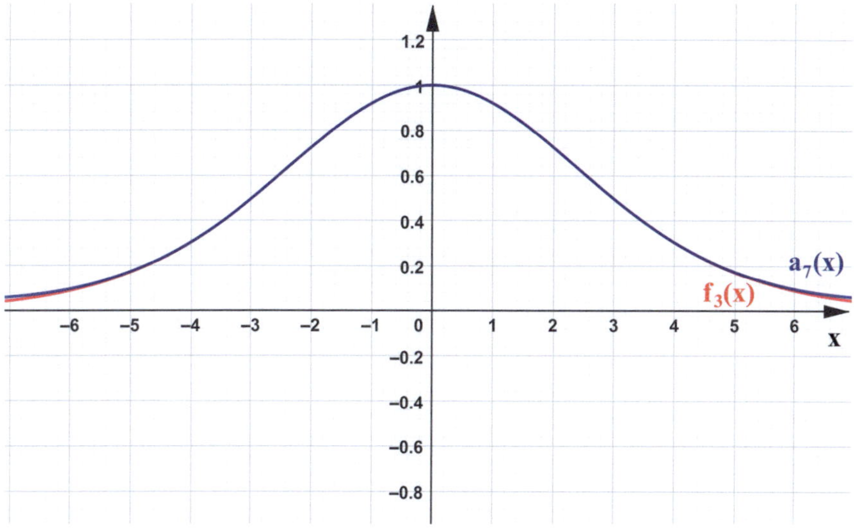

Fig. 1.22 First Einstein function $f_3(x)$ and the second approximation function $a_7(x)$

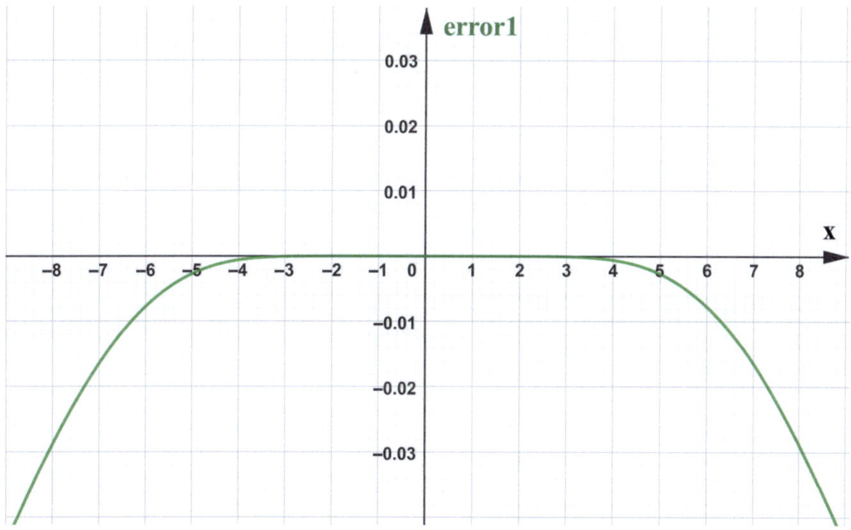

Fig. 1.23 Absolute error of approximation of the first Einstein function using second approximation function $a_7(x)$

1.4.4 Third Approximation Function for Generating the First Einstein Activation Function

The general form of the third proposed approximation function $a_8(x)$ is the following:

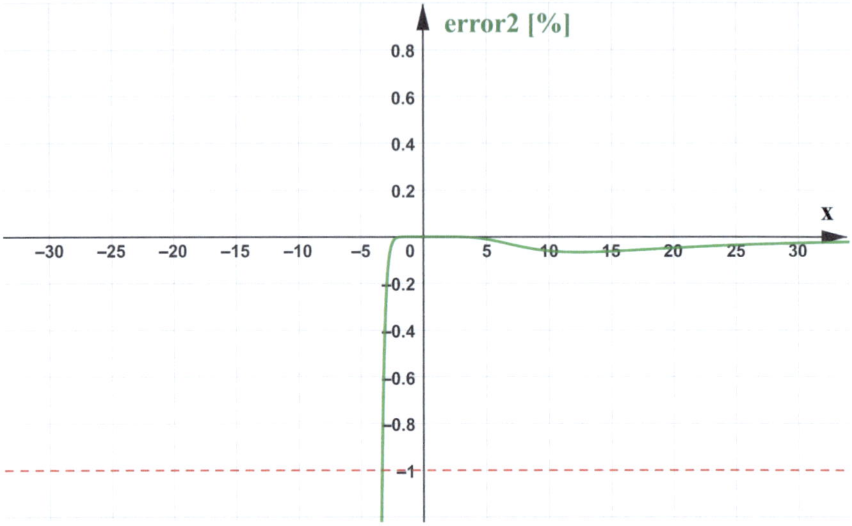

Fig. 1.24 Relative error of approximation of the first Einstein activation function $f_3(x)$ using approximation function $a_7(x)$

$$a_{8\mathrm{GEN}}(x) = \frac{c_2x^2 + c_0}{d_4x^4 + d_2x^2 + d_0}, \tag{1.18}$$

c_0, c_2, d_0, d_2 and d_4 being constant coefficients having particular values that can be determined by the condition that the superior-order Taylor series expansion of $f_3(x)$ first Einstein function, expressed by (1.13), to be identical with the superior-order Taylor series expansion of the general form of the approximation function $a_{8\mathrm{GEN}}(x)$, expressed by (1.18). The values of these coefficients, obtained considering the previous conditions, can be written as: $c_0 = 1$, $c_2 = -\frac{1}{56}$, $d_0 = 1$, $d_2 = \frac{11}{168}$ and $d_4 = \frac{13}{10{,}080}$. Replacing these particular values in relation (1.18), the third approximation function $a_8(x)$ has the following mathematical expression:

$$a_8(x) = \frac{1 - \frac{x^2}{56}}{\frac{13x^4}{10{,}080} + \frac{11x^2}{168} + 1}. \tag{1.19}$$

The graphical representation of the first Einstein function $f_3(x)$ and of the third approximation function $a_8(x)$ is presented in Fig. 1.25.

The absolute error of approximation of the first Einstein activation function $f_3(x)$ using the approximation function $a_8(x)$ is illustrated in Fig. 1.26. It is defined as the difference between the previous functions, error1 $= f_3(x) - a_8(x)$.

For more accurately evaluating the error of approximation, the relative error of approximation is shown in Fig. 1.27. This error is defined as error2 $= 100\frac{f_3(x) - a_8(x)}{f_3(x)}$ (in percents). Considering as acceptable a maximal approximation error of $\pm 1\%$, the maximal range of the input variable x is $[-2.7\,;\ \infty]$.

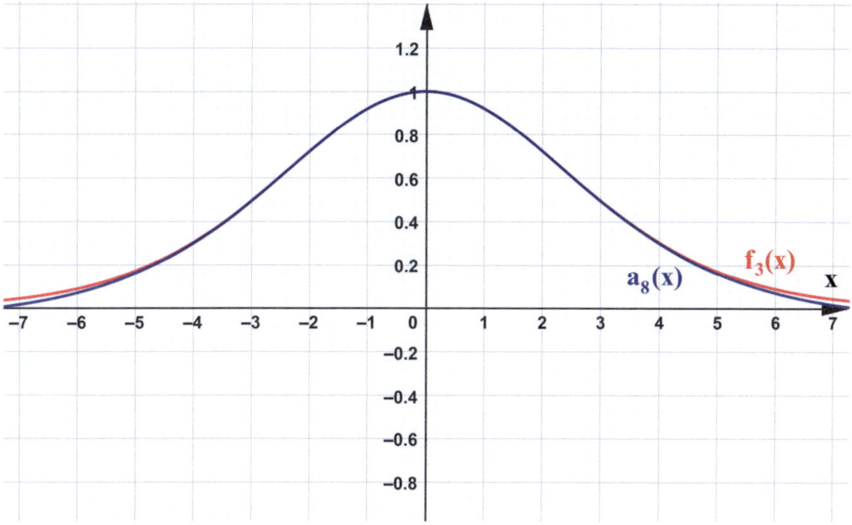

Fig. 1.25 First Einstein function $f_3(x)$ and the third approximation function $a_8(x)$

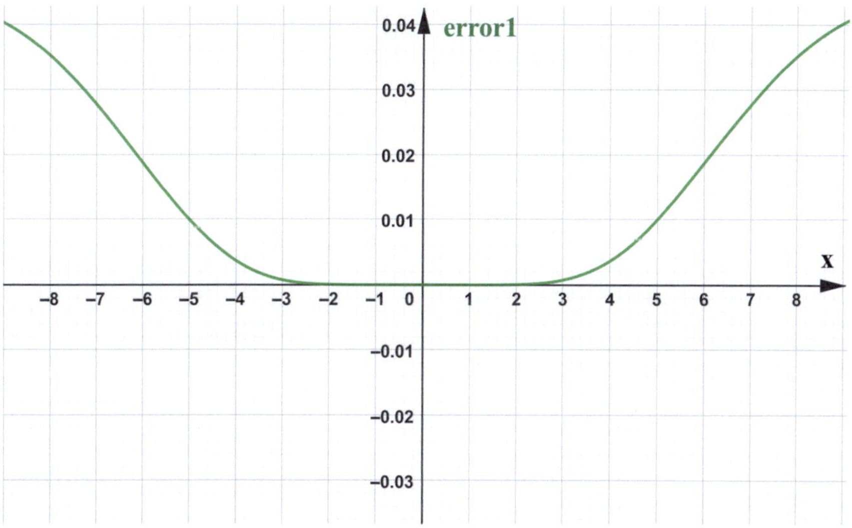

Fig. 1.26 Absolute error of approximation of the first Einstein function using third approximation function $a_8(x)$

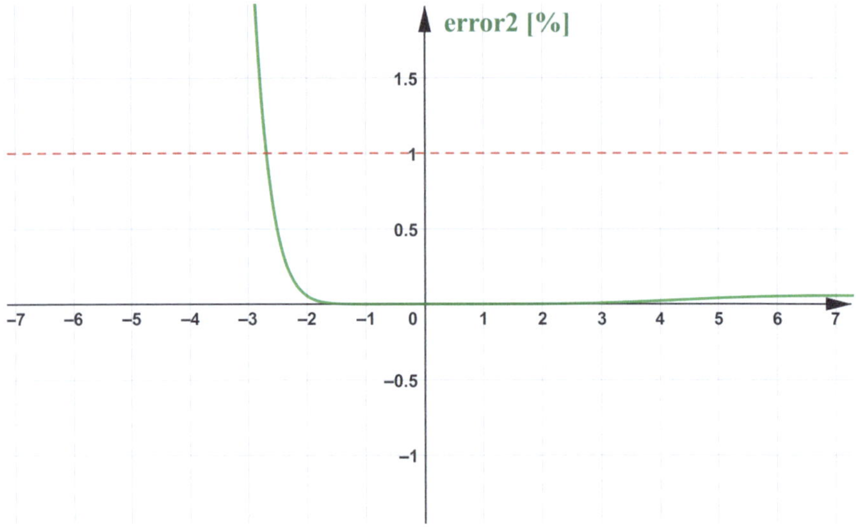

Fig. 1.27 Relative error of approximation of the first Einstein activation function $f_3(x)$ using approximation function $a_8(x)$

1.4.5 *General Form of the Second Einstein Activation Function*

The mathematical form of the second Einstein activation function can be expressed as follows:

$$f_4(x) = \frac{x}{\exp(x) - 1}. \tag{1.20}$$

The graphical representation of the second Einstein activation function $f_4(x)$ is presented in Fig. 1.28.

In order to approximate the second Einstein activation function, two new approximation functions $a_9(x)$ and $a_{10}(x)$ are presented, able to accurately approximate $f_4(x)$ function.

1.4.6 *First Approximation Function for Generating the Second Einstein Activation Function*

The general form of the first proposed approximation function $a_9(x)$ is the following:

$$a_{9\text{GEN}}(x) = \frac{c_3x^3 + c_2x^2 + c_1x + c_0}{d_2x^2 + d_0}, \tag{1.21}$$

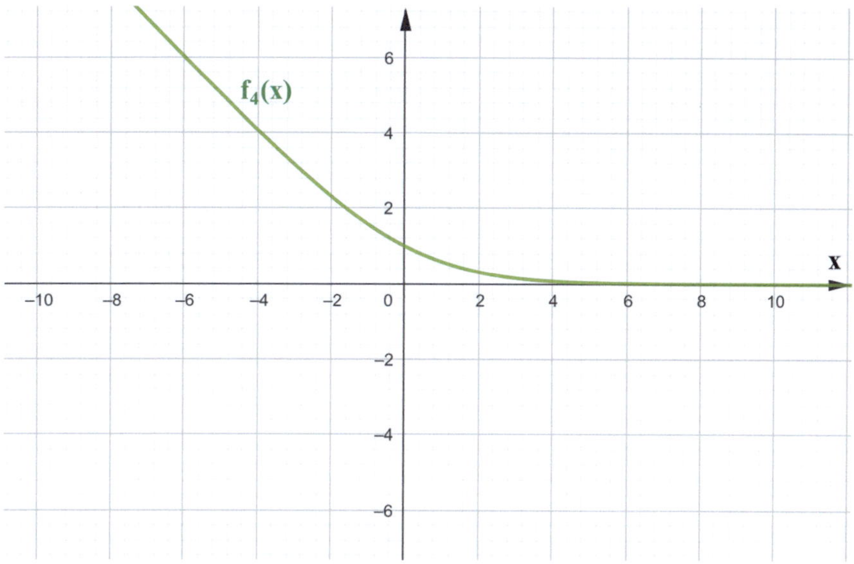

Fig. 1.28 Graphical representation of the second Einstein activation function

c_0, c_1, c_2, c_3, d_0 and d_2 being constant coefficients having particular values that can be determined by the condition that the superior-order Taylor series expansion of $f_4(x)$ second Einstein activation function, expressed by (1.20), to be identical with the superior-order Taylor series expansion of the general form of the approximation function $a_{9\mathrm{GEN}}(x)$, expressed by (1.21). The values of these coefficients, obtained considering the previous conditions, can be written as: $c_0 = 1$, $c_1 = -\frac{1}{2}$, $c_2 = \frac{1}{10}$, $c_3 = -\frac{1}{120}$, $d_0 = 1$ and $d_2 = \frac{1}{60}$. Replacing these particular values in relation (1.21), the first approximation function $a_9(x)$ has the following mathematical expression:

$$a_9(x) = \frac{-\frac{x^3}{120} + \frac{x^2}{10} - \frac{x}{2} + 1}{\frac{x^2}{60} + 1}. \tag{1.22}$$

The graphical representation of the second Einstein activation function $f_4(x)$ and of the first approximation function $a_9(x)$ is presented in Fig. 1.29.

The absolute error of approximation of the second Einstein activation function $f_4(x)$ using the approximation function $a_9(x)$ is illustrated in Fig. 1.30. It is defined as the difference between the previous functions, error1 $= f_4(x) - a_9(x)$.

For more accurately evaluating the error of approximation, the relative error of approximation is shown in Fig. 1.31. This error is defined as error2 $= 100\frac{f_4(x) - a_9(x)}{f_4(x)}$ (in percents). Considering as acceptable a maximal approximation error of $\pm 1\%$, the maximal range of the input variable x is $[-4.7\,;\,2.6]$.

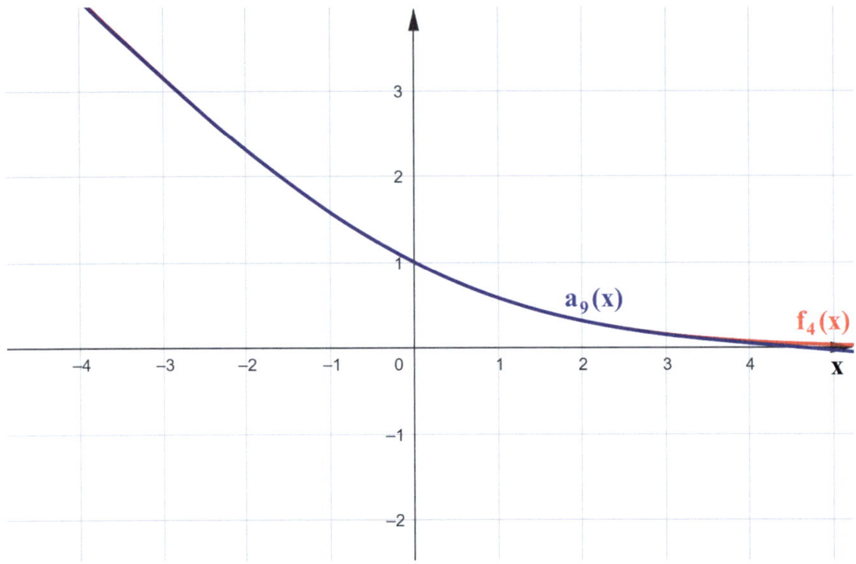

Fig. 1.29 Second Einstein activation function $f_4(x)$ and the first approximation function $a_9(x)$

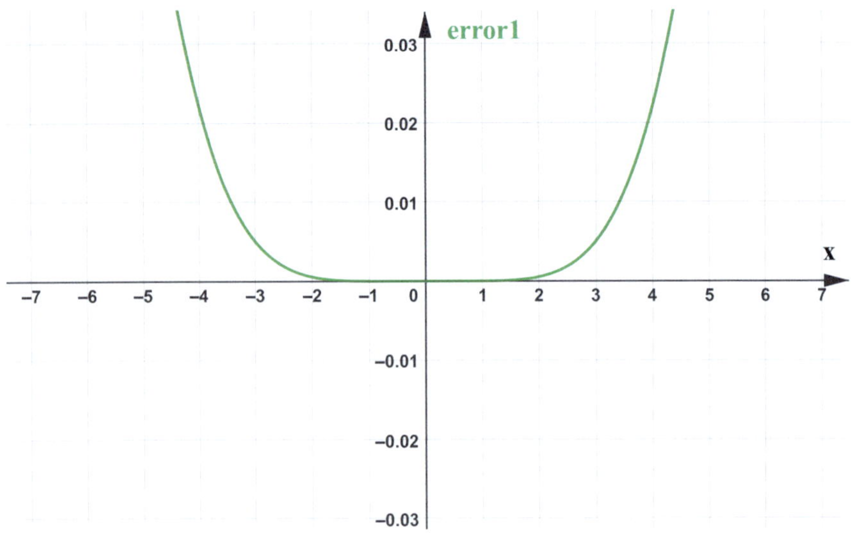

Fig. 1.30 Absolute error of approximation of the second Einstein activation function using first approximation function $a_9(x)$

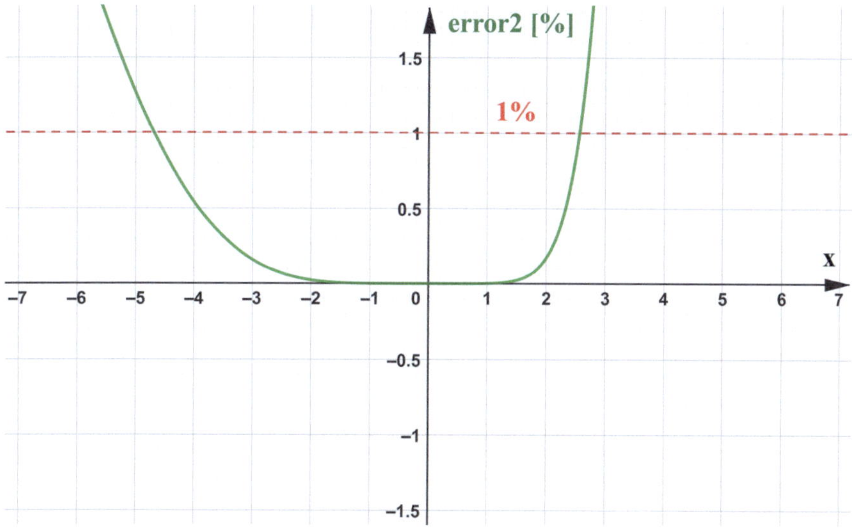

Fig. 1.31 Relative error of approximation of the second Einstein activation function $f_4(x)$ using approximation function $a_9(x)$

1.4.7 Second Approximation Function for Generating the Second Einstein Activation Function

The general form of the second proposed approximation function $a_{10}(x)$ is the following:

$$a_{10\mathrm{GEN}}(x) = \frac{c_3x^3 + c_2x^2 + c_1x + c_0}{d_3x^3 + d_2x^2 + d_1x + d_0}, \tag{1.23}$$

$c_0, c_1, c_2, c_3, d_0, d_1, d_2$ and d_3 being constant coefficients having particular values that can be determined by the condition that the superior-order Taylor series expansion of $f_4(x)$ second Einstein activation function, expressed by (1.20), to be identical with the superior-order Taylor series expansion of the general form of the approximation function $a_{10\mathrm{GEN}}(x)$, expressed by (1.23). The values of these coefficients, obtained considering the previous conditions, can be written as: $c_0 = 1$, $c_1 = -\frac{3}{7}$, $c_2 = \frac{1}{14}$, $c_3 = -\frac{1}{210}$, $d_0 = 1$, $d_1 = \frac{1}{14}$, $d_2 = \frac{1}{42}$ and $d_3 = \frac{1}{840}$. Replacing these particular values in relation (1.23), the second approximation function $a_{10}(x)$ has the following mathematical expression:

$$a_{10}(x) = \frac{-\frac{x^3}{210} + \frac{x^2}{14} - \frac{3x}{7} + 1}{\frac{x^3}{840} + \frac{x^2}{42} + \frac{x}{14} + 1}. \tag{1.24}$$

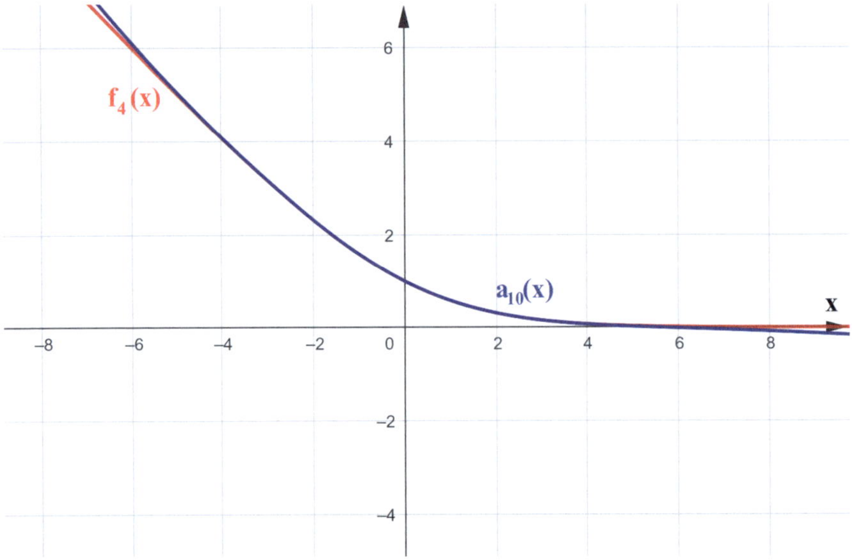

Fig. 1.32 Second Einstein activation function $f_4(x)$ and the second approximation function $a_{10}(x)$

The graphical representation of the second Einstein activation function $f_4(x)$ and of the second approximation function $a_{10}(x)$ is presented in Fig. 1.32.

The absolute error of approximation of the second Einstein activation function $f_4(x)$ using the approximation function $a_{10}(x)$ is illustrated in Fig. 1.33. It is defined as the difference between the previous functions, error1 $= f_4(x) - a_{10}(x)$.

For more accurately evaluating the error of approximation, the relative error of approximation is shown in Fig. 1.34. This error is defined as error2 $= 100\frac{f_4(x) - a_{10}(x)}{f_4(x)}$ (in percents). Considering as acceptable a maximal approximation error of $\pm 1\%$, the maximal range of the input variable x is $[-5.3\,;\ 3.3]$.

1.5 Approximation Functions for Hyperbolic Tangent (Tan-Sigmoid) Activation Function

The hyperbolic tangent activation function is a commonly used sigmoidal function in neural networks. This function is capable to map input values to the range from -1 and 1 and it is zero-centered, which provides several practical advantages. The area of utilization of the hyperbolic tangent activation function is:

a. ***Hidden layers in neural networks (especially recurrent neural networks—RNNs)***

   - widely used in RNNs.

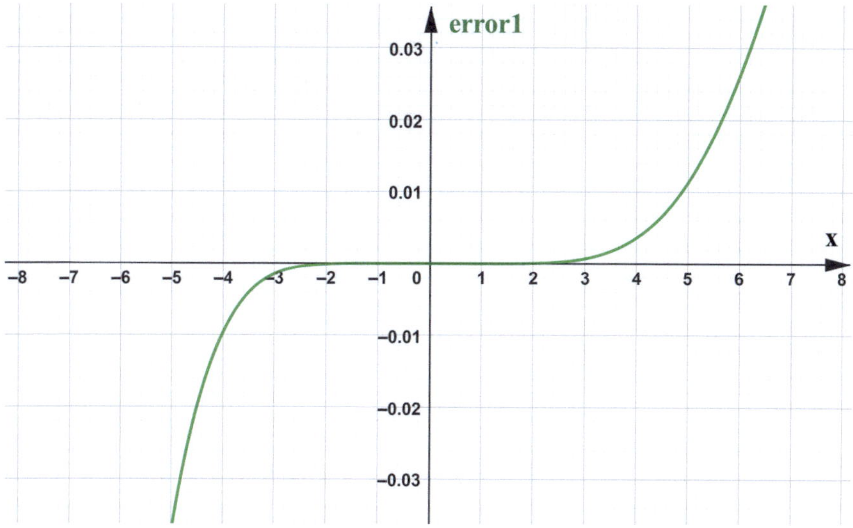

Fig. 1.33 Absolute error of approximation of the second Einstein activation function using second approximation function $a_{10}(x)$

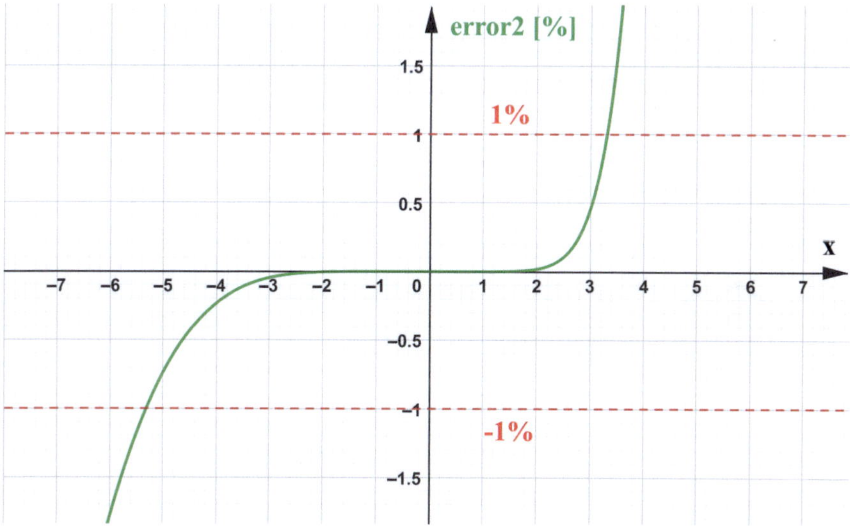

Fig. 1.34 Relative error of approximation of the second Einstein activation function $f_4(x)$ using approximation function $a_{10}(x)$

- suitable for hidden layers because it is zero-centered, helping gradient-based optimization algorithms likes stochastic gradient descent (SGD) converge faster than unipolar functions.

b. ***Deep learning architectures***

- prior to rectified linear unit (ReLU), hyperbolic tangent activation function was a standard activation function for multilayer perceptrons (MLPs).
- still found in models requiring bounded output with both positive and negative signal propagation.

c. ***Autoencoders***

- used in encoding and decoding layers to compress and reconstruct data.
- especially helpful when input data is normalized or centered around zero.

d. ***Normalization-sensitive networks***

- because it outputs values centered around 0, it works well in networks where input features are normalized (mean ≈ 0).

e. ***Control systems and dynamical models***

- used in neural controllers and dynamic systems modeling where smooth transitions between states (positive and negative) are required.

f. ***Bio-inspired neural models***

- hyperbolic tangent activation functions mimics biological neuron responses more closely than ReLU in some systems.
- used in computational neuroscience and psychological modeling.

1.5.1 General Form of the Hyperbolic Tangent Activation Function

The mathematical form of this function can be expressed as follows:

$$f_5(x) = \frac{\exp(x) - \exp(-x)}{\exp(x) + \exp(-x)}. \tag{1.25}$$

The graphical representation of the hyperbolic tangent activation function $f_5(x)$ is presented in Fig. 1.35.

In order to approximate the hyperbolic tangent activation function, four new approximation functions $a_{11}(x)$, $a_{12}(x)$, $a_{13}(x)$ and $a_{14}(x)$ are presented, able to accurately approximate the previous activation function. Their particular mathematical forms are developed having two very important objectives. The first objective is represented by the accuracy of the approximation that could be obtained using these approximation functions. The second objective is related to the possibility of their

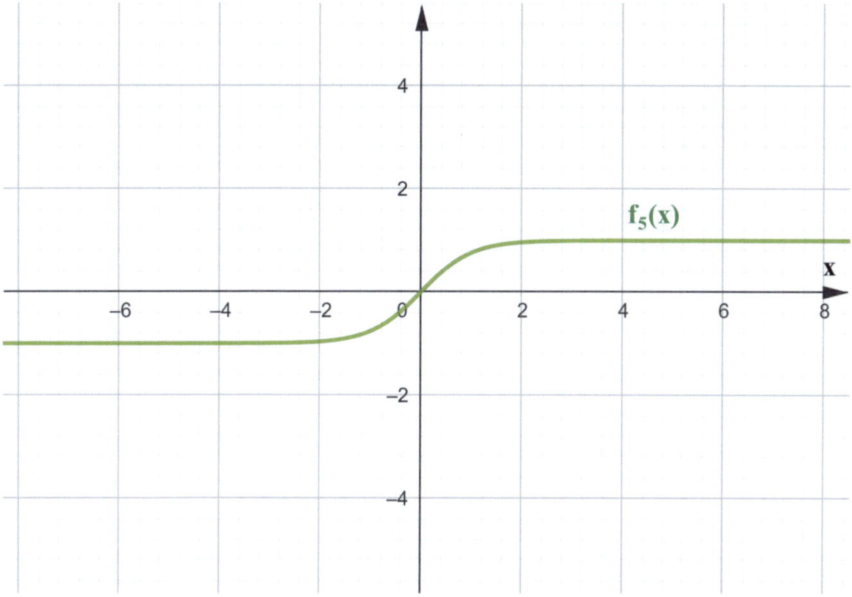

Fig. 1.35 Graphical representation of the hyperbolic tangent activation function

facile analog implementation in CMOS technology using as reasonable as possible hardware resources (see Chaps. 4 and 5).

1.5.2 *First Approximation Function for Generating the Hyperbolic Tangent Activation Function*

The general form of the first proposed approximation function $a_{11}(x)$ is the following:

$$a_{11\text{GEN}}(x) = \frac{c_3 x^3 + c_1 x}{d_2 x^2 + d_0}, \tag{1.26}$$

c_1, c_3, d_0 and d_2 being constant coefficients having particular values that can be determined by the condition that the superior-order Taylor series expansion of $f_5(x)$ hyperbolic tangent function, expressed by (1.25), to be identical with the superior-order Taylor series expansion of the general form of the approximation function $a_{11\text{GEN}}(x)$, expressed by (1.26). The values of these coefficients, obtained considering the previous conditions, can be written as: $c_1 = 1$, $c_3 = \frac{1}{15}$, $d_0 = 1$ and $d_2 = \frac{2}{5}$. Replacing these particular values in relation (1.26), the first approximation function $a_{11}(x)$ has the following mathematical expression:

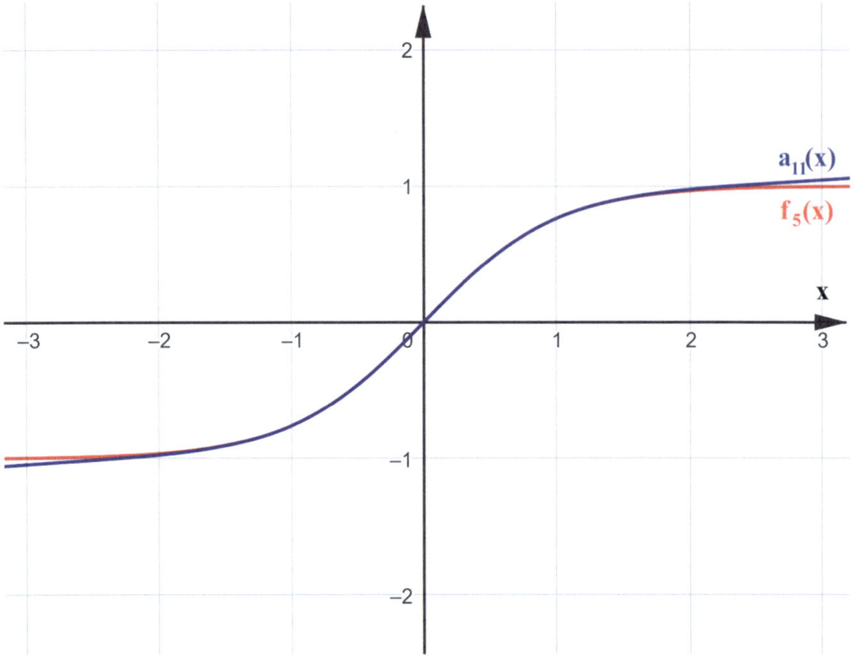

Fig. 1.36 Hyperbolic tangent activation function $f_5(x)$ and the first approximation function $a_{11}(x)$

$$a_{11}(x) = \frac{\frac{x^3}{15} + x}{\frac{2x^2}{5} + 1}. \tag{1.27}$$

The graphical representation of the hyperbolic tangent activation function $f_5(x)$ and of the first approximation function $a_{11}(x)$ is presented in Fig. 1.36.

The absolute error of approximation of the hyperbolic tangent activation function $f_5(x)$ using the approximation function $a_{11}(x)$ is illustrated in Fig. 1.37. It is defined as the difference between the previous functions, error1 $= f_5(x) - a_{11}(x)$.

For more accurately evaluating the error of approximation, the relative error of approximation is shown in Fig. 1.38. This error is defined as error2 $= 100\frac{f_5(x) - a_{11}(x)}{f_5(x)}$ (in percents). Considering as acceptable a maximal approximation error of $\pm 1\%$, the maximal range of the input variable x is $[-2\,;\ 2]$.

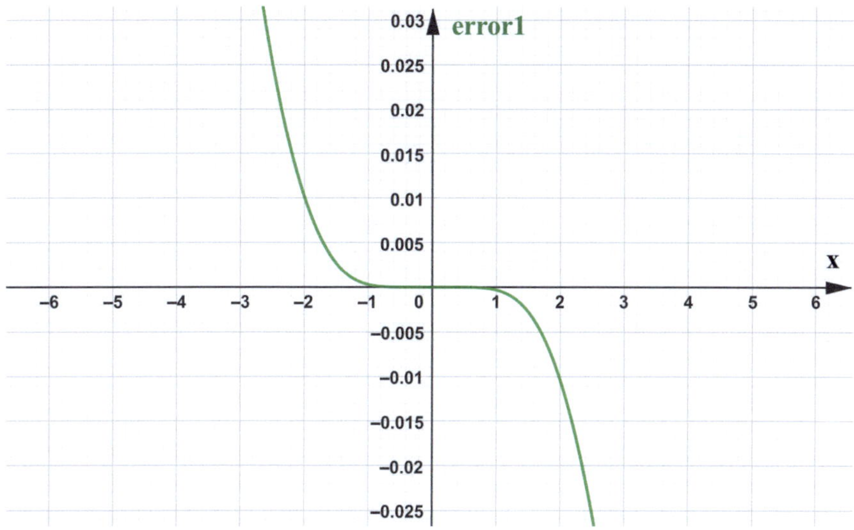

Fig. 1.37 Absolute error of approximation of the hyperbolic tangent activation function using approximation function $a_{11}(x)$

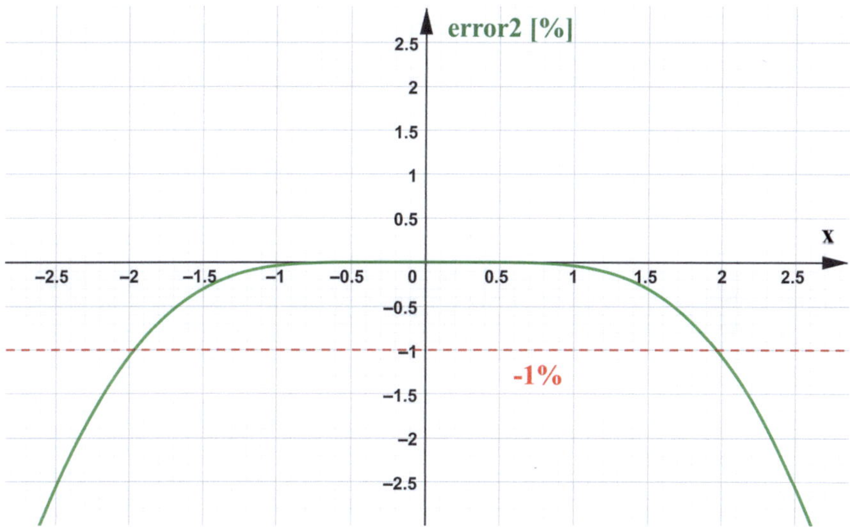

Fig. 1.38 Relative error of approximation of the hyperbolic tangent activation function $f_5(x)$ using approximation function $a_{11}(x)$

1.5.3 Second Approximation Function for Generating the Hyperbolic Tangent Activation Function

The general form of the second proposed approximation function $a_{12}(x)$ is the following:

$$a_{12\text{GEN}}(x) = \frac{c_3 x^3 + c_1 x}{d_4 x^4 + d_2 x^2 + d_0}, \tag{1.28}$$

c_1, c_3, d_0, d_2 and d_4 being constant coefficients having particular values that can be determined by the condition that the superior-order Taylor series expansion of $f_5(x)$ hyperbolic tangent activation function, expressed by (1.25), to be identical with the superior-order Taylor series expansion of the general form of the approximation function $a_{12\text{GEN}}(x)$, expressed by (1.28). The values of these coefficients, obtained considering the previous conditions, can be written as: $c_1 = 1$, $c_3 = \frac{2}{21}$, $d_0 = 1$, $d_2 = \frac{3}{7}$ and $d_4 = \frac{1}{105}$. Replacing these particular values in relation (1.28), the second approximation function $a_{12}(x)$ has the following mathematical expression:

$$a_{12}(x) = \frac{10x^3 + 105x}{x^4 + 45x^2 + 105}. \tag{1.29}$$

The graphical representation of the hyperbolic tangent activation function $f_5(x)$ and of the second approximation function $a_{12}(x)$ is presented in Fig. 1.39.

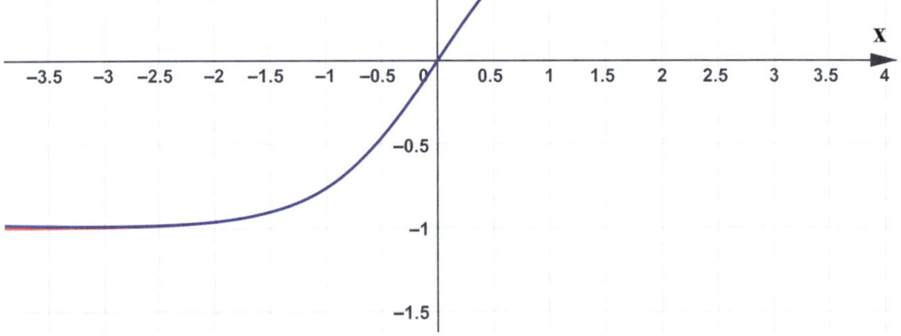

Fig. 1.39 Hyperbolic tangent activation function $f_5(x)$ and the second approximation function $a_{12}(x)$

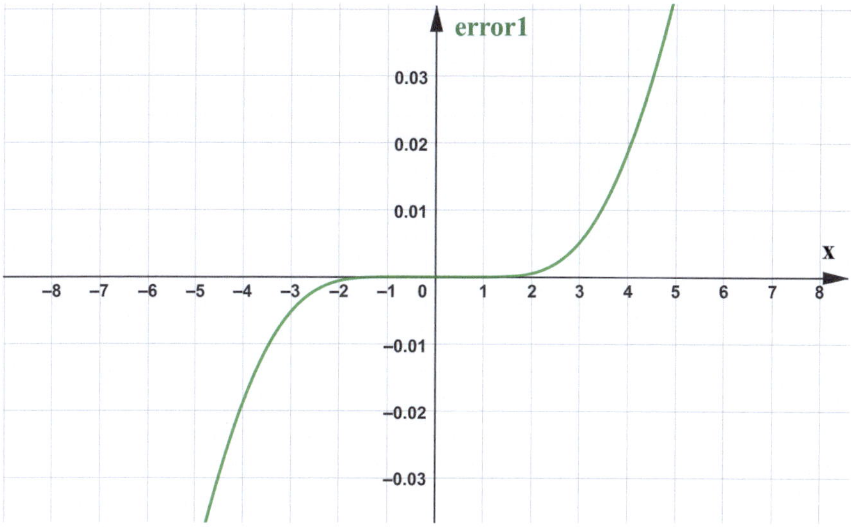

Fig. 1.40 Absolute error of approximation of the hyperbolic tangent activation function using approximation function $a_{12}(x)$

The absolute error of approximation of the hyperbolic tangent activation function $f_5(x)$ using the approximation function $a_{12}(x)$ is illustrated in Fig. 1.40. It is defined as the difference between the previous functions, error1 $= f_5(x) - a_{12}(x)$.

For more accurately evaluating the error of approximation, the relative error of approximation is shown in Fig. 1.41. This error is defined as error2 $= 100\frac{f_5(x)-a_{12}(x)}{f_5(x)}$ (in percents). Considering as acceptable a maximal approximation error of $\pm 1\%$, the maximal range of the input variable x is $[-3.5\,;\,3.5]$.

1.5.4 Third Approximation Function for Generating the Hyperbolic Tangent Activation Function

The general form of the third proposed approximation function $a_{13}(x)$ is the following:

$$a_{13\mathrm{GEN}}(x) = \frac{c_5x^5 + c_3x^3 + c_1x}{d_4x^4 + d_2x^2 + d_0}, \tag{1.30}$$

c_1, c_3, c_5, d_0, d_2 and d_4 being constant coefficients having particular values that can be determined by the condition that the superior-order Taylor series expansion of $f_5(x)$ hyperbolic tangent activation function, expressed by (1.25), to be identical with the superior-order Taylor series expansion of the general form of the approximation function $a_{13\mathrm{GEN}}(x)$, expressed by (1.30). The values of these coefficients, obtained

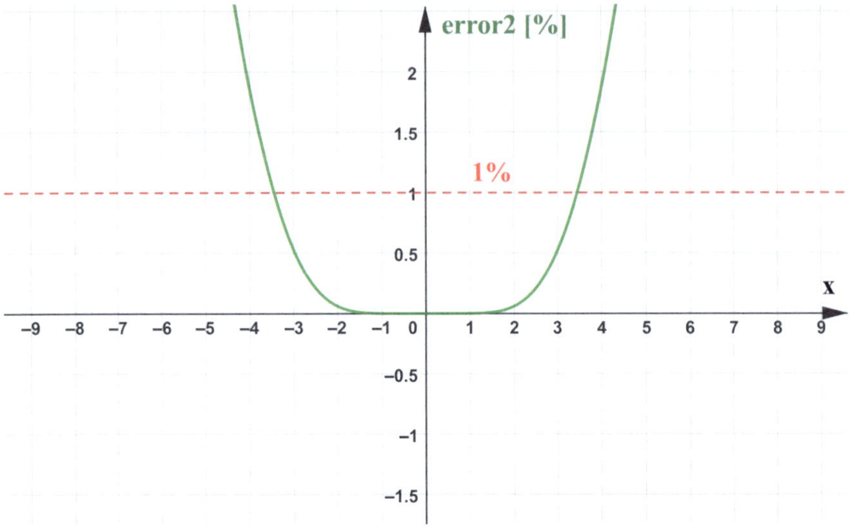

Fig. 1.41 Relative error of approximation of the hyperbolic tangent activation function $f_5(x)$ using approximation function $a_{12}(x)$

considering the previous conditions, can be written as: $c_1 = 1$, $c_3 = \frac{1}{9}$, $c_5 = \frac{1}{945}$ $d_0 = 1$, $d_2 = \frac{4}{9}$ and $d_4 = \frac{1}{63}$. Replacing these particular values in relation (1.30), the third approximation function $a_{13}(x)$ has the following mathematical expression:

$$a_{13}(x) = \frac{x}{15}\frac{x^4 + 105x^2 + 945}{x^4 + 28x^2 + 63}. \tag{1.31}$$

The graphical representation of the hyperbolic tangent activation function $f_5(x)$ and of the third approximation function $a_{13}(x)$ is presented in Fig. 1.42.

The absolute error of approximation of the hyperbolic tangent activation function $f_5(x)$ using the approximation function $a_{13}(x)$ is illustrated in Fig. 1.43. It is defined as the difference between the previous functions, $\text{error1} = f_5(x) - a_{13}(x)$.

For more accurately evaluating the error of approximation, the relative error of approximation is shown in Fig. 1.44. This error is defined as $\text{error2} = 100\frac{f_5(x) - a_{13}(x)}{f_5(x)}$ (in percents). Considering as acceptable a maximal approximation error of $\pm 1\%$, the maximal range of the input variable x is $[-5.2\,;\ 5.2]$.

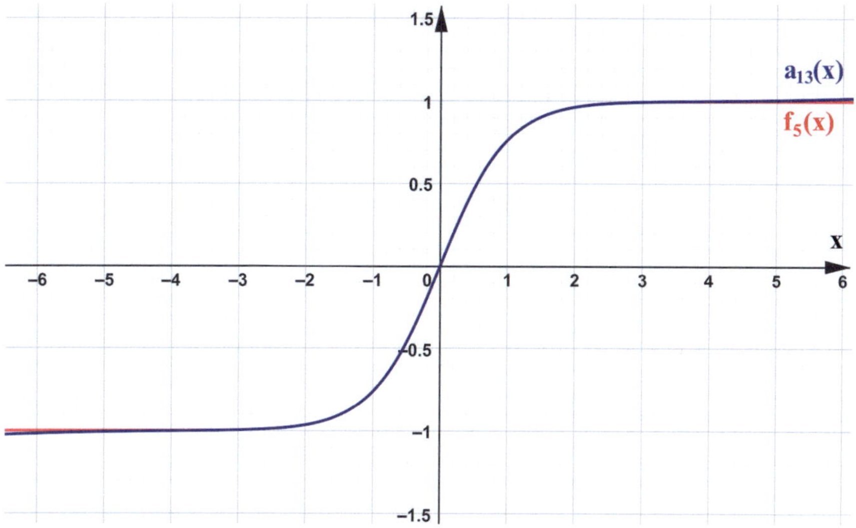

Fig. 1.42 Hyperbolic tangent activation function $f_5(x)$ and the third approximation function $a_{13}(x)$

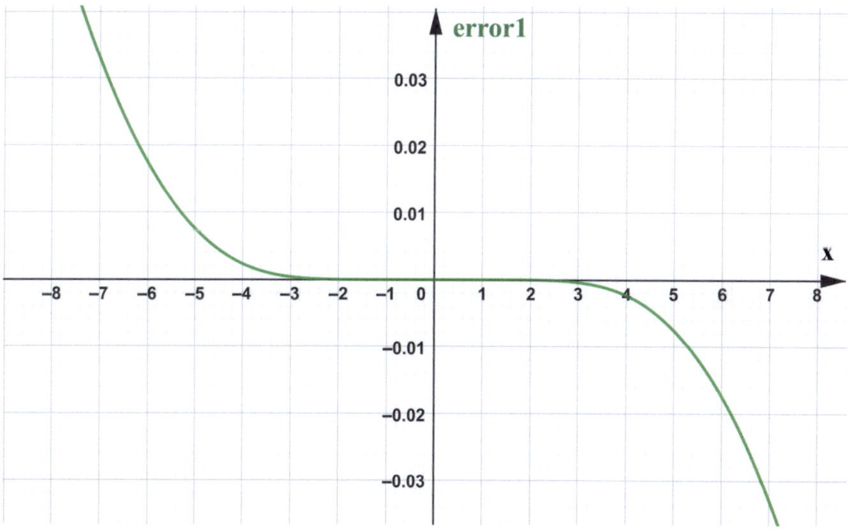

Fig. 1.43 Absolute error of approximation of the hyperbolic tangent activation function using approximation function $a_{13}(x)$

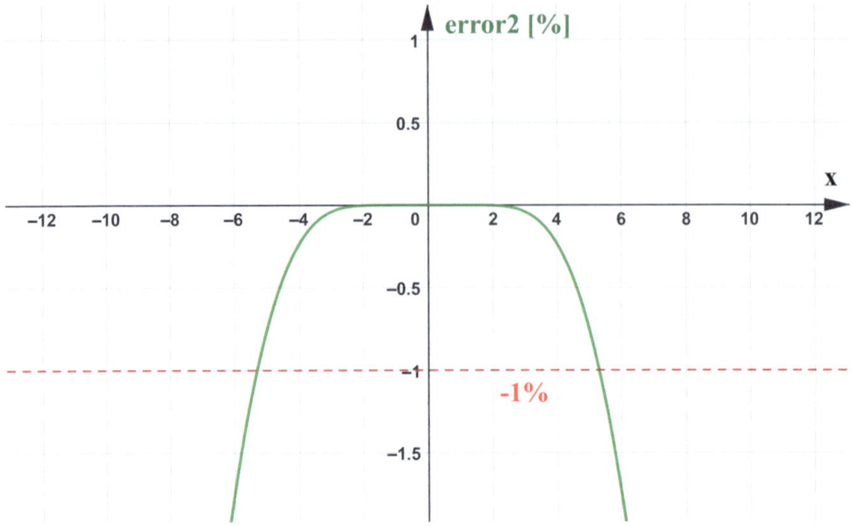

Fig. 1.44 Relative error of approximation of the hyperbolic tangent activation function $f_5(x)$ using approximation function $a_{13}(x)$

1.5.5 Fourth Approximation Function for Generating the Hyperbolic Tangent Activation Function

The general form of the fourth proposed approximation function $a_{14}(x)$ is the following:

$$a_{14\mathrm{GEN}}(x) = \frac{c_5x^5 + c_3x^3 + c_1x}{d_6x^6 + d_4x^4 + d_2x^2 + d_0}, \tag{1.32}$$

c_1, c_3, c_5, d_0, d_2, d_4 and d_6 being constant coefficients having particular values that can be determined by the condition that the superior-order Taylor series expansion of $f_5(x)$ hyperbolic tangent activation function, expressed by (1.25), to be identical with the superior-order Taylor series expansion of the general form of the approximation function $a_{14GEN}(x)$, expressed by (1.32). The values of these coefficients, obtained considering the previous conditions, can be written as: $c_1 = 1$, $c_3 = \frac{4}{33}$, $c_5 = \frac{1}{495}$, $d_0 = 1$, $d_2 = \frac{5}{11}$, $d_4 = \frac{2}{99}$ and $d_6 = \frac{1}{10{,}395}$. Replacing these particular values in relation (1.32), the third approximation function $a_{14}(x)$ has the following mathematical expression:

$$a_{14}(x) = \frac{\frac{x^5}{495} + \frac{4x^3}{33} + x}{\frac{x^6}{10{,}395} + \frac{2x^4}{99} + \frac{5x^2}{11} + 1}. \tag{1.33}$$

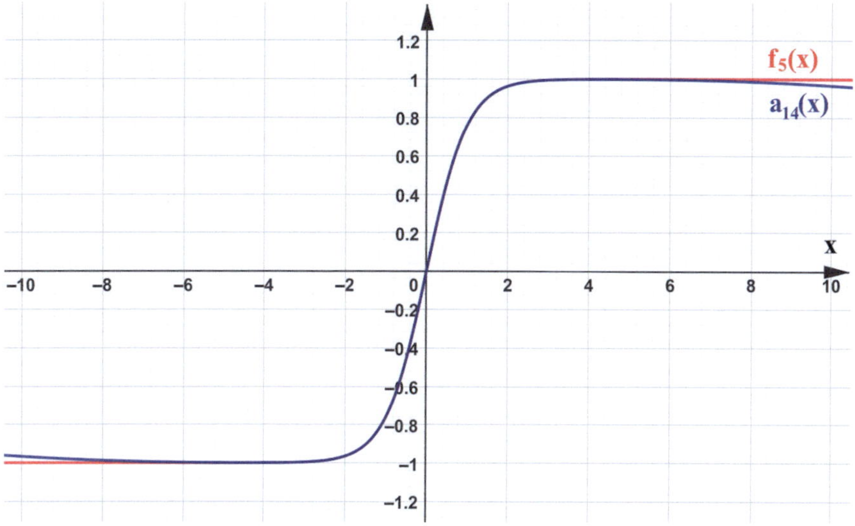

Fig. 1.45 Hyperbolic tangent activation function $f_5(x)$ and the fourth approximation function $a_{14}(x)$

The graphical representation of the hyperbolic tangent activation function $f_5(x)$ and of the fourth approximation function $a_{14}(x)$ is presented in Fig. 1.45.

The absolute error of approximation of the hyperbolic tangent activation function $f_5(x)$ using the approximation function $a_{14}(x)$ is illustrated in Fig. 1.46. It is defined as the difference between the previous functions, error1 $= f_5(x) - a_{14}(x)$.

For more accurately evaluating the error of approximation, the relative error of approximation is shown in Fig. 1.47. This error is defined as error2 $= 100\frac{f_5(x)-a_{14}(x)}{f_5(x)}$ (in percents). Considering as acceptable a maximal approximation error of $\pm 1\%$, the maximal range of the input variable x is $[-7.6\,;\ 7.6]$.

1.6 Approximation Functions for Continuous-Log Sigmoid Activation Function

The continuous-log activation function is a nonlinear function, used primarily in contexts where a smooth, compressive nonlinearity is needed that behaves similarly to logarithmic scaling, but remains continuous and differentiable across its entire domain.

The advantages of the continuous-log activation function are the following: it grows slowly for large inputs (like a logarithm), it is continuous and differentiable across all real numbers, it is zero-centered and it provides symmetric compression for both positive and negative inputs. The areas of utilization of the continuous-log activation function are:

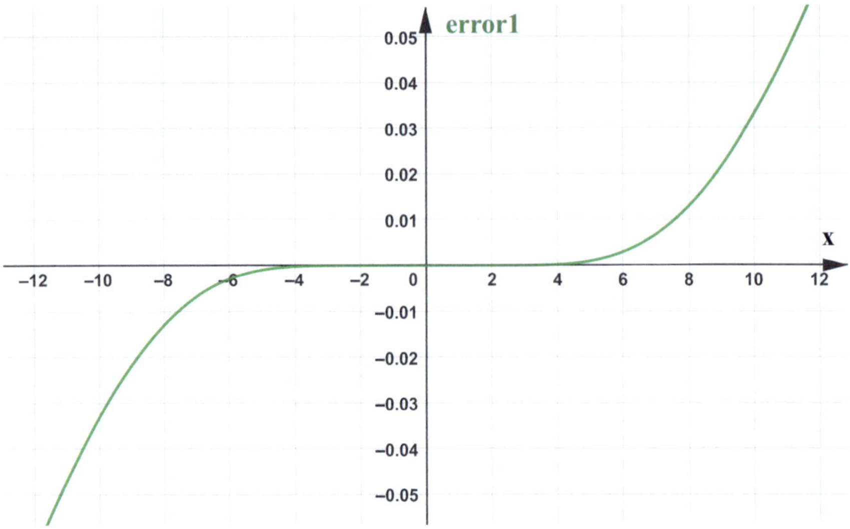

Fig. 1.46 Absolute error of approximation of the hyperbolic tangent activation function using approximation function $a_{14}(x)$

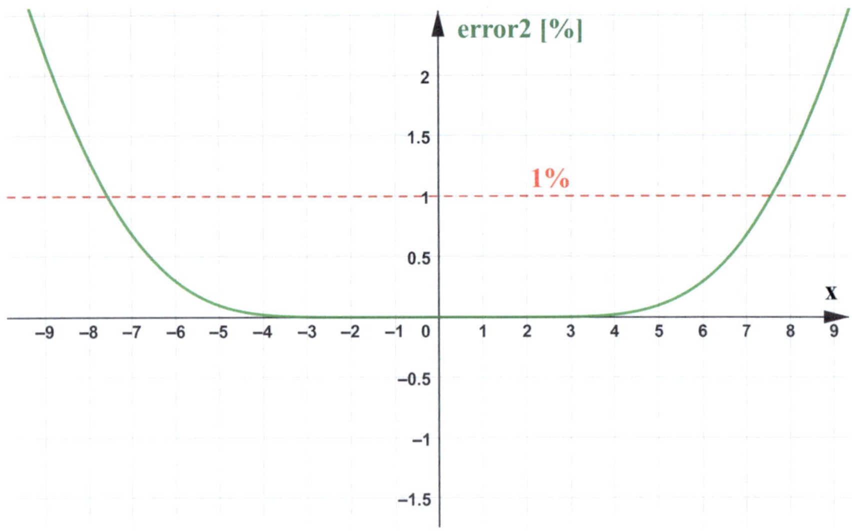

Fig. 1.47 Relative error of approximation of the hyperbolic tangent activation function $f_5(x)$ using approximation function $a_{14}(x)$

a. ***Robust neural networks***

- it helps prevent exploding activations by compressing large values smoothly.
- useful in networks where input magnitudes vary greatly and stable and bounded activations are needed.

b. ***Noise-resilient feature extraction***

- in image processing or signal processing, continuous-log activations can suppress the impact of outliers and high-frequency noise.
- useful in autoencoders or denoising networks.

c. ***Financial and scientific modeling***

- where inputs may span orders of magnitude, like stock returns, earthquake magnitudes, or biological measurements, and you want scale-insensitive responses.
- the log-like behavior ensures smoother transitions and avoids hard clipping.

d. ***Sparse and compressive representations***

- used in sparse coding or information-theoretic neural models, where the goal is to compress information content while preserving key features.
- encourages small activations and sparsity due to the sublinear growth.

1.6.1 General Form of the Continuous-Log Sigmoid Activation Function

The mathematical form of this function can be expressed as follows:

$$f_6(x) = \frac{1}{1 + \exp(-bx)}. \tag{1.34}$$

b being a parameter that models the form of the continuous-log sigmoid activation function. The graphical representation of the continuous-log sigmoid activation function $f_6(x)$, for multiple values of b parameter ($b = 1, b = 2$ and $b = 3$, respectively), is presented in Fig. 1.48.

Because the case $b = 1$ has been already treated as unipolar sigmoidal function, which represents a particular case of the continuous-log sigmoid activation function (see paragraph 1.2), the following approximation functions are developed for generating the continuous-log sigmoidal approximation functions for $b = 2$ and $b = 3$ cases.

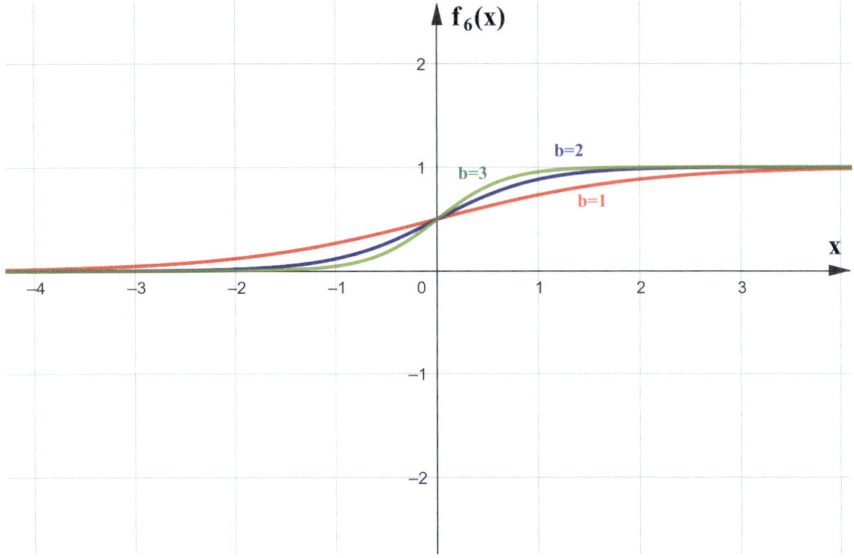

Fig. 1.48 Graphical representation of the continuous-log activation function for ($b = 1, b = 2$ and $b = 3$)

1.6.2 First Approximation Function for Generating the Continuous-Log Sigmoid Activation Function (for b = 2)

The general expression $f_6(x)$ of the continuous-log sigmoid activation function has the following mathematical expression for the particular value of the parameter, $b = 2$:

$$f_8(x) = \frac{1}{1 + \exp(-2x)}. \tag{1.35}$$

In order to approximate the $f_8(x)$ continuous-log sigmoid activation function, two new approximation functions $a_{15}(x)$ and $a_{16}(x)$ are presented, able to accurately approximate the previous activation function.

The general form of the first proposed approximation function $a_{15}(x)$ is the following:

$$a_{15\text{GEN}}(x) = \frac{c_3x^3 + c_2x^2 + c_1x + c_0}{d_2x^2 + d_0}, \tag{1.36}$$

c_0, c_1, c_2, c_3, d_0 and d_2 being constant coefficients having particular values that can be determined by the condition that the superior-order Taylor series expansion of $f_8(x)$ continuous-log sigmoid activation function, expressed by (1.35), to be identical with the superior-order Taylor series expansion of the general form of the approximation function $a_{15GEN}(x)$, expressed by (1.36). The values of these coefficients, obtained considering the previous conditions, can be written as: $c_0 = \frac{1}{2}$, $c_1 = \frac{1}{2}$, $c_2 = \frac{1}{5}$, $c_3 = \frac{1}{30}$, $d_0 = 1$ and $d_2 = \frac{2}{5}$. Replacing these particular values in relation (1.36), the first approximation function $a_{15}(x)$ has the following mathematical expression:

$$a_{15}(x) = \frac{\frac{x^3}{30} + \frac{x^2}{5} + \frac{x}{2} + \frac{1}{2}}{\frac{2x^2}{5} + 1}. \tag{1.37}$$

The graphical representation of the continuous-log sigmoid activation function $f_8(x)$ and of the first approximation function $a_{15}(x)$ is presented in Fig. 1.49.

The absolute error of approximation of the continuous-log sigmoidal activation function $f_8(x)$ using the approximation function $a_{15}(x)$ is illustrated in Fig. 1.50. It is defined as the difference between the previous functions, error1 $= f_8(x) - a_{15}(x)$.

For more accurately evaluating the error of approximation, the relative error of approximation is shown in Fig. 1.51. This error is defined as error2 $= 100\frac{f_8(x) - a_{15}(x)}{f_8(x)}$ (in percents). Considering as acceptable a maximal approximation error of $\pm 1\%$, the maximal range of the input variable x is $[-1.3\,;\ 2.35]$.

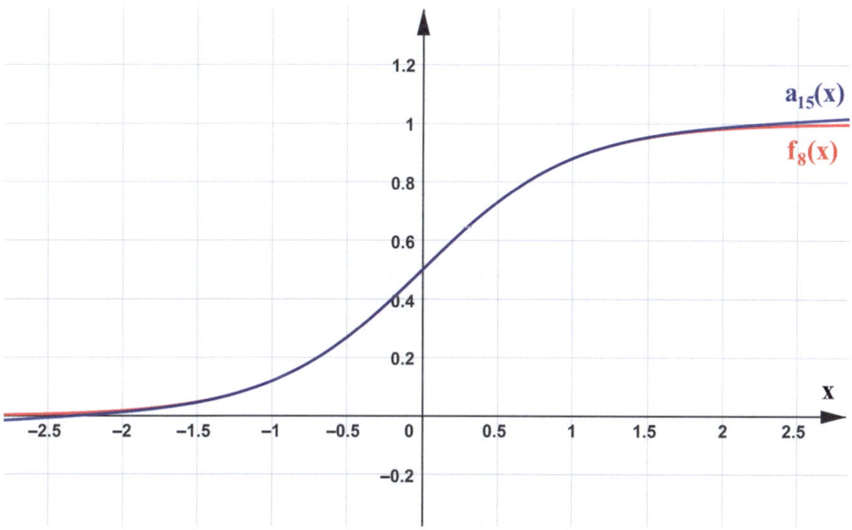

Fig. 1.49 Continuous-log sigmoid activation function $f_8(x)$ and the first approximation function $a_{15}(x)$

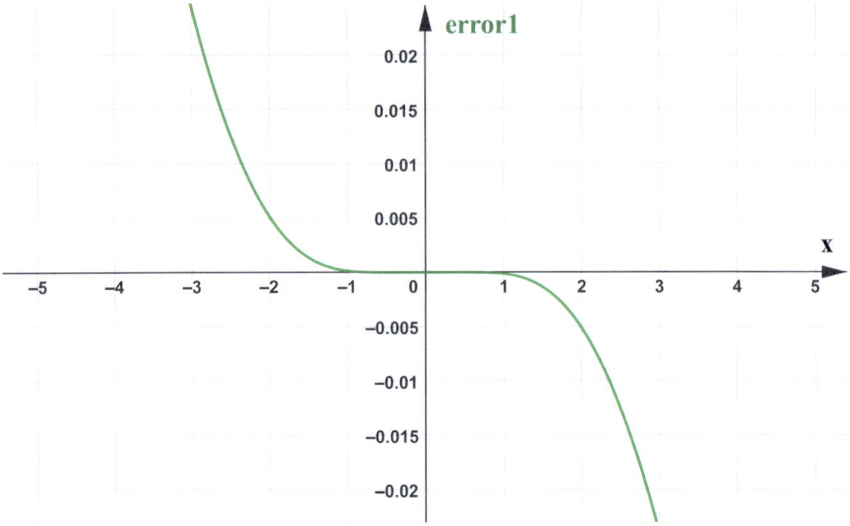

Fig. 1.50 Absolute error of approximation of the continuous-log sigmoid activation function $f_8(x)$ using first approximation function $a_{15}(x)$

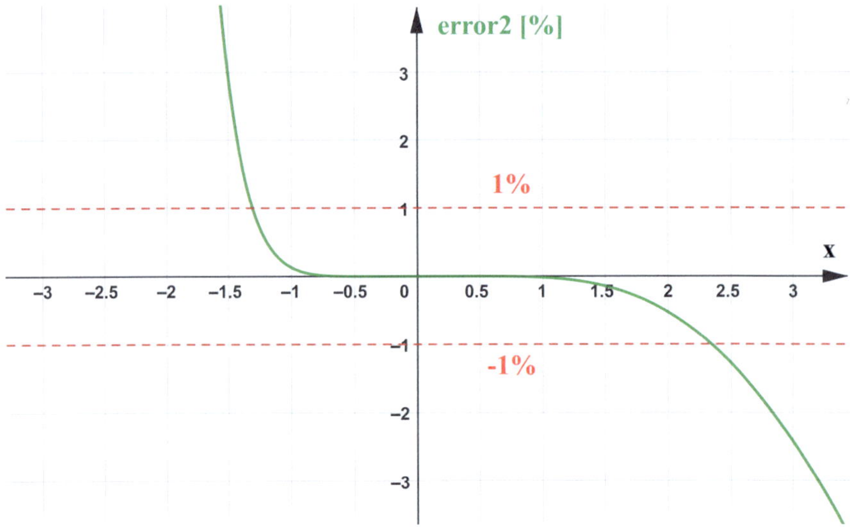

Fig. 1.51 Relative error of approximation of the continuous-log sigmoidal activation function $f_8(x)$ using approximation function $a_{15}(x)$

1.6.3 Second Approximation Function for Generating the Continuous-Log Sigmoid Activation Function (for b = 2)

The general form of the second proposed approximation function $a_{16}(x)$ is the following:

$$a_{16\text{GEN}}(x) = \frac{c_4x^4 + c_3x^3 + c_2x^2 + c_1x + c_0}{d_4x^4 + d_2x^2 + d_0}, \tag{1.38}$$

$c_0, c_1, c_2, c_3, c_4, d_0, d_2$ and d_4 being constant coefficients having particular values that can be determined by the condition that the superior-order Taylor series expansion of $f_8(x)$ continuous-log sigmoid activation function, expressed by (1.35), to be identical with the superior-order Taylor series expansion of the general form of the approximation function $a_{16\text{GEN}}(x)$, expressed by (1.38). The values of these coefficients, obtained considering the previous conditions, can be written as: $c_0 = \frac{1}{2}$, $c_1 = \frac{1}{2}$, $c_2 = \frac{3}{14}$, $c_3 = \frac{1}{21}$, $c_4 = \frac{1}{210}$, $d_0 = 1$, $d_2 = \frac{3}{7}$ and $d_4 = \frac{1}{105}$. Replacing these particular values in relation (1.43), the second approximation function $a_{16}(x)$ has the following mathematical expression:

$$a_{16}(x) = \frac{\frac{x^4}{210} + \frac{x^3}{21} + \frac{3x^2}{14} + \frac{x}{2} + \frac{1}{2}}{\frac{x^4}{105} + \frac{3x^2}{7} + 1}. \tag{1.39}$$

The graphical representation of the continuous-log sigmoid activation function $f_8(x)$ and of the second approximation function $a_{16}(x)$ is presented in Fig. 1.52.

The absolute error of approximation of the continuous-log sigmoidal activation function $f_8(x)$ using the approximation function $a_{16}(x)$ is illustrated in Fig. 1.53. It is defined as the difference between the previous functions, error1 $= f_8(x) - a_{16}(x)$.

For more accurately evaluating the error of approximation, the relative error of approximation is shown in Fig. 1.54. This error is defined as error2 $= 100\frac{f_8(x) - a_{16}(x)}{f_8(x)}$ (in percents). Considering as acceptable a maximal approximation error of $\pm 1\%$, the maximal range of the input variable x is $[-1.9\,;\ 4.1]$.

1.6.4 First Approximation Function for Generating the Continuous-Log Sigmoid Activation Function (for b = 3)

The general expression $f_6(x)$ of the continuous-log sigmoid activation function has the following mathematical expression for the particular value of the parameter, $b = 3$:

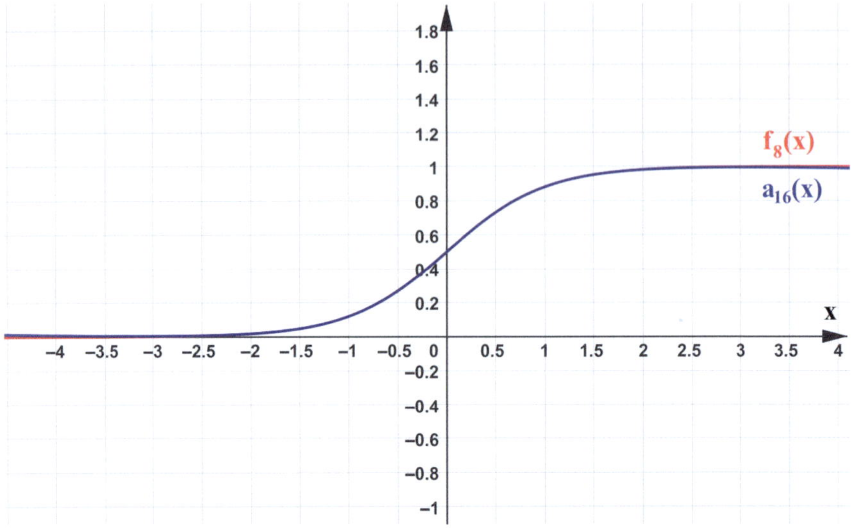

Fig. 1.52 Continuous-log sigmoid activation function $f_8(x)$ and the second approximation function $a_{16}(x)$

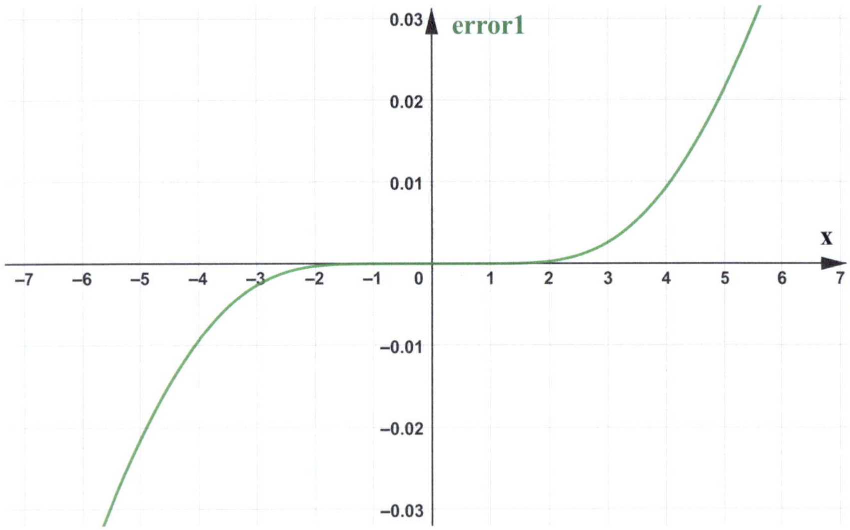

Fig. 1.53 Absolute error of approximation of the continuous-log sigmoid activation function $f_8(x)$ using second approximation function $a_{16}(x)$

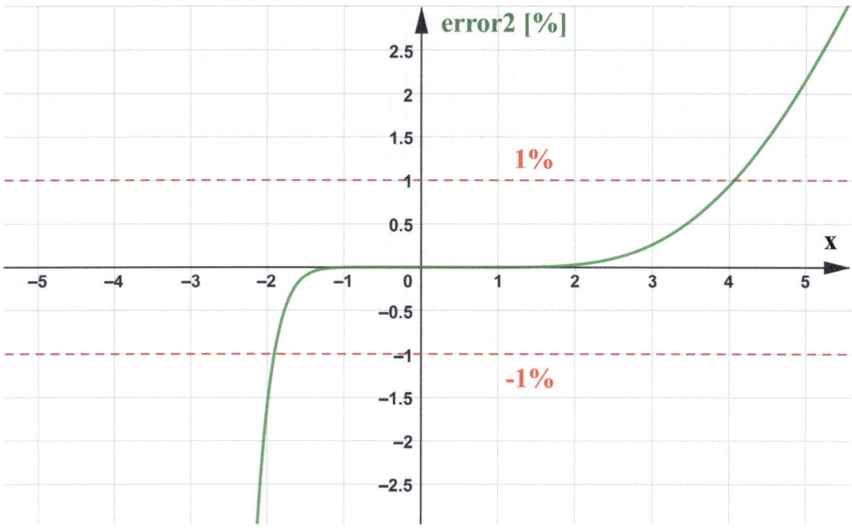

Fig. 1.54 Relative error of approximation of the continuous-log sigmoidal activation function $f_8(x)$ using approximation function $a_{18}(x)$

$$f_9(x) = \frac{1}{1 + \exp(-3x)}. \tag{1.40}$$

In order to approximate the $f_9(x)$ continuous-log sigmoid activation function, two new approximation functions $a_{17}(x)$ and $a_{18}(x)$ are presented, able to accurately approximate the previous activation function.

The general form of the first proposed approximation function $a_{17}(x)$ is the following:

$$a_{17\text{GEN}}(x) = \frac{c_3x^3 + c_2x^2 + c_1x + c_0}{d_2x^2 + d_0}, \tag{1.41}$$

c_0, c_1, c_2, c_3, d_0 and d_2 being constant coefficients having particular values that can be determined by the condition that the superior-order Taylor series expansion of $f_9(x)$ continuous-log sigmoid activation function, expressed by (1.40), to be identical with the superior-order Taylor series expansion of the general form of the approximation function $a_{17\text{GEN}}(x)$, expressed by (1.41). The values of these coefficients, obtained considering the previous conditions, can be written as: $c_0 = \frac{1}{2}$, $c_1 = \frac{3}{4}$, $c_2 = \frac{9}{20}$, $c_3 = \frac{9}{80}$, $d_0 = 1$ and $d_2 = \frac{9}{10}$. Replacing these particular values in relation (1.41), the first approximation function $a_{19}(x)$ has the following mathematical expression:

$$a_{17}(x) = \frac{\frac{9x^3}{80} + \frac{9x^2}{20} + \frac{3x}{4} + \frac{1}{2}}{\frac{9x^2}{10} + 1}. \tag{1.42}$$

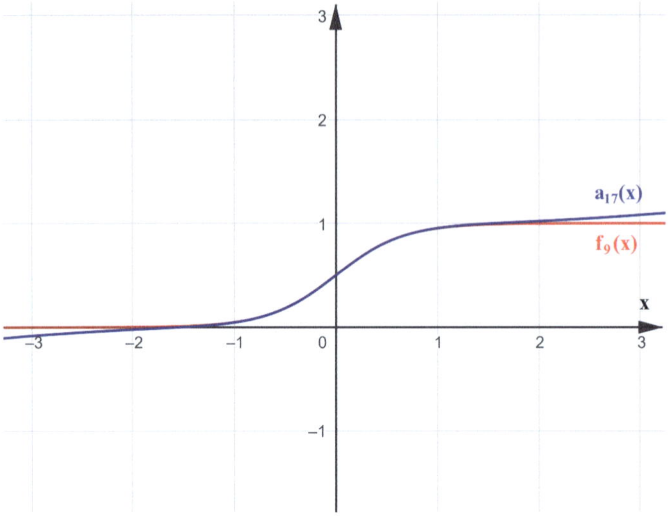

Fig. 1.55 Continuous-log activation sigmoid activation function $f_9(x)$

The graphical representation of the continuous-log sigmoid activation function $f_9(x)$ and of the first approximation function $a_{17}(x)$ is presented in Fig. 1.55.

and the first approximation function $a_{17}(x)$.

The absolute error of approximation of the continuous-log sigmoidal activation function $f_9(x)$ using the approximation function $a_{17}(x)$ is illustrated in Fig. 1.56. It is defined as the difference between the previous functions, $\text{error1} = f_9(x) - a_{17}(x)$.

For more accurately evaluating the error of approximation, the relative error of approximation is shown in Fig. 1.57. This error is defined as $\text{error2} = 100\frac{f_9(x)-a_{17}(x)}{f_9(x)}$ (in percents). Considering as acceptable a maximal approximation error of $\pm 1\%$, the maximal range of the input variable x is $[-0.9\,;\ 1.6]$.

1.6.5 Second Approximation Function for Generating the Continuous-Log Sigmoid Activation Function (for b = 3)

The general form of the second proposed approximation function $a_{18}(x)$ is the following:

$$a_{18\text{GEN}}(x) = \frac{c_4x^4 + c_3x^3 + c_2x^2 + c_1x + c_0}{d_4x^4 + d_2x^2 + d_0}, \qquad (1.43)$$

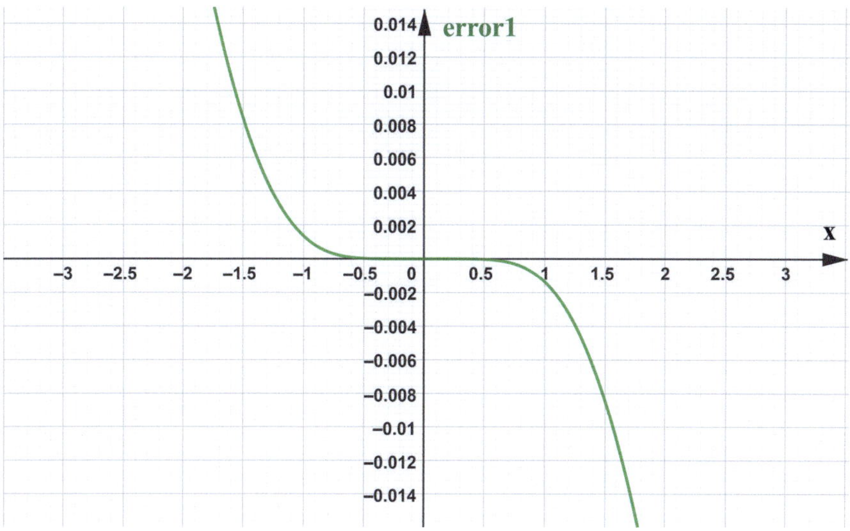

Fig. 1.56 Absolute error of approximation of the continuous-log sigmoid activation $f_9(x)$ function using first approximation function $a_{17}(x)$

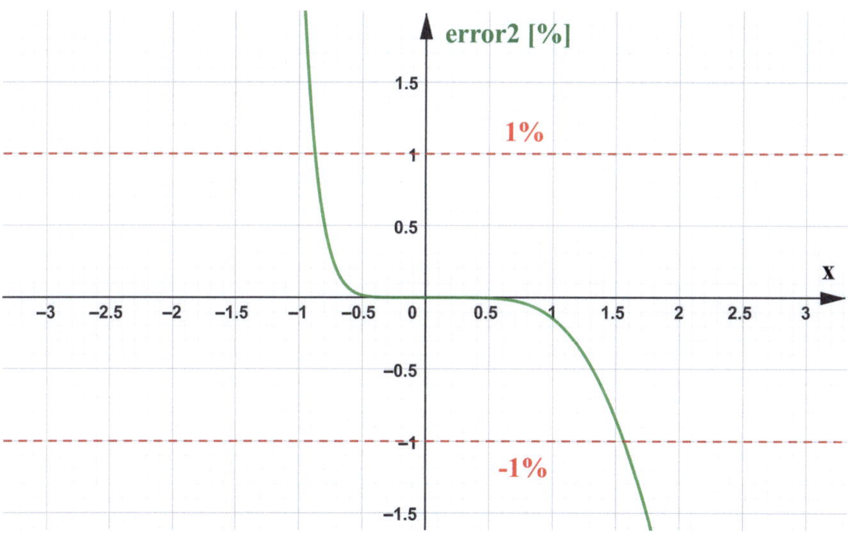

Fig. 1.57 Relative error of approximation of the continuous-log sigmoidal activation function $f_9(x)$ using approximation function $a_{17}(x)$

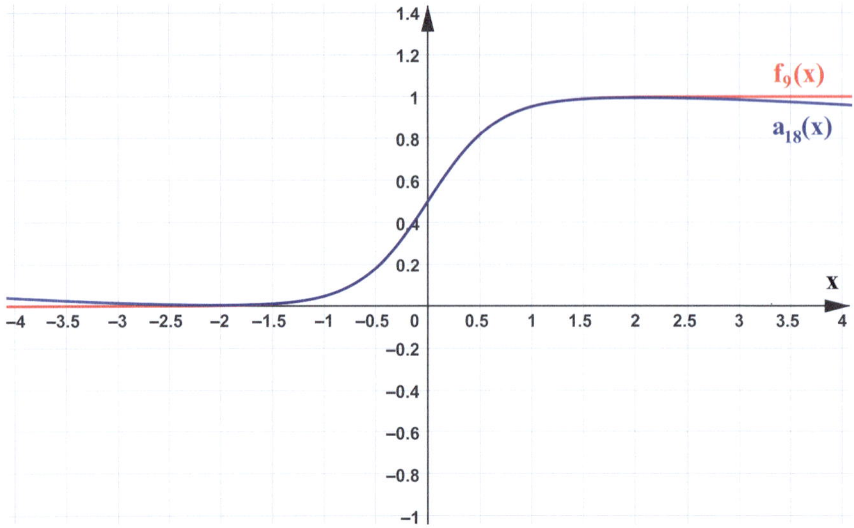

Fig. 1.58 Continuous-log sigmoid activation function $f_9(x)$ and the second approximation function $a_{18}(x)$

c_0, c_1, c_2, c_3, c_4, d_0, d_2 and d_4 being constant coefficients having particular values that can be determined by the condition that the superior-order Taylor series expansion of $f_9(x)$ continuous-log sigmoid activation function, expressed by (1.40), to be identical with the superior-order Taylor series expansion of the general form of the approximation function $a_{18\text{GEN}}(x)$, expressed by (1.43). The values of these coefficients, obtained considering the previous conditions, can be written as: $c_0 = \frac{1}{2}$, $c_1 = \frac{3}{4}$, $c_2 = \frac{27}{56}$, $c_3 = \frac{9}{56}$, $c_4 = \frac{27}{1120}$, $d_0 = 1$, $d_2 = \frac{27}{28}$ and $d_4 = \frac{27}{560}$. Replacing these particular values in relation (1.43), the second approximation function $a_{18}(x)$ has the following mathematical expression:

$$a_{18}(x) = \frac{\frac{27x^4}{1120} + \frac{9x^3}{56} + \frac{27x^2}{56} + \frac{3x}{4} + \frac{1}{2}}{\frac{27x^4}{560} + \frac{27x^2}{28} + 1}. \tag{1.44}$$

The graphical representation of the continuous-log sigmoid activation function $f_9(x)$ and of the second approximation function $a_{18}(x)$ is presented in Fig. 1.58.

The absolute error of approximation of the continuous-log sigmoidal activation function $f_9(x)$ using the second approximation function $a_{18}(x)$ is illustrated in Fig. 1.59. It is defined as the difference between the previous functions, $\text{error1} = f_9(x) - a_{18}(x)$.

For more accurately evaluating the error of approximation, the relative error of approximation is shown in Fig. 1.60. This error is defined as $\text{error2} = 100\frac{f_9(x)-a_{18}(x)}{f_9(x)}$ (in percents). Considering as acceptable a maximal approximation error of $\pm 1\%$, the maximal range of the input variable x is $[-1.3;\ 2.7]$.

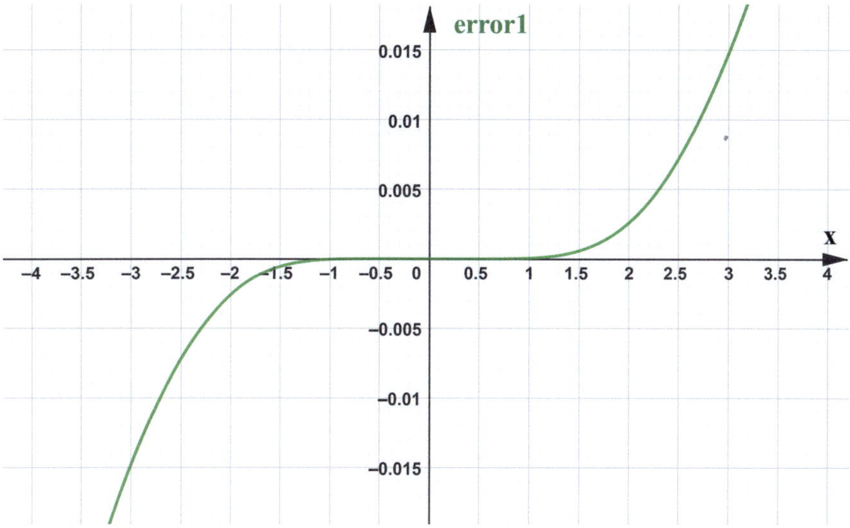

Fig. 1.59 Absolute error of approximation of the continuous-log sigmoid activation function $f_9(x)$ using the second approximation function $a_{18}(x)$

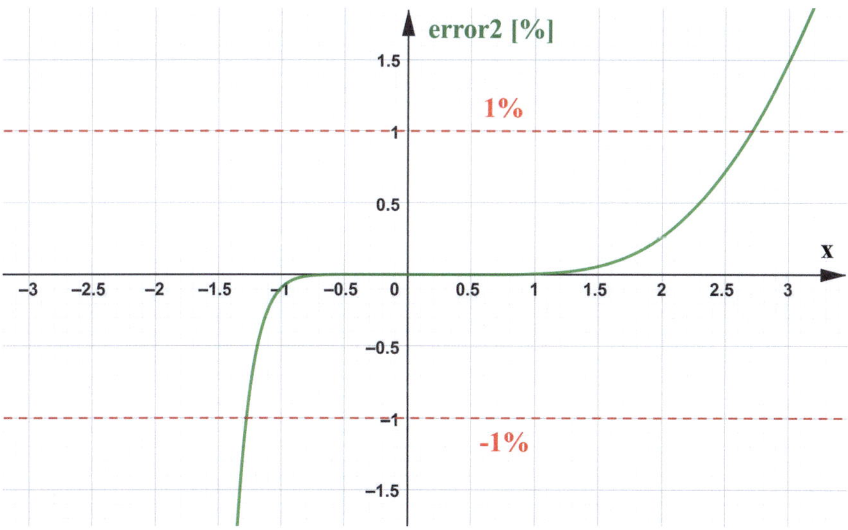

Fig. 1.60 Relative error of approximation of the continuous-log sigmoidal activation function $f_9(x)$ using the second approximation function $a_{18}(x)$

Reference

Popa CR (2023) High-accuracy Gaussian function generator for neural networks. Electronics 12(1):24. https://doi.org/10.3390/electronics12010024

Chapter 2
Superior-Order Approximation Functions for Generating Radial Basis Activation Functions

2.1 Introduction

The neural networks based on radial basis functions (RBF) represent a class of neural networks that are especially used for approximating functions, interpolations, regressions and classifications. There are feedforward networks, but their operation is different. The RBF neural networks have a multitude of applications: models recognition, medical classifications (very useful for deciding the diagnostic), fuzzy systems or function interpolation.

2.2 Approximation Functions for Cubic Activation Function

The cubic activation function is extremely useful in deep neural networks. It can be used as activation function, as an alternative to other activation functions, such as ReLU, hyperbolic tangent or Swish functions. The small disadvantage of the cubic function is a possible instability of deep neural networks that are based on it (because the cubic function is strongly increasing for large values of the input variable). However, the cubic function is very useful in shallow networks with few layers and in networks with batch normalization, as they are able to prevent the strong increasing of its values.

The mathematical form of this function can be expressed as follows:

$$f_1(x) = x^3. \quad (2.1)$$

The graphical representation of the cubic activation function $f_1(x)$ is presented in Fig. 2.1.

C. R. Popa, *Analog Current-Mode Computational Circuits for Artificial Neural Networks*, Analog Circuits and Signal Processing,
https://doi.org/10.1007/978-3-032-03989-7_2

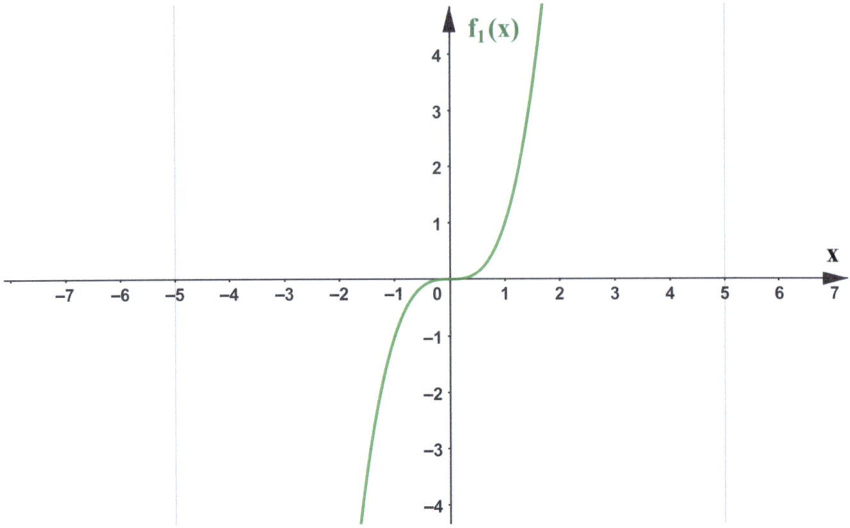

Fig. 2.1 Graphical representation of the cubic activation function

In order to obtain the cubic activation function using convenient elementary functions that could be very easily generated using simple and accurate CMOS computational circuits, the cubic function can be expressed as a product of two functions, $a_1(x)$ and $a_2(x)$:

$$f_1(x) = a_1(x)a_2(x), \tag{2.2}$$

where $a_1(x)$ and $a_2(x)$ have the following expressions:

$$a_1(x) = x \tag{2.3}$$

and:

$$a_2(x) = x^2. \tag{2.4}$$

2.3 Approximation Functions for Gaussian Activation Function

The Gaussian function (or normal distribution function in probability applications) (Popa 2012, 2023) represents a bell-shaped curve. It has a multitude of applications in machine learning, statistics and neural networks, most of its applications being related to radial basis function (RBF) networks: in machine learning, RBF networks use Gaussian functions as activation function in hidden neurons; in statistics, the probability density function (PDF) has a normal distribution; in image processing, it is possible to smooth images by averaging pixels with Gaussian weights.

Because the Gaussian activation strongly responds only to inputs close to the center of the curve, it is excellent for clustering and interpolation. Additionally, its characteristics of being smooth and differentiable makes the Gaussian function a very good choice for gradient-based learning. The Gaussian function is very useful for modeling nonlinear decision boundaries when combined with linear output layers. In consequence, the areas of utilization of the Gaussian activation functions are:

a. ***Statistics and probability***

- normal distribution: the Gaussian function describes the normal distribution, fundamental in statistical modeling.
- central limit theorem: many distributions tend toward a Gaussian as sample size increases.
- error analysis: measurement errors often follow a normal distribution.

b. ***Machine learning and AI***

- Gaussian Naive Bayes classifier.
- Gaussian mixture models (GMMs): for clustering and density estimation.
- kernel functions: in support vector machines (SVM), Gaussian (RBF) kernels are common.
- Gaussian processes: for non-parametric regression and probabilistic modeling.

c. ***Signal processing and image processing***

- Gaussian blur: used in image filtering to reduce noise and detail.
- edge detection: preprocessing step with Gaussian smoothing.
- noise modeling: assumes Gaussian noise in many signal acquisition systems.

d. ***Physics***

- quantum mechanics: the wavefunctions of ground states (like harmonic oscillators) are Gaussian.
- optics: laser beam profiles are often Gaussian.
- heat diffusion: solutions to the diffusion equation are Gaussian.

e. ***Chemistry and spectroscopy***

- spectral line shapes: many absorption/emission lines are Gaussian-shaped.
- molecular modeling: Gaussian functions are used as basis functions in quantum chemistry (for example, Gaussian software).

f. ***Finance***

- asset return modeling: assumes Gaussian distribution in basic models (for example, Black–Scholes).
- risk assessment: value at risk (VaR) and other metrics often assume normality.

g. ***Biology and medicine***

- population genetics: traits often follow a Gaussian distribution.
- medical imaging: Gaussian filters used in preprocessing MRI/CT data.

h. ***Mathematics***

- Fourier transform: the Fourier transform of a Gaussian is another Gaussian.
- probability theory: as the basis of many theorems and approximations.

2.3.1 General Form of the Gaussian Activation Function

The mathematical form of this function can be expressed as follows:

$$f_2(x, a, b, c) = a \exp\left[-\frac{(x-b)^2}{2\sigma^2}\right], \tag{2.5}$$

where

a is the amplitude (height of the curve's peak),
b is the mean (position of the peak center) and
σ is the standard deviation.

The graphical representation of the Gaussian activation function $f_2(x)$ for $a = 1$ and $\sigma = \frac{1}{\sqrt{2}}$, considering three values of b parameter ($b = -1$, $b = 0$ and $b = 1$), is presented in Fig. 2.2.

The graphical representation of the Gaussian activation function $f_2(x)$ for $a = 1$ and $\sigma = 2$, considering three values of b parameter ($b = -1$, $b = 0$ and $b = 1$), is presented in Fig. 2.3.

The graphical representation of the Gaussian activation function $f_2(x)$ for $a = 1$ and $\sigma = 1/4$, considering three values of b parameter ($b = -1$, $b = 0$ and $b = 1$), is presented in Fig. 2.4.

The graphical representation of the Gaussian activation function $f_2(x)$ for $a = 5$ and $\sigma = \frac{1}{\sqrt{2}}$, considering three values of b parameter ($b = -3$, $b = 0$ and $b = 3$), is presented in Fig. 2.5.

The graphical representation of the Gaussian activation function $f_2(x)$ for $a = 5$ and $\sigma = 2$, considering three values of b parameter ($b = -3$, $b = 0$ and $b = 3$), is presented in Fig. 2.6.

The graphical representation of the Gaussian activation function $f_2(x)$ for $a = 5$ and $\sigma = 1/4$, considering three values of b parameter ($b = -3$, $b = 0$ and $b = 3$), is presented in Fig. 2.7.

Considering one of the simplest cases $a = 1, b = 0$ and $\sigma = \frac{1}{\sqrt{2}}$, the mathematical form (2.5) of the Gaussian activation function becomes:

$$f_3(x) = \exp\left(-x^2\right). \tag{2.6}$$

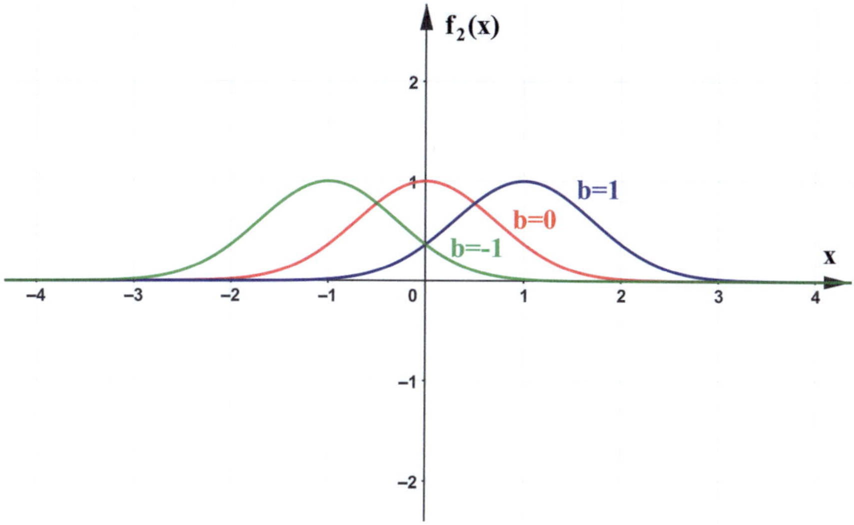

Fig. 2.2 Gaussian activation function $f_2(x)$ for $a = 1, \sigma = 1/\sqrt{2}$ and b parameter ($b = -1, b = 0$ and $b = 1$)

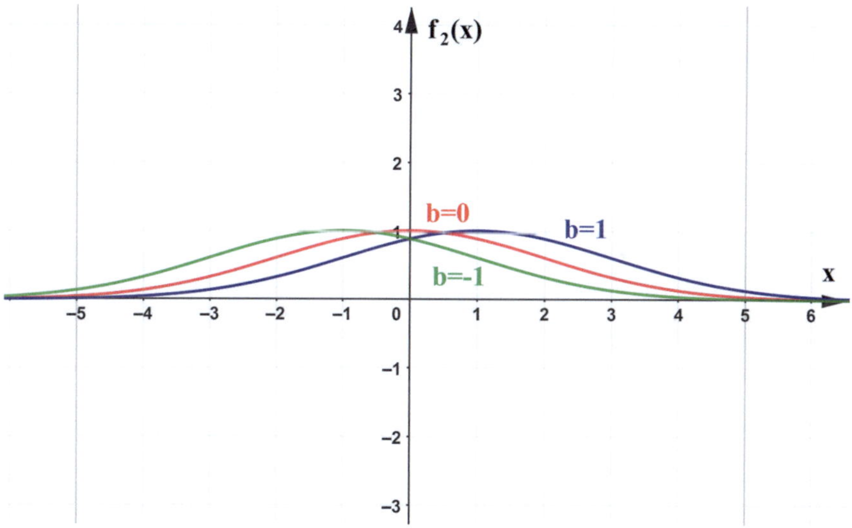

Fig. 2.3 Gaussian activation function $f_2(x)$ for $a = 1, \sigma = 2$ and b parameter ($b = -1, b = 0$ and $b = 1$)

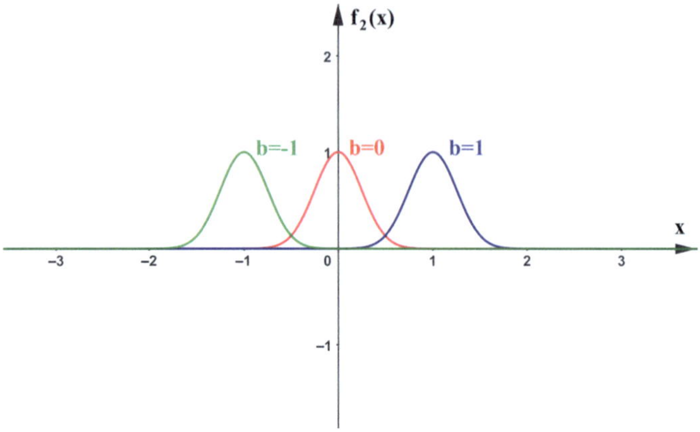

Fig. 2.4 Gaussian activation function $f_2(x)$ for $a = 1$, $\sigma = 1/4$ and b parameter ($b = -1$, $b = 0$ and $b = 1$)

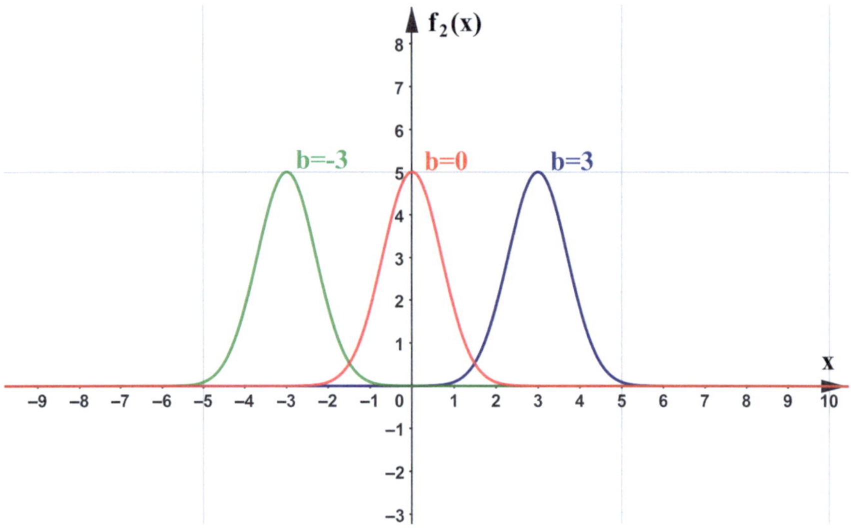

Fig. 2.5 Gaussian activation function $f_2(x)$ for $a = 5$, $\sigma = 1/\sqrt{2}$ and b parameter ($b = -3$, $b = 0$ and $b = 3$)

The more general form of the Gaussian activation function, often used in a multitude of practical applications, is also derived from the most general form (2.5) of the Gaussian activation function, particularized for $a = 1$ and $b = 0$:

$$f_3(x, \sigma) = \exp\left(-\frac{x^2}{2\sigma^2}\right). \tag{2.7}$$

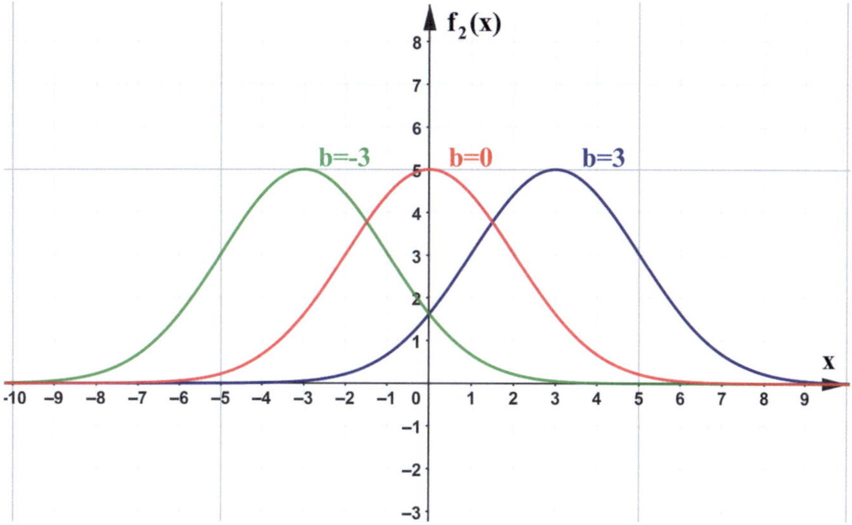

Fig. 2.6 Gaussian activation function $f_2(x)$ for $a = 5$, $\sigma = 2$ and b parameter ($b = -3$, $b = 0$ and $b = 3$)

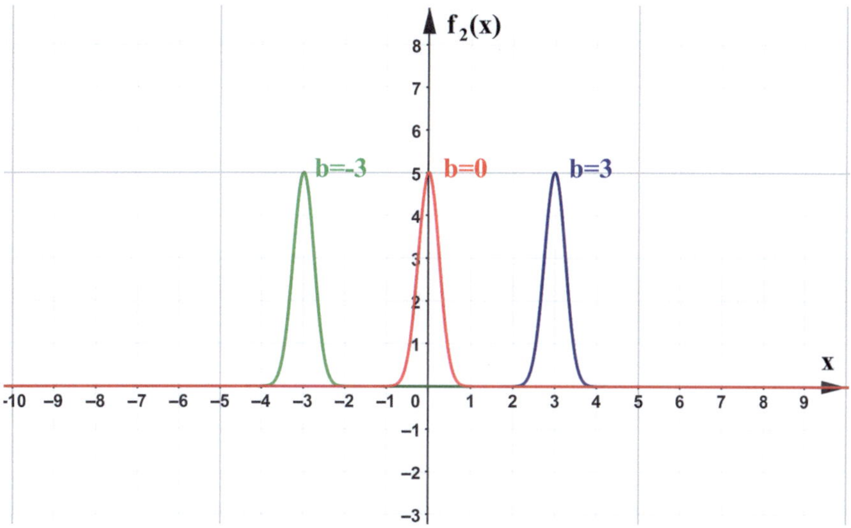

Fig. 2.7 Gaussian activation function $f_2(x)$ for $a = 5$, $\sigma = 1/4$ and b parameter ($b = -1$, $b = 0$ and $b = 1$)

In order to obtain an analog implementation of the Gaussian activation function, a multitude of superior-order approximation functions is presented. The most important differences between these functions are mainly related to the accuracy of the approximation, as well as to the harware resources required for their implementations in CMOS technology. In order to simplify the mathematical analysis, these superior-order approximation functions are developed for generating the simplest Gaussian activation function, expressed by relation (2.6). The generate of the more general mathematical form of the Gaussian activation function is posssible introducing small changes in the previous mentioned superior-order approximation functions.

2.3.2 *First Approximation Function for Generating the Gaussian Activation Function*

In order to approximate the Gaussian activation function $f_3(x)$, a new approximation function $a_3(x)$ is presented. Its particular mathematical form is developed having two very important objectives. The first objective is represented by the accuracy of the approximation that could be obtained using this approximation function. The second objective is related to the possibility of its facile analog implementation in CMOS technology using as reasonable as possible hardware resources (see Chaps. 4 and 6).

Because the Gaussian function is an even mathematical function, the general form of the first proposed approximation function $a_3(x)$ is the following:

$$a_{3\mathrm{GEN}}(x) = \frac{c_2 x^2 + c_0}{d_2 x^2 + d_0}, \tag{2.8}$$

c_0, c_2, d_0 and d_2 being constant coefficients having particular values that can be determined by the condition that the superior-order Taylor series expansion of $f_3(x)$ Gaussian activation function, expressed by (2.6), to be identical with the superior-order Taylor series expansion of the general form of the approximation function $a_{3\mathrm{GEN}}(x)$, expressed by (2.8).

The particular form of the approximation function is developed as a ratio of two polynomial functions: the numerator degree is two, while the denominator degree is also two. The values of these coefficients, obtained considerring the previous conditions, can be witten as: $c_0 = 1$, $c_2 = -\frac{1}{2}$, $d_0 = 1$ and $d_2 = \frac{1}{2}$. Replacing these particular values in relation (2.8), the first approximation function $a_3(x)$ has the following mathematical expression:

$$a_3(x) = \frac{1 - \frac{x^2}{2}}{1 + \frac{x^2}{2}}. \tag{2.9}$$

The graphical representation of the Gaussian activation function $f_3(x)$ and of the first approximation function $a_3(x)$ is presented in Fig. 2.8.

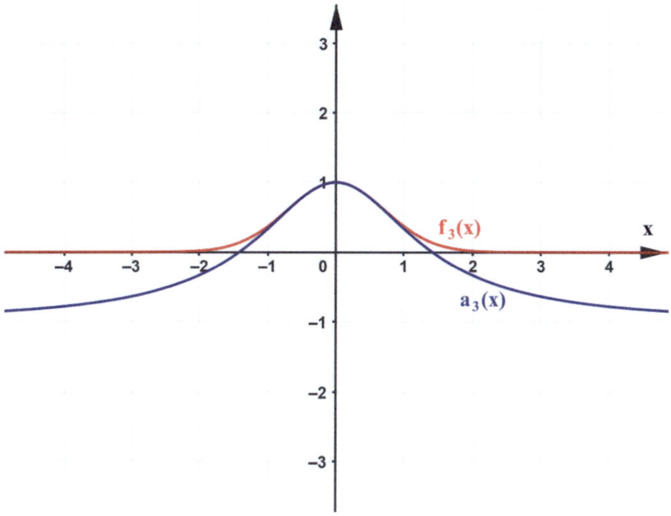

Fig. 2.8 Gaussian activation function $f_3(x)$ and the first approximation function $a_3(x)$

The absolute error of approximation of the Gaussian activation function $f_3(x)$ using the approximation function $a_3(x)$ is illustrated in Fig. 2.9. It is defined as the difference between the previous functions, error1 $= f_3(x) - a_3(x)$.

For more accurately evaluating the error of approximation, the relative error of approximation is shown in Fig. 2.10. This error is defined as error2 $= 100\frac{f_3(x)-a_3(x)}{f_3(x)}$

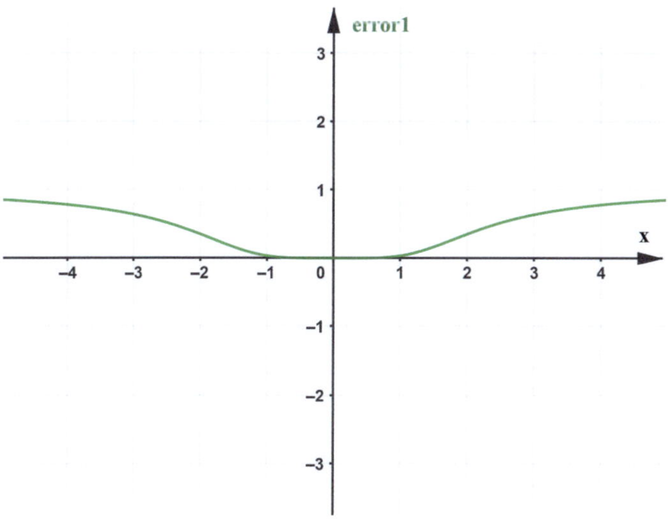

Fig. 2.9 Absolute error of approximation of the Gaussian activation function $f_3(x)$ using approximation function $a_3(x)$

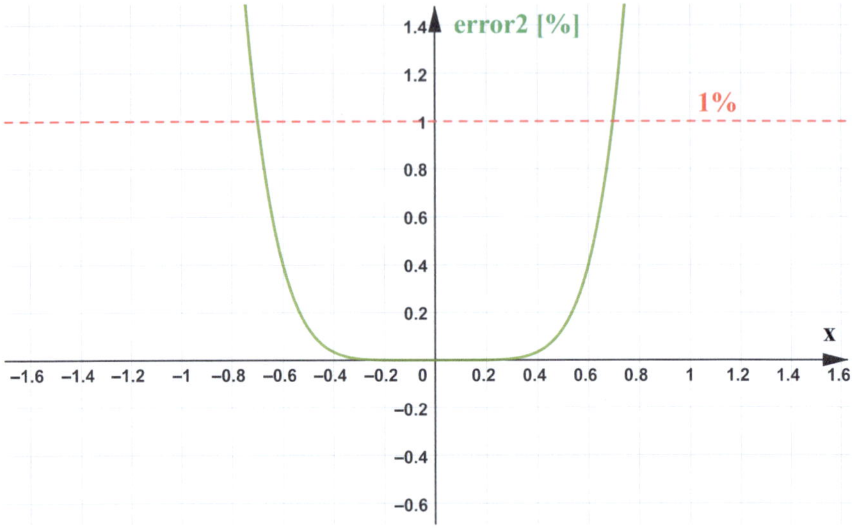

Fig. 2.10 Relative error of approximation of the Gaussian activation function $f_3(x)$ using approximation function $a_3(x)$

(in percents). Considering as acceptable a maximal approximation error of $\pm 1\%$, the maximal range of the input variable x is $[-0.7\,;\ 0.7]$.

In order to approximate the $f_3(x, \sigma)$ more general Gaussian activation function expressed by (2.7), $x \to \frac{x}{\sigma\sqrt{2}}$ variable changing must be used in the previous relation (2.9) of the $a_3(x)$ particular approximation function, resulting the following mathematical form of the general approximation function $a_3(x, \sigma)$:

$$a_3(x, \sigma) = \frac{1 - \frac{x^2}{4\sigma^2}}{1 + \frac{x^2}{4\sigma^2}}. \tag{2.10}$$

2.3.3 *Second Approximation Function for Generating the Gaussian Activation Function*

In order to approximate the Gaussian activation function $f_3(x)$, a new approximation function $a_4(x)$ is presented. Its mathematical form is obtained exploiting the following particularity of the Gaussian activation function, based on the variable changing $x \to \frac{x}{2}$:

$$\exp\left(-x^2\right) = \exp\left[-\left(\frac{x}{2}\right)^2\right]^4. \tag{2.11}$$

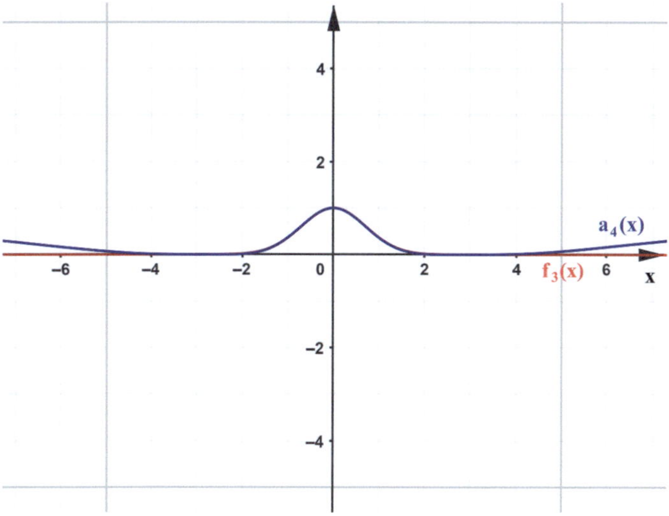

Fig. 2.11 Gaussian activation function $f_3(x)$ and the second approximation function $a_4(x)$

This variable changing strongly increases the accuracy of approximating the Gaussian activation function, by decreasing the equivalent variable range of the input variable x. Using this variable changing method, the new improved accuracy approximation function $a_4(x)$, based on the previous approximation function $a_3(x)$, has the following expression:

$$a_4(x) = \left(\frac{1 - \frac{x^2}{8}}{1 + \frac{x^2}{8}} \right)^4. \tag{2.12}$$

The graphical representation of the Gaussian activation function $f_3(x)$ and of the second approximation function $a_4(x)$ is presented in Fig. 2.11.

The absolute error of approximation of the Gaussian activation function $f_3(x)$ using the approximation function $a_4(x)$ is illustrated in Fig. 2.12. It is defined as the difference between the previous functions, error1 $= f_3(x) - a_4(x)$.

For more accurately evaluating the error of approximation, the relative error of approximation is shown in Fig. 2.13. This error is defined as error2 $= 100\frac{f_3(x) - a_4(x)}{f_3(x)}$ (in percents). Considering as acceptable a maximal approximation error of $\pm 1\%$, the maximal range of the input variable x is $[-1.1\,;\ 1.1]$.

In order to approximate the $f_3(x, \sigma)$ more general Gaussian activation function expressed by (2.7), $x \to \frac{x}{\sigma\sqrt{2}}$ variable changing must be used in the previous relation (2.12) of the $a_4(x)$ particular approximation function, resulting the following mathematical form of the general approximation function $a_4(x, \sigma)$:

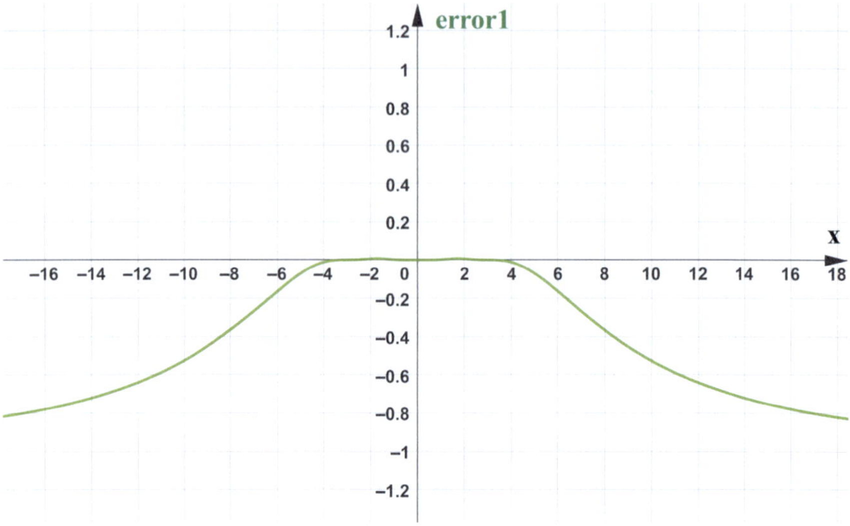

Fig. 2.12 Absolute error of approximation of the Gaussian activation function $f_3(x)$ using approximation function $a_4(x)$

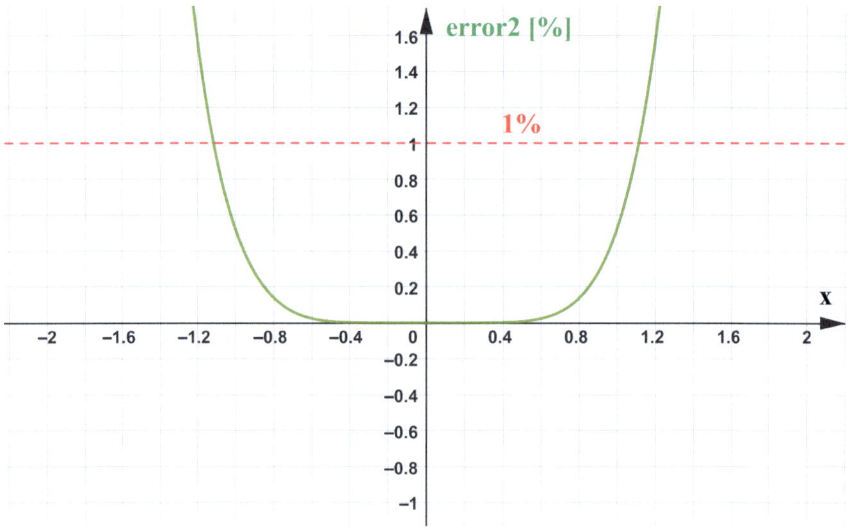

Fig. 2.13 Relative error of approximation of the Gaussian activation function $f_3(x)$ using approximation function $a_4(x)$

$$a_4(x,\sigma) = \left(\frac{1 - \frac{x^2}{16\sigma^2}}{1 + \frac{x^2}{16\sigma^2}} \right)^4. \tag{2.13}$$

2.3.4 Third Approximation Function for Generating the Gaussian Activation Function

In order to approximate the Gaussian activation function $f_3(x)$, a new approximation function $a_5(x)$ is presented. Because the Gaussian function is an even mathematical function, the general form of the proposed third approximation function $a_5(x)$ could be expressed as follows:

$$a_{5\mathrm{GEN}}(x) = \frac{c_4x^4 + c_2x^2 + c_0}{d_2x^2 + d_0}, \tag{2.14}$$

c_0, c_2, c_4, d_0 and d_2 being constant coefficients having particular values that can be determined by the condition that the superior-order Taylor series expansion of $f_3(x)$ Gaussian activation function, expressed by (2.6), to be identical with the superior-order Taylor series expansion of the general form of the approximation function $a_{5\mathrm{GEN}}(x)$, expressed by (2.14).

The particular form of the approximation function is developed as a ratio of two polynomial functions: the numerator degree is four, while the denominator degree is two. The values of these coefficients, obtained considerring the previous conditions, can be witten as: $c_0 = 1$, $c_2 = -\frac{2}{3}$, $c_4 = \frac{1}{6}$, $d_0 = 1$ and $d_2 = \frac{1}{3}$. Replacing these particular values in relation (2.14), the third approximation function $a_5(x)$ has the following mathematical expression:

$$a_5(x) = \frac{\frac{x^4}{6} - \frac{2x^2}{3} + 1}{\frac{x^2}{3} + 1}. \tag{2.15}$$

The graphical representation of the Gaussian activation function $f_3(x)$ and of the third approximation function $a_5(x)$ is presented in Fig. 2.14.

The absolute error of approximation of the Gaussian activation function $f_3(x)$ using the approximation function $a_5(x)$ is illustrated in Fig. 2.15. It is defined as the difference between the previous functions, error1 $= f_3(x) - a_5(x)$.

For more accurately evaluating the error of approximation, the relative error of approximation is shown in Fig. 2.16. This error is defined as error2 $= 100\frac{f_3(x) - a_5(x)}{f_3(x)}$ (in percents). Considering as acceptable a maximal approximation error of $\pm 1\%$, the maximal range of the input variable x is $[-0.92\,;\,0.92]$.

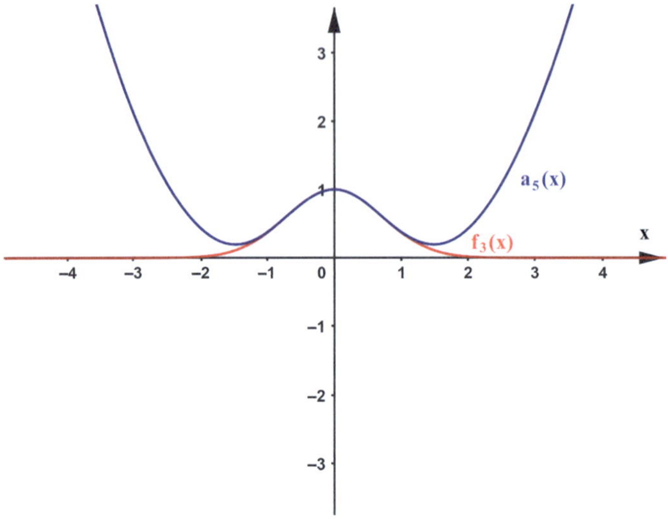

Fig. 2.14 Gaussian activation function $f_3(x)$ and the third approximation function $a_5(x)$

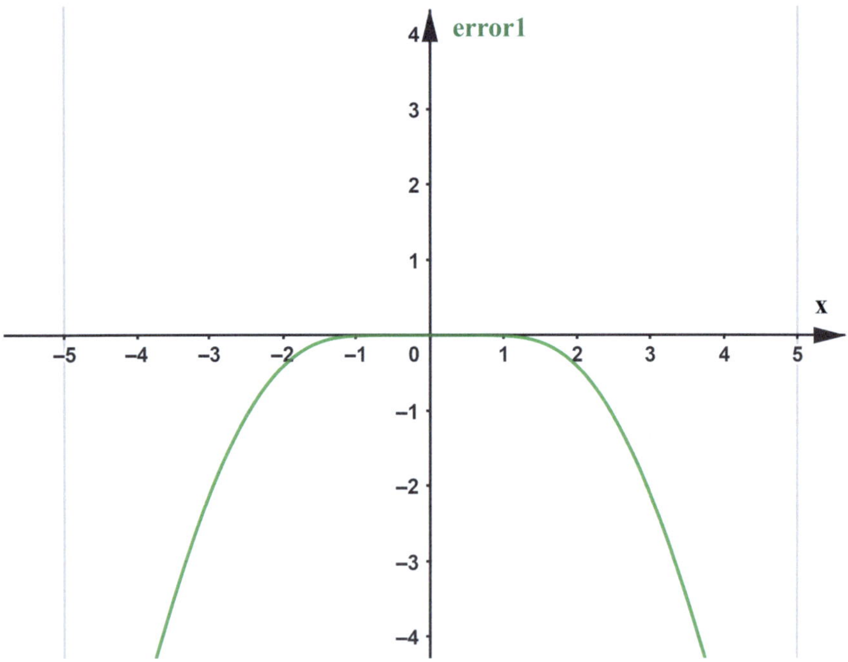

Fig. 2.15 Absolute error of approximation of the Gaussian activation function $f_3(x)$ using approximation function $a_5(x)$

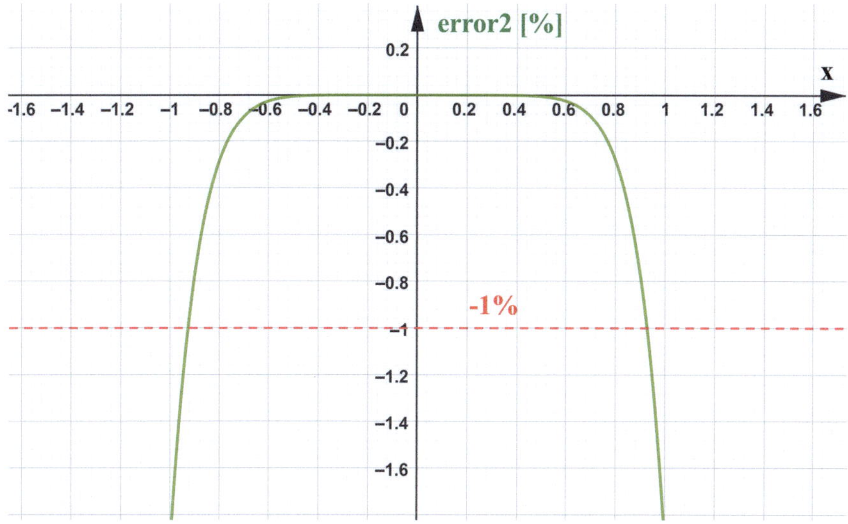

Fig. 2.16 Relative error of approximation of the Gaussian activation function $f_3(x)$ using approximation function $a_5(x)$

In order to approximate the $f_3(x, \sigma)$ more general Gaussian activation function expressed by (2.7), $x \rightarrow \frac{x}{\sigma\sqrt{2}}$ variable changing must be used in the previous relation (2.15) of the $a_5(x)$ particular approximation function, resulting the following mathematical form of the general approximation function $a_5(x, \sigma)$:

$$a_5(x, \sigma) = \frac{\frac{x^4}{24\sigma^4} - \frac{x^2}{3\sigma^2} + 1}{\frac{x^2}{6\sigma^2} + 1}. \tag{2.16}$$

2.3.5 *Fourth Approximation Function for Generating the Gaussian Activation Function*

The variable changing method, expressed by relation (2.11), is used to approximate the Gaussian activation function $f_3(x)$. The new improved accuracy approximation function $a_6(x)$, based on the previous approximation function $a_5(x)$, has the following expression:

$$a_6(x) = \left(\frac{\frac{x^4}{96} - \frac{x^2}{6} + 1}{\frac{x^2}{12} + 1} \right)^4. \tag{2.17}$$

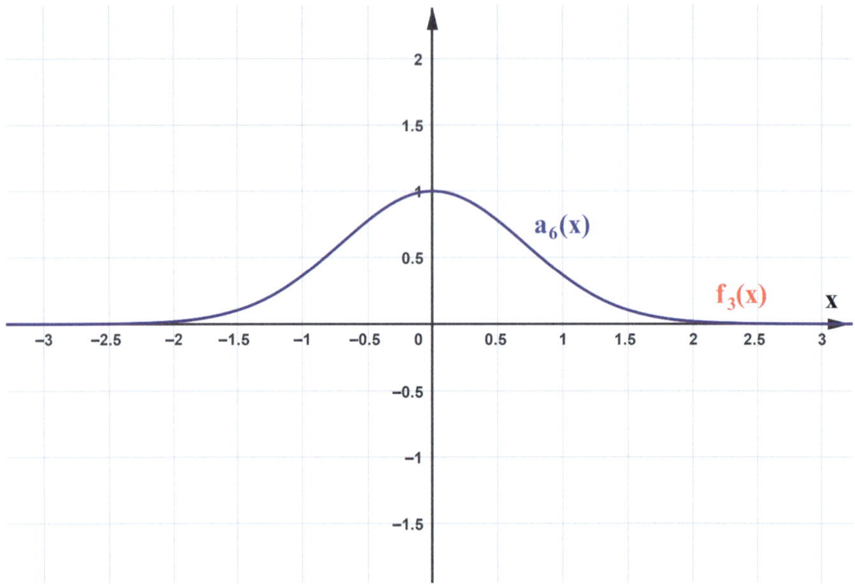

Fig. 2.17 Gaussian activation function $f_3(x)$ and the fourth approximation function $a_6(x)$

The graphical representation of the Gaussian activation function $f_3(x)$ and of the fourth approximation function $a_6(x)$ is presented in Fig. 2.17.

The absolute error of approximation of the Gaussian activation function $f_3(x)$ using the approximation function $a_3(x)$ is illustrated in Fig. 2.18. It is defined as the difference between the previous functions, error1 $= f_3(x) - a_6(x)$.

For more accurately evaluating the error of approximation, the relative error of approximation is shown in Fig. 2.19. This error is defined as error2 $= 100\frac{f_3(x)-a_6(x)}{f_3(x)}$ (in percents). Considering as acceptable a maximal approximation error of $\pm 1\%$, the maximal range of the input variable x is $[-1.6;\ 1.6]$.

In order to approximate the $f_3(x,\sigma)$ more general Gaussian activation function expressed by (2.7), $x \to \frac{x}{\sigma\sqrt{2}}$ variable changing must be used in the previous relation (2.17) of the $a_6(x)$ particular approximation function, resulting the following mathematical form of the general approximation function $a_6(x,\sigma)$:

$$a_6(x,\sigma) = \left(\frac{\frac{x^4}{384\sigma^4} - \frac{x^2}{12\sigma^2} + 1}{\frac{x^2}{24\sigma^2} + 1}\right)^4. \tag{2.18}$$

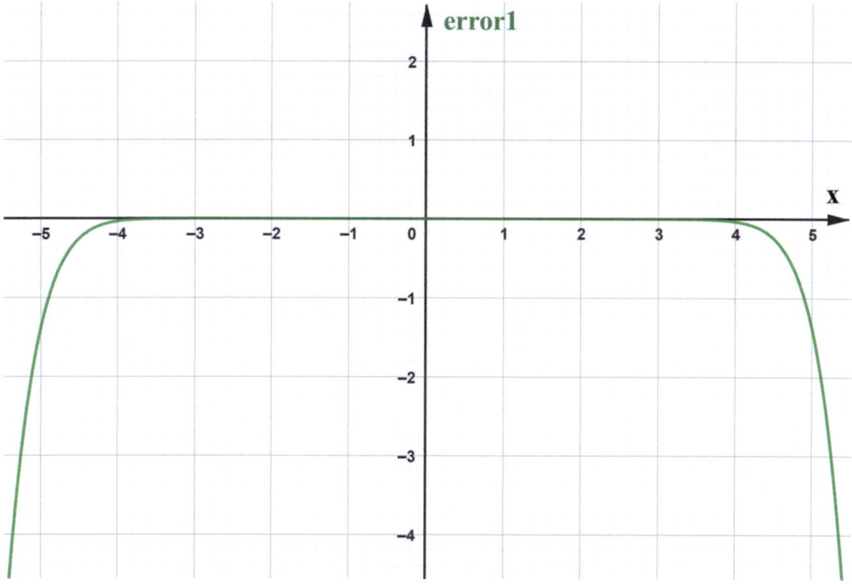

Fig. 2.18 Absolute error of approximation of the Gaussian activation function $f_3(x)$ using approximation function $a_6(x)$

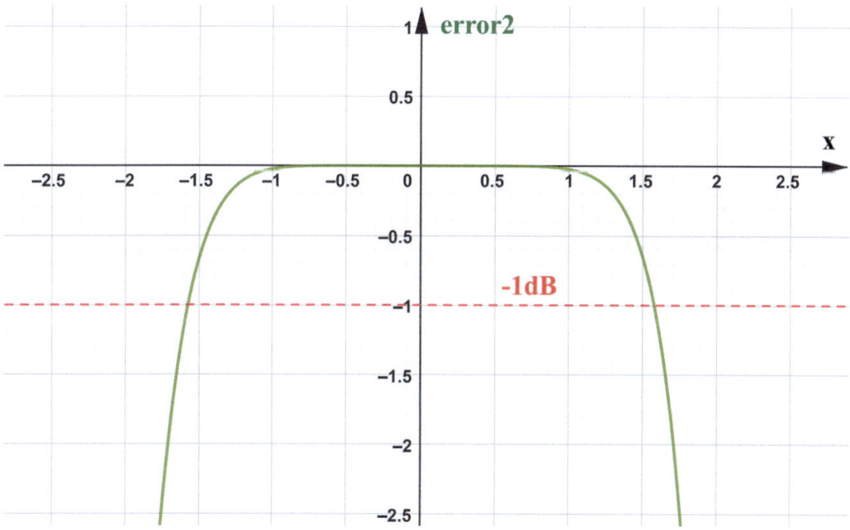

Fig. 2.19 Relative error of approximation of the Gaussian activation function $f_3(x)$ using approximation function $a_6(x)$

2.3.6 Fiveth Approximation Function for Generating the Gaussian Activation Function

In order to approximate the Gaussian activation function $f_3(x)$, a new approximation function $a_7(x)$ is presented. Because the Gaussian function is an even mathematical function, the general form of the proposed fiveth approximation function $a_7(x)$ can be expressed as follows:

$$a_{7\text{GEN}}(x) = \frac{c_2x^2 + c_0}{d_4x^4 + d_2x^2 + d_0}, \tag{2.19}$$

c_0, c_2, d_0, d_2 and d_4 being constant coefficients having particular values that can be determined by the condition that the superior-order Taylor series expansion of $f_3(x)$ Gaussian activation function, expressed by (2.6), to be identical with the superior-order Taylor series expansion of the general form of the approximation function $a_{7\text{GEN}}(x)$, expressed by (2.19).

The particular form of the approximation function is developed as a ratio of two polynomial functions: the numerator degree is two, while the denominator degree is four. The values of these coefficients, obtained considerring the previous conditions, can be witten as: $c_0 = 1$, $c_2 = -\frac{1}{3}$, $d_0 = 1$, $d_2 = \frac{2}{3}$ and $d_4 = \frac{1}{6}$. Replacing these particular values in relation (2.19), the fiveth approximation function $a_7(x)$ has the following mathematical expression:

$$a_7(x) = \frac{1 - \frac{x^2}{3}}{1 + \frac{2x^2}{3} + \frac{x^4}{6}}. \tag{2.20}$$

The graphical representation of the Gaussian activation function $f_3(x)$ and of the fiveth approximation function $a_7(x)$ is presented in Fig. 2.20.

The absolute error of approximation of the Gaussian activation function $f_3(x)$ using the approximation function $a_7(x)$ is illustrated in Fig. 2.21. It is defined as the difference between the previous functions, $\text{error1} = f_3(x) - a_7(x)$.

For more accurately evaluating the error of approximation, the relative error of approximation is shown in Fig. 2.22. This error is defined as $\text{error2} = 100\frac{f_3(x) - a_7(x)}{f_7(x)}$ (in percents). Considering as acceptable a maximal approximation error of $\pm 1\%$, the maximal range of the input variable x is $[-1\,;\ 1]$.

In order to approximate the $f_3(x, \sigma)$ more general Gaussian activation function expressed by (2.7), $x \to \frac{x}{\sigma\sqrt{2}}$ variable changing must be used in the previous relation (2.20) of the $a_7(x)$ particular approximation function, resulting the following mathematical form of the general approximation function $a_7(x, \sigma)$:

$$a_7(x, \sigma) = \frac{1 - \frac{x^2}{6\sigma^2}}{1 + \frac{x^2}{3\sigma^2} + \frac{x^4}{24\sigma^4}}. \tag{2.21}$$

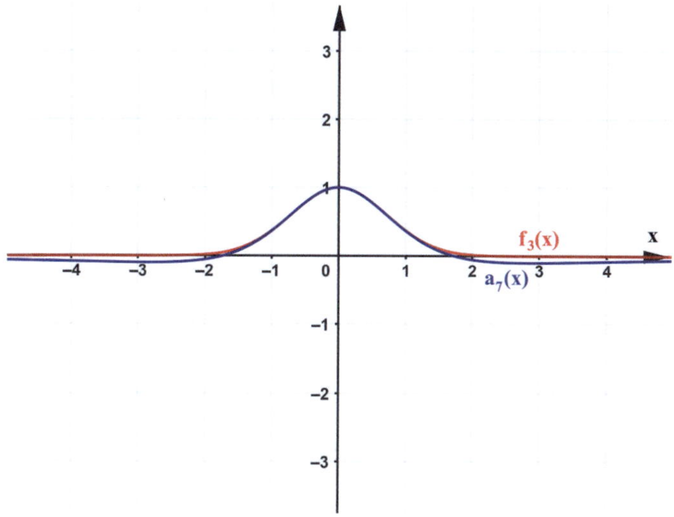

Fig. 2.20 Gaussian activation function $f_3(x)$ and the fiveth approximation function $a_7(x)$

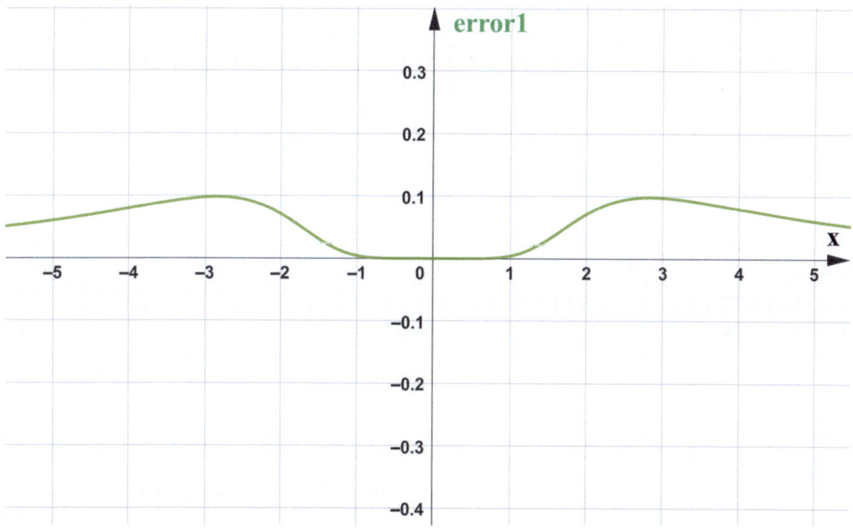

Fig. 2.21 Absolute error of approximation of the Gaussian activation function $f_3(x)$ using approximation function $a_7(x)$

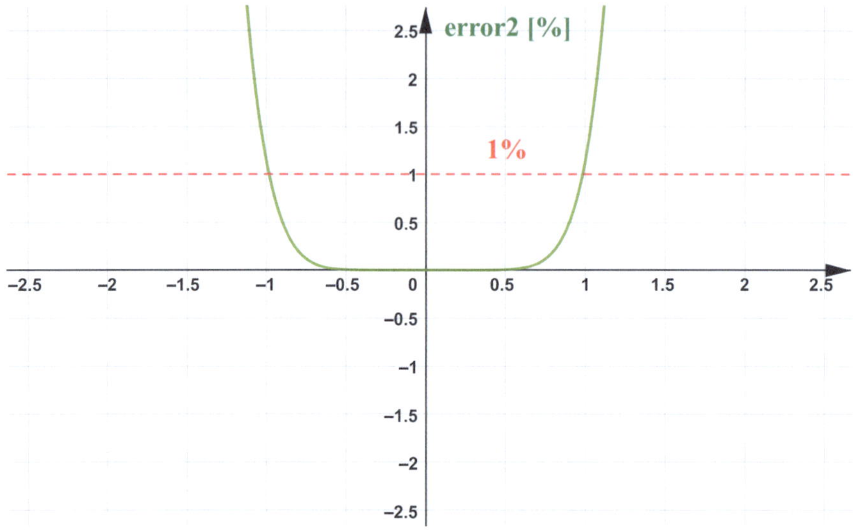

Fig. 2.22 Relative error of approximation of the Gaussian activation function $f_3(x)$ using approximation function $a_7(x)$

2.3.7 *Sixth Approximation Function for Generating the Gaussian Activation Function*

The variable changing method, expressed by relation (2.11), is used to approximate the Gaussian activation function $f_3(x)$. The new improved accuracy approximation function $a_8(x)$, based on the previous approximation function $a_7(x)$, has the following expression:

$$a_8(x) = \left(\frac{1 - \frac{x^2}{12}}{1 + \frac{x^2}{6} + \frac{x^4}{96}} \right)^4. \tag{2.22}$$

The graphical representation of the Gaussian activation function $f_3(x)$ and of the sixth approximation function $a_8(x)$ is presented in Fig. 2.23.

The absolute error of approximation of the Gaussian activation function $f_3(x)$ using the approximation function $a_8(x)$ is illustrated in Fig. 2.24. It is defined as the difference between the previous functions, $\text{error1} = f_3(x) - a_8(x)$.

For more accurately evaluating the error of approximation, the relative error of approximation is shown in Fig. 2.25. This error is defined as $\text{error2} = 100\frac{f_3(x)-a_8(x)}{f_3(x)}$ (in percents). Considering as acceptable a maximal approximation error of $\pm 1\%$, the maximal range of the input variable x is $[-1.7\,;\ 1.7]$.

In order to approximate the $f_3(x, \sigma)$ more general Gaussian activation function expressed by (2.7), $x \to \frac{x}{\sigma\sqrt{2}}$ variable changing must be used in the previous relation (2.22) of the $a_8(x)$ particular approximation function, resulting the following

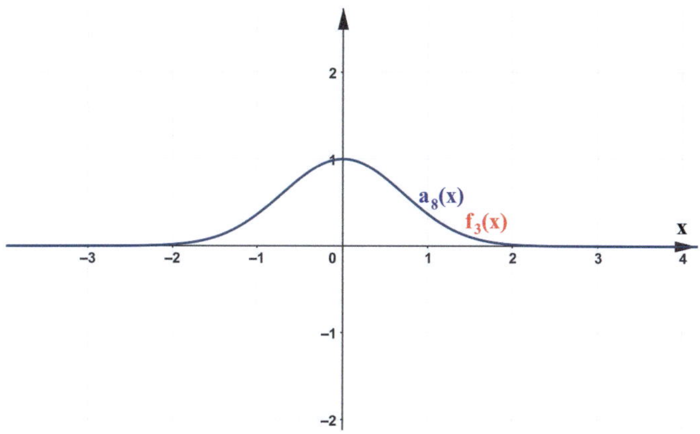

Fig. 2.23 Gaussian activation function $f_3(x)$ and the sixth approximation function $a_8(x)$

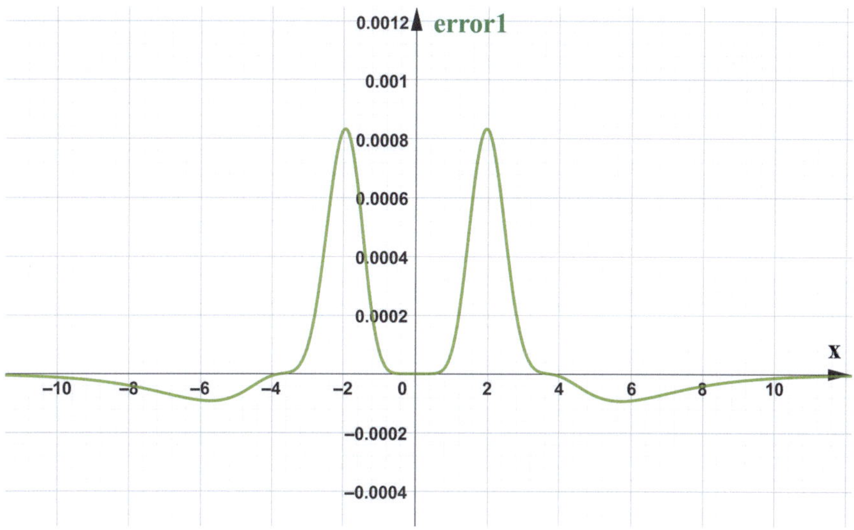

Fig. 2.24 Absolute error of approximation of the Gaussian activation function $f_3(x)$ using approximation function $a_8(x)$

mathematical form of the general approximation function $a_8(x, \sigma)$:

$$a_8(x, \sigma) = \left(\frac{1 - \frac{x^2}{24\sigma^2}}{1 + \frac{x^2}{12\sigma^2} + \frac{x^4}{384\sigma^4}} \right)^4. \tag{2.23}$$

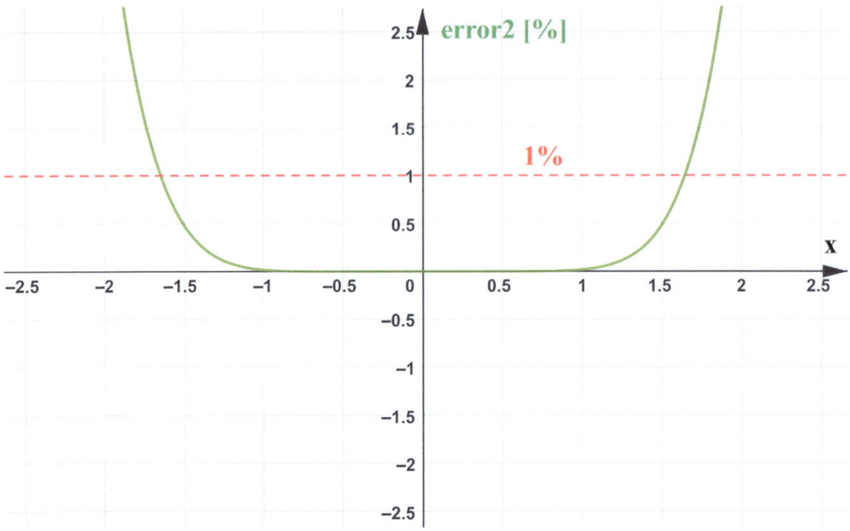

Fig. 2.25 Relative error of approximation of the Gaussian activation function $f_3(x)$ using approximation function $a_8(x)$

2.3.8 Seventh Approximation Function for Generating the Gaussian Activation Function

In order to approximate the Gaussian activation function $f_3(x)$, a new approximation function $a_9(x)$ is presented. Because the Gaussian function is an even mathematical function, the general form of the proposed seventh approximation function $a_9(x)$ can be expressed as follows:

$$a_{9\text{GEN}}(x) = \frac{c_4x^4 + c_2x^2 + c_0}{d_4x^4 + d_2x^2 + d_0}, \tag{2.24}$$

c_0, c_2, c_4, d_0, d_2 and d_4 being constant coefficients having particular values that can be determined by the condition that the superior-order Taylor series expansion of $f_3(x)$ Gaussian activation function, expressed by (2.6), to be identical with the superior-order Taylor series expansion of the general form of the approximation function $a_{9\text{GEN}}(x)$, expressed by (2.24).

The particular form of the approximation function is developed as a ratio of two polynomial functions: the numerator degree is four, while the denominator degree is also four. The values of these coefficients, obtained considerring the previous conditions, can be witten as: $c_0 = 1$, $c_2 = -\frac{1}{2}$, $c_4 = \frac{1}{12}$, $d_0 = 1$, $d_2 = \frac{1}{2}$ and $d_4 = \frac{1}{12}$. Replacing these particular values in relation (2.24), the seventh approximation function $a_9(x)$ has the following mathematical expression:

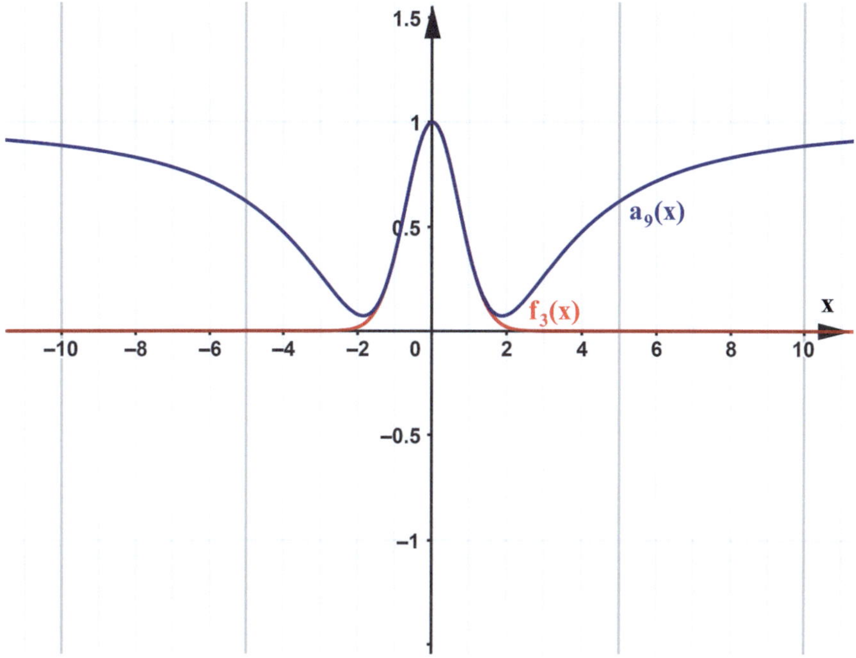

Fig. 2.26 Gaussian activation function $f_3(x)$ and the seventh approximation function $a_9(x)$

$$a_9(x) = \frac{1 - \frac{x^2}{2} + \frac{x^4}{12}}{1 + \frac{x^2}{2} + \frac{x^4}{12}}. \tag{2.25}$$

The graphical representation of the Gaussian activation function $f_3(x)$ and of the seventh approximation function $a_9(x)$ is presented in Fig. 2.26.

The absolute error of approximation of the Gaussian activation function $f_3(x)$ using the approximation function $a_9(x)$ is illustrated in Fig. 2.27. It is defined as the difference between the previous functions, error1 $= f_3(x) - a_9(x)$.

For more accurately evaluating the error of approximation, the relative error of approximation is shown in Fig. 2.28. This error is defined as error2 $= 100\frac{f_3(x) - a_9(x)}{f_3(x)}$ (in percents). Considering as acceptable a maximal approximation error of $\pm 1\%$, the maximal range of the input variable x is $[-1.2\,;\ 1.2]$.

In order to approximate the $f_3(x, \sigma)$ more general Gaussian activation function expressed by (2.7), $x \rightarrow \frac{x}{\sigma\sqrt{2}}$ variable changing must be used in the previous relation (2.25) of the $a_9(x)$ particular approximation function, resulting the following mathematical form of the general approximation function $a_9(x, \sigma)$:

$$a_9(x, \sigma) = \frac{1 - \frac{x^2}{4\sigma^2} + \frac{x^4}{48\sigma^4}}{1 + \frac{x^2}{4\sigma^2} + \frac{x^4}{48\sigma^4}}. \tag{2.26}$$

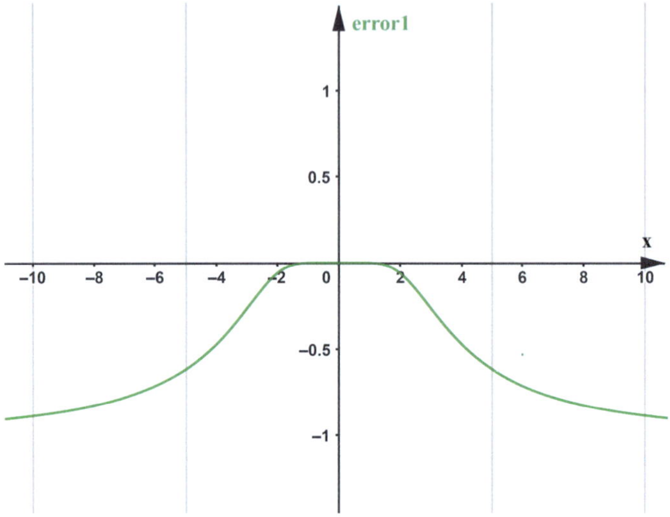

Fig. 2.27 Absolute error of approximation of the Gaussian activation function $f_3(x)$ using approximation function $a_9(x)$

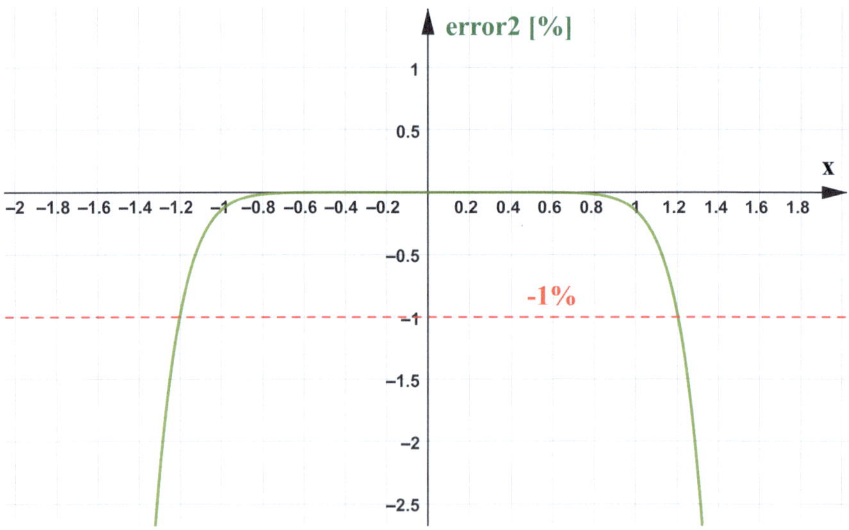

Fig. 2.28 Relative error of approximation of the Gaussian activation function $f_3(x)$ using approximation function $a_9(x)$

2.3.9 *Eighth Approximation Function for Generating the Gaussian Activation Function*

The variable changing method, expressed by relation (2.11), is used to approximate the Gaussian activation function $f_3(x)$. The new improved accuracy approximation function $a_{10}(x)$, based on the previous approximation function $a_9(x)$, has the following expression:

$$a_{10}(x) = \left(\frac{1 - \frac{x^2}{8} + \frac{x^4}{192}}{1 + \frac{x^2}{8} + \frac{x^4}{192}} \right)^4. \tag{2.27}$$

The graphical representation of the Gaussian activation function $f_3(x)$ and of the eighth approximation function $a_{10}(x)$ is presented in Fig. 2.29.

The absolute error of approximation of the Gaussian activation function $f_3(x)$ using the approximation function $a_{10}(x)$ is illustrated in Fig. 2.30. It is defined as the difference between the previous functions, error1 $= f_3(x) - a_{10}(x)$.

For more accurately evaluating the error of approximation, the relative error of approximation is shown in Fig. 2.31. This error is defined as error2 $= 100\frac{f_3(x) - a_{10}(x)}{f_3(x)}$

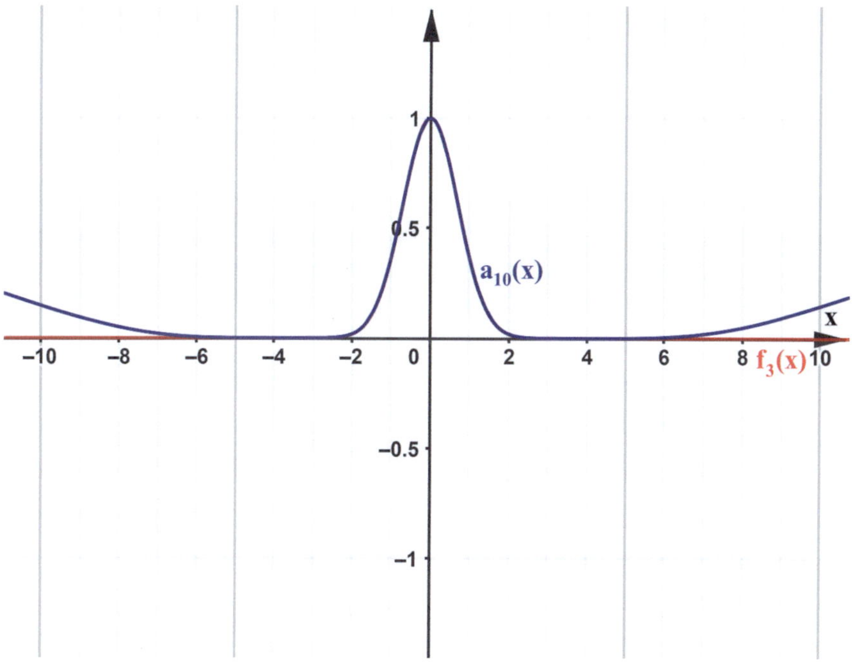

Fig. 2.29 Gaussian activation function $f_3(x)$ and the eighth approximation function $a_{10}(x)$

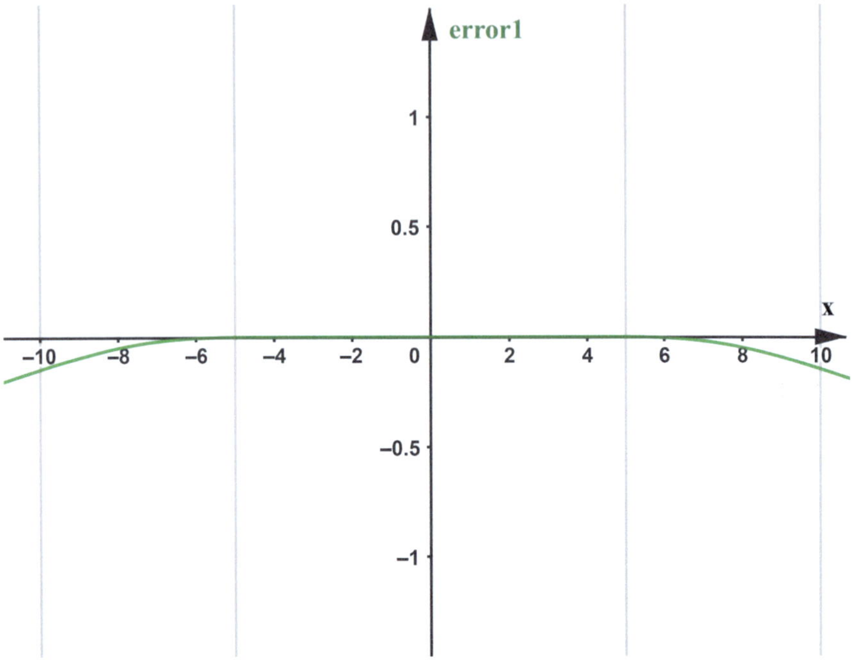

Fig. 2.30 Absolute error of approximation of the Gaussian activation function $f_3(x)$ using approximation function $a_{10}(x)$

(in percents). Considering as acceptable a maximal approximation error of $\pm 1\%$, the maximal range of the input variable x is $[-2.1\ ;\ 2.1]$.

In order to approximate the $f_3(x, \sigma)$ more general Gaussian activation function expressed by (2.7), $x \rightarrow \frac{x}{\sigma\sqrt{2}}$ variable changing must be used in the previous relation (2.27) of the $a_{10}(x)$ particular approximation function, resulting the following mathematical form of the general approximation function $a_{10}(x, \sigma)$:

$$a_{10}(x, \sigma) = \left(\frac{1 - \frac{x^2}{16\sigma^2} + \frac{x^4}{768\sigma^4}}{1 + \frac{x^2}{16\sigma^2} + \frac{x^4}{768\sigma^4}} \right)^4. \tag{2.28}$$

2.3.10 *Nineth Approximation Function for Generating the Gaussian Activation Function*

In order to approximate the Gaussian activation function $f_3(x)$, a new approximation function $a_{11}(x)$ is presented. Because the Gaussian function is an even mathematical function, the general form of the proposed nineth approximation function $a_{11}(x)$ can

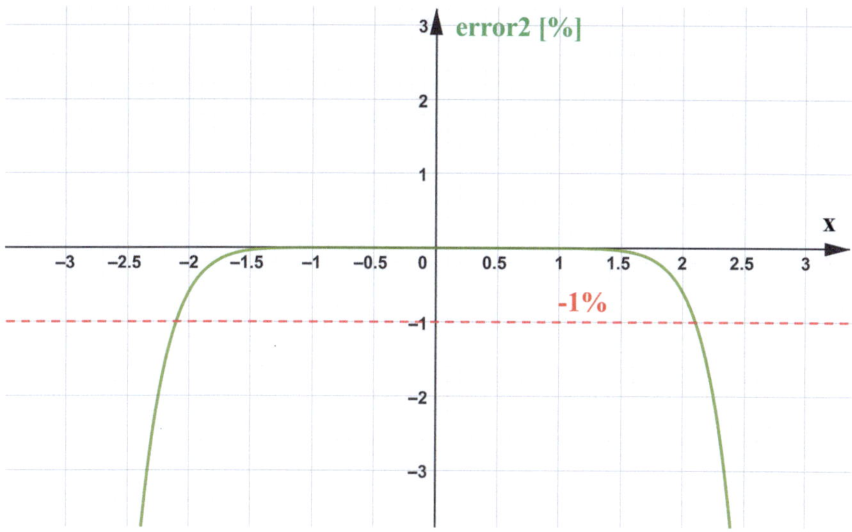

Fig. 2.31 Relative error of approximation of the Gaussian activation function $f_3(x)$ using approximation function $a_{10}(x)$

be expressed as follows:

$$a_{11\text{GEN}}(x) = \frac{c_6x^6 + c_4x^4 + c_2x^2 + c_0}{d_2x^2 + d_0}. \tag{2.29}$$

c_0, c_2, c_4, c_6, d_0 and d_2 being constant coefficients having particular values that can be determined by the condition that the superior-order Taylor series expansion of $f_3(x)$ Gaussian activation function, expressed by (2.6), to be identical with the superior-order Taylor series expansion of the general form of the approximation function $a_{11\text{GEN}}(x)$, expressed by (2.29).

The particular form of the approximation function is developed as a ratio of two polynomial functions: the numerator degree is six, while the denominator degree is two. The values of these coefficients, obtained considerring the previous conditions, can be witten as: $c_0 = 1, c_2 = -\frac{3}{4}, c_4 = \frac{1}{4}, c_6 = -\frac{1}{24}, d_0 = 1$ and $d_2 = \frac{1}{4}$. Replacing these particular values in relation (2.29), the nineth approximation function $a_{11}(x)$ has the following mathematical expression:

$$a_{11}(x) = \frac{1 - \frac{3x^2}{4} + \frac{x^4}{4} - \frac{x^6}{24}}{1 + \frac{x^2}{4}}. \tag{2.30}$$

The graphical representation of the Gaussian activation function $f_3(x)$ and of the nineth approximation function $a_{11}(x)$ is presented in Fig. 2.32.

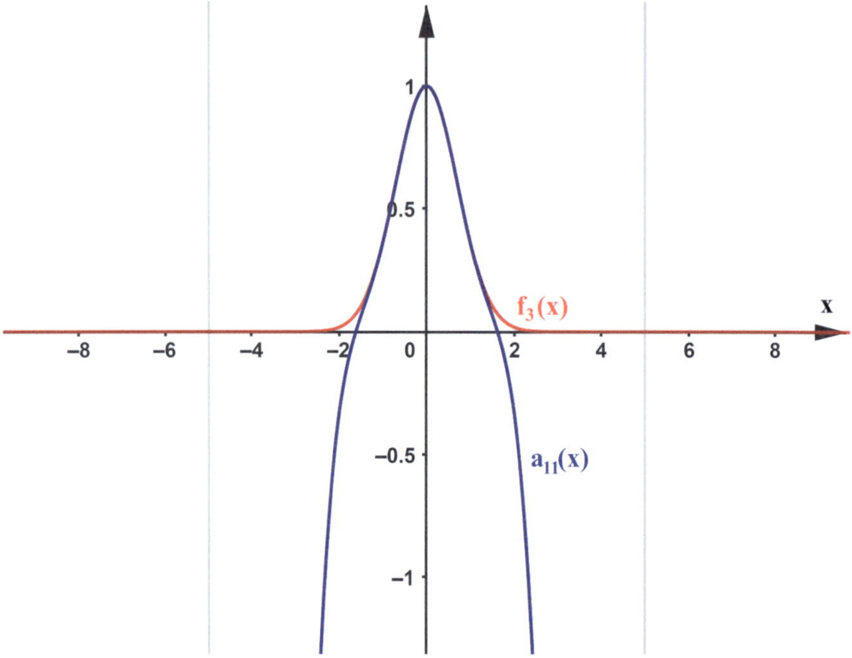

Fig. 2.32 Gaussian activation function $f_3(x)$ and the nineth approximation function $a_{11}(x)$

The absolute error of approximation of the Gaussian activation function $f_3(x)$ using the approximation function $a_{11}(x)$ is illustrated in Fig. 2.33. It is defined as the difference between the previous functions, error1 $= f_3(x) - a_{11}(x)$.

For more accurately evaluating the error of approximation, the relative error of approximation is shown in Fig. 2.34. This error is defined as error2 $= 100\frac{f_3(x)-a_{11}(x)}{f_3(x)}$ (in percents). Considering as acceptable a maximal approximation error of $\pm 1\%$, the maximal range of the input variable x is $[-1.1\,;\ 1.1]$.

In order to approximate the $f_3(x, \sigma)$ more general Gaussian activation function expressed by (2.7), $x \rightarrow \frac{x}{\sigma\sqrt{2}}$ variable changing must be used in the previous relation (2.30) of the $a_{11}(x)$ particular approximation function, resulting the following mathematical form of the general approximation function $a_{11}(x, \sigma)$:

$$a_{11}(x, \sigma) = \frac{1 - \frac{3x^2}{8\sigma^2} + \frac{x^4}{16\sigma^4} - \frac{x^6}{192\sigma^6}}{1 + \frac{x^2}{8\sigma^2}}. \tag{2.31}$$

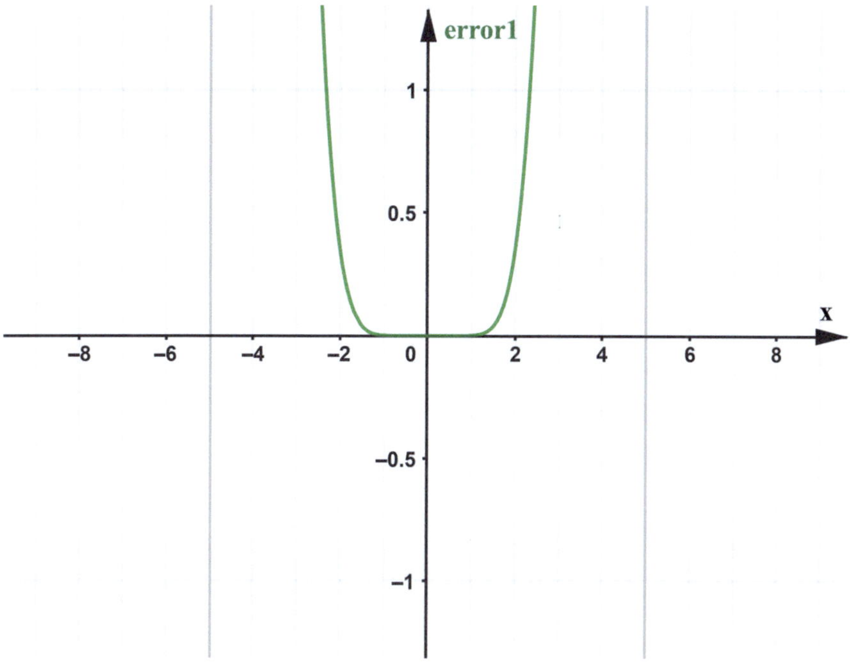

Fig. 2.33 Absolute error of approximation of the Gaussian activation function $f_3(x)$ using approximation function $a_{11}(x)$

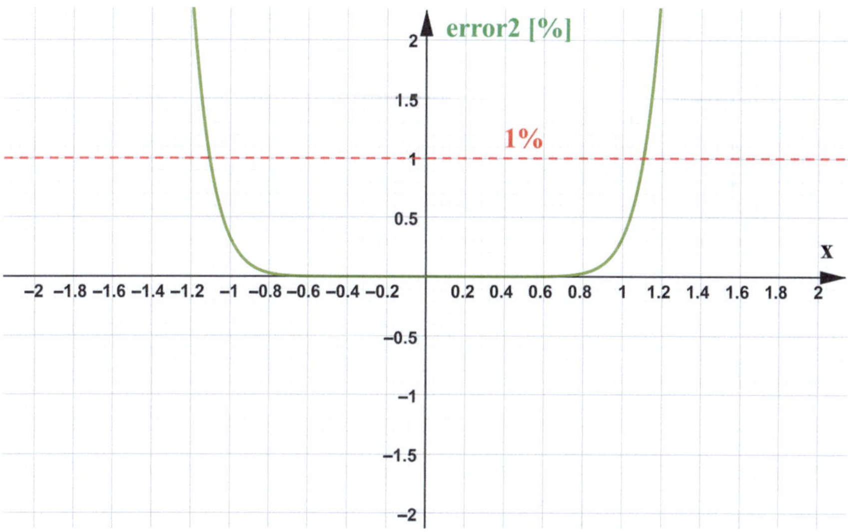

Fig. 2.34 Relative error of approximation of the Gaussian activation function $f_3(x)$ using $a_{11}(x)$ approximation function

2.3.11 Tenth Approximation Function for Generating the Gaussian Activation Function

The variable changing method, expressed by relation (2.11), is used to approximate the Gaussian activation function $f_3(x)$. The new improved accuracy approximation function $a_{12}(x)$, based on the previous approximation function $a_{11}(x)$, has the following expression:

$$a_{12}(x) = \left(\frac{1 - \frac{3x^2}{16} + \frac{x^4}{64} - \frac{x^6}{1536}}{1 + \frac{x^2}{16}} \right)^4. \tag{2.32}$$

The graphical representation of the Gaussian activation function $f_3(x)$ and of the tenth approximation function $a_{12}(x)$ is presented in Fig. 2.35.

The absolute error of approximation of the Gaussian activation function $f_3(x)$ using the approximation function $a_{12}(x)$ is illustrated in Fig. 2.36. It is defined as the difference between the previous functions, $\text{error1} = f_3(x) - a_{12}(x)$.

For more accurately evaluating the error of approximation, the relative error of approximation is shown in Fig. 2.37. This error is defined as $\text{error2} = 100\frac{f_3(x) - a_{12}(x)}{f_3(x)}$

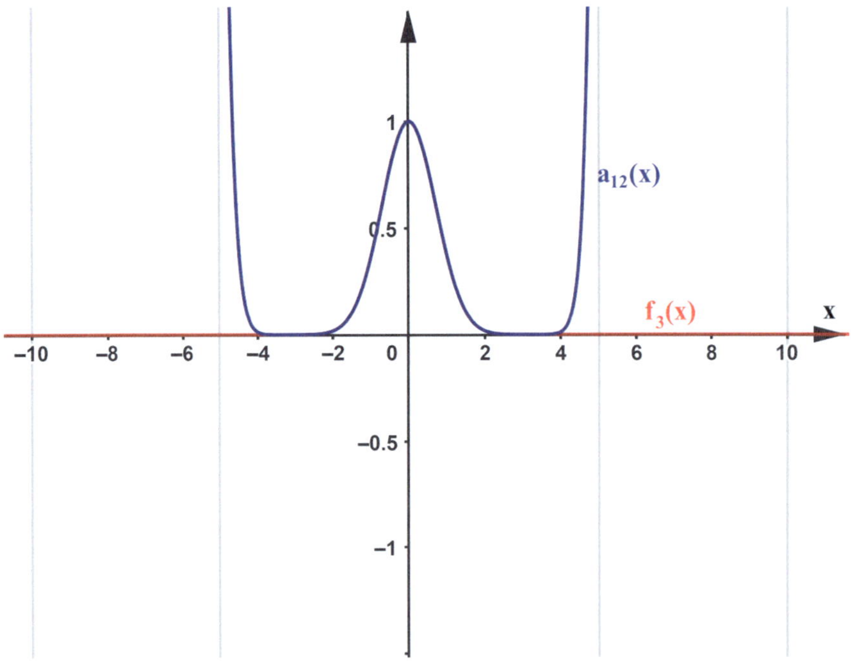

Fig. 2.35 Gaussian activation function $f_3(x)$ and the tenth approximation function $a_{12}(x)$

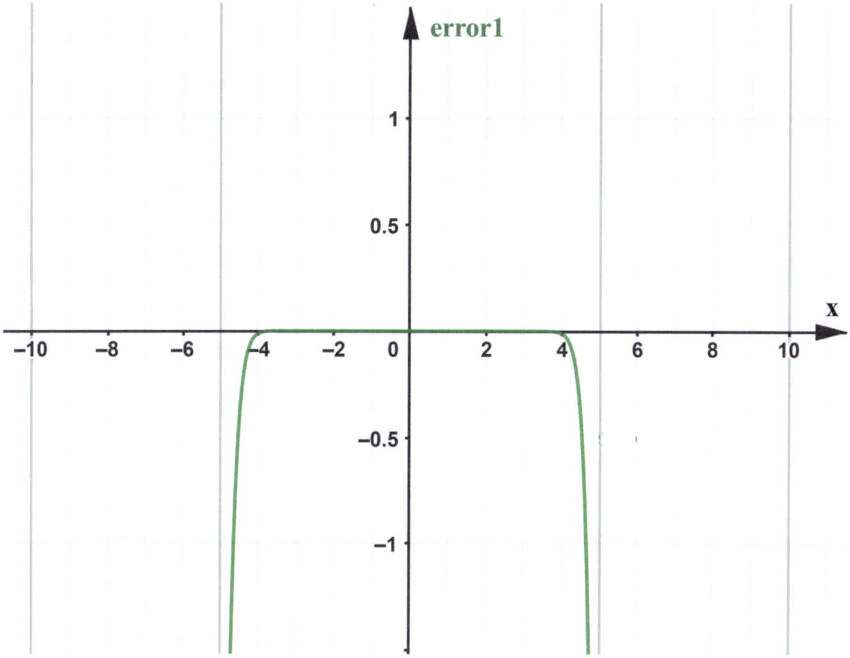

Fig. 2.36 Absolute error of approximation of the Gaussian activation function $f_3(x)$ using approximation function $a_{12}(x)$

(in percents). Considering as acceptable a maximal approximation error of $\pm 1\%$, the maximal range of the input variable x is $[-1.9;\ 1.9]$.

In order to approximate the $f_3(x, \sigma)$ more general Gaussian activation function expressed by (2.7), $x \to \frac{x}{\sigma\sqrt{2}}$ variable changing must be used in the previous relation (2.32) of the $a_{12}(x)$ particular approximation function, resulting the following mathematical form of the general approximation function $a_{12}(x, \sigma)$:

$$a_{12}(x, \sigma) = \left(\frac{1 - \frac{3x^2}{32\sigma^2} + \frac{x^4}{256\sigma^4} - \frac{x^6}{12288\sigma^6}}{1 + \frac{x^2}{32\sigma^2}} \right)^4. \tag{2.33}$$

2.3.12 11th Approximation Function for Generating the Gaussian Activation Function

In order to approximate the Gaussian activation function $f_3(x)$, a new approximation function $a_{13}(x)$ is presented. Because the Gaussian function is an even mathematical function, the general form of the proposed first approximation function $a_{13}(x)$ can

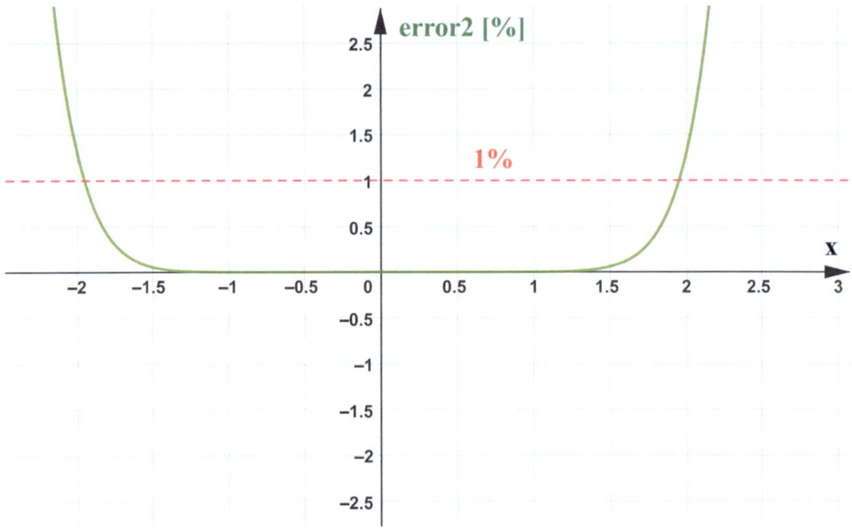

Fig. 2.37 Relative error of approximation of the Gaussian activation function $f_3(x)$ using approximation function $a_{12}(x)$

be expressed as follows:

$$a_{13\text{GEN}}(x) = \frac{c_6x^6 + c_4x^4 + c_2x^2 + c_0}{d_6x^6 + d_4x^4 + d_2x^2 + d_0}, \tag{2.34}$$

$c_0, c_2, c_4, c_6, d_0, d_2, d_4$ and d_6 being constant coefficients having particular values that can be determined by the condition that the superior-order Taylor series expansion of $f_3(x)$ Gaussian activation function, expressed by (2.6), to be identical with the superior-order Taylor series expansion of the general form of the approximation function $a_{13\text{GEN}}(x)$, expressed by (2.34).

The particular form of the approximation function is developed as a ratio of two polynomial functions: the numerator degree is six, while the denominator degree is also six. The values of these coefficients, obtained considerring the previous conditions, can be witten as: $c_0 = 1$, $c_2 = -\frac{1}{2}$, $c_4 = \frac{1}{10}$, $c_6 = -\frac{1}{120}$, $d_0 = 1$, $d_2 = \frac{1}{2}$, $d_4 = \frac{1}{10}$ and $d_6 = \frac{1}{120}$. Replacing these particular values in relation (2.34), the 11th approximation function $a_{13}(x)$ has the following mathematical expression:

$$a_{13}(x) = \frac{1 - \frac{x^2}{2} + \frac{x^4}{10} - \frac{x^6}{120}}{1 + \frac{x^2}{2} + \frac{x^4}{10} + \frac{x^6}{120}}. \tag{2.35}$$

The graphical representation of the Gaussian activation function $f_3(x)$ and of the 11th approximation function $a_{13}(x)$ is presented in Fig. 2.38.

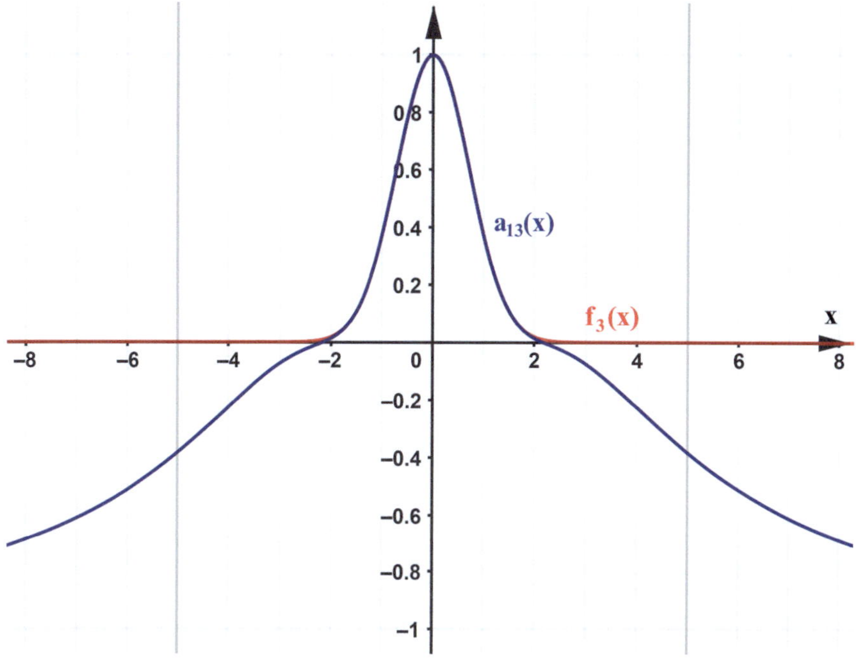

Fig. 2.38 Gaussian activation function $f_3(x)$ and the 11th approximation function $a_{13}(x)$

The absolute error of approximation of the Gaussian activation function $f_3(x)$ using the approximation function $a_{13}(x)$ is illustrated in Fig. 2.39. It is defined as the difference between the previous functions, error1 $= f_3(x) - a_{13}(x)$.

For more accurately evaluating the error of approximation, the relative error of approximation is shown in Fig. 2.40. This error is defined as error2 $= 100\frac{f_3(x) - a_{13}(x)}{f_3(x)}$ (in percents). Considering as acceptable a maximal approximation error of $\pm 1\%$, the maximal range of the input variable x is $[-1.65\,;\ 1.65]$.

In order to approximate the $f_3(x, \sigma)$ more general Gaussian activation function expressed by (2.7), $x \to \frac{x}{\sigma\sqrt{2}}$ variable changing must be used in the previous relation (2.35) of the $a_{13}(x)$ particular approximation function, resulting the following mathematical form of the general approximation function $a_{13}(x, \sigma)$:

$$a_{13}(x, \sigma) = \frac{1 - \frac{x^2}{4\sigma^2} + \frac{x^4}{40\sigma^4} - \frac{x^6}{960\sigma^6}}{1 + \frac{x^2}{4\sigma^2} + \frac{x^4}{40\sigma^4} + \frac{x^6}{960\sigma^6}}. \tag{2.36}$$

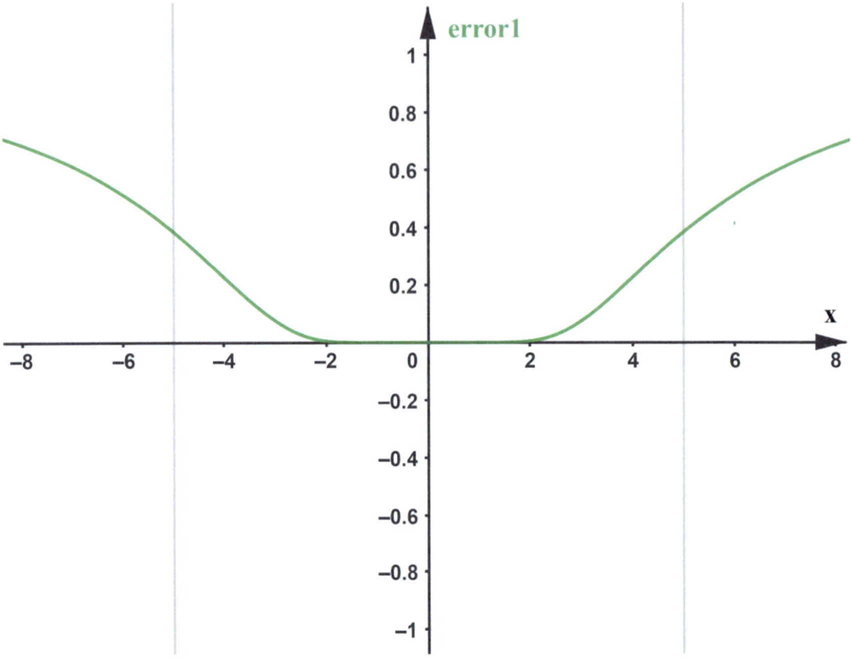

Fig. 2.39 Absolute error of approximation of the Gaussian activation function $f_3(x)$ using approximation function $a_{13}(x)$

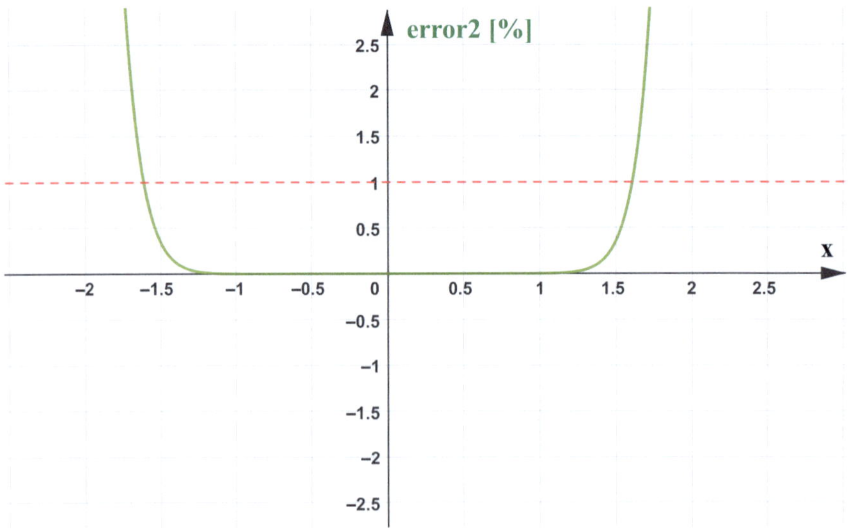

Fig. 2.40 Relative error of approximation of the Gaussian activation function $f_3(x)$ using approximation function $a_{13}(x)$

2.3.13 12th Approximation Function for Generating the Gaussian Activation Function

The variable changing method, expressed by relation (2.11), is used to approximate the Gaussian activation function $f_3(x)$. The new improved accuracy approximation function $a_{14}(x)$, based on the previous approximation function $a_{13}(x)$, has the following expression:

$$a_{14}(x) = \left(\frac{1 - \frac{x^2}{8} + \frac{x^4}{160} - \frac{x^6}{7680}}{1 + \frac{x^2}{8} + \frac{x^4}{160} + \frac{x^6}{7680}} \right)^4. \tag{2.37}$$

The graphical representation of the Gaussian activation function $f_3(x)$ and of the 12th approximation function $a_{14}(x)$ is presented in Fig. 2.41.

The absolute error of approximation of the Gaussian activation function $f_3(x)$ using the approximation function $a_{14}(x)$ is illustrated in Fig. 2.42. It is defined as the difference between the previous functions, error1 $= f_3(x) - a_{14}(x)$.

For more accurately evaluating the error of approximation, the relative error of approximation is shown in Fig. 2.43. This error is defined as error2 $= 100\frac{f_3(x) - a_{14}(x)}{f_3(x)}$ (in percents). Considering as acceptable a maximal approximation error of $\pm 1\%$, the maximal range of the input variable x is $[-1.4;\ 1.4]$.

In order to approximate the $f_3(x, \sigma)$ more general Gaussian activation function expressed by (2.7), $x \to \frac{x}{\sigma\sqrt{2}}$ variable changing must be used in the previous relation (2.37) of the $a_{14}(x)$ particular approximation function, resulting the following mathematical form of the general approximation function $a_{14}(x, \sigma)$:

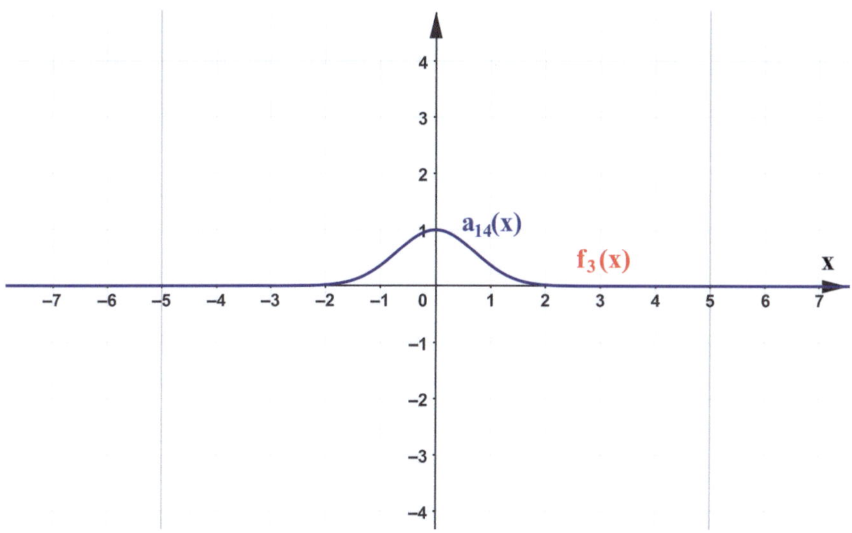

Fig. 2.41 Gaussian activation function $f_3(x)$ and the 12th approximation function $a_{14}(x)$

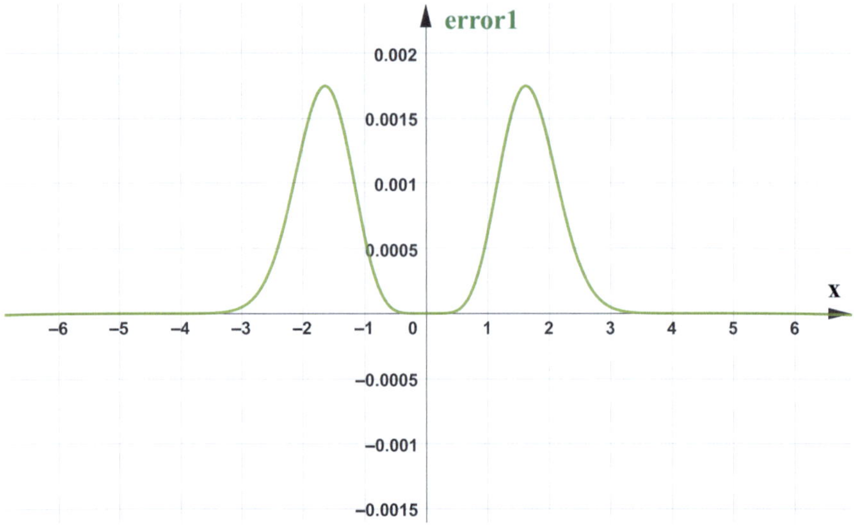

Fig. 2.42 Absolute error of approximation of the Gaussian activation function $f_3(x)$ using approximation function $a_{14}(x)$

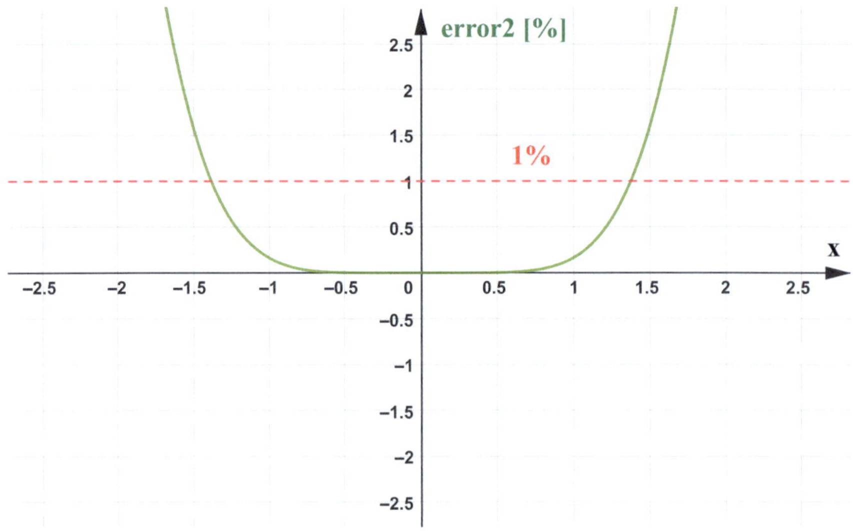

Fig. 2.43 Relative error of approximation of the Gaussian activation function $f_3(x)$ using approximation function $a_{14}(x)$

$$a_{14}(x,\sigma) = \left(\frac{1 - \frac{x^2}{16\sigma^2} + \frac{x^4}{640\sigma^4} - \frac{x^6}{61{,}440\sigma^6}}{1 + \frac{x^2}{16\sigma^2} + \frac{x^4}{640\sigma^4} + \frac{x^6}{61{,}440\sigma^6}} \right)^4. \tag{2.38}$$

2.3.14 13th Approximation Function for Generating the Gaussian Activation Function

In order to approximate the Gaussian activation function $f_3(x)$, a new approximation function $a_{15}(x)$ is presented. Because the Gaussian function is an even mathematical function, the general form of the proposed 13th approximation function $a_{15}(x)$ can be expressed as follows:

$$a_{15\text{GEN}}(x) = \frac{c_6x^6 + c_4x^4 + c_2x^2 + c_0}{d_4x^4 + d_2x^2 + d_0}, \tag{2.39}$$

c_0, c_2, c_4, c_6, d_0, d_2 and d_4 being constant coefficients having particular values that can be determined by the condition that the superior-order Taylor series expansion of $f_3(x)$ Gaussian activation function, expressed by (2.6), to be identical with the superior-order Taylor series expansion of the general form of the approximation function $a_{15\text{GEN}}(x)$, expressed by (2.39).

The particular form of the approximation function is developed as a ratio of two polynomial functions: the numerator degree is six, while the denominator degree is four. The values of these coefficients, obtained considerring the previous conditions, can be witten as: $c_0 = 1$, $c_2 = -\frac{3}{5}$, $c_4 = \frac{3}{20}$, $c_6 = -\frac{1}{60}$, $d_0 = 1$, $d_2 = \frac{2}{5}$ and $d_4 = \frac{1}{20}$. Replacing these particular values in relation (2.38), the 13th approximation function $a_{15\text{GEN}}(x)$ has the following mathematical expression:

$$a_{15}(x) = \frac{1 - \frac{3x^2}{5} + \frac{3x^4}{20} - \frac{x^6}{60}}{1 + \frac{2x^2}{5} + \frac{x^4}{20}}. \tag{2.40}$$

The graphical representation of the Gaussian activation function $f_3(x)$ and of the 13th approximation function $a_{15}(x)$ is presented in Fig. 2.44.

The absolute error of approximation of the Gaussian activation function $f_3(x)$ using the approximation function $a_{15}(x)$ is illustrated in Fig. 2.45. It is defined as the difference between the previous functions, error1 $= f_3(x) - a_{15}(x)$.

For more accurately evaluating the error of approximation, the relative error of approximation is shown in Fig. 2.46. This error is defined as error2 $= 100\frac{f_3(x) - a_{15}(x)}{f_3(x)}$ (in percents). Considering as acceptable a maximal approximation error of $\pm 1\%$, the maximal range of the input variable x is $[-1.4;\ 1.4]$.

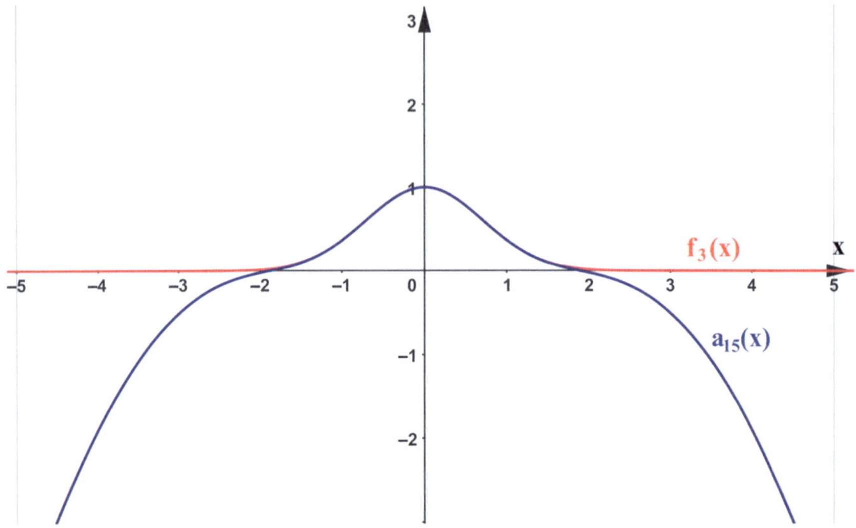

Fig. 2.44 Gaussian activation function $f_3(x)$ and the 13th approximation function $a_{15}(x)$

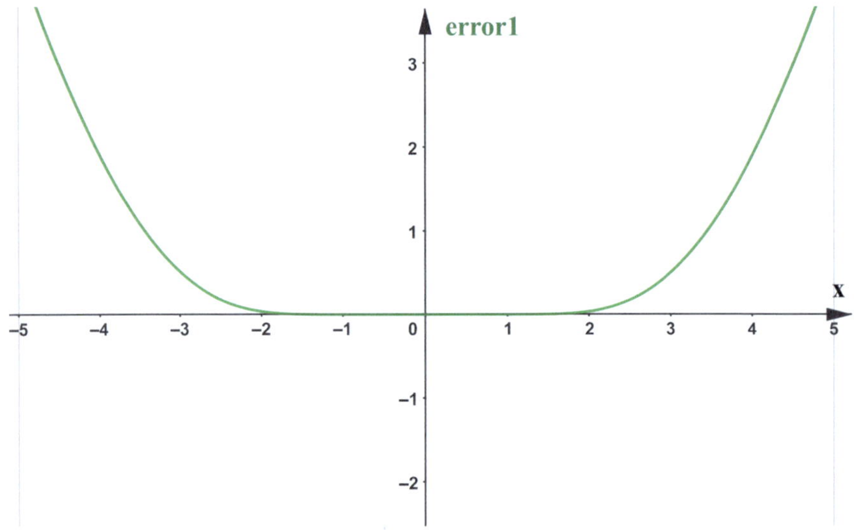

Fig. 2.45 Absolute error of approximation of the Gaussian activation function $f_3(x)$ using approximation function $a_{15}(x)$

In order to approximate the $f_3(x, \sigma)$ more general Gaussian activation function expressed by (2.7), $x \to \frac{x}{\sigma\sqrt{2}}$ variable changing must be used in the previous relation (2.40) of the $a_{15}(x)$ particular approximation function, resulting the following mathematical form of the general approximation function $a_{15}(x, \sigma)$:

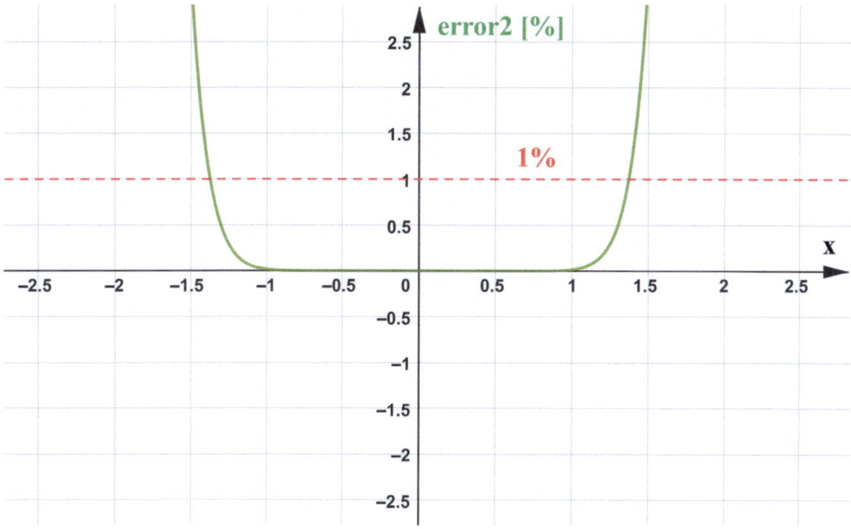

Fig. 2.46 Relative error of approximation of the Gaussian activation function $f_3(x)$ using approximation function $a_{15}(x)$

$$a_{15}(x,\sigma) = \frac{1 - \frac{3x^2}{10\sigma^2} + \frac{3x^4}{80\sigma^4} - \frac{x^6}{480\sigma^6}}{1 + \frac{x^2}{5\sigma^2} + \frac{x^4}{80\sigma^4}}. \tag{2.41}$$

2.3.15 *14th Approximation Function for Generating the Gaussian Activation Function*

The variable changing method, expressed by relation (2.11), is used to approximate the Gaussian activation function $f_3(x)$. The new improved accuracy approximation function $a_{16}(x)$, based on the previous approximation function $a_{15}(x)$, has the following expression:

$$a_{16}(x) = \left(\frac{1 - \frac{3x^2}{20} + \frac{3x^4}{320} - \frac{x^6}{3840}}{1 + \frac{x^2}{10} + \frac{x^4}{320}}\right)^4. \tag{2.42}$$

The graphical representation of the Gaussian activation function $f_3(x)$ and of the 14th approximation function $a_{16}(x)$ is presented in Fig. 2.47.

The absolute error of approximation of the Gaussian activation function $f_3(x)$ using the approximation function $a_{16}(x)$ is illustrated in Fig. 2.48. It is defined as the difference between the previous functions, error1 $= f_3(x) - a_{16}(x)$.

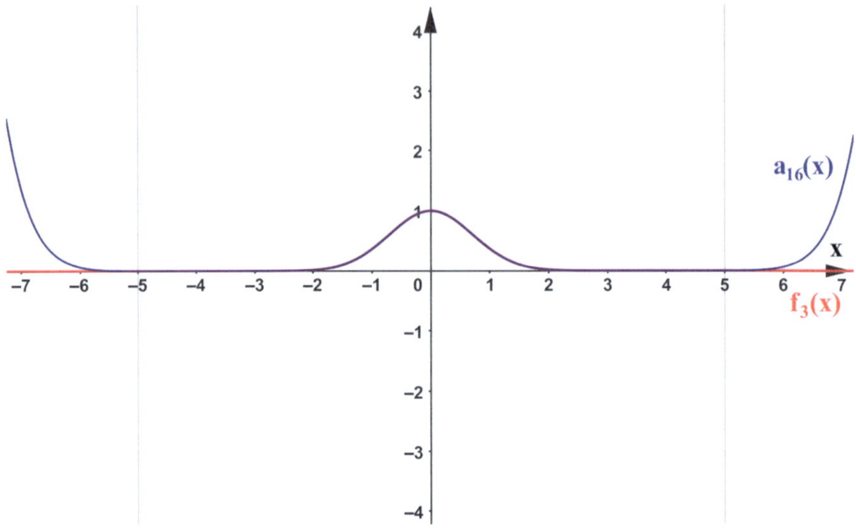

Fig. 2.47 Gaussian activation function $f_3(x)$ and the 14th approximation function $a_{16}(x)$

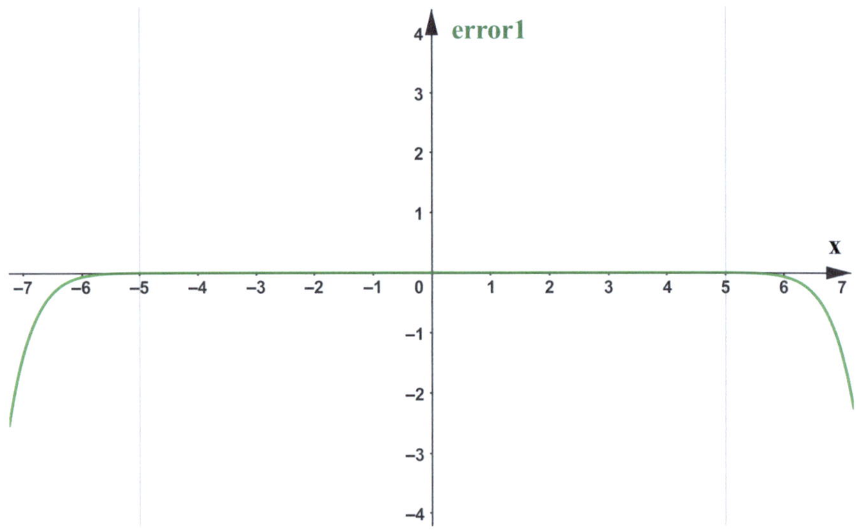

Fig. 2.48 Absolute error of approximation of the Gaussian activation function $f_3(x)$ using approximation function $a_{16}(x)$

For more accurately evaluating the error of approximation, the relative error of approximation is shown in Fig. 2.49. This error is defined as error2 $= 100\frac{f_3(x)-a_{16}(x)}{f_3(x)}$ (in percents). Considering as acceptable a maximal approximation error of $\pm 1\%$, the maximal range of the input variable x is $[-2.5\,;\ 2.5]$.

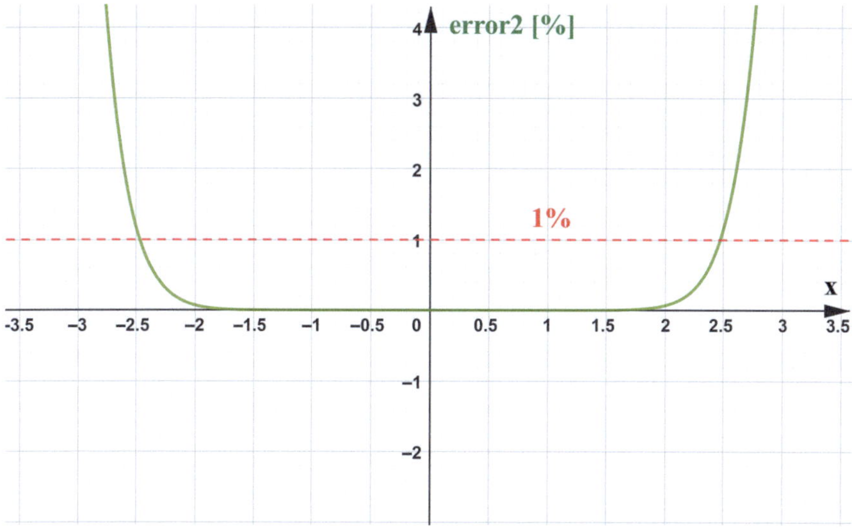

Fig. 2.49 Relative error of approximation of the Gaussian activation function $f_3(x)$ using approximation function $a_{16}(x)$

In order to approximate the $f_3(x, \sigma)$ more general Gaussian activation function expressed by (2.7), $x \to \frac{x}{\sigma\sqrt{2}}$ variable changing must be used in the previous relation (2.42) of the $a_{16}(x)$ particular approximation function, resulting the following mathematical form of the general approximation function $a_{16}(x, \sigma)$:

$$a_{16}(x,\sigma) = \left(\frac{1 - \frac{3x^2}{40\sigma^2} + \frac{3x^4}{1280\sigma^4} - \frac{x^6}{30720\sigma^6}}{1 + \frac{x^2}{20\sigma^2} + \frac{x^4}{1280\sigma^4}} \right)^4. \tag{2.43}$$

2.3.16 15th Approximation Function for Generating the Gaussian Activation Function

In order to approximate the Gaussian activation function $f_3(x)$, a new approximation function $a_{17}(x)$ is presented. Because the Gaussian function is an even mathematical function, the general form of the proposed 15th approximation function $a_{17}(x)$ can be expressed as follows:

$$a_{17\text{GEN}}(x) = \left(\frac{c_8x^2 + c_6x^6 + c_4x^4 + c_2x^2 + c_0}{d_4x^4 + d_2x^2 + d_0} \right)^4, \tag{2.44}$$

$c_0, c_2, c_4, c_6, c_8, d_0, d_2$ and d_4 being constant coefficients having particular values that can be determined by the condition that the superior-order Taylor series expansion of $f_3(x)$ Gaussian activation function, expressed by (2.6), to be identical with the superior-order Taylor series expansion of the general form of the approximation function $a_{17\mathrm{GEN}}(x)$, expressed by (2.44).

The values of these coefficients, obtained considerring the previous conditions, can be witten as: $c_0 = 1$, $c_2 = -\frac{1}{20}$, $c_4 = -\frac{1}{160}$, $c_6 = \frac{1}{1920}$, $c_8 = \frac{1}{30720}$, $d_0 = 1$, $d_2 = \frac{1}{5}$ and $d_4 = \frac{1}{80}$. Replacing these particular values in relation (2.44), the 15th approximation function $a_{17}(x)$ has the following mathematical expression:

$$a_{17}(x) = \left(\frac{1 - \frac{x^2}{20} - \frac{x^4}{160} + \frac{x^6}{1920} + \frac{x^8}{30720}}{1 + \frac{x^2}{5} + \frac{x^4}{80}} \right)^4. \tag{2.45}$$

The graphical representation of the Gaussian activation function $f_3(x)$ and of the 15th approximation function $a_{17}(x)$ is presented in Fig. 2.50.

The absolute error of approximation of the Gaussian activation function $f_3(x)$ using the approximation function $a_{17}(x)$ is illustrated in Fig. 2.51. It is defined as the difference between the previous functions, error1 $= f_3(x) - a_{17}(x)$.

For more accurately evaluating the error of approximation, the relative error of approximation is shown in Fig. 2.52. This error is defined as error2 $= 100\frac{f_3(x) - a_{17}(x)}{f_3(x)}$ (in percents). Considering as acceptable a maximal approximation error of $\pm 1\%$, the maximal range of the input variable x is $[-1.8\,;\ 1.8]$.

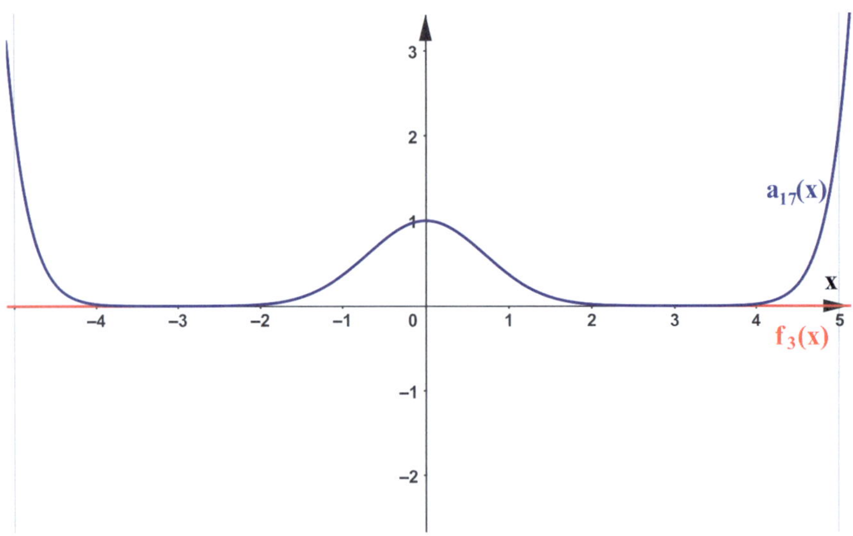

Fig. 2.50 Gaussian activation function $f_3(x)$ and the 15th approximation function $a_{17}(x)$

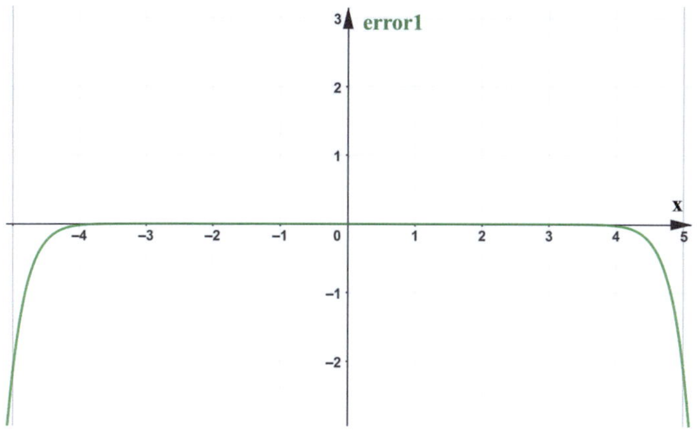

Fig. 2.51 Absolute error of approximation of the Gaussian activation function $f_3(x)$ using approximation function $a_{17}(x)$

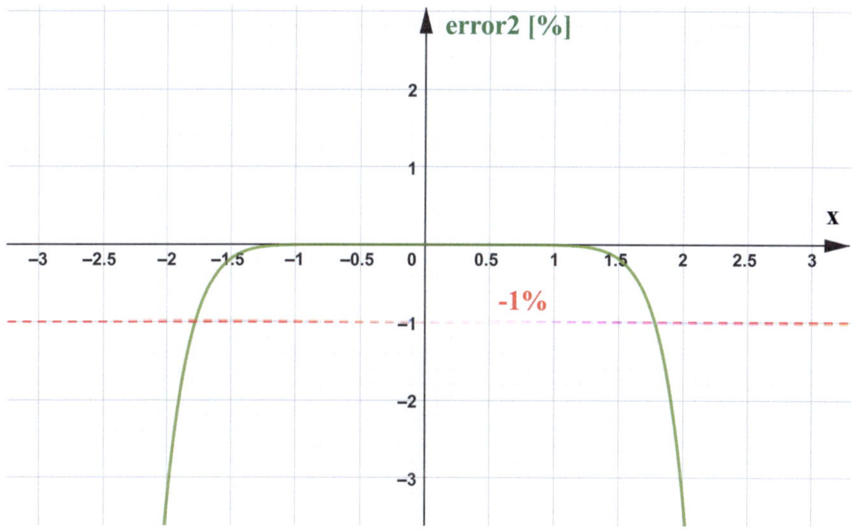

Fig. 2.52 Relative error of approximation of the Gaussian activation function $f_3(x)$ using approximation function $a_{17}(x)$

2.3.17 16th Approximation Function for Generating the Gaussian Activation Function

In order to approximate the Gaussian activation function $f_3(x)$, a new approximation function $a_{18}(x)$ is presented. Because the Gaussian function is an even mathematical function, the general form of the proposed 16th approximation function $a_{18}(x)$ can be expressed as follows:

$$a_{18\mathrm{GEN}}(x) = \left(c_0 + c_2x^2 + c_4x^4 + c_6x^6\right)^4, \tag{2.46}$$

c_0, c_2, c_4 and c_6 being constant coefficients having particular values that can be determined by the condition that the superior-order Taylor series expansion of $f_3(x)$ Gaussian activation function, expressed by (2.6), to be identical with the superior-order Taylor series expansion of the general form of the approximation function $a_{18\mathrm{GEN}}(x)$, expressed by (2.46).

The values of these coefficients, obtained considerring the previous conditions, can be witten as: $c_0 = 1$, $c_2 = -\frac{1}{4}$, $c_4 = \frac{1}{32}$ and $c_6 = -\frac{1}{384}$. Replacing these particular values in relation (2.46), the 16th approximation function $a_{18}(x)$ has the following mathematical expression:

$$a_{18}(x) = \left(1 - \frac{x^2}{4} + \frac{x^4}{32} - \frac{x^6}{384}\right)^4. \tag{2.47}$$

The graphical representation of the Gaussian activation function $f_3(x)$ and of the 16th approximation function $a_{18}(x)$ is presented in Fig. 2.53.

The absolute error of approximation of the Gaussian activation function $f_3(x)$ using the approximation function $a_{18}(x)$ is illustrated in Fig. 2.54. It is defined as the difference between the previous functions, error1 $= f_3(x) - a_{18}(x)$.

For more accurately evaluating the error of approximation, the relative error of approximation is shown in Fig. 2.55. This error is defined as error2 $= 100\frac{f_3(x)-a_{18}(x)}{f_3(x)}$ (in percents). Considering as acceptable a maximal approximation error of $\pm 1\%$, the maximal range of the input variable x is $[-1.35\,;\ 1.35]$.

In order to approximate the $f_3(x, \sigma)$ more general Gaussian activation function expressed by (2.7), $x \to \frac{x}{\sigma\sqrt{2}}$ variable changing must be used in the previous relation (2.47) of the $a_{18}(x)$ particular approximation function, resulting the following mathematical form of the general approximation function $a_{18}(x, \sigma)$:

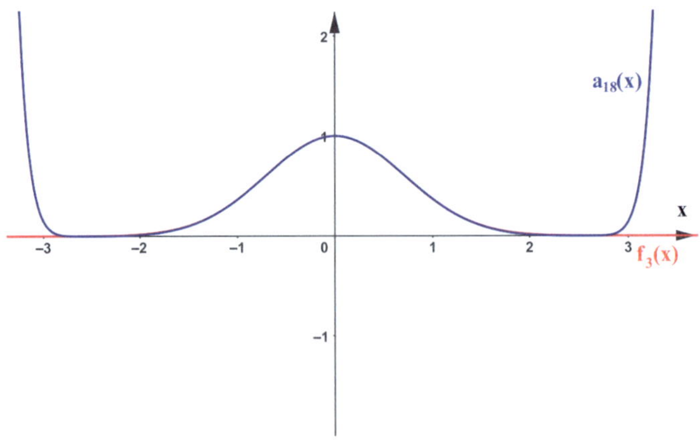

Fig. 2.53 Gaussian activation function $f_3(x)$ and the 16th approximation function $a_{18}(x)$

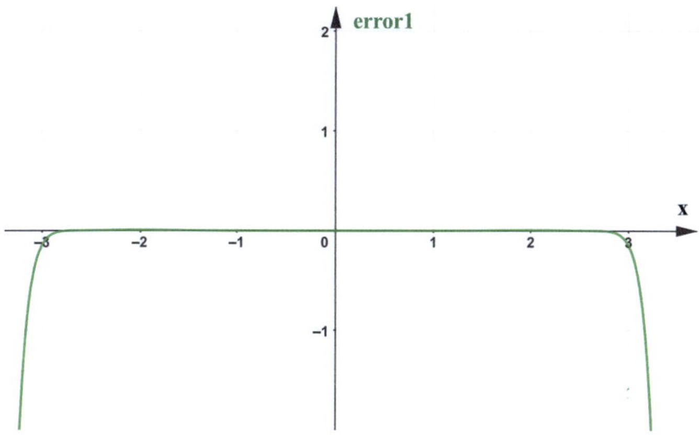

Fig. 2.54 Absolute error of approximation of the Gaussian activation function $f_3(x)$ using approximation function $a_{18}(x)$

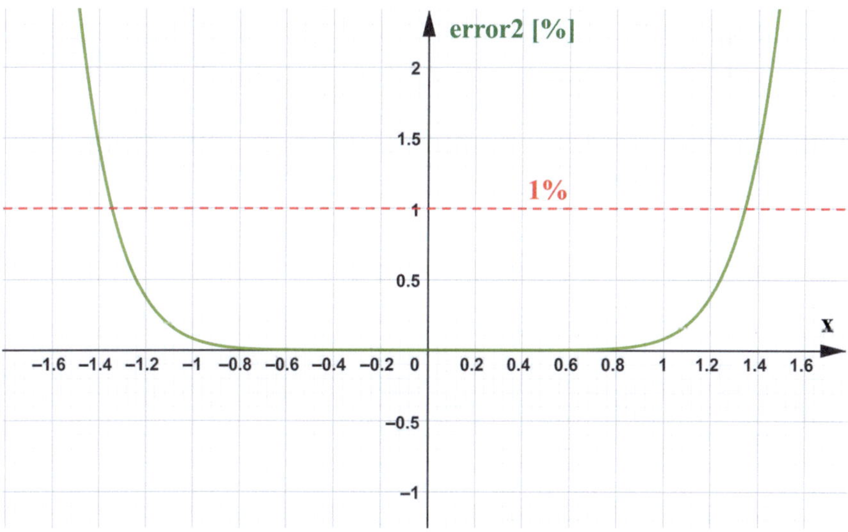

Fig. 2.55 Relative error of approximation of the Gaussian activation function $f_3(x)$ using approximation function $a_{18}(x)$

$$a_{18}(x, \sigma) = \left(1 - \frac{x^4}{8\sigma^2} + \frac{x^4}{128\sigma^4} - \frac{x^6}{3072\sigma^6}\right)^4. \tag{2.48}$$

2.4 Quadratic and Inverse Quadratic Activation Functions

Quadratic activation functions can be used as alternatives to other types of activation functions, such as ReLU or hyperbolic tangent activation functions. They are very useful in quadratic layers (second-order neurons) and have the advantages of modeling more complex functions with fewer layers and, also, of capturing feature interactions. The areas of utilization of the quadratic functions are:

a. ***Statistics and data fitting***

- regression analysis: quadratic regression models relationships that are curved, not linear.
- least squares minimization: the objective function in least squares is quadratic.

b. ***Physics and engineering***

- projectile motion: position vs. time graphs are quadratic due to constant acceleration.
- potential energy curves.
- structural engineering: stress/strain relationships and deformation curves.

c. ***Optimization***

- quadratic programming (QP): optimization problems with a quadratic objective and linear constraints-common in operations research and machine learning.
- convex optimization: many convex problems involve minimizing quadratic functions.

d. ***Computer graphics and geometry***

- Bezier curves: quadratic Bezier curves are used in drawing smooth paths.
- surface modeling: parabolic surfaces in modeling and simulation.

e. ***Economics and finance***

- profit/cost modeling: many cost functions are modeled as quadratic (for example, economies of scale).
- portfolio optimization: the variance (risk) is a quadratic function of asset weights.

The areas of utilization of the inverse quadratic functions are:

a. ***Machine learning***

- kernel methods: multiquadratic radial basis functions are used in support vector machines (SVMs), kernel regression, and function interpolation.
- Gaussian process regression: sometimes uses multiquadratic-type covariance functions.

b. ***Approximation theory***

- scattered data interpolation: multuquadratic functions are effective for high-dimensional, scattered datasets.
- radial basis function (RBF) networks: neural network models with multi-quadratic basis functions for function approximation.

c. ***Numerical methods***

- meshfree methods: used in solving partial differential equations (PDEs), where grid-based methods are difficult.
- finite point method / radial basis functions-finite differences (RBF-FD): solving differential equations using multiquadratic interpolation.

d. ***Geostatistics and environmental modeling***

- surface modeling: elevation, temperature, pollution data can be interpolated using multiquadratic radial basis functions (RBFs).

2.4.1 General Form of the Quadratic Function

The mathematical form of this function can be expressed as follows:

$$f_4(x) = x^2 + a^2. \tag{2.49}$$

The graphical representation of the quadratic function $f_4(x)$, for multiple values of a parameter ($a = 0$, $a = 1$ and $a = 2$), is presented in Fig. 2.56.

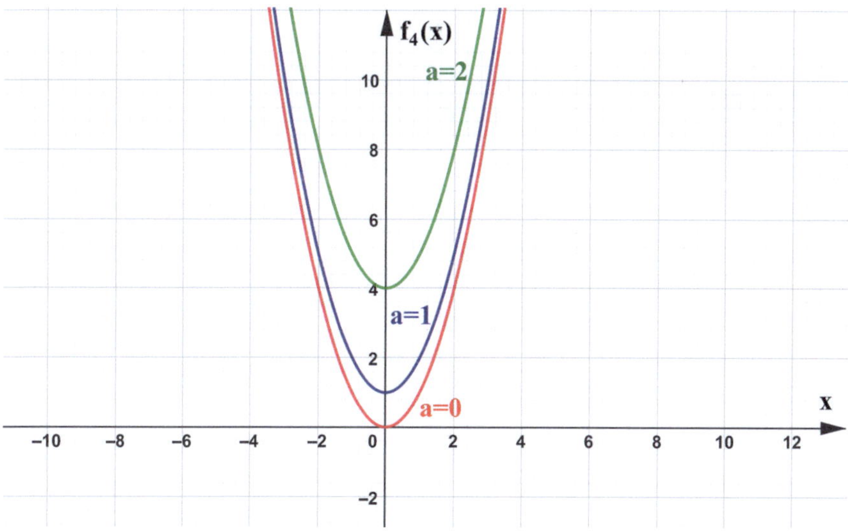

Fig. 2.56 Graphical representation of the quadratic function for $a = 0$, $a = 1$ and $a = 2$

Because of the capability of facile implementation in CMOS technology of the squaring circuit, this function can be very easily obtained, without using any approximation function (see Chaps. 4 and 6).

2.4.2 *General Form of the Inverse Quadratic Function*

The mathematical form of this function can be expressed as follows:

$$f_5(x) = \frac{1}{x^2 + a^2}. \tag{2.50}$$

The graphical representation of the quadratic function $f_5(x)$, for multiple values of a parameter ($a = 0$, $a = 1$ and $a = 2$), is presented in Fig. 2.57.

Because of the capability of facile implementation in CMOS technology of the squaring and of the multiplier/divider circuits, this function can be very easily obtained, without using any approximation function (see Chaps. 4 and 6).

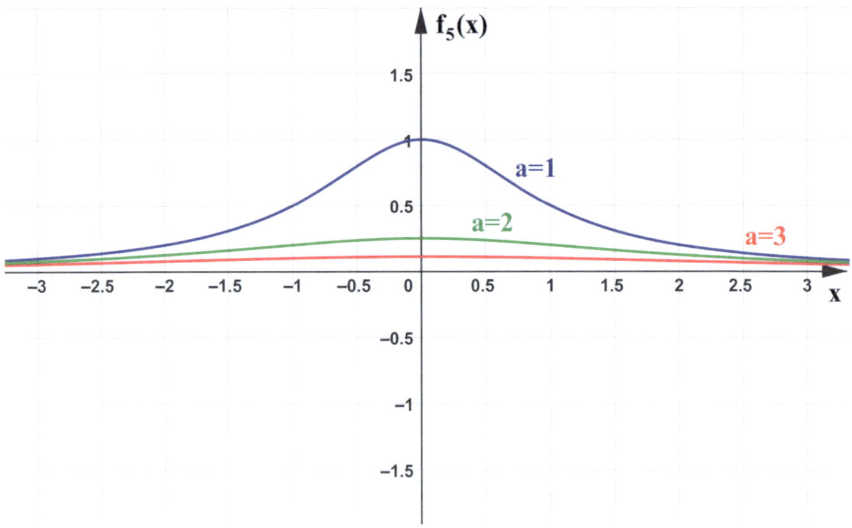

Fig. 2.57 Graphical representation of the inverse quadratic function for $a = 0$, $a = 1$ and $a = 2$

2.5 Approximation Functions for Multiquadratic Activation Functions

The multiquadratic functions represent an alternative of the utilization of Gaussian functions as activation functions in neural networks. They have the advantage of a strong nonlinearity, being, also, derivable and smooth. Their applications include function interpolation, regression and classification neural networks, hybrid networks—radial basis function (RBFs) and perceptron, or pattern recognition systems. The areas of utilization of the multiquadratic and inverse multiquadratic functions (see the following paragraph) are the following:

a. ***Scattered Data Interpolation***

- interpolating data points that are irregularly spaced in 2D or 3D.
- geographic information systems (GIS)—elevation, rainfall, pollution maps, meteorology (weather data interpolation), or medical imaging (reconstructing surfaces from point clouds—for example MRI, CT scans).

b. ***Numerical solution of PDEs (meshfree methods)***

- solving partial differential equations without a structured grid.
- techniques: radial basis function—finite difference (RBF-FD), or Kansa's method.
- heat transfer, diffusion problems, electromagnetic wave propagation or groundwater or contaminant transport modeling.

c. ***Machine learning and neural networks***

- function approximation using RBF neural networks.
- time series prediction, function regression from sparse data or classification (with RBF kernels).

d. ***Image and signal processing***

- surface and shape reconstruction, edge smoothing.
- surface fitting for 3D objects (for example, face models, organ modeling), interpolation in computer graphics or image warping and morphing.

e. ***Physics and engineering simulations***

- modeling physical systems with limited data.
- structural mechanics (modeling stress–strain from scattered sensors), vibration and wave simulation in complex geometries, or electromagnetics (for example, antenna modeling).

f. ***Biological and environmental modeling***

- modeling biological processes or environmental fields where measurements are sparse.

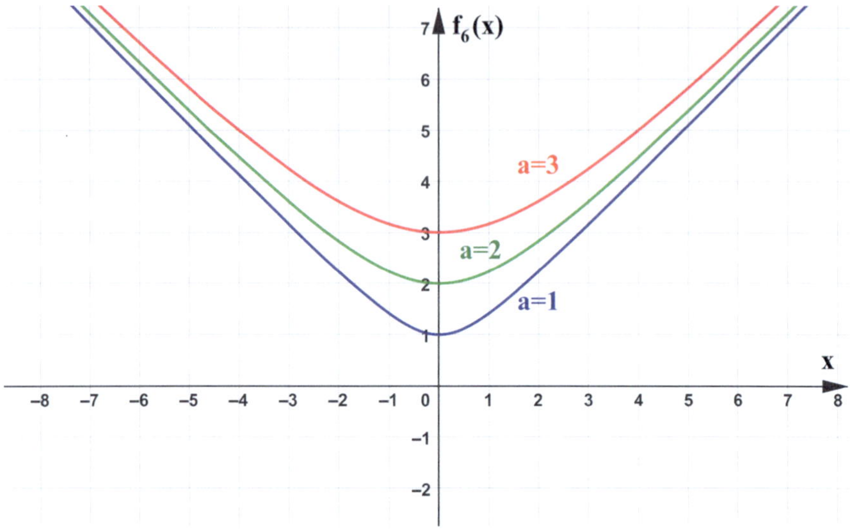

Fig. 2.58 Graphical representation of the multiquadratic function for $a = 0$, $a = 1$ and $a = 2$

- growth of tumors or organs from sampled data, pollution dispersion in air or water, or oceanographic or climate modeling.

2.5.1 General Form of the Multiquadratic Function

The mathematical form of the multiquadratic function can be expressed as follows:

$$f_6(x) = \sqrt{x^2 + a^2}. \tag{2.51}$$

The graphical representation of the multiquadratic function $f_6(x)$, for multiple values of a parameter ($a = 0$, $a = 1$ and $a = 2$) is presented in Fig. 2.58.

2.5.2 First Approximation Function for Generating the Multiquadratic Function for a = 1

In order to approximate the multiquadratic function $f_6(x)$ (for $a = 1$), a new approximation function $a_{19}(x)$ is presented. Because the multiquadratic function is an even mathematical function, the general form of the proposed first approximation function $a_{19}(x)$ can be expressed as follows:

$$a_{19GEN}(x) = \frac{c_4 x^4 + c_2 x^2 + c_0}{d_2 x^2 + d_0}, \tag{2.52}$$

c_0, c_2, c_4, d_0 and d_2 being constant coefficients having particular values that can be determined by the condition that the superior-order Taylor series expansion of $f_6(x)$ multiquadratic function (for $a = 1$), expressed by (2.51), to be identical with the superior-order Taylor series expansion of the general form of the approximation function $a_{19GEN}(x)$, expressed by (2.52).

The values of these coefficients, obtained considerring the previous conditions, can be witten as: $c_0 = 1$, $c_2 = 1$, $c_4 = \frac{1}{8}$, $d_0 = 1$ and $d_2 = \frac{1}{2}$. Replacing these particular values in relation (2.52), the first approximation function $a_{19}(x)$ has the following mathematical expression:

$$a_{19}(x) = \frac{\frac{x^4}{8} + x^2 + 1}{\frac{x^2}{2} + 1}. \tag{2.53}$$

The graphical representation of the multiquadratic function $f_6(x)$ (for $a = 1$) and of the first approximation function $a_{19}(x)$ is presented in Fig. 2.59.

The absolute error of approximation of the multiquadratic function $f_6(x)$ (for $a = 1$) using the approximation function $a_{19}(x)$ is illustrated in Fig. 2.60. It is defined as the difference between the previous functions, error1 $= f_6(x) - a_{19}(x)$.

For more accurately evaluating the error of approximation, the relative error of approximation is shown in Fig. 2.61. This error is defined as error2 $= 100\frac{f_6(x) - a_{19}(x)}{f_6(x)}$ (in percents). Considering as acceptable a maximal approximation error of $\pm 1\%$, the maximal range of the input variable x is $[-0.55\,;\ 0.55]$.

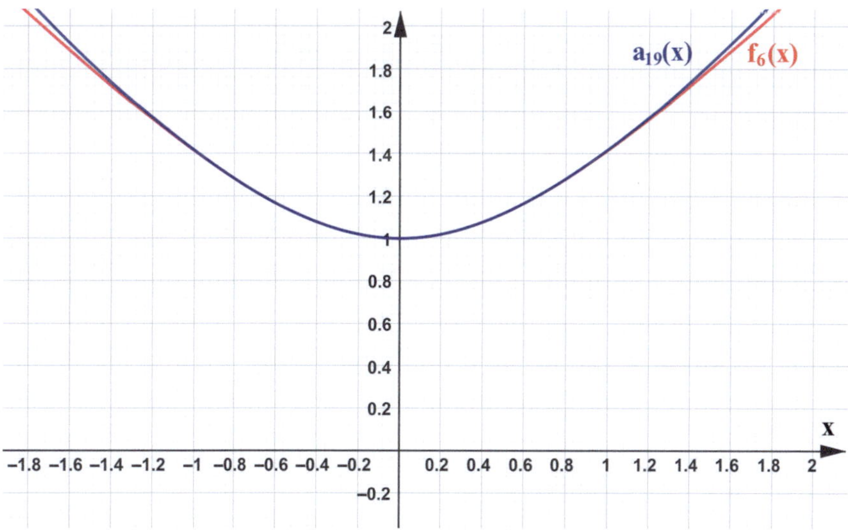

Fig. 2.59 Multiquadratic function $f_6(x)$ (for $a = 1$) and the first approximation function $a_{19}(x)$

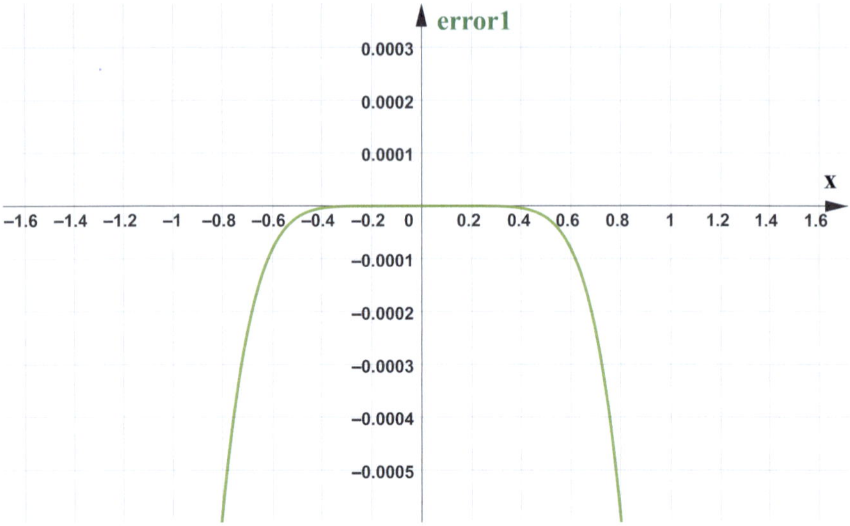

Fig. 2.60 Absolute error of approximation of the multiquadratic function $f_6(x)$ (for $a = 1$) using approximation function $a_{19}(x)$

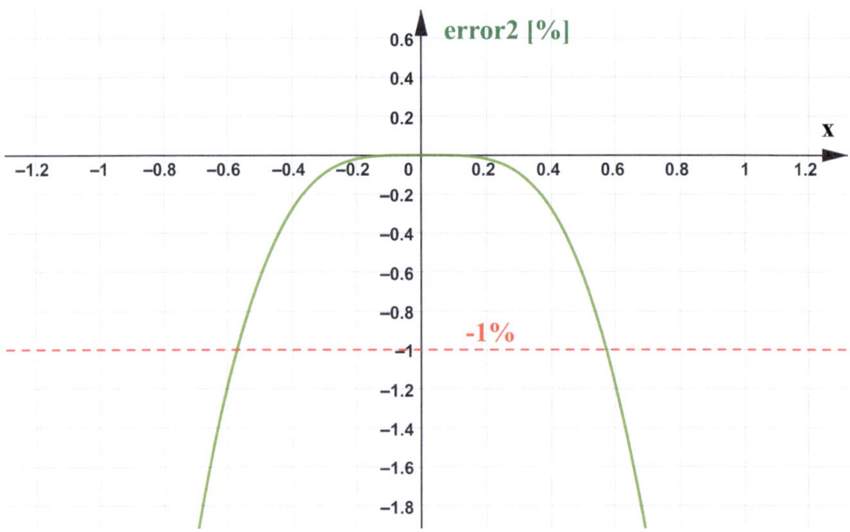

Fig. 2.61 Relative error of approximation of the multiquadratic function $f_6(x)$ (for $a = 1$) using approximation function $a_{19}(x)$

2.5.3 Second Approximation Function for Generating the Multiquadratic Function for a = 1

In order to approximate the multiquadratic function $f_6(x)$ (for $a = 1$), a new approximation function $a_{20}(x)$ is presented. Because the multiquadratic function is an even mathematical function, the general form of the proposed second approximation function $a_{20}(x)$ can be expressed as follows:

$$a_{20\mathrm{GEN}}(x) = \frac{c_4x^4 + c_2x^2 + c_0}{d_4x^4 + d_2x^2 + d_0}, \tag{2.54}$$

c_0, c_2, c_4, d_0, d_2 and d_4 being constant coefficients having particular values that can be determined by the condition that the superior-order Taylor series expansion of $f_6(x)$ multiquadratic function (for $a = 1$), expressed by (2.51), to be identical with the superior-order Taylor series expansion of the general form of the approximation function $a_{20\mathrm{GEN}}(x)$, expressed by (2.54).

The values of these coefficients, obtained considerring the previous conditions, can be witten as: $c_0 = 1$, $c_2 = \frac{5}{4}$, $c_4 = \frac{5}{16}$, $d_0 = 1$, $d_2 = \frac{3}{4}$ and $d_4 = \frac{1}{16}$. Replacing these particular values in relation (2.54), the second approximation function $a_{20}(x)$ has the following mathematical expression:

$$a_{20}(x) = \frac{\frac{5x^4}{16} + \frac{5x^2}{4} + 1}{\frac{x^4}{16} + \frac{3x^2}{4} + 1}. \tag{2.55}$$

The graphical representation of the multiquadratic function $f_6(x)$ (for $a = 1$) and of the second approximation function $a_{20}(x)$ is presented in Fig. 2.62.

The absolute error of approximation of the multiquadratic function $f_6(x)$ (for $a = 1$) using the approximation function $a_{20}(x)$ is illustrated in Fig. 2.63. It is defined as the difference between the previous functions, error1 $= f_6(x) - a_{20}(x)$.

For more accurately evaluating the error of approximation, the relative error of approximation is shown in Fig. 2.64. This error is defined as error2 $= 100\frac{f_6(x) - a_{20}(x)}{f_6(x)}$ (in percents). Considering as acceptable a maximal approximation error of $\pm 1\%$, the maximal range of the input variable x is $[-1.8; 1.8]$.

2.5.4 Third Approximation Function for Generating the Multiquadratic Function for a = 1

In order to approximate the multiquadratic function $f_6(x)$ (for $a = 1$), a new approximation function $a_{21}(x)$ is presented. Because the multiquadratic function is an even mathematical function, the general form of the proposed third approximation function $a_{21}(x)$ can be expressed as follows:

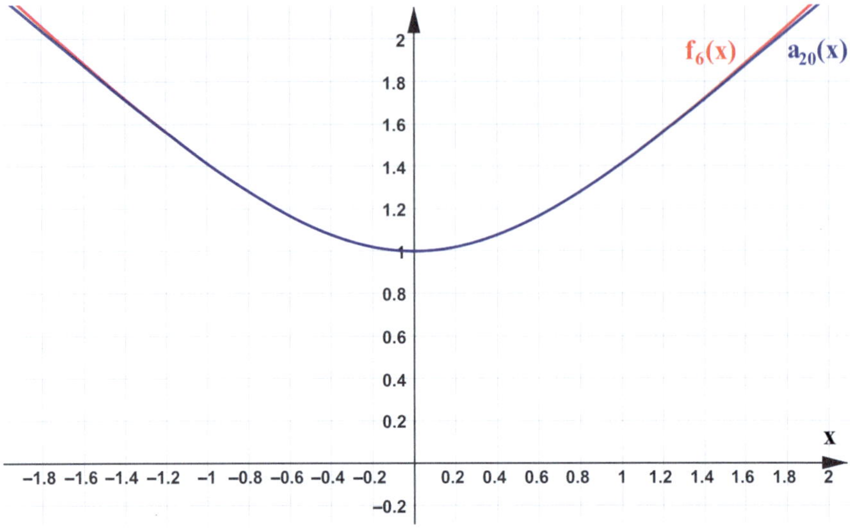

Fig. 2.62 Multiquadratic function $f_6(x)$ (for $a = 1$) and the second approximation function $a_{20}(x)$

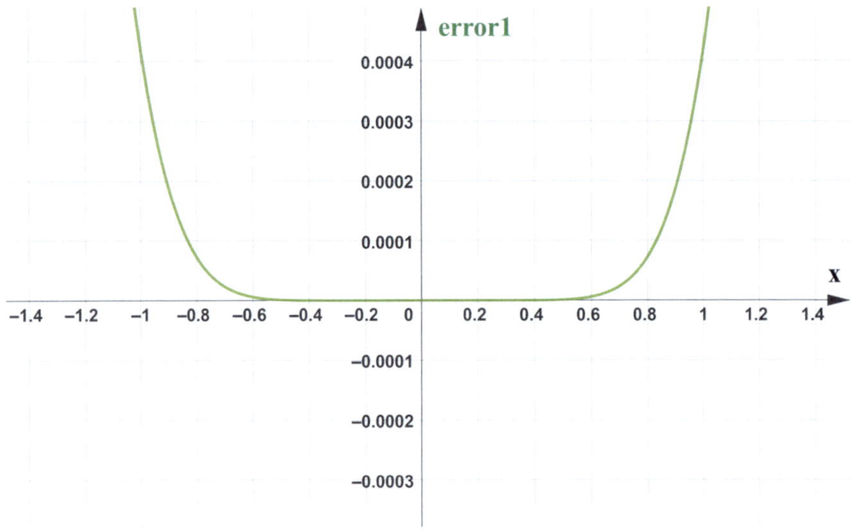

Fig. 2.63 Absolute error of approximation of the multiquadratic function $f_6(x)$ (for $a = 1$) using approximation function $a_{20}(x)$

$$a_{21\mathrm{GEN}}(x) = \frac{c_6 x^6 + c_4 x^4 + c_2 x^2 + c_0}{d_4 x^4 + d_2 x^2 + d_0}, \tag{2.56}$$

c_0, c_2, c_4, c_6, d_0, d_2 and d_4 being constant coefficients having particular values that can be determined by the condition that the superior-order Taylor series expansion

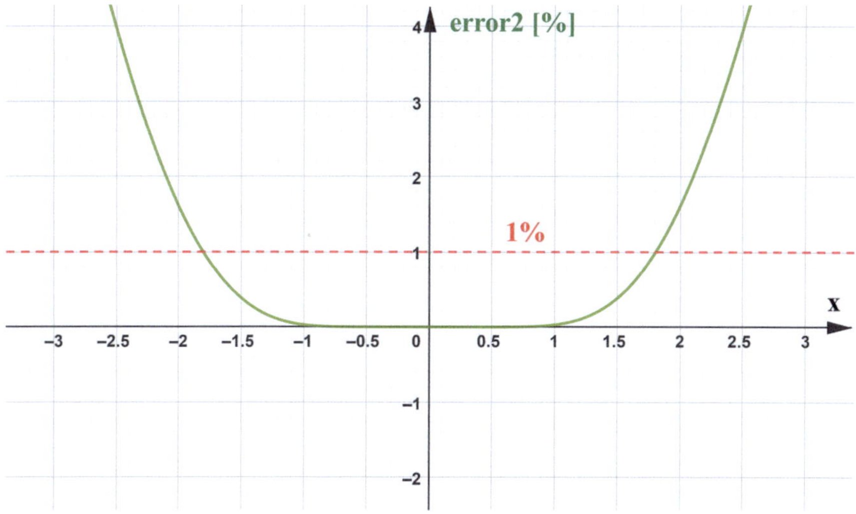

Fig. 2.64 Relative error of approximation of the multiquadratic function $f_6(x)$ (for $a = 1$) using approximation function $a_{20}(x)$

of $f_6(x)$ multiquadratic function (for $a = 1$), expressed by (2.51), to be identical with the superior-order Taylor series expansion of the general form of the approximation function $a_{21\text{GEN}}(x)$, expressed by (2.56).

The values of these coefficients, obtained considerring the previous conditions, can be witten as: $c_0 = 1$, $c_2 = \frac{3}{2}$, $c_4 = \frac{9}{16}$, $c_6 = \frac{1}{32}$, $d_0 = 1$, $d_2 = 1$ and $d_4 = \frac{3}{16}$. Replacing these particular values in relation (2.56), the third approximation function $a_{21}(x)$ has the following mathematical expression:

$$a_{21}(x) = \frac{\frac{x^6}{32} + \frac{9x^4}{16} + \frac{3x^2}{2} + 1}{\frac{3x^4}{16} + x^2 + 1}. \tag{2.57}$$

The graphical representation of the multiquadratic function $f_6(x)$ (for $a = 1$) and of the third approximation function $a_{21}(x)$ is presented in Fig. 2.65.

The absolute error of approximation of the multiquadratic function $f_6(x)$ (for $a = 1$) using the approximation function $a_{21}(x)$ is illustrated in Fig. 2.66. It is defined as the difference between the previous functions, error1 $= f_6(x) - a_{21}(x)$.

For more accurately evaluating the error of approximation, the relative error of approximation is shown in Fig. 2.67. This error is defined as error2 $= 100\frac{f_6(x) - a_{21}(x)}{f_6(x)}$ (in percents). Considering as acceptable a maximal approximation error of $\pm 1\%$, the maximal range of the input variable x is $[-2.2\,;\ 2.2]$.

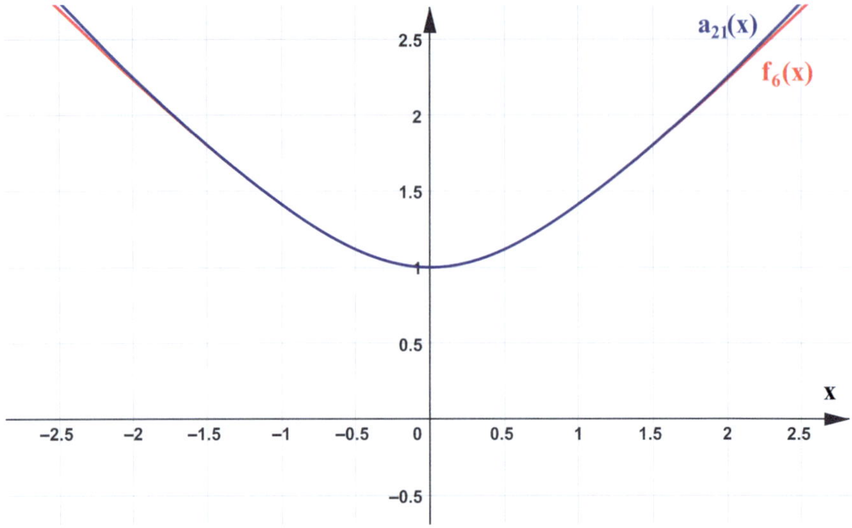

Fig. 2.65 Multiquadratic function $f_6(x)$ (for $a = 1$) and the third approximation function $a_{21}(x)$

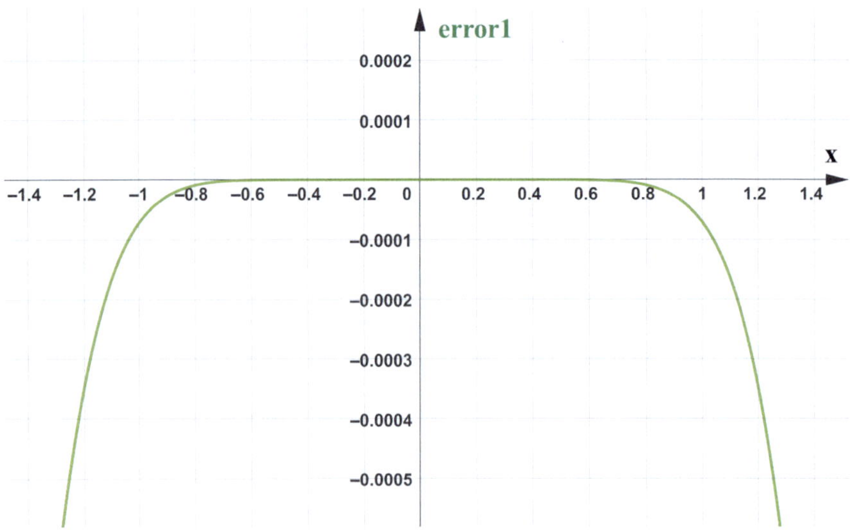

Fig. 2.66 Absolute error of approximation of the multiquadratic function $f_6(x)$ (for $a = 1$) using approximation function $a_{21}(x)$

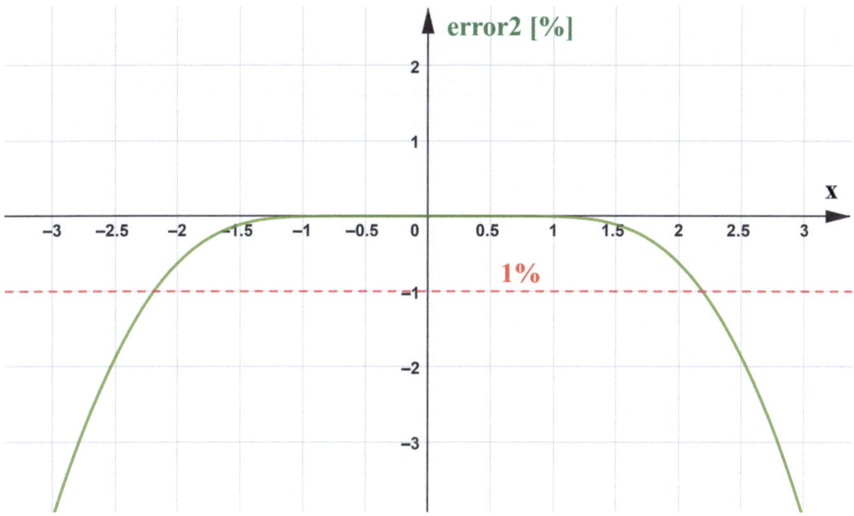

Fig. 2.67 Relative error of approximation of the multiquadratic function $f_6(x)$ using approximation function $a_{21}(x)$

2.5.5 *Fourth Approximation Function for Generating the Multiquadratic Function for a = 1*

In order to approximate the multiquadratic function $f_6(x)$ (for $a = 1$), a new approximation function $a_{22}(x)$ is presented. Because the multiquadratic function is an even mathematical function, the general form of the proposed fourth approximation function $a_{22}(x)$ can be expressed as follows:

$$a_{22\text{GEN}}(x) = \frac{c_6x^6 + c_4x^4 + c_2x^2 + c_0}{d_6x^6 + d_4x^4 + d_2x^2 + d_0}, \tag{2.58}$$

$c_0, c_2, c_4, c_6, d_0, d_2, d_4$ and d_6 being constant coefficients having particular values that can be determined by the condition that the superior-order Taylor series expansion of $f_6(x)$ multiquadratic function (for $a = 1$), expressed by (2.51), to be identical with the superior-order Taylor series expansion of the general form of the approximation function $a_{22\text{GEN}}(x)$, expressed by (2.58).

The values of these coefficients, obtained considerring the previous conditions, can be witten as: $c_0 = 1, c_2 = \frac{7}{4}, c_4 = \frac{7}{8}, c_6 = \frac{7}{64}, d_0 = 1, d_2 = \frac{5}{4}, d_4 = \frac{3}{8}$ and $d_6 = \frac{1}{64}$. Replacing these particular values in relation (2.58), the fourth approximation function $a_{22}(x)$ has the following mathematical expression:

$$a_{22}(x) = \frac{\frac{7x^6}{64} + \frac{7x^4}{8} + \frac{7x^2}{4} + 1}{\frac{x^6}{64} + \frac{3x^4}{8} + \frac{5x^2}{4} + 1}. \tag{2.59}$$

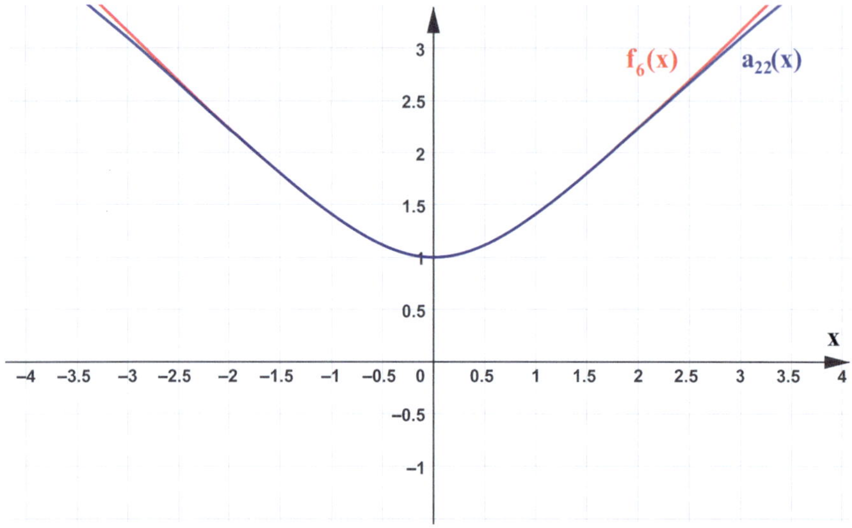

Fig. 2.68 Multiquadratic function $f_6(x)$ (for $a = 1$) and the fourth approximation function $a_{22}(x)$

The graphical representation of the multiquadratic function $f_6(x)$ (for $a = 1$) and of the fourth approximation function $a_{22}(x)$ is presented in Fig. 2.68.

The absolute error of approximation of the multiquadratic function $f_6(x)$ (for $a = 1$) using the approximation function $a_{22}(x)$ is illustrated in Fig. 2.69. It is defined as the difference between the previous functions, error1 $= f_6(x) - a_{22}(x)$.

For more accurately evaluating the error of approximation, the relative error of approximation is shown in Fig. 2.70. This error is defined as error2 $= 100\frac{f_6(x)-a_{22}(x)}{f_6(x)}$ (in percents). Considering as acceptable a maximal approximation error of $\pm 1\%$, the maximal range of the input variable x is $[-2.6;\ 2.6]$.

2.5.6 *Fifth Approximation Function for Generating the Multiquadratic Function for a = 1*

In order to approximate the multiquadratic function $f_6(x)$ (for $a = 1$), a new approximation function $a_{23}(x)$ is presented. Because the multiquadratic function is an even mathematical function, the general form of the proposed fifth approximation function $a_{23}(x)$ can be expressed as follows:

$$a_{23\mathrm{GEN}}(x) = \frac{c_6x^6 + c_4x^4 + c_2x^2 + c_0}{d_8x^8 + d_6x^6 + d_4x^4 + d_2x^2 + d_0}, \qquad (2.60)$$

c_0, c_2, c_4, c_6, d_0, d_2, d_4, d_6 and d_8 being constant coefficients having particular values that can be determined by the condition that the superior-order Taylor series

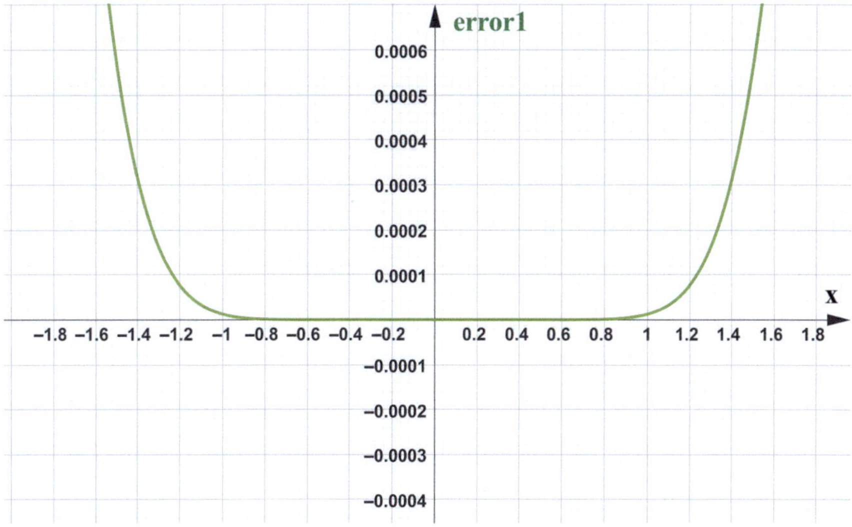

Fig. 2.69 Absolute error of approximation of the multiquadratic function $f_6(x)$ (for $a = 1$) using approximation function $a_{22}(x)$

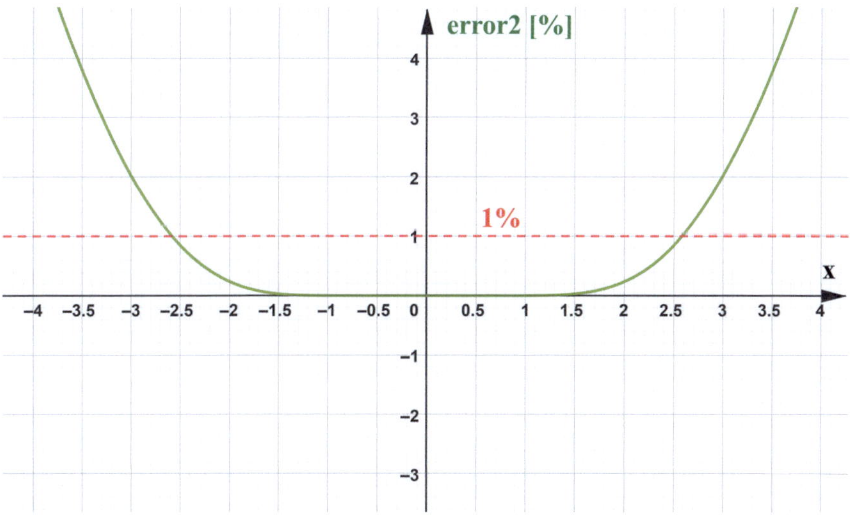

Fig. 2.70 Relative error of approximation of the multiquadratic function $f_6(x)$ using approximation function $a_{22}(x)$

expansion of $f_6(x)$ multiquadratic function (for $a = 1$), expressed by (2.51), to be identical with the superior-order Taylor series expansion of the general form of the approximation function $a_{23GEN}(x)$, expressed by (2.60).

The values of these coefficients, obtained considerring the previous conditions, can be witten as: $c_0 = 1$, $c_2 = \frac{27}{14}$, $c_4 = \frac{9}{8}$, $c_6 = \frac{3}{16}$, $d_0 = 1$, $d_2 = \frac{10}{7}$, $d_4 = \frac{15}{28}$, $d_6 = \frac{1}{28}$ and $d_8 = -\frac{1}{896}$. Replacing these particular values in relation (2.60), the fifth approximation function $a_{23}(x)$ has the following mathematical expression:

$$a_{23}(x) = \frac{\frac{3x^6}{16} + \frac{9x^4}{8} + \frac{27x^2}{14} + 1}{-\frac{x^8}{896} + \frac{x^6}{28} + \frac{15x^4}{28} + \frac{10x^2}{7} + 1}. \tag{2.61}$$

The graphical representation of the multiquadratic function $f_6(x)$ (for $a = 1$) and of the fifth approximation function $a_{23}(x)$ is presented in Fig. 2.71.

The absolute error of approximation of the multiquadratic function $f_6(x)$ (for $a = 1$) using the approximation function $a_{23}(x)$ is illustrated in Fig. 2.72. It is defined as the difference between the previous functions, error1 $= f_6(x) - a_{23}(x)$.

For more accurately evaluating the error of approximation, the relative error of approximation is shown in Fig. 2.73. This error is defined as error2 $= 100\frac{f_6(x) - a_{23}(x)}{f_6(x)}$ (in percents). Considering as acceptable a maximal approximation error of $\pm 1\%$, the maximal range of the input variable x is $[-2.6;\ 2.6]$.

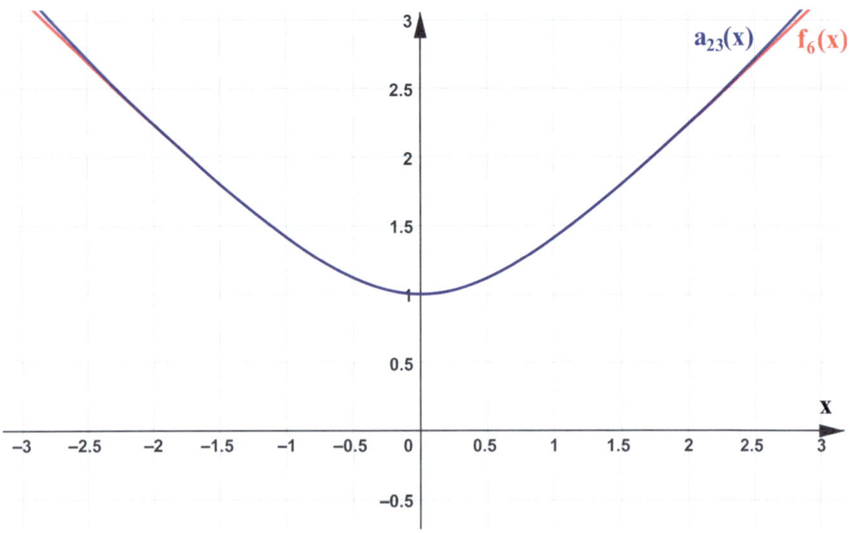

Fig. 2.71 Multiquadratic function $f_6(x)$ (for $a = 1$) and the fifth approximation function $a_{23}(x)$

Fig. 2.72 Absolute error of approximation of the multiquadratic function $f_6(x)$ (for $a = 1$) using approximation function $a_{23}(x)$

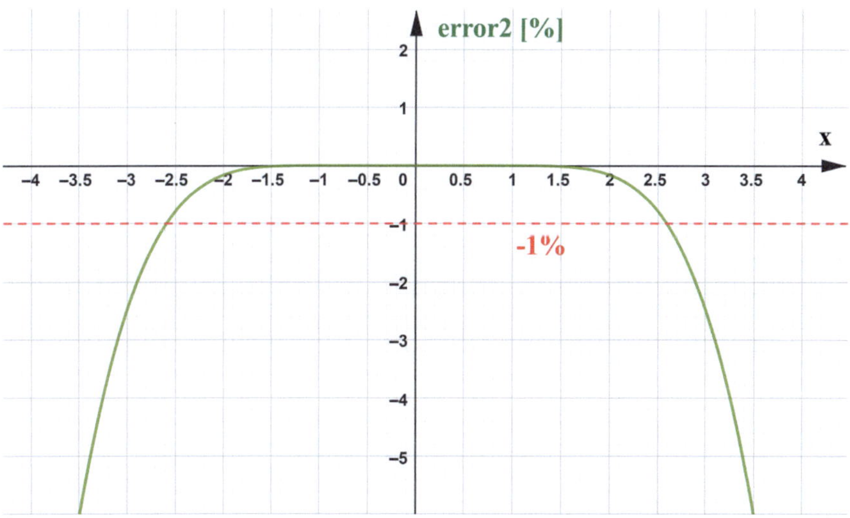

Fig. 2.73 Relative error of approximation of the multiquadratic function $f_6(x)$ (for $a = 1$) using approximation function $a_{23}(x)$

2.5.7 Sixth Approximation Function for Generating the Multiquadratic Function for a = 1

In order to approximate the multiquadratic function $f_6(x)$ (for $a = 1$), a new approximation function $a_{24}(x)$ is presented. Because the multiquadratic function is an even mathematical function, the general form of the proposed sixth approximation function $a_{24}(x)$ can be expressed as follows:

$$a_{24\mathrm{GEN}}(x) = \frac{c_8x^8 + c_6x^6 + c_4x^4 + c_2x^2 + c_0}{d_8x^8 + d_6x^6 + d_4x^4 + d_2x^2 + d_0}, \tag{2.62}$$

c_0, c_2, c_4, c_6, c_8, d_0, d_2, d_4, d_6 and d_8 being constant coefficients having particular values that can be determined by the condition that the superior-order Taylor series expansion of $f_6(x)$ multiquadratic function (for $a = 1$), expressed by (2.51), to be identical with the superior-order Taylor series expansion of the general form of the approximation function $a_{24\mathrm{GEN}}(x)$, expressed by (2.62).

The values of these coefficients, obtained considerring the previous conditions, can be witten as: $c_0 = 1$, $c_2 = \frac{9}{4}$, $c_4 = \frac{27}{16}$, $c_6 = \frac{15}{32}$, $c_8 = \frac{9}{256}$, $d_0 = 1$, $d_2 = \frac{7}{4}$, $d_4 = \frac{15}{16}$, $d_6 = \frac{5}{32}$ and $d_8 = \frac{1}{256}$. Replacing these particular values in relation (2.62), the sixth approximation function $a_{24}(x)$ has the following mathematical expression:

$$a_{24}(x) = \frac{\frac{9x^8}{256} + \frac{15x^6}{32} + \frac{27x^4}{16} + \frac{9x^2}{4} + 1}{\frac{x^8}{256} + \frac{5x^6}{32} + \frac{15x^4}{16} + \frac{7x^2}{4} + 1}. \tag{2.63}$$

The graphical representation of the multiquadratic function $f_6(x)$ (for $a = 1$) and of the sixth approximation function $a_{24}(x)$ is presented in Fig. 2.74.

The absolute error of approximation of the multiquadratic function $f_6(x)$ (for $a = 1$) using the approximation function $a_{24}(x)$ is illustrated in Fig. 2.75. It is defined as the difference between the previous functions, error1 $= f_6(x) - a_{24}(x)$.

For more accurately evaluating the error of approximation, the relative error of approximation is shown in Fig. 2.76. This error is defined as error2 $= 100\frac{f_6(x) - a_{24}(x)}{f_6(x)}$ (in percents). Considering as acceptable a maximal approximation error of $\pm 1\%$, the maximal range of the input variable x is $[-3.4\,;\,3.4]$.

2.5.8 First Approximation Function for Generating the Multiquadratic Function for a = 2

In order to approximate the multiquadratic function $f_6(x)$ (for $a = 2$), a new approximation function $a_{25}(x)$ is presented. Because the multiquadratic function is an even mathematical function, the general form of the proposed first approximation function $a_{25}(x)$ can be expressed as follows:

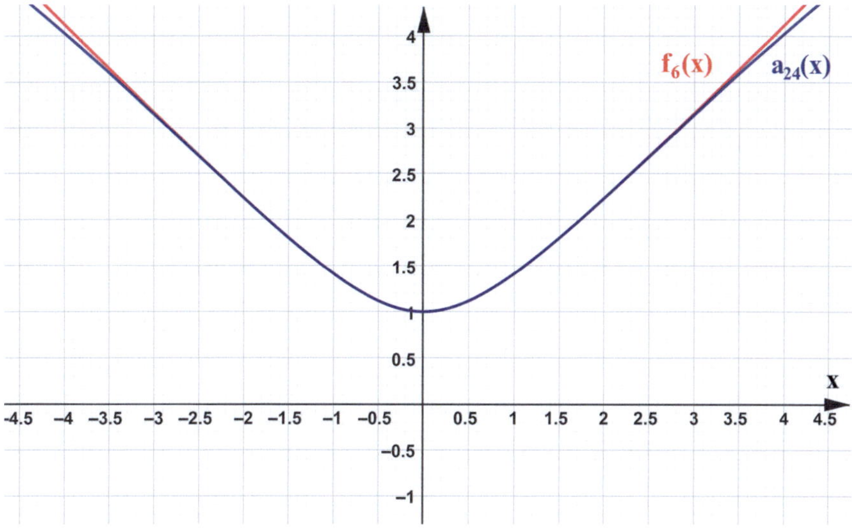

Fig. 2.74 Multiquadratic function $f_6(x)$ (for $a = 1$) and the sixth approximation function $a_{24}(x)$

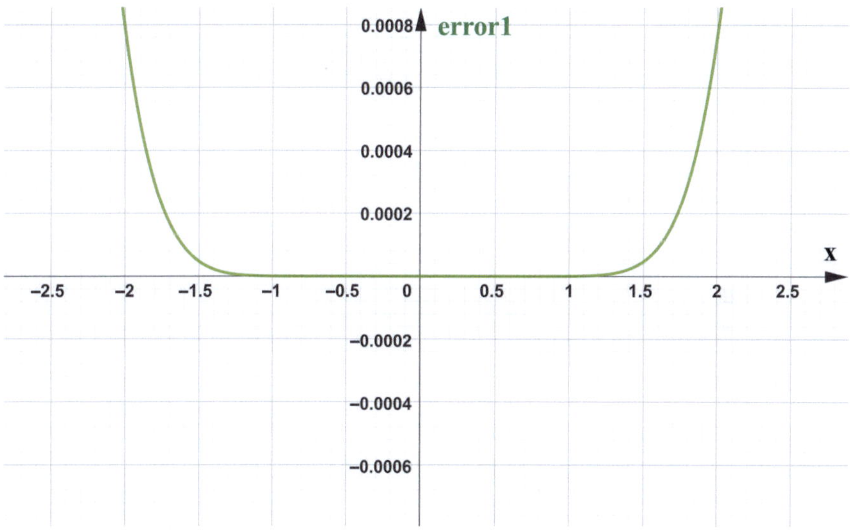

Fig. 2.75 Absolute error of approximation of the multiquadratic function $f_6(x)$ (for $a = 1$) using approximation function $a_{24}(x)$

$$a_{25\mathrm{GEN}}(x) = \frac{c_4x^4 + c_2x^2 + c_0}{d_2x^2 + d_0}, \tag{2.64}$$

c_0, c_2, c_4, d_0 and d_2 being constant coefficients having particular values that can be determined by the condition that the superior-order Taylor series expansion of

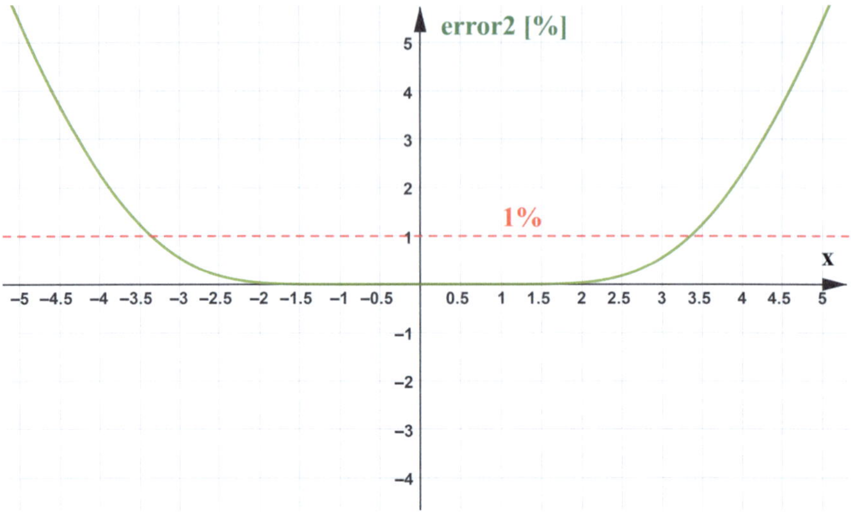

Fig. 2.76 Relative error of approximation of the multiquadratic function $f_6(x)$ (for $a = 1$) using $a_{24}(x)$ approximation function

$f_6(x)$ multiquadratic function (for $a = 2$), expressed by (2.51), to be identical with the superior-order Taylor series expansion of the general form of the approximation function $a_{25\text{GEN}}(x)$, expressed by (2.64).

The values of these coefficients, obtained considerring the previous conditions, can be witten as: $c_0 = 2$, $c_2 = \frac{1}{2}$, $c_4 = \frac{1}{64}$, $d_0 = 1$ and $d_2 = \frac{1}{8}$. Replacing these particular values in relation (2.64), the first approximation function $a_{25}(x)$ has the following mathematical expression:

$$a_{25}(x) = \frac{\frac{x^4}{64} + \frac{x^2}{2} + 2}{\frac{x^2}{8} + 1}. \tag{2.65}$$

The graphical representation of the multiquadratic function $f_6(x)$ (for $a = 2$) and of the first approximation function $a_{25}(x)$ is presented in Fig. 2.77.

The absolute error of approximation of the multiquadratic function $f_6(x)$ using the approximation function $a_{25}(x)$ is illustrated in Fig. 2.78. It is defined as the difference between the previous functions, error1 $= f_6(x) - a_{25}(x)$.

For more accurately evaluating the error of approximation, the relative error of approximation is shown in Fig. 2.79. This error is defined as error2 $= 100\frac{f_6(x) - a_{25}(x)}{f_6(x)}$ (in percents). Considering as acceptable a maximal approximation error of $\pm 1\%$, the maximal range of the input variable x is $[-2.75\,;\ 2.75]$.

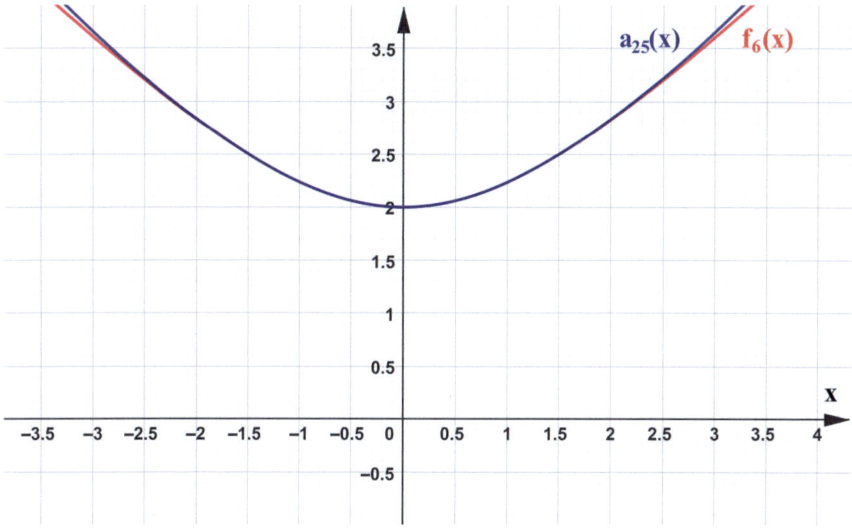

Fig. 2.77 Multiquadratic function $f_6(x)$ (for $a = 2$) and the first approximation function $a_{25}(x)$

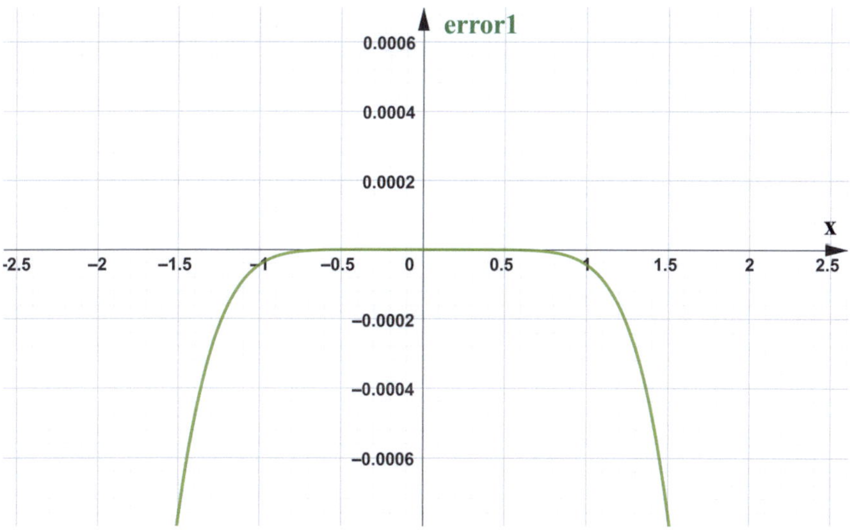

Fig. 2.78 Absolute error of approximation of the multiquadratic function $f_6(x)$ (for $a = 2$) using approximation function $a_{25}(x)$

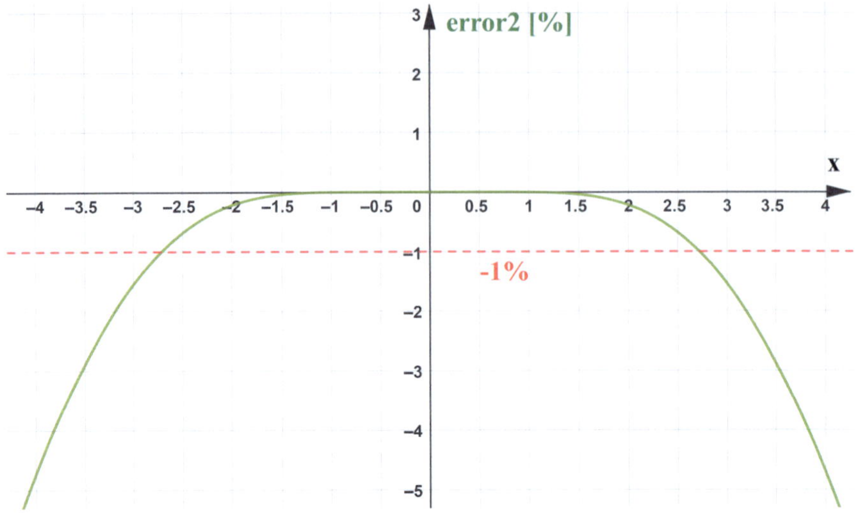

Fig. 2.79 Relative error of approximation of the multiquadratic function $f_6(x)$ (for $a = 2$) using approximation function $a_{25}(x)$

2.5.9 Second Approximation Function for Generating the Multiquadratic Function for a = 2

In order to approximate the multiquadratic function $f_6(x)$ (for $a = 2$), a new approximation function $a_{26}(x)$ is presented. Because the multiquadratic function is an even mathematical function, the general form of the proposed second approximation function $a_{26}(x)$ can be expressed as follows:

$$a_{26\text{GEN}}(x) = \frac{c_4 x^4 + c_2 x^2 + c_0}{d_4 x^4 + d_2 x^2 + d_0}, \tag{2.66}$$

c_0, c_2, c_4, d_0, d_2 and d_4 being constant coefficients having particular values that can be determined by the condition that the superior-order Taylor series expansion of $f_6(x)$ multiquadratic function (for $a = 2$), expressed by (2.51), to be identical with the superior-order Taylor series expansion of the general form of the approximation function $a_{24\text{GEN}}(x)$, expressed by (2.66).

The values of these coefficients, obtained considerring the previous conditions, can be witten as: $c_0 = 2, c_2 = \frac{5}{8}, c_4 = \frac{5}{128}, d_0 = 1, d_2 = \frac{3}{16}$ and $d_4 = \frac{1}{256}$. Replacing these particular values in relation (2.66), the second approximation function $a_{26}(x)$ has the following mathematical expression:

$$a_{26}(x) = \frac{\frac{5x^4}{128} + \frac{5x^2}{8} + 2}{\frac{x^4}{256} + \frac{3x^2}{16} + 1}. \tag{2.67}$$

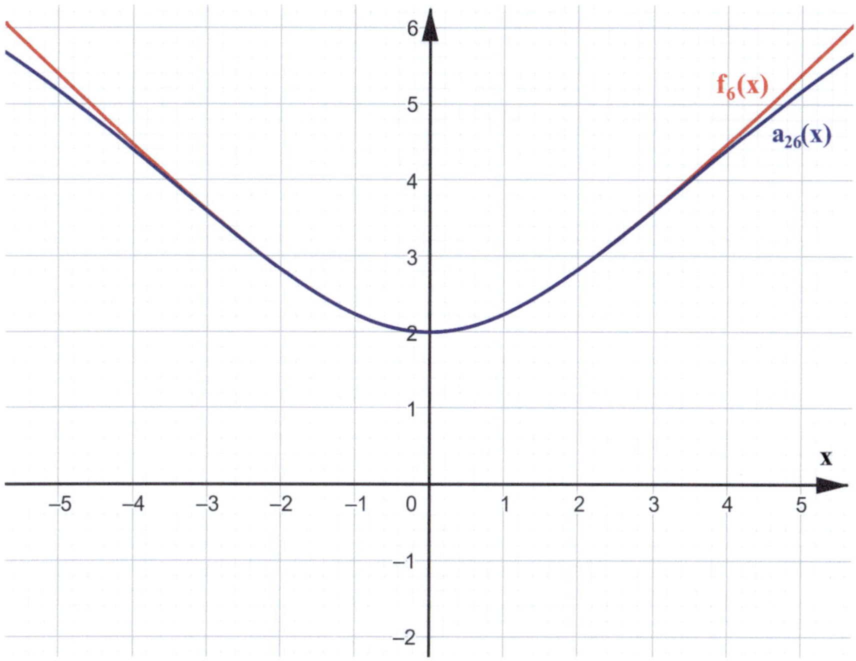

Fig. 2.80 Multiquadratic function $f_6(x)$ (for $a = 2$) and the second approximation function $a_{26}(x)$

The graphical representation of the multiquadratic function $f_6(x)$ (for $a = 2$) and of the second approximation function $a_{26}(x)$ is presented in Fig. 2.80.

The absolute error of approximation of the multiquadratic function $f_6(x)$ (for $a = 2$) using the approximation function $a_{26}(x)$ is illustrated in Fig. 2.81. It is defined as the difference between the previous functions, error1 $= f_6(x) - a_{26}(x)$.

For more accurately evaluating the error of approximation, the relative error of approximation is shown in Fig. 2.82. This error is defined as error2 $= 100\frac{f_6(x)-a_{26}(x)}{f_6(x)}$ (in percents). Considering as acceptable a maximal approximation error of $\pm 1\%$, the maximal range of the input variable x is $[-3.6\,;\ 3.6]$.

2.5.10 *Third Approximation Function for Generating the Multiquadratic Function for a = 2*

In order to approximate the multiquadratic function $f_6(x)$ (for $a = 2$), a new approximation function $a_{27}(x)$ is presented. Because the multiquadratic function is an even mathematical function, the general form of the proposed third approximation function $a_{27}(x)$ can be expressed as follows:

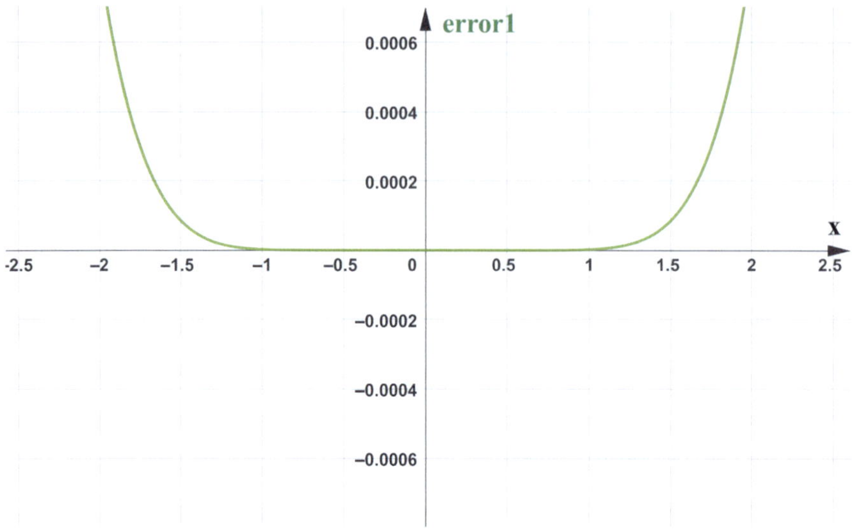

Fig. 2.81 Absolute error of approximation of the multiquadratic function $f_6(x)$ (for $a = 2$) using approximation function $a_{26}(x)$

$$a_{27\text{GEN}}(x) = \frac{c_6 x^6 + c_4 x^4 + c_2 x^2 + c_0}{d_4 x^4 + d_2 x^2 + d_0}, \tag{2.68}$$

c_0, c_2, c_4, c_6, d_0, d_2 and d_4 being constant coefficients having particular values that can be determined by the condition that the superior-order Taylor series expansion of $f_6(x)$ multiquadratic function (for $a = 2$), expressed by (2.51), to be identical with the superior-order Taylor series expansion of the general form of the approximation function $a_{27\text{GEN}}(x)$, expressed by (2.68).

The values of these coefficients, obtained considerring the previous conditions, can be witten as: $c_0 = 2$, $c_2 = \frac{3}{4}$, $c_4 = \frac{9}{128}$, $c_6 = \frac{1}{1024}$, $d_0 = 1$, $d_2 = \frac{1}{4}$ and $d_4 = \frac{3}{256}$. Replacing these particular values in relation (2.68), the third approximation function $a_{27}(x)$ has the following mathematical expression:

$$a_{27}(x) = \frac{\frac{x^6}{1024} + \frac{9x^4}{128} + \frac{3x^2}{4} + 2}{\frac{3x^4}{256} + \frac{x^2}{4} + 1}. \tag{2.69}$$

The graphical representation of the multiquadratic function $f_6(x)$ (for $a = 2$) and of the third approximation function $a_{27}(x)$ is presented in Fig. 2.83.

The absolute error of approximation of the multiquadratic function $f_6(x)$ (for $a = 2$) using the approximation function $a_{27}(x)$ is illustrated in Fig. 2.84. It is defined as the difference between the previous functions, error1 $= f_6(x) - a_{27}(x)$.

For more accurately evaluating the error of approximation, the relative error of approximation is shown in Fig. 2.85. This error is defined as error2 $= 100\frac{f_6(x) - a_{27}(x)}{f_6(x)}$

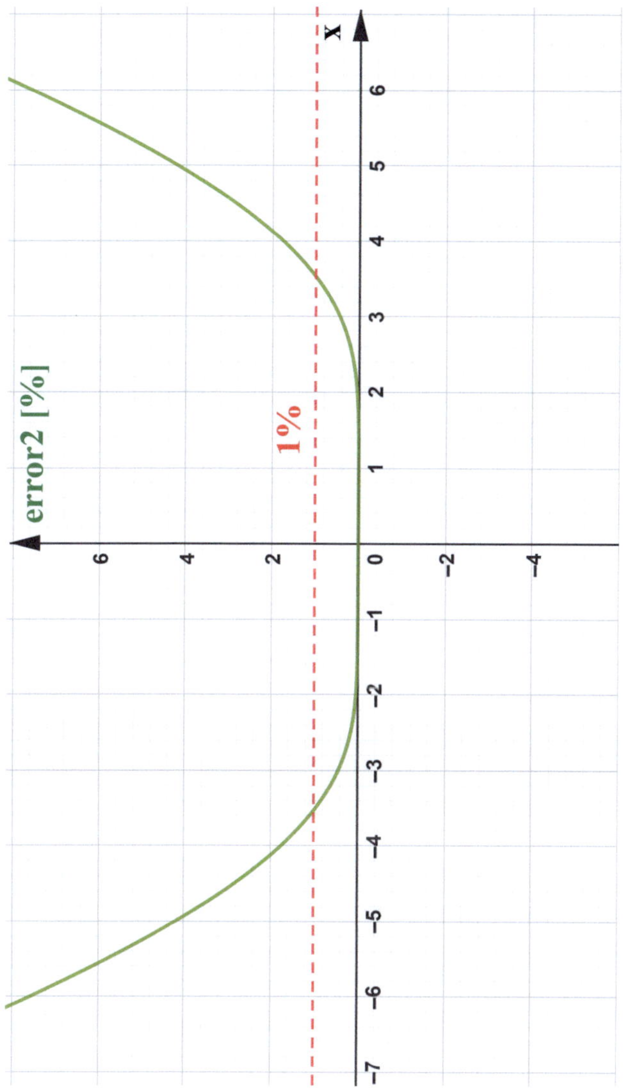

Fig. 2.82 Relative error of approximation of the multiquadratic function $f_6(x)$ (for $a = 2$) using approximation function $a_{26}(x)$

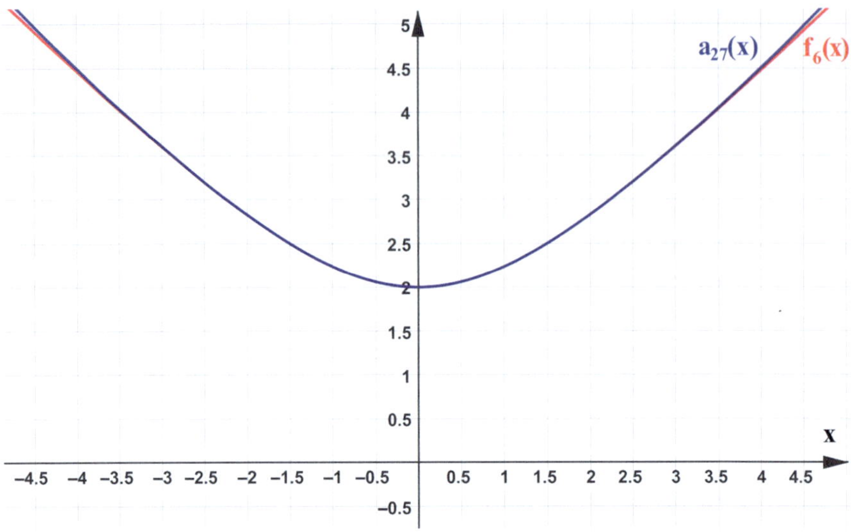

Fig. 2.83 Multiquadratic function $f_6(x)$ (for $a = 2$) and the third approximation function $a_{27}(x)$

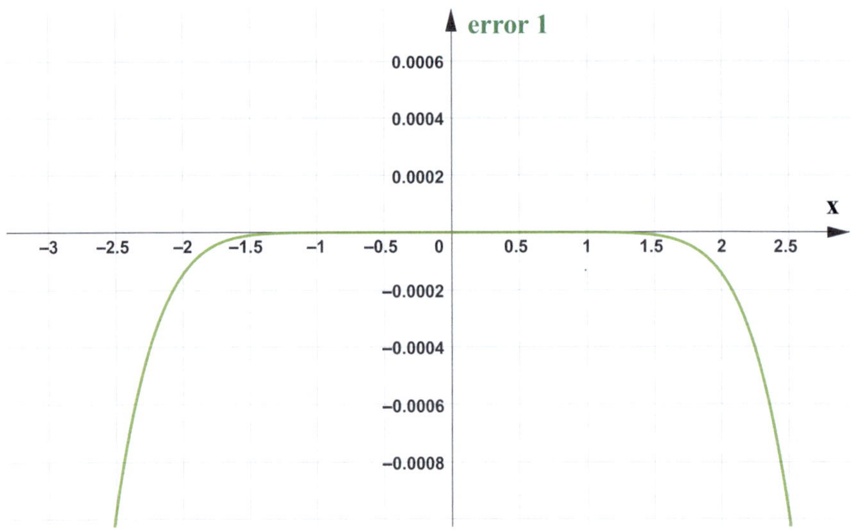

Fig. 2.84 Absolute error of approximation of the multiquadratic function $f_6(x)$ (for $a = 2$) using approximation function $a_{27}(x)$

(in percents). Considering as acceptable a maximal approximation error of $\pm 1\%$, the maximal range of the input variable x is $[-4.3\,;\ 4.3]$.

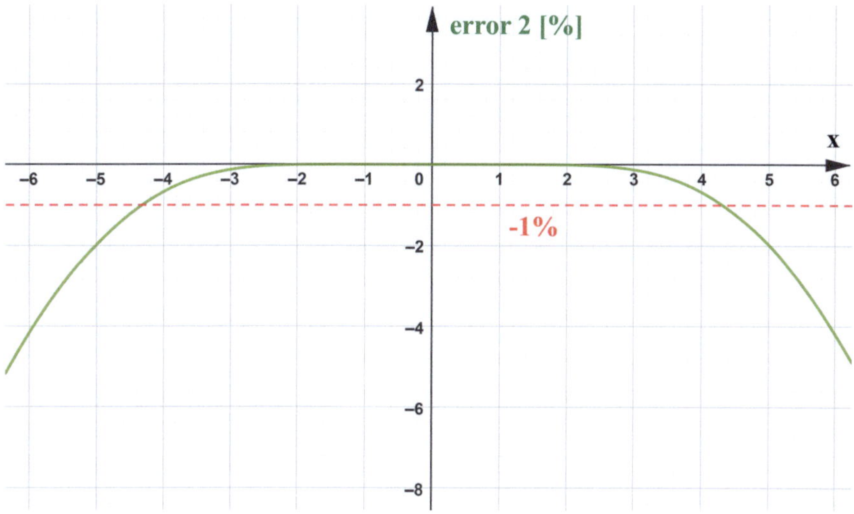

Fig. 2.85 Relative error of approximation of the multiquadratic function $f_6(x)$ (for $a = 2$) using approximation function $a_{27}(x)$

2.5.11 Fourth Approximation Function for Generating the Multiquadratic Function for a = 2

In order to approximate the multiquadratic function $f_6(x)$ (for $a = 2$), a new approximation function $a_{28}(x)$ is presented. Because the multiquadratic function is an even mathematical function, the general form of the proposed fourth approximation function $a_{28}(x)$ can be expressed as follows:

$$a_{28\text{GEN}}(x) = \frac{c_6x^6 + c_4x^4 + c_2x^2 + c_0}{d_6x^6 + d_4x^4 + d_2x^2 + d_0}, \tag{2.70}$$

$c_0, c_2, c_4, c_6, d_0, d_2, d_4$ and d_6 being constant coefficients having particular values that can be determined by the condition that the superior-order Taylor series expansion of $f_6(x)$ multiquadratic function (for $a = 2$), expressed by (2.51), to be identical with the superior-order Taylor series expansion of the general form of the approximation function $a_{28\text{GEN}}(x)$, expressed by (2.70).

The values of these coefficients, obtained considerring the previous conditions, can be witten as: $c_0 = 2$, $c_2 = \frac{7}{8}$, $c_4 = \frac{7}{64}$, $c_6 = \frac{7}{2048}$, $d_0 = 1$, $d_2 = \frac{5}{16}$, $d_4 = \frac{3}{128}$ and $d_6 = \frac{1}{4096}$. Replacing these particular values in relation (2.70), the fourth approximation function $a_{28}(x)$ has the following mathematical expression:

$$a_{28}(x) = \frac{\frac{7x^6}{2048} + \frac{7x^4}{64} + \frac{7x^2}{8} + 2}{\frac{x^6}{4096} + \frac{3x^4}{128} + \frac{5x^2}{16} + 1}. \tag{2.71}$$

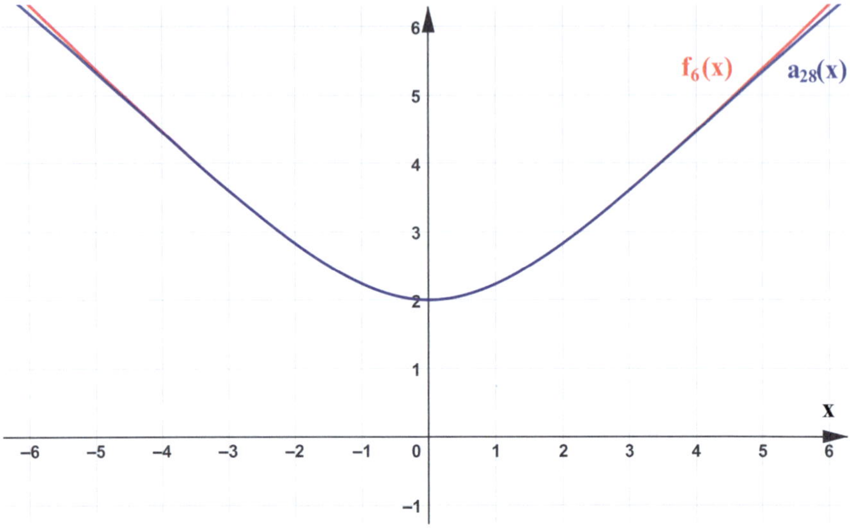

Fig. 2.86 Multiquadratic function $f_6(x)$ (for $a = 2$) and the fourth approximation function $a_{28}(x)$

The graphical representation of the multiquadratic function $f_6(x)$ (for $a = 2$) and of the fourth approximation function $a_{28}(x)$ is presented in Fig. 2.86.

The absolute error of approximation of the multiquadratic function $f_6(x)$ (for $a = 2$) using the approximation function $a_{28}(x)$ is illustrated in Fig. 2.87. It is defined as the difference between the previous functions, $\text{error1} = f_6(x) - a_{28}(x)$.

For more accurately evaluating the error of approximation, the relative error of approximation is shown in Fig. 2.88. This error is defined as $\text{error2} = 100\frac{f_6(x)-a_{28}(x)}{f_6(x)}$ (in percents). Considering as acceptable a maximal approximation error of $\pm 1\%$, the maximal range of the input variable x is $[-5.2\,;\ 5.2]$.

2.5.12 First Approximation Function for Generating the Multiquadratic Function for a = 3

In order to approximate the multiquadratic function $f_6(x)$ (for $a = 3$), a new approximation function $a_{29}(x)$ is presented. Because the multiquadratic function is an even mathematical function, the general form of the proposed first approximation function $a_{29}(x)$ can be expressed as follows:

$$a_{29\text{GEN}}(x) = \frac{c_4x^4 + c_2x^2 + c_0}{d_2x^2 + d_0}, \tag{2.72}$$

c_0, c_2, c_4, d_0 and d_2 being constant coefficients having particular values that can be determined by the condition that the superior-order Taylor series expansion of

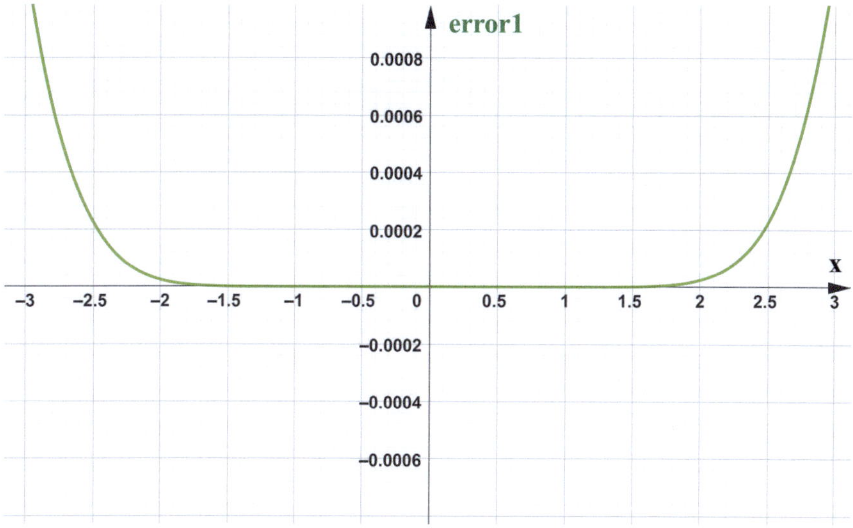

Fig. 2.87 Absolute error of approximation of the multiquadratic function $f_6(x)$ (for $a = 2$) using approximation function $a_{28}(x)$

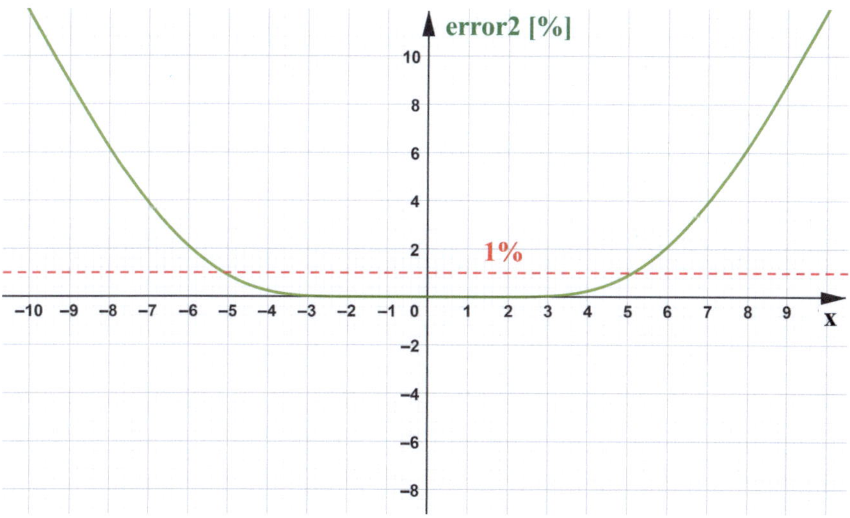

Fig. 2.88 Relative error of approximation of the multiquadratic function $f_6(x)$ using approximation function $a_{28}(x)$

$f_6(x)$ multiquadratic function (for $a = 3$), expressed by (2.51), to be identical with the superior-order Taylor series expansion of the general form of the approximation function $a_{29\text{GEN}}(x)$, expressed by (2.72).

The values of these coefficients, obtained considerring the previous conditions, can be witten as: $c_0 = 3$, $c_2 = \frac{1}{3}$, $c_4 = \frac{1}{216}$, $d_0 = 1$ and $d_2 = \frac{1}{18}$. Replacing these particular values in relation (2.72), the first approximation function $a_{29}(x)$ has the following mathematical expression:

$$a_{29}(x) = \frac{\frac{x^4}{216} + \frac{x^2}{3} + 3}{\frac{x^2}{18} + 1}. \tag{2.73}$$

The graphical representation of the multiquadratic function $f_6(x)$ (for $a = 3$) and of the first approximation function $a_{29}(x)$ is presented in Fig. 2.89.

The absolute error of approximation of the multiquadratic function $f_6(x)$ (for $a = 3$) using the approximation function $a_{29}(x)$ is illustrated in Fig. 2.90. It is defined as the difference between the previous functions, error1 $= f_6(x) - a_{29}(x)$.

For more accurately evaluating the error of approximation, the relative error of approximation is shown in Fig. 2.91. This error is defined as error2 $= 100\frac{f_6(x) - a_{29}(x)}{f_6(x)}$ (in percents). Considering as acceptable a maximal approximation error of $\pm 1\%$, the maximal range of the input variable x is $[-4.2\,;\,4.2]$.

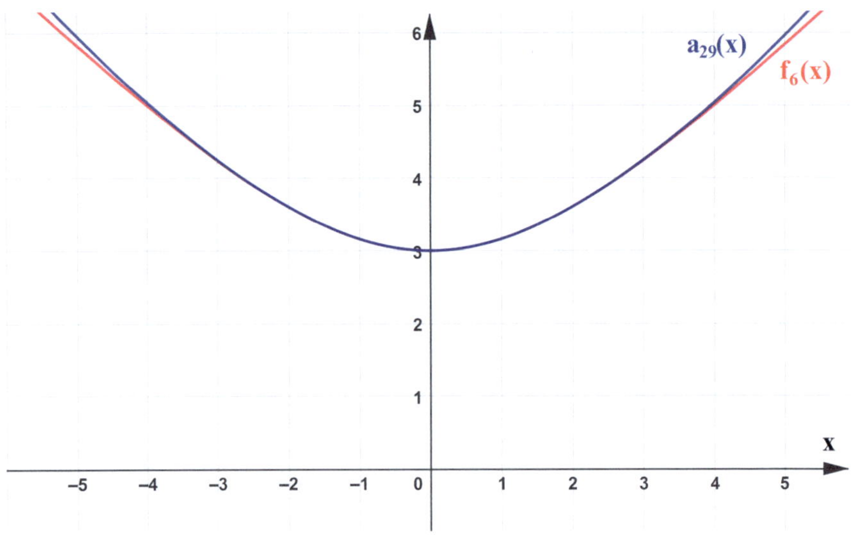

Fig. 2.89 Multiquadratic function $f_6(x)$ (for $a = 3$) and the first approximation function $a_{29}(x)$

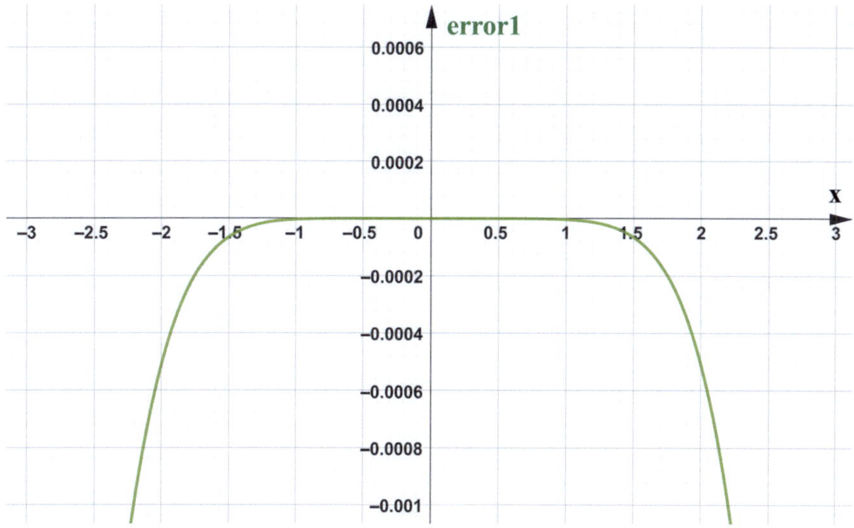

Fig. 2.90 Error of approximation of the multiquadratic function $f_6(x)$ (for $a = 3$) using approximation function $a_{29}(x)$

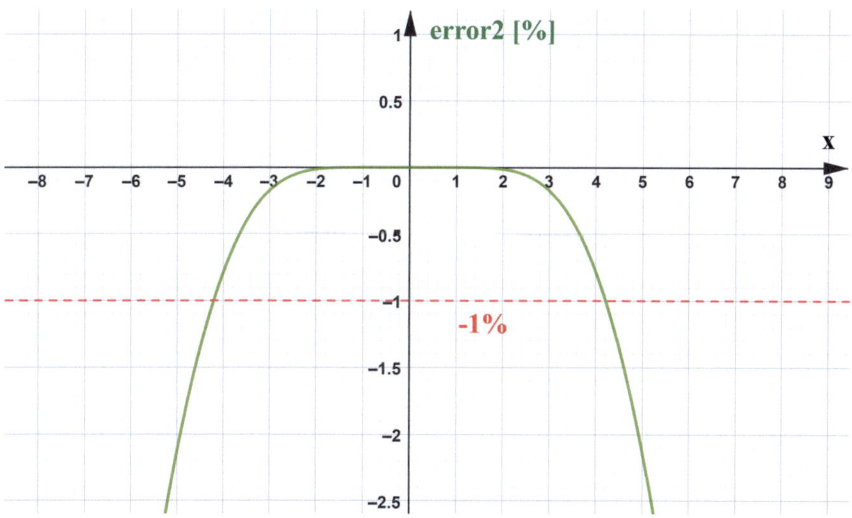

Fig. 2.91 Relative error of approximation of the multiquadratic function $f_6(x)$ (for $a = 3$) using approximation function $a_{29}(x)$

2.5.13 Second Approximation Function for Generating the Multiquadratic Function for a = 3

In order to approximate the multiquadratic function $f_6(x)$ (for $a = 3$), a new approximation function $a_{30}(x)$ is presented. Because the multiquadratic function is an even mathematical function, the general form of the proposed second approximation function $a_{30}(x)$ can be expressed as follows:

$$a_{30\mathrm{GEN}}(x) = \frac{c_4x^4 + c_2x^2 + c_0}{d_4x^4 + d_2x^2 + d_0}, \tag{2.74}$$

c_0, c_2, c_4, d_0, d_2 and d_4 being constant coefficients having particular values that can be determined by the condition that the superior-order Taylor series expansion of $f_6(x)$ multiquadratic function (for $a = 3$), expressed by (2.51), to be identical with the superior-order Taylor series expansion of the general form of the approximation function $a_{30\mathrm{GEN}}(x)$, expressed by (2.74).

The values of these coefficients, obtained considerring the previous conditions, can be witten as: $c_0 = 3$, $c_2 = \frac{5}{12}$, $c_4 = \frac{5}{432}$, $d_0 = 1$, $d_2 = \frac{1}{12}$ and $d_4 = \frac{1}{1296}$. Replacing these particular values in relation (2.74), the second approximation function $a_{30}(x)$ has the following mathematical expression:

$$a_{30}(x) = \frac{\frac{5x^4}{432} + \frac{5x^2}{12} + 3}{\frac{x^4}{1296} + \frac{x^2}{12} + 1}. \tag{2.75}$$

The graphical representation of the multiquadratic function $f_6(x)$ (for $a = 3$) and of the second approximation function $a_{30}(x)$ is presented in Fig. 2.92.

The absolute error of approximation of the multiquadratic function $f_6(x)$ (for $a = 3$) using the approximation function $a_{30}(x)$ is illustrated in Fig. 2.93. It is defined as the difference between the previous functions, error1 $= f_6(x) - a_{30}(x)$.

For more accurately evaluating the error of approximation, the relative error of approximation is shown in Fig. 2.94. This error is defined as error2 $= 100\frac{f_6(x) - a_{30}(x)}{f_6(x)}$ (in percents). Considering as acceptable a maximal approximation error of $\pm 1\%$, the maximal range of the input variable x is $[-5.4\,;\ 5.4]$.

2.5.14 Third Approximation Function for Generating the Multiquadratic Function for a = 3

In order to approximate the multiquadratic function $f_6(x)$ (for $a = 3$), a new approximation function $a_{31}(x)$ is presented. Because the multiquadratic function is an even mathematical function, the general form of the proposed third approximation function $a_{31}(x)$ can be expressed as follows:

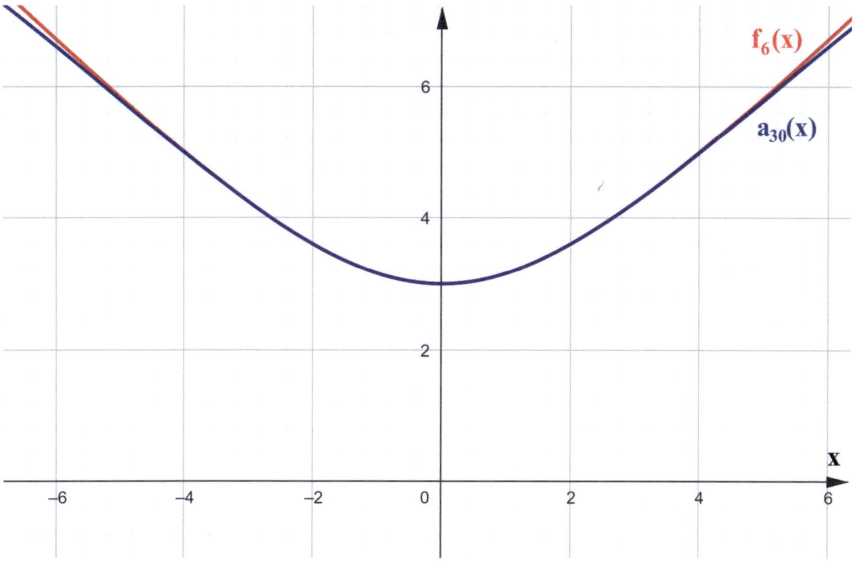

Fig. 2.92 Multiquadratic function $f_6(x)$ (for $a = 3$) and the second approximation function $a_{30}(x)$

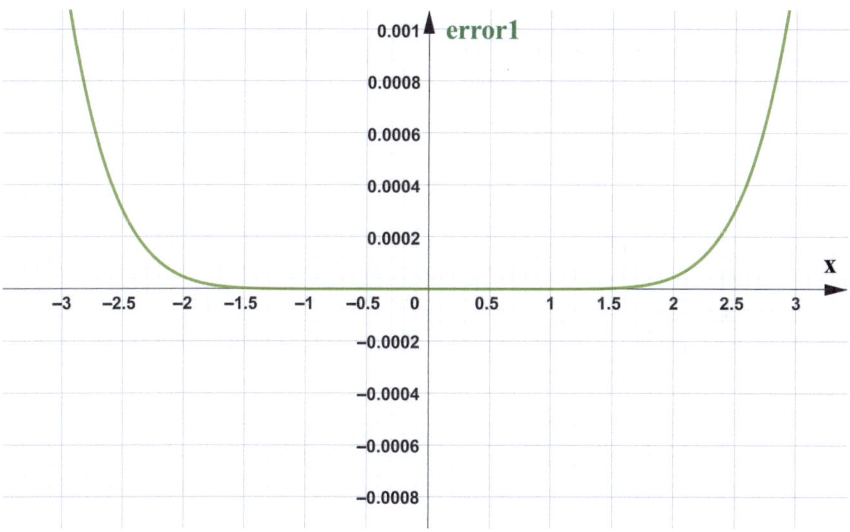

Fig. 2.93 Absolute error of approximation of the multiquadratic function $f_6(x)$ (for $a = 3$) using $a_{30}(x)$ approximation function

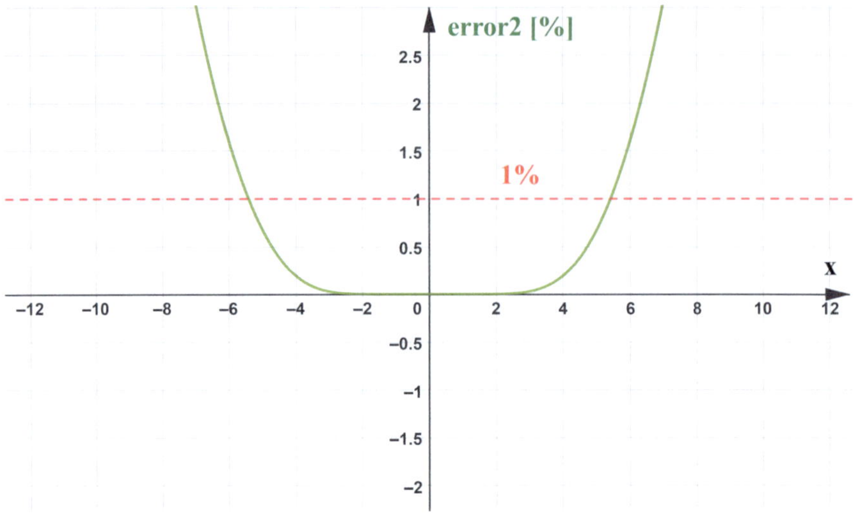

Fig. 2.94 Relative error of approximation of the multiquadratic function $f_6(x)$ (for $a = 3$) using approximation function $a_{30}(x)$

$$a_{31\mathrm{GEN}}(x) = \frac{c_6x^6 + c_4x^4 + c_2x^2 + c_0}{d_4x^4 + d_2x^2 + d_0}, \tag{2.76}$$

c_0, c_2, c_4, c_6, d_0, d_2 and d_4 being constant coefficients having particular values that can be determined by the condition that the superior-order Taylor series expansion of $f_6(x)$ multiquadratic function (for $a = 3$), expressed by (2.51), to be identical with the superior-order Taylor series expansion of the general form of the approximation function $a_{31\mathrm{GEN}}(x)$, expressed by (2.76).

The values of these coefficients, obtained considerring the previous conditions, can be witten as: $c_0 = 3$, $c_2 = \frac{1}{2}$, $c_4 = \frac{1}{48}$, $c_6 = \frac{1}{7776}$, $d_0 = 1$, $d_2 = \frac{1}{9}$ and $d_4 = \frac{1}{432}$. Replacing these particular values in relation (2.76), the third approximation function $a_{27}(x)$ has the following mathematical expression:

$$a_{31}(x) = \frac{\frac{x^6}{7776} + \frac{x^4}{48} + \frac{x^2}{2} + 3}{\frac{x^4}{432} + \frac{x^2}{9} + 1}. \tag{2.77}$$

The graphical representation of the multiquadratic function $f_6(x)$ (for $a = 3$) and of the third approximation function $a_{31}(x)$ is presented in Fig. 2.95.

The absolute error of approximation of the multiquadratic function $f_6(x)$ (for $a = 3$) using the approximation function $a_{31}(x)$ is illustrated in Fig. 2.96. It is defined as the difference between the previous functions, error1 $= f_6(x) - a_{31}(x)$.

For more accurately evaluating the error of approximation, the relative error of approximation is shown in Fig. 2.97. This error is defined as error2 $= 100\frac{f_6(x) - a_{31}(x)}{f_6(x)}$

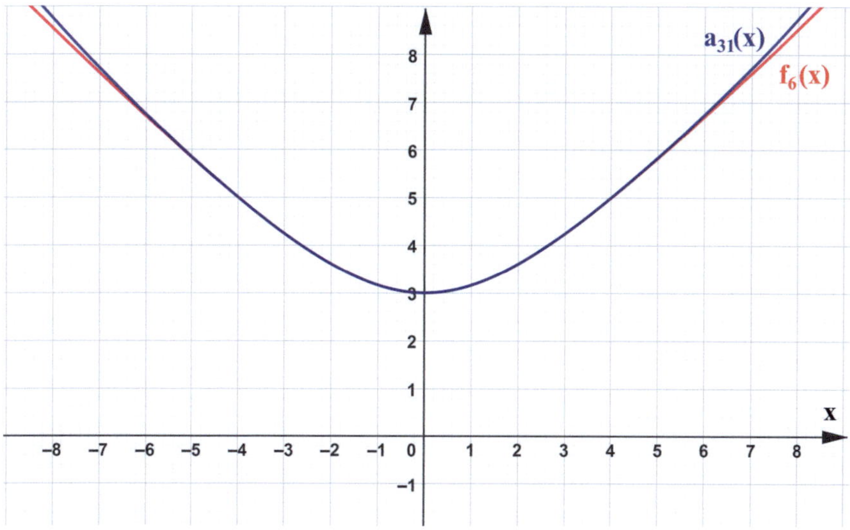

Fig. 2.95 Multiquadratic function $f_6(x)$ (for $a = 3$) and the third approximation function $a_{31}(x)$

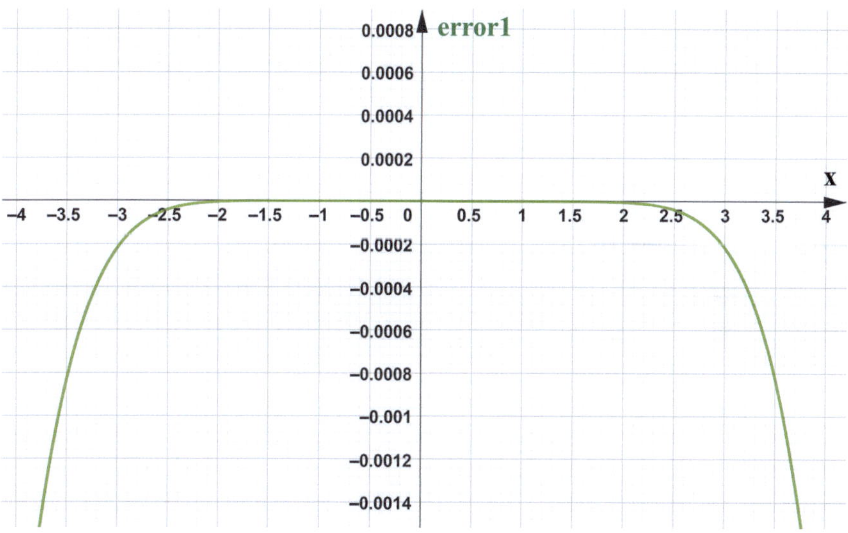

Fig. 2.96 Absolute error of approximation of the multiquadratic function $f_6(x)$ (for $a = 3$) using approximation function $a_{31}(x)$

(in percents). Considering as acceptable a maximal approximation error of $\pm 1\%$, the maximal range of the input variable x is $[-6.4\,;\ 6.4]$.

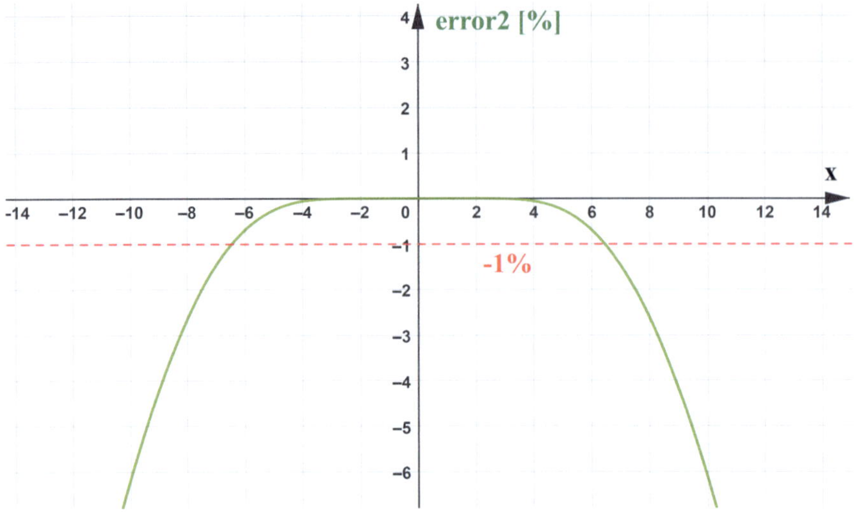

Fig. 2.97 Relative error of approximation of the multiquadratic function $f_6(x)$ (for $a = 3$) using approximation function $a_{31}(x)$

2.5.15 *Fourth Approximation Function for Generating the Multiquadratic Function for a = 3*

In order to approximate the multiquadratic function $f_6(x)$ (for $a = 3$), a new approximation function $a_{32}(x)$ is presented. Because the multiquadratic function is an even mathematical function, the general form of the proposed fourth approximation function $a_{32}(x)$ can be expressed as follows:

$$a_{32\text{GEN}}(x) = \frac{c_6x^6 + c_4x^4 + c_2x^2 + c_0}{d_6x^6 + d_4x^4 + d_2x^2 + d_0}, \tag{2.78}$$

$c_0, c_2, c_4, c_6, d_0, d_2, d_4$ and d_6 being constant coefficients having particular values that can be determined by the condition that the superior-order Taylor series expansion of $f_6(x)$ multiquadratic function (for $a = 3$), expressed by (2.51), to be identical with the superior-order Taylor series expansion of the general form of the approximation function $a_{32\text{GEN}}(x)$, expressed by (2.78).

The values of these coefficients, obtained considerring the previous conditions, can be witten as: $c_0 = 3$, $c_2 = \frac{7}{12}$, $c_4 = \frac{7}{216}$, $c_6 = \frac{7}{15{,}552}$, $d_0 = 1$, $d_2 = \frac{5}{36}$, $d_4 = \frac{1}{216}$ and $d_6 = \frac{1}{46{,}656}$. Replacing these particular values in relation (2.78), the fourth approximation function $a_{32}(x)$ has the following mathematical expression:

$$a_{32}(x) = \frac{\frac{7x^6}{15{,}552} + \frac{7x^4}{216} + \frac{7x^2}{12} + 3}{\frac{x^6}{46{,}656} + \frac{x^4}{216} + \frac{5x^2}{36} + 1}. \tag{2.79}$$

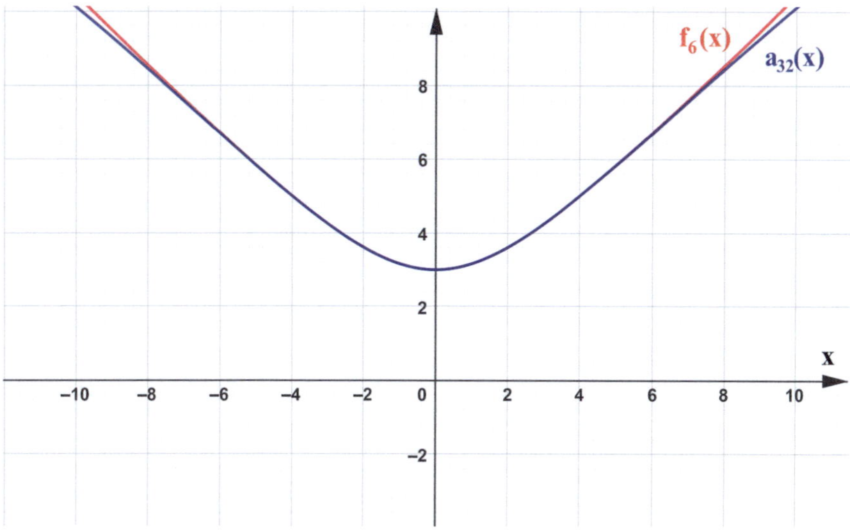

Fig. 2.98 Multiquadratic function $f_6(x)$ (for $a = 3$) and the fourth approximation function $a_{32}(x)$

The graphical representation of the multiquadratic function $f_6(x)$ (for $a = 3$) and of the fourth approximation function $a_{32}(x)$ is presented in Fig. 2.98.

The absolute error of approximation of the multiquadratic function $f_6(x)$ (for $a = 3$) using the approximation function $a_{32}(x)$ is illustrated in Fig. 2.99. It is defined as the difference between the previous functions, error1 $= f_6(x) - a_{32}(x)$.

For more accurately evaluating the error of approximation, the relative error of approximation is shown in Fig. 2.100. This error is defined as error2 $= 100\frac{f_6(x)-a_{32}(x)}{f_6(x)}$ (in percents). Considering as acceptable a maximal approximation error of $\pm 1\%$, the maximal range of the input variable x is $[-7.7\,;\ 7.7]$.

2.6 Approximation Functions for Inverse Multiquadratic Functions

The inverse multiquadratic function represents a type of radial basis function (RBF), mostly used in RBF neural networks and kernel methods. It provides smooth, localized responses and decays slowly as the input moves away from the center, making it very valuable in modeling complex patterns.

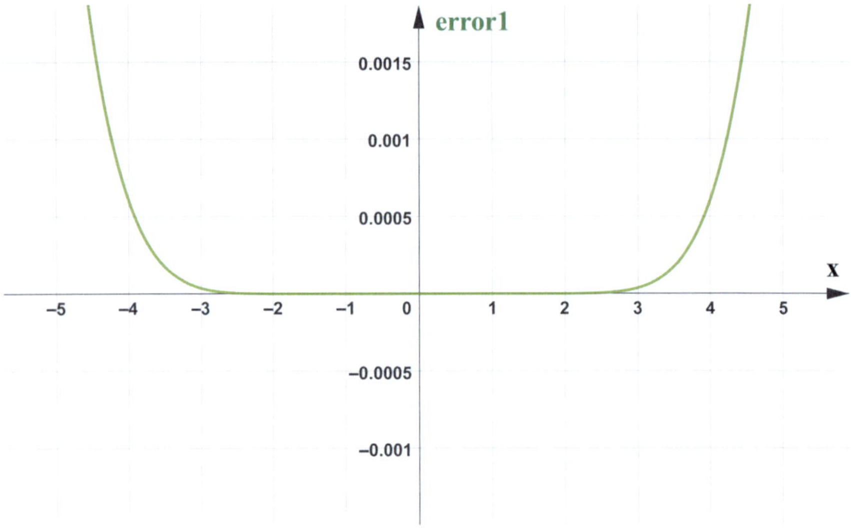

Fig. 2.99 Absolute error of approximation of the multiquadratic function $f_6(x)$ (for $a = 3$) using approximation function $a_{32}(x)$

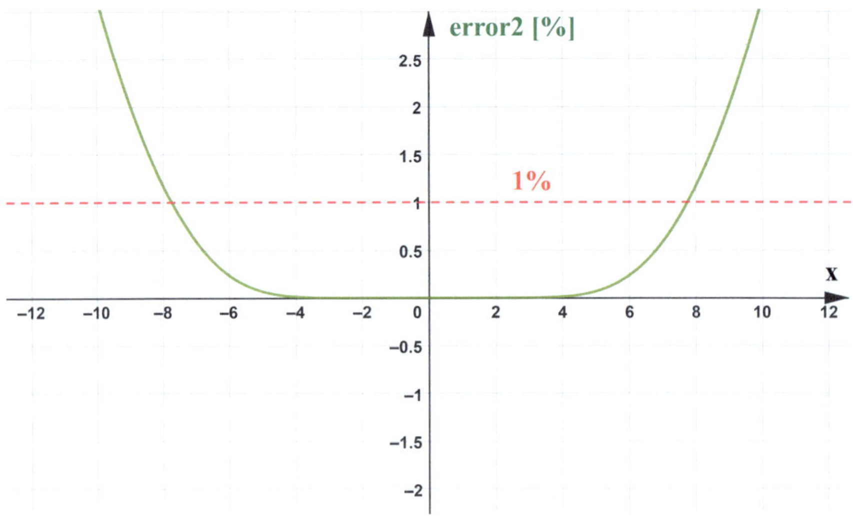

Fig. 2.100 Relative error of approximation of the multiquadratic function $f_6(x)$ (for $a = 3$) using approximation function $a_{32}(x)$

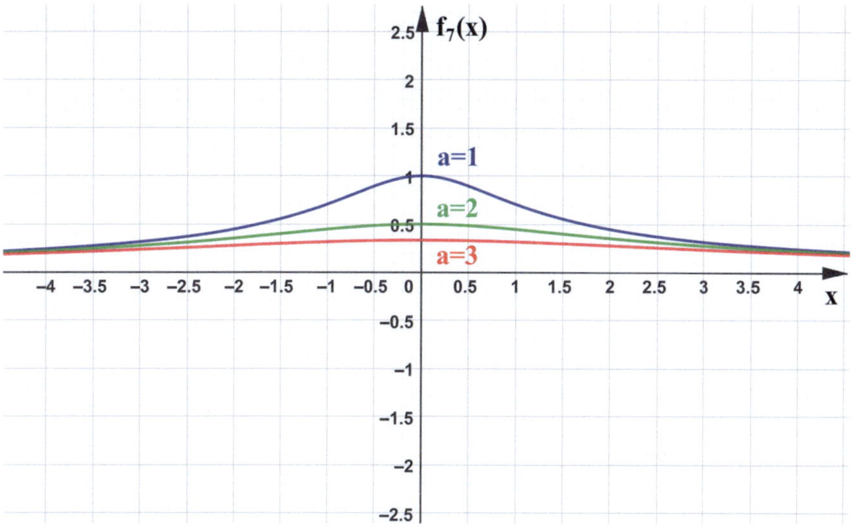

Fig. 2.101 Graphical representation of the inverse multiquadratic function for $a = 0$, $a = 1$ and $a = 2$

2.6.1 *General Form of the Inverse Multiquadratic Function*

The mathematical form of the inverse inverse multiquadratic function can be expressed as follows:

$$f_7(x) = \frac{1}{\sqrt{x^2 + a^2}}. \tag{2.80}$$

The graphical representation of the inverse multiquadratic function $f_7(x)$, for multiple values of a parameter ($a = 0$, $a = 1$ and $a = 2$) is presented in Fig. 2.101.

2.6.2 *First Approximation Function for Generating the Inverse Multiquadratic Function for a = 1*

In order to approximate the inverse multiquadratic function $f_7(x)$ (for $a = 1$), a new approximation function $a_{33}(x)$ is presented. Its particular mathematical form is developed having two very important objectives. The first objective is represented by the accuracy of the approximation that could be obtained using this approximation function. The second objective is related to the possibility of its facile analog implementation in CMOS technology using as reasonable as possible hardware resources (see Chaps. 4 and 6).

Because the inverse multiquadratic function is an even mathematical function, the general form of the proposed first approximation function $a_{33}(x)$ can be expressed as follows:

$$a_{33\text{GEN}}(x) = \frac{c_4x^4 + c_2x^2 + c_0}{d_6x^6 + d_4x^4 + d_2x^2 + d_0}, \tag{2.81}$$

c_0, c_2, c_4, d_0, d_2, d_4 and d_6 being constant coefficients having particular values that can be determined by the condition that the superior-order Taylor series expansion of $f_7(x)$ inverse multiquadratic function (for $a = 1$), expressed by (2.80), to be identical with the superior-order Taylor series expansion of the general form of the approximation function $a_{33\text{GEN}}(x)$, expressed by (2.81).

The values of these coefficients, obtained considerring the previous conditions, can be witten as: $c_0 = 1$, $c_2 = 1$, $c_4 = \frac{3}{16}$, $d_0 = 1$, $d_2 = \frac{3}{2}$, $d_4 = \frac{9}{16}$ and $d_6 = \frac{1}{32}$. Replacing these particular values in relation (2.81), the first approximation function $a_{33}(x)$ has the following mathematical expression:

$$a_{33}(x) = \frac{\frac{3x^4}{16} + x^2 + 1}{\frac{x^6}{32} + \frac{9x^4}{16} + \frac{3x^2}{2} + 1}. \tag{2.82}$$

The graphical representation of the inverse multiquadratic function $f_7(x)$ (for $a = 1$) and of the first approximation function $a_{33}(x)$ is presented in Fig. 2.102.

The absolute error of approximation of the multiquadratic function $f_7(x)$ (for $a = 1$) using the approximation function $a_{33}(x)$ is illustrated in Fig. 2.103. It is defined as the difference between the previous functions, error1 $= f_6(x) - a_{33}(x)$.

For more accurately evaluating the error of approximation, the relative error of approximation is shown in Fig. 2.104. This error is defined as error2 $= 100\frac{f_7(x) - a_{33}(x)}{f_6(x)}$ (in percents). Considering as acceptable a maximal approximation error of $\pm 1\%$, the maximal range of the input variable x is $[-2.2\,;\ 2.2]$.

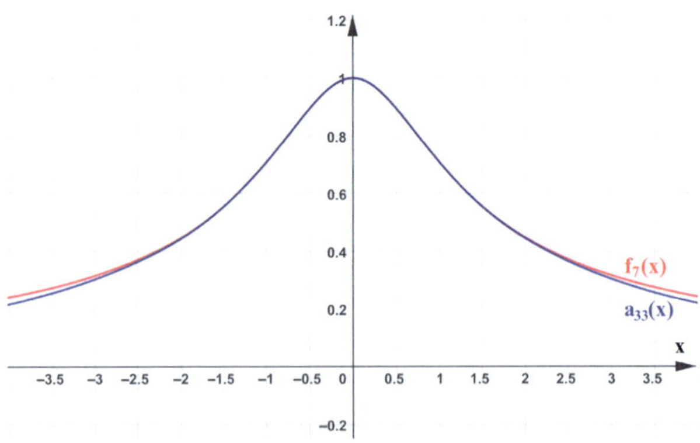

Fig. 2.102 Inverse multiquadratic function $f_7(x)$ (for $a = 1$) and the first approximation function $a_{33}(x)$

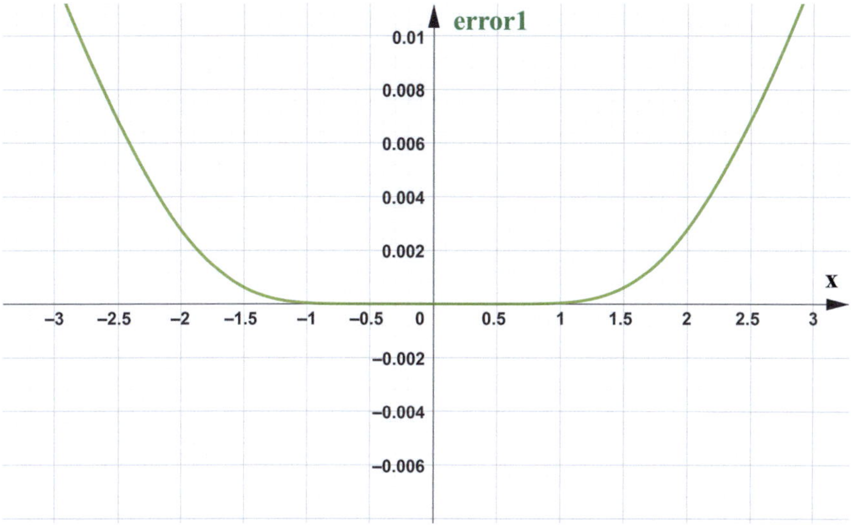

Fig. 2.103 Absolute error of approximation of the inverse multiquadratic function $f_7(x)$ (for $a = 1$) using approximation function $a_{33}(x)$

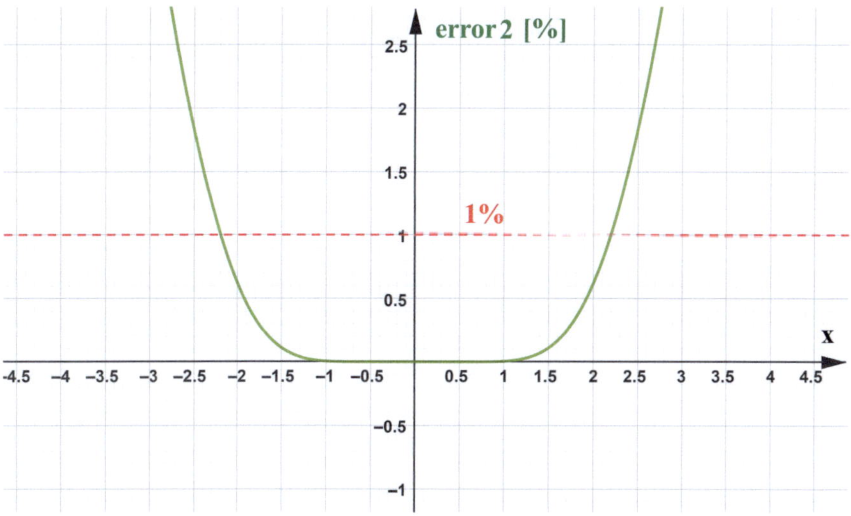

Fig. 2.104 Relative error of approximation of the multiquadratic function $f_7(x)$ (for $a = 1$) using approximation function $a_{33}(x)$

2.6.3 Second Approximation Function for Generating the Inverse Multiquadratic Function for a = 1

In order to approximate the inverse multiquadratic function $f_7(x)$ (for $a = 1$), a new approximation function $a_{34}(x)$ is presented. Because the inverse multiquadratic function is an even mathematical function, the general form of the proposed second approximation function $a_{34}(x)$ can be expressed as follows:

$$a_{34\mathrm{GEN}}(x) = \frac{c_6x^6 + c_4x^4 + c_2x^2 + c_0}{d_6x^6 + d_4x^4 + d_2x^2 + d_0}, \tag{2.83}$$

$c_0, c_2, c_4, c_6, d_0, d_2, d_4$ and d_6 being constant coefficients having particular values that can be determined by the condition that the superior-order Taylor series expansion of $f_7(x)$ inverse multiquadratic function (for $a = 1$), expressed by (2.80), to be identical with the superior-order Taylor series expansion of the general form of the approximation function $a_{34\mathrm{GEN}}(x)$, expressed by (2.83).

The values of these coefficients, obtained considerring the previous conditions, can be witten as: $c_0 = 1$, $c_2 = \frac{5}{4}$, $c_4 = \frac{3}{8}$, $c_6 = \frac{1}{64}$, $d_0 = 1$, $d_2 = \frac{7}{4}$, $d_4 = \frac{7}{8}$ and $d_6 = \frac{7}{64}$. Replacing these particular values in relation (2.83), the second approximation function $a_{34}(x)$ has the following mathematical expression:

$$a_{34}(x) = \frac{\frac{x^6}{64} + \frac{3x^4}{8} + \frac{5x^2}{4} + 1}{\frac{7x^6}{64} + \frac{7x^4}{8} + \frac{7x^2}{4} + 1}. \tag{2.84}$$

The graphical representation of the inverse multiquadratic function $f_7(x)$ (for $a = 1$) and of the second approximation function $a_{34}(x)$ is presented in Fig. 2.105.

The absolute error of approximation of the multiquadratic function $f_7(x)$ (for $a = 1$) using the approximation function $a_{34}(x)$ is illustrated in Fig. 2.106. It is defined as the difference between the previous functions, error1 $= f_7(x) - a_{34}(x)$.

For more accurately evaluating the error of approximation, the relative error of approximation is shown in Fig. 2.107. This error is defined as error2 $= 100\frac{f_7(x) - a_{34}(x)}{f_6(x)}$ (in percents). Considering as acceptable a maximal approximation error of $\pm 1\%$, the maximal range of the input variable x is $[-2.6\,;\,2.6]$.

2.6.4 First Approximation Function for Generating the Inverse Multiquadratic Function for a = 2

In order to approximate the inverse multiquadratic function $f_7(x)$ (for $a = 2$), a new approximation function $a_{35}(x)$ is presented. Because the inverse multiquadratic function is an even mathematical function, the general form of the proposed first approximation function $a_{35}(x)$ can be expressed as follows:

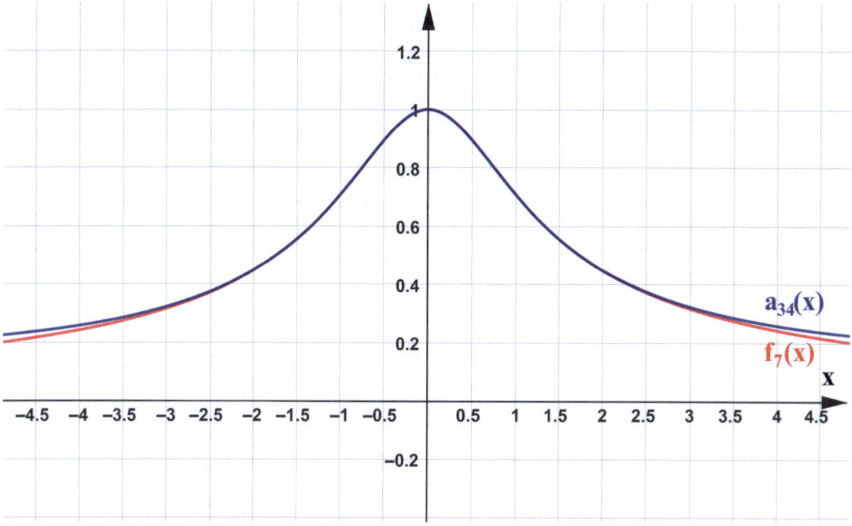

Fig. 2.105 Inverse multiquadratic function $f_7(x)$ (for $a = 1$) and the second approximation function $a_{34}(x)$

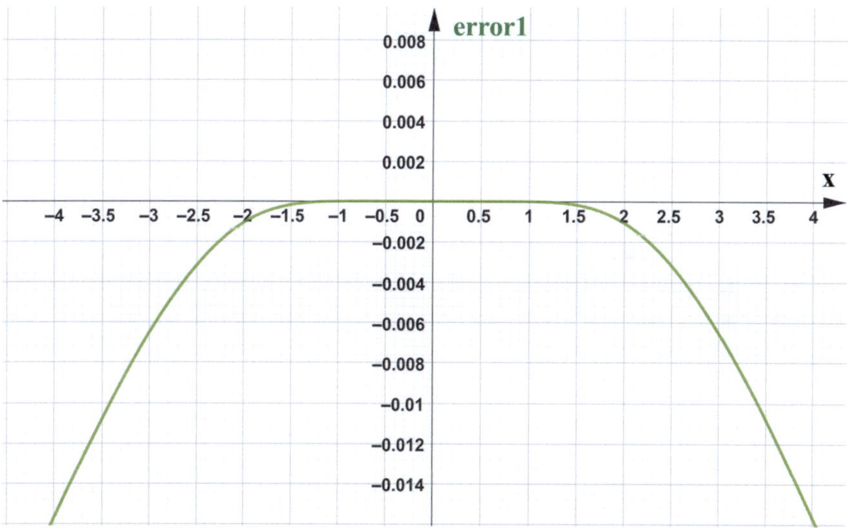

Fig. 2.106 Absolute error of approximation of the inverse multiquadratic function $f_7(x)$ (for $a = 1$) using approximation function $a_{34}(x)$

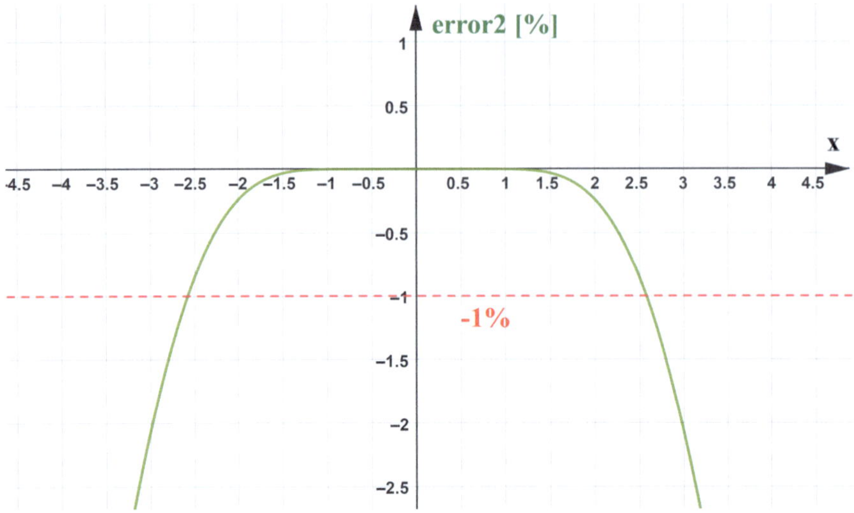

Fig. 2.107 Relative error of approximation of the multiquadratic function $f_7(x)$ (for $a = 1$) using approximation function $a_{34}(x)$

$$a_{35\text{GEN}}(x) = \frac{c_4x^4 + c_2x^2 + c_0}{d_4x^4 + d_2x^2 + d_0}, \tag{2.85}$$

c_0, c_2, c_4, d_0, d_2 and d_4 being constant coefficients having particular values that can be determined by the condition that the superior-order Taylor series expansion of $f_7(x)$ inverse multiquadratic function (for $a = 2$), expressed by (2.80), to be identical with the superior-order Taylor series expansion of the general form of the approximation function $a_{35\text{GEN}}(x)$, expressed by (2.85).

The values of these coefficients, obtained considerring the previous conditions, can be witten as: $c_0 = \frac{1}{2}$, $c_2 = \frac{3}{32}$, $c_4 = \frac{1}{512}$, $d_0 = 1$, $d_2 = \frac{5}{16}$ and $d_4 = \frac{5}{256}$. Replacing these particular values in relation (2.85), the first approximation function $a_{35}(x)$ has the following mathematical expression:

$$a_{35}(x) = \frac{\frac{x^4}{512} + \frac{3x^2}{32} + \frac{1}{2}}{\frac{5x^4}{256} + \frac{5x^2}{16} + 1}. \tag{2.86}$$

The graphical representation of the inverse multiquadratic function $f_7(x)$ (for $a = 2$) and of the first approximation function $a_{35}(x)$ is presented in Fig. 2.108.

The absolute error of approximation of the multiquadratic function $f_7(x)$ (for $a = 2$) using the approximation function $a_{35}(x)$ is illustrated in Fig. 2.109. It is defined as the difference between the previous functions, error1 $= f_7(x) - a_{35}(x)$.

For more accurately evaluating the error of approximation, the relative error of approximation is shown in Fig. 2.110. This error is defined as error2 $= 100\frac{f_7(x) - a_{35}(x)}{f_6(x)}$

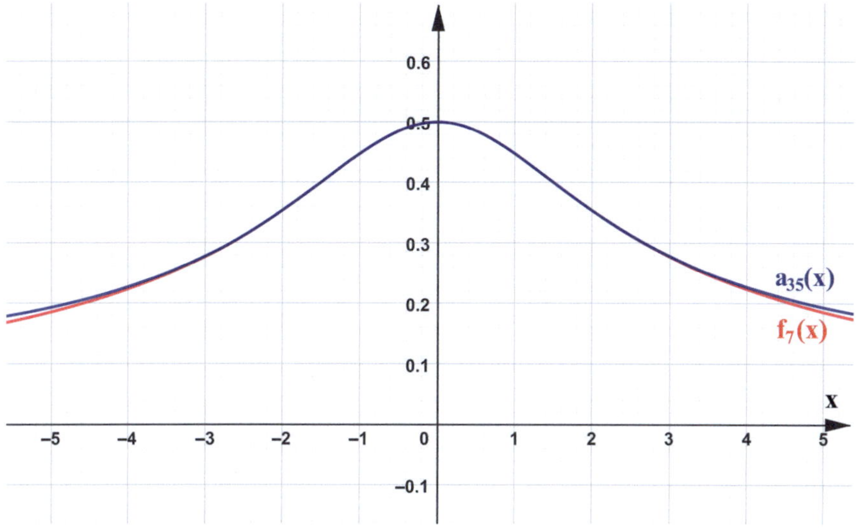

Fig. 2.108 Inverse multiquadratic function $f_7(x)$ (for $a = 2$) and the first approximation function $a_{35}(x)$

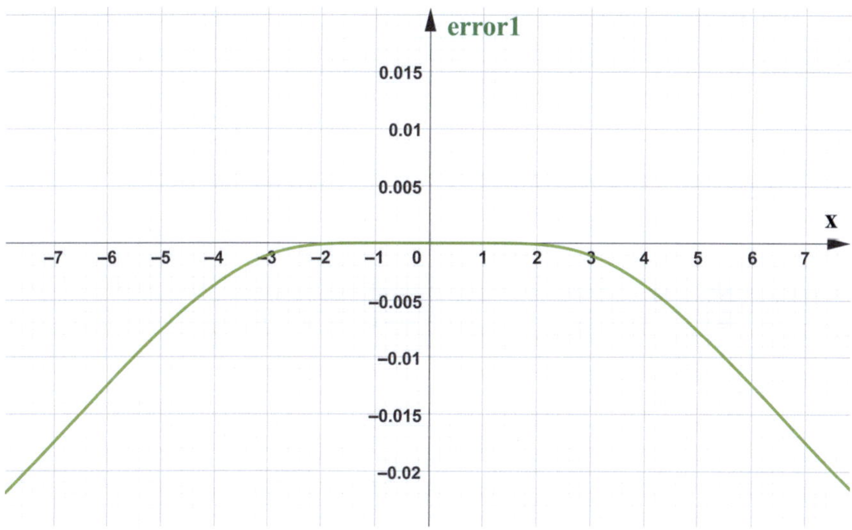

Fig. 2.109 Absolute error of approximation of the inverse multiquadratic function $f_7(x)$ (for $a = 2$) using approximation function $a_{35}(x)$

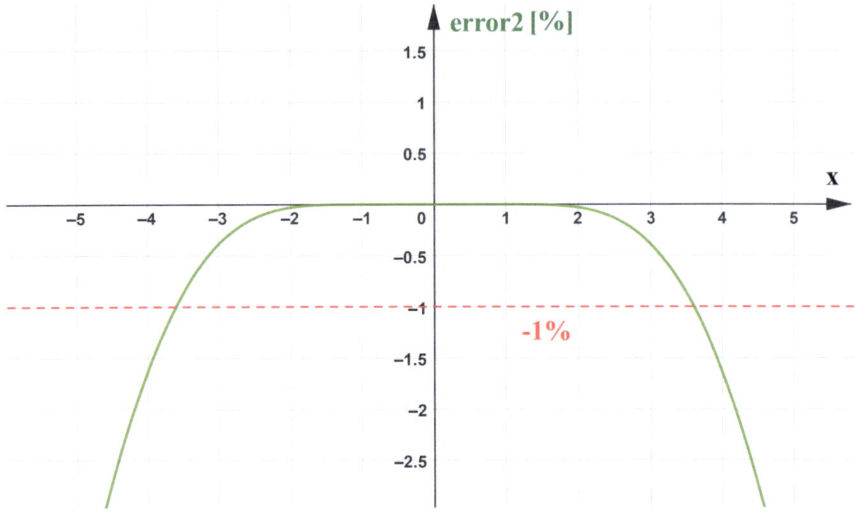

Fig. 2.110 Relative error of approximation of the multiquadratic function $f_7(x)$ (for $a = 2$) using approximation function $a_{35}(x)$

(in percents). Considering as acceptable a maximal approximation error of $\pm 1\%$, the maximal range of the input variable x is $[-3.6\,;\ 3.6]$.

2.6.5 Second Approximation Function for Generating the Inverse Multiquadratic Function for a = 2

In order to approximate the inverse multiquadratic function $f_7(x)$ (for $a = 2$), a new approximation function $a_{36}(x)$ is presented. Because the inverse multiquadratic function is an even mathematical function, the general form of the proposed second approximation function $a_{36}(x)$ can be expressed as follows:

$$a_{36\mathrm{GEN}}(x) = \frac{c_4 x^4 + c_2 x^2 + c_0}{d_6 x^6 + d_4 x^4 + d_2 x^2 + d_0}, \tag{2.87}$$

c_0, c_2, c_4, d_0, d_2, d_4 and d_6 being constant coefficients having particular values that can be determined by the condition that the superior-order Taylor series expansion of $f_7(x)$ inverse multiquadratic function (for $a = 2$), expressed by (2.80), to be identical with the superior-order Taylor series expansion of the general form of the approximation function $a_{36\mathrm{GEN}}(x)$, expressed by (2.87).

The values of these coefficients, obtained considerring the previous conditions, can be witten as: $c_0 = \frac{1}{2}$, $c_2 = \frac{1}{8}$, $c_4 = \frac{3}{512}$, $d_0 = 1$, $d_2 = \frac{3}{8}$, $d_4 = \frac{9}{256}$ and $d_6 = \frac{1}{2048}$. Replacing these particular values in relation (2.87), the second approximation

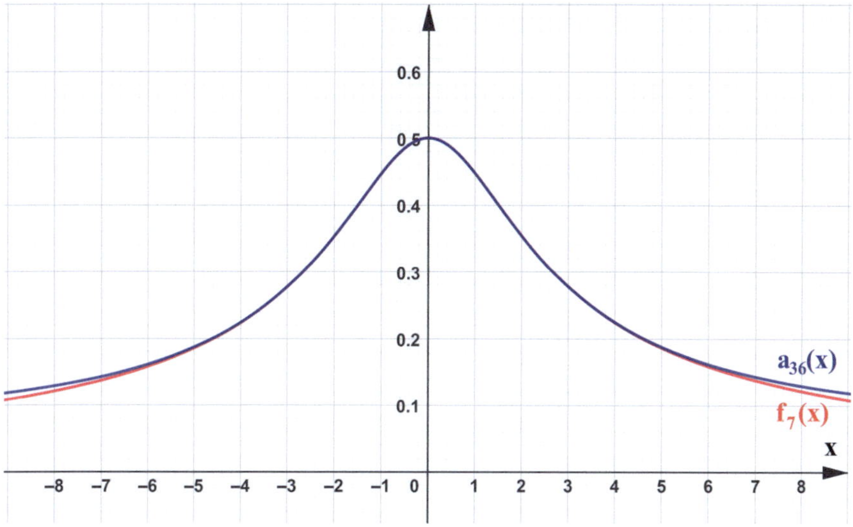

Fig. 2.111 Inverse multiquadratic function $f_7(x)$ (for $a = 2$) and the second approximation function $a_{36}(x)$

function $a_{36}(x)$ has the following mathematical expression:

$$a_{36}(x) = \frac{\frac{3x^4}{512} + \frac{x^2}{8} + \frac{1}{2}}{\frac{x^6}{2048} + \frac{9x^4}{256} + \frac{3x^2}{8} + 1}. \tag{2.88}$$

The graphical representation of the inverse multiquadratic function $f_7(x)$ (for $a = 2$) and of the second approximation function $a_{36}(x)$ is presented in Fig. 2.111.

The absolute error of approximation of the multiquadratic function $f_7(x)$ (for $a = 2$) using the approximation function $a_{36}(x)$ is illustrated in Fig. 2.112. It is defined as the difference between the previous functions, error1 $= f_7(x) - a_{36}(x)$.

For more accurately evaluating the error of approximation, the relative error of approximation is shown in Fig. 2.113. This error is defined as error2 $= 100\frac{f_7(x) - a_{36}(x)}{f_6(x)}$ (in percents). Considering as acceptable a maximal approximation error of $\pm 1\%$, the maximal range of the input variable x is $[-4.4\,;\,4.4]$.

2.6.6 *Third Approximation Function for Generating the Inverse Multiquadratic Function for a = 2*

In order to approximate the inverse multiquadratic function $f_7(x)$ (for $a = 2$), a new approximation function $a_{37}(x)$ is presented. Because the inverse multiquadratic function is an even mathematical function, the general form of the proposed third

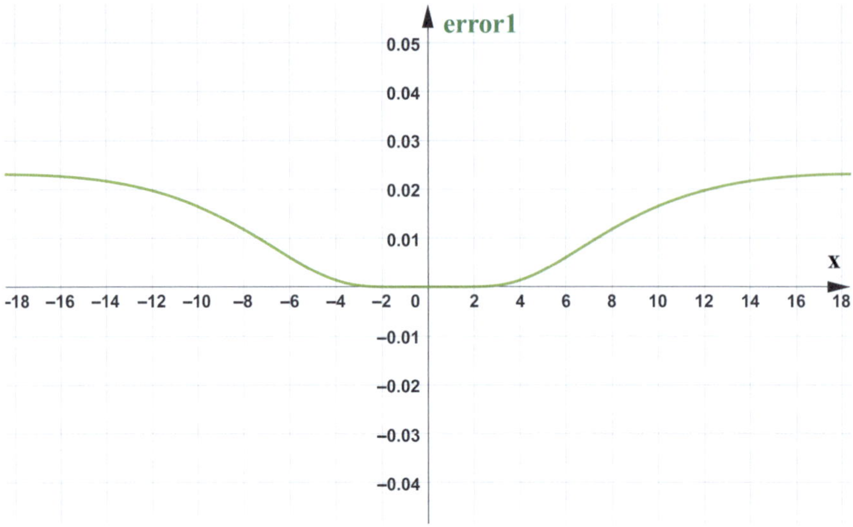

Fig. 2.112 Error of approximation of the inverse multiquadratic function $f_7(x)$ (for $a = 2$) using approximation function $a_{36}(x)$

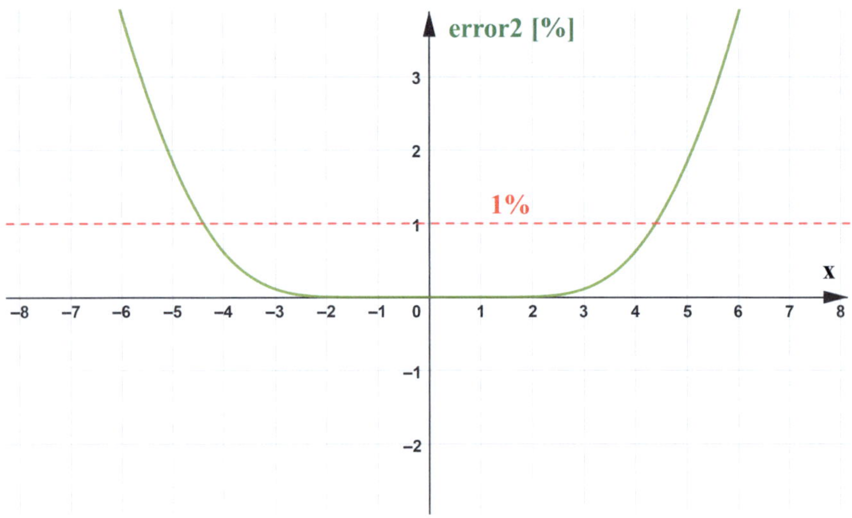

Fig. 2.113 Relative error of approximation of the multiquadratic function $f_7(x)$ (for $a = 2$) using approximation function $a_{36}(x)$

approximation function $a_{37}(x)$ can be expressed as follows:

$$a_{37\mathrm{GEN}}(x) = \frac{c_6x^6 + c_4x^4 + c_2x^2 + c_0}{d_6x^6 + d_4x^4 + d_2x^2 + d_0}, \tag{2.89}$$

$c_0, c_2, c_4, c_6, d_0, d_2, d_4$ and d_6 being constant coefficients having particular values that can be determined by the condition that the superior-order Taylor series expansion of $f_7(x)$ inverse multiquadratic function (for $a = 2$), expressed by (2.80), to be identical with the superior-order Taylor series expansion of the general form of the approximation function $a_{37\mathrm{GEN}}(x)$, expressed by (2.89).

The values of these coefficients, obtained considerring the previous conditions, can be witten as: $c_0 = \frac{1}{2}$, $c_2 = \frac{5}{32}$, $c_4 = \frac{3}{256}$, $c_6 = \frac{1}{8192}$, $d_0 = 1$, $d_2 = \frac{7}{16}$, $d_4 = \frac{7}{128}$ and $d_6 = \frac{7}{4096}$. Replacing these particular values in relation (2.89), the third approximation function $a_{37}(x)$ has the following mathematical expression:

$$a_{37}(x) = \frac{\frac{x^6}{8192} + \frac{3x^4}{256} + \frac{5x^2}{32} + \frac{1}{2}}{\frac{7x^6}{4096} + \frac{7x^4}{128} + \frac{7x^2}{16} + 1}. \tag{2.90}$$

The graphical representation of the inverse multiquadratic function $f_7(x)$ (for $a = 2$) and of the third approximation function $a_{37}(x)$ is presented in Fig. 2.114.

The absolute error of approximation of the multiquadratic function $f_7(x)$ (for $a = 2$) using the approximation function $a_{37}(x)$ is illustrated in Fig. 2.115. It is defined as the difference between the previous functions, error1 $= f_7(x) - a_{37}(x)$.

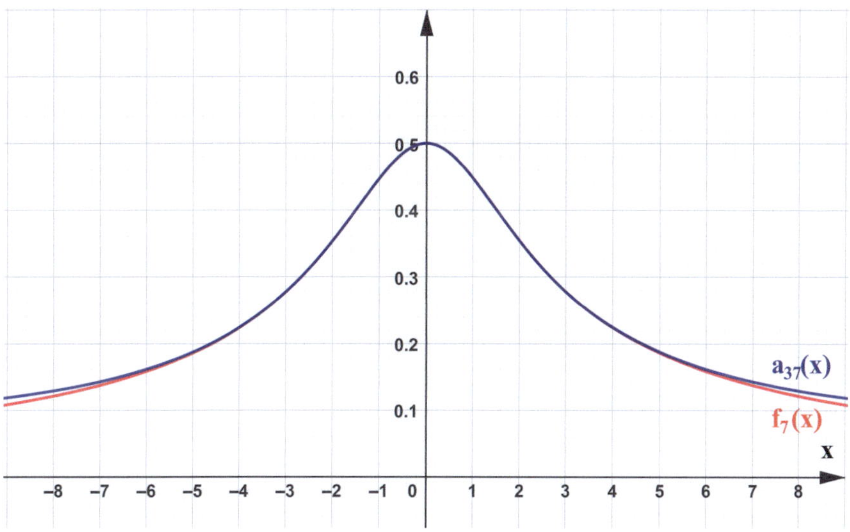

Fig. 2.114 Inverse multiquadratic function $f_7(x)$ (for $a = 2$) and the third approximation function $a_{37}(x)$

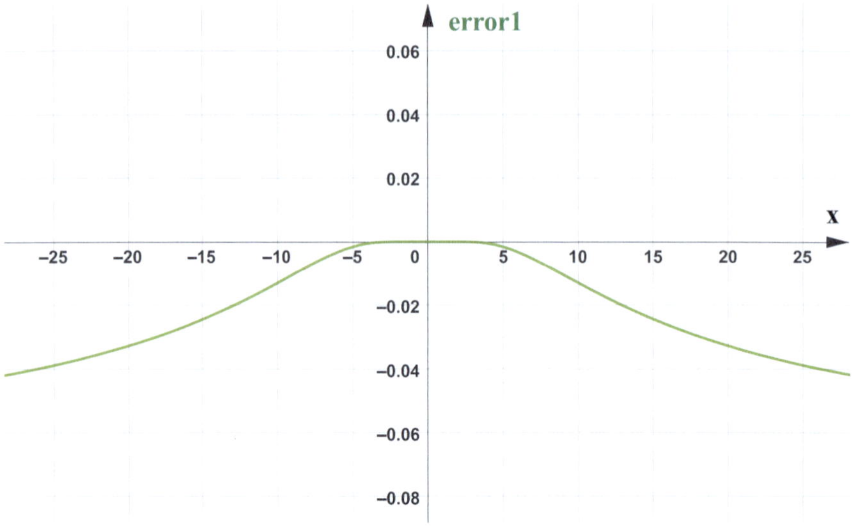

Fig. 2.115 Absolute error of approximation of the inverse multiquadratic function $f_7(x)$ (for $a = 2$) using approximation function $a_{37}(x)$

For more accurately evaluating the error of approximation, the relative error of approximation is shown in Fig. 2.116. This error is defined as error2 $= 100\frac{f_7(x)-a_{37}(x)}{f_6(x)}$ (in percents). Considering as acceptable a maximal approximation error of $\pm 1\%$, the maximal range of the input variable x is $[-5.2\,;\ 5.2]$.

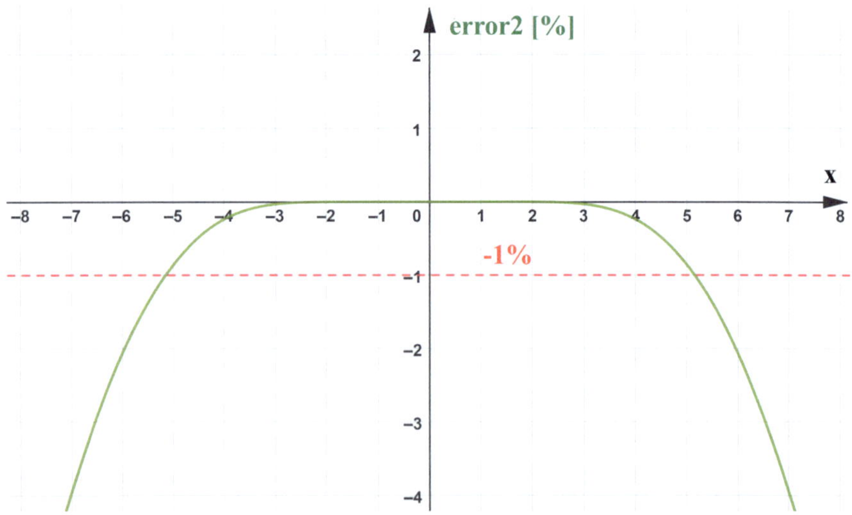

Fig. 2.116 Relative error of approximation of the multiquadratic function $f_7(x)$ (for $a = 2$) using approximation function $a_{37}(x)$

2.6.7 First Approximation Function for Generating the Inverse Multiquadratic Function for a = 3

In order to approximate the inverse multiquadratic function $f_7(x)$ (for $a = 3$), a new approximation function $a_{38}(x)$ is presented. Because the inverse multiquadratic function is an even mathematical function, the general form of the proposed first approximation function $a_{38}(x)$ can be expressed as follows:

$$a_{38\text{GEN}}(x) = \frac{c_4 x^4 + c_2 x^2 + c_0}{d_4 x^4 + d_2 x^2 + d_0}, \tag{2.91}$$

c_0, c_2, c_4, d_0, d_2 and d_4 being constant coefficients having particular values that can be determined by the condition that the superior-order Taylor series expansion of $f_7(x)$ inverse multiquadratic function (for $a = 3$), expressed by (2.80), to be identical with the superior-order Taylor series expansion of the general form of the approximation function $a_{38\text{GEN}}(x)$, expressed by (2.91).

The values of these coefficients, obtained considerring the previous conditions, can be witten as: $c_0 = \frac{1}{3}$, $c_2 = \frac{1}{36}$, $c_4 = \frac{1}{3888}$, $d_0 = 1$, $d_2 = \frac{5}{36}$ and $d_6 = \frac{5}{1296}$. Replacing these particular values in relation (2.91), the first approximation function $a_{38}(x)$ has the following mathematical expression:

$$a_{38}(x) = \frac{\frac{x^4}{3888} + \frac{x^2}{36} + \frac{1}{3}}{\frac{5x^4}{1296} + \frac{5x^2}{36} + 1}. \tag{2.92}$$

The graphical representation of the inverse multiquadratic function $f_7(x)$ (for $a = 3$) and of the first approximation function $a_{38}(x)$ is presented in Fig. 2.117.

The absolute error of approximation of the multiquadratic function $f_7(x)$ (for $a = 3$) using the approximation function $a_{38}(x)$ is illustrated in Fig. 2.118. It is defined as the difference between the previous functions, error1 $= f_7(x) - a_{38}(x)$.

For more accurately evaluating the error of approximation, the relative error of approximation is shown in Fig. 2.119. This error is defined as error2 $= 100\frac{f_7(x) - a_{38}(x)}{f_6(x)}$ (in percents). Considering as acceptable a maximal approximation error of $\pm 1\%$, the maximal range of the input variable x is $[-5.4\,;\ 5.4]$.

2.6.8 Second Approximation Function for Generating the Inverse Multiquadratic Function for a = 3

In order to approximate the inverse multiquadratic function $f_7(x)$ (for $a = 3$), a new approximation function $a_{39}(x)$ is presented. Because the inverse multiquadratic function is an even mathematical function, the general form of the proposed second approximation function $a_{39}(x)$ can be expressed as follows:

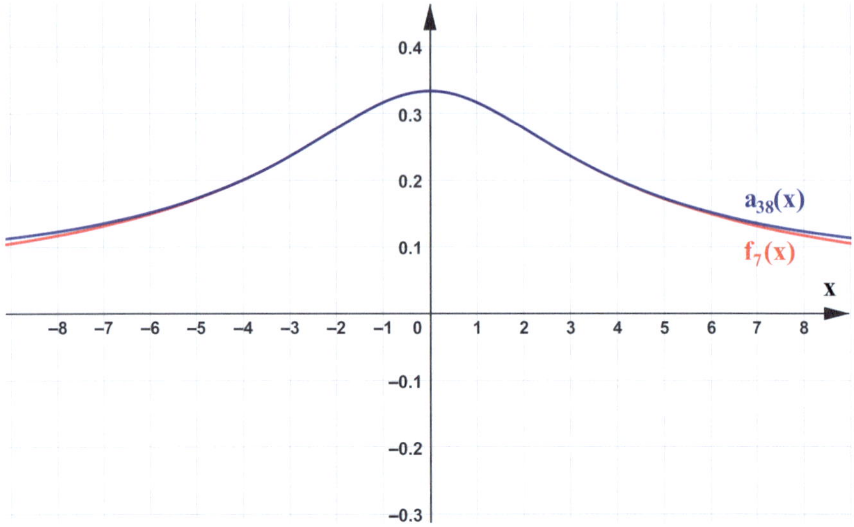

Fig. 2.117 Inverse multiquadratic function $f_7(x)$ (for $a = 3$) and the first approximation function $a_{38}(x)$

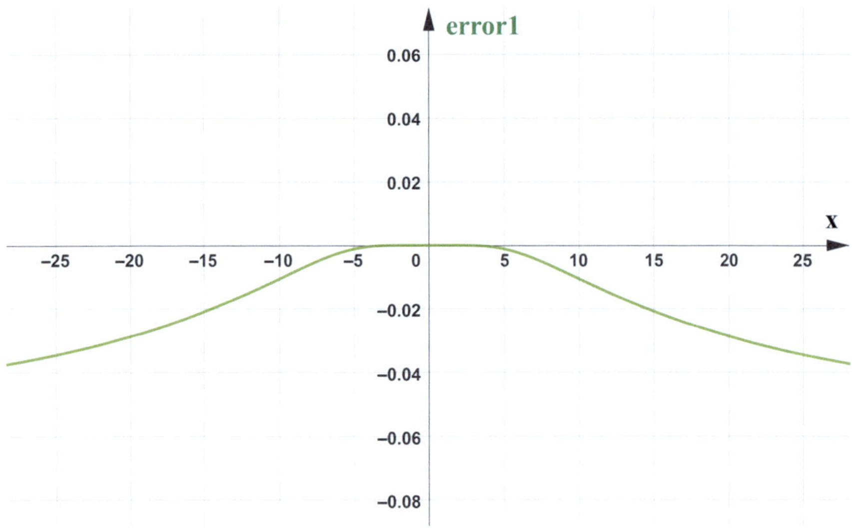

Fig. 2.118 Absolute error of approximation of the inverse multiquadratic function $f_7(x)$ (for $a = 3$) using approximation function $a_{38}(x)$

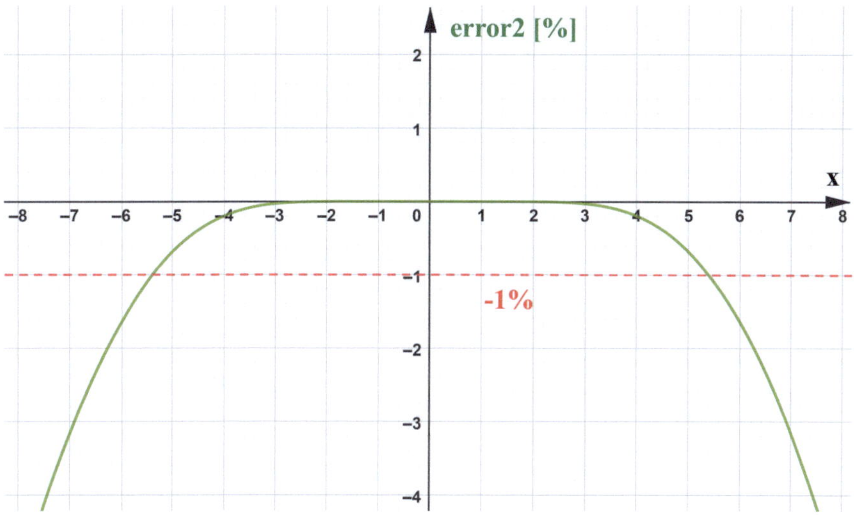

Fig. 2.119 Relative error of approximation of the multiquadratic function $f_7(x)$ (for $a = 3$) using approximation function $a_{38}(x)$

$$a_{39\mathrm{GEN}}(x) = \frac{c_4x^4 + c_2x^2 + c_0}{d_6x^6 + d_4x^4 + d_2x^2 + d_0}, \tag{2.93}$$

c_0, c_2, c_4, d_0, d_2, d_4 and d_6 being constant coefficients having particular values that can be determined by the condition that the superior-order Taylor series expansion of $f_7(x)$ inverse multiquadratic function (for $a = 3$), expressed by (2.80), to be identical with the superior-order Taylor series expansion of the general form of the approximation function $a_{39\mathrm{GEN}}(x)$, expressed by (2.93).

The values of these coefficients, obtained considerring the previous conditions, can be witten as: $c_0 = \frac{1}{3}$, $c_2 = \frac{1}{27}$, $c_4 = \frac{1}{1296}$, $d_0 = 1$, $d_2 = \frac{1}{6}$, $d_4 = \frac{1}{144}$ and $d_6 = \frac{1}{23328}$. Replacing these particular values in relation (2.93), the second approximation function $a_{39}(x)$ has the following mathematical expression:

$$a_{39}(x) = \frac{\frac{x^4}{1296} + \frac{x^2}{27} + \frac{1}{3}}{\frac{x^6}{23328} + \frac{x^4}{144} + \frac{x^2}{6} + 1}. \tag{2.94}$$

The graphical representation of the inverse multiquadratic function $f_7(x)$ (for $a = 3$) and of the second approximation function $a_{39}(x)$ is presented in Fig. 2.120.

The absolute error of approximation of the multiquadratic function $f_7(x)$ (for $a = 3$) using the approximation function $a_{39}(x)$ is illustrated in Fig. 2.121. It is defined as the difference between the previous functions, error1 $= f_7(x) - a_{39}(x)$.

For more accurately evaluating the error of approximation, the relative error of approximation is shown in Fig. 2.122. This error is defined as error2 $= 100\frac{f_7(x) - a_{39}(x)}{f_6(x)}$

Fig. 2.120 Inverse multiquadratic function $f_7(x)$ (for $a = 3$) and the second approximation function $a_{39}(x)$

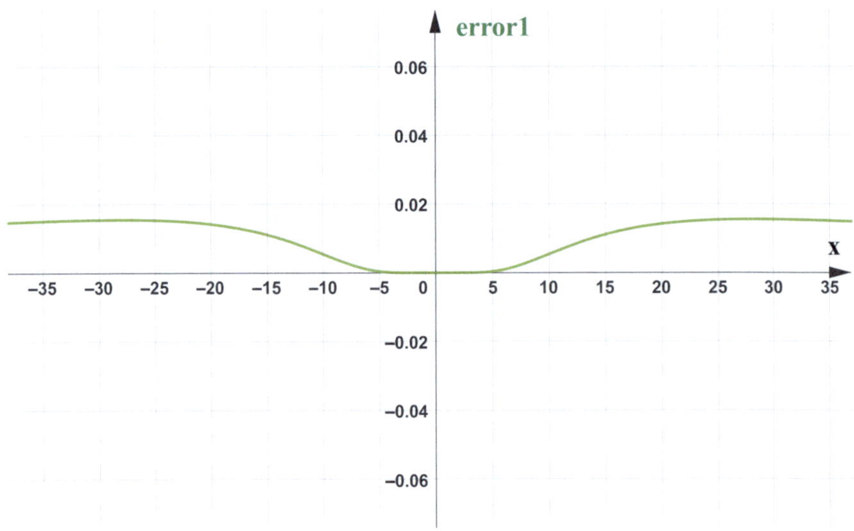

Fig. 2.121 Absolute error of approximation of the inverse multiquadratic function $f_7(x)$ (for $a = 3$) using approximation function $a_{39}(x)$

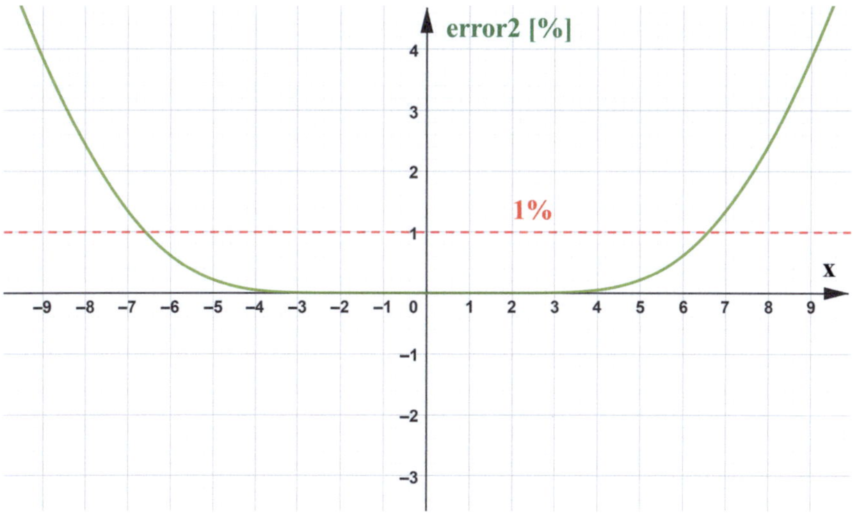

Fig. 2.122 Relative error of approximation of the multiquadratic function $f_7(x)$ (for $a = 3$) using approximation function $a_{39}(x)$

(in percents). Considering as acceptable a maximal approximation error of $\pm 1\%$, the maximal range of the input variable x is $[-6.6\,;\ 6.6]$.

References

Popa C (2012) High-accuracy function synthesizer circuit with applications in signal processing. EURASIP J Adv Signal Process 2012:129. https://doi.org/10.1186/1687-6180-2012-129

Popa CR (2023) High-accuracy Gaussian function generator for neural networks. Electronics 12(1):24. https://doi.org/10.3390/electronics12010024

Chapter 3
Superior-Order Approximation Functions for Artificial Neural Networks Applications

3.1 Introduction

In addition to the functions from previous two chapters (sigmoidal activation functions, presented in Chap. 1 and radial basis activation functions described in Chap. 2), it exists other important mathematical functions, having a multitude of applications in analog signal processing and artificial neural networks (Popa 2012, 2023). These functions are detailed analyzed in this chapter and the most accurate approximation functions are proposed, in order to generate them. Two important objectives are considered: as good as possible accuracy of the approximation and reasonable hardware resources required for their implementation in CMOS technology using fundamental CMOS computational circuits (further described in Chap. 4).

3.2 Approximation Functions for Exponential Function

The exponential function is a fundamental mathematical function with a wide range of applications across science, engineering and machine learning. In neural networks and computational models, variants of the exponential function play a crucial role. The areas of utilization of the exponential function are:

a. ***Activation and Output Functions in Neural Networks***

- Softmax function uses exponentials to convert logits into probabilities in classification tasks.

b. ***Loss Functions***

- Exponential loss is used in models like adaptive boosting, where the loss penalizes misclassified points exponentially.

C. R. Popa, *Analog Current-Mode Computational Circuits for Artificial Neural Networks*, Analog Circuits and Signal Processing,
https://doi.org/10.1007/978-3-032-03989-7_3

- Exponential decay functions are used in regularization and learning rate schedules.

c. ***Probability and Statistics***

- In probabilistic modeling, exponential functions are central to exponential distributions (for example, modeling time between events in Poisson processes), log-linear models and Gibbs distributions and Bayesian inference and Boltzmann machines.

d. ***Attention Mechanisms in Transformers***

- Exponentials are key in computing attention weights.
- The softmax uses exponentials to emphasize higher scores while keeping the output normalized.

e. ***Physics-Inspired and Bio-inspired Models***

- Exponential functions appear in models of radioactive decay, population growth, synaptic decay and activation energy.

3.2.1 General Form of the Exponential Function

The mathematical form of this function can be expressed as follows:

$$f_1(x, a) = \exp(ax). \tag{3.1}$$

a is being a parameter. The graphical representation of the exponential function $f_1(x, a)$, considering three values of a parameter ($a = 1$, $a = 2$ and $a = 3$), is presented in Fig. 3.1.

Considering the simplest case $a = 1$, the mathematical form (3.1) of the exponential function becomes:

$$f_1(x) \overset{\text{not}}{=} f_1(x, 1) = \exp(x). \tag{3.2}$$

In order to obtain an analog implementation of the exponential function, a multitude of superior-order approximation functions is presented. The most important differences between these functions are mainly related to the accuracy of their approximation, as well as to the hardware resources required for their implementations in CMOS technology. In order to simplify the mathematical analysis, these superior-order approximation functions are developed for generating the simplest exponential function, expressed by relation (3.2). The obtaining of the more general mathematical forms of the exponential function is possible introducing small changes in the previous mentioned superior-order approximation functions.

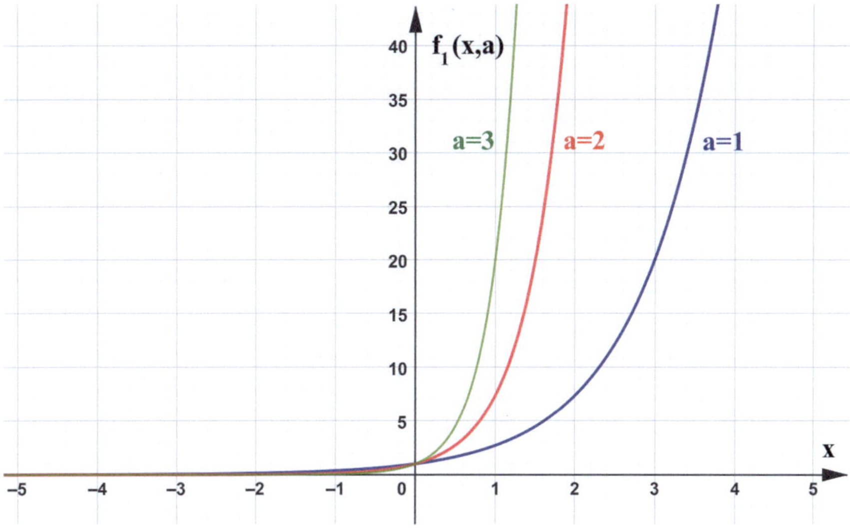

Fig. 3.1 Exponential function $f_1(x, a)$ for $a = 1$, $a = 2$ and $a = 3$

3.2.2 *First Approximation Function for Generating the Exponential Function*

In order to approximate the exponential function $f_1(x)$, a new approximation function $a_1(x)$ is presented. Its particular mathematical form is developed having two very important objectives. The first objective is represented by the accuracy of the approximation that could be obtained using this approximation function. The second objective is related to the possibility of its facile analog implementation in CMOS technology using as reasonable as possible hardware resources (see Chaps. 4 and 7).

The general form of the proposed first approximation function $a_1(x)$ is the following:

$$a_{1\text{GEN}}(x) = \left(\frac{c_1 x + c_0}{d_1 x + d_0}\right)^2, \tag{3.3}$$

c_0, c_1, d_0 and d_1 being constant coefficients having particular values that can be determined by the condition that the superior-order Taylor series expansion of $f_1(x)$ exponential function, expressed by (3.2), must be identical with the superior-order Taylor series expansion of the general form of the approximation function $a_{1\text{GEN}}(x)$, expressed by (3.3).

The values of these coefficients, obtained considering the previous conditions, can be written as: $c_0 = 1$, $c_1 = \frac{1}{4}$, $d_0 = 1$ and $d_1 = -\frac{1}{4}$. Replacing these particular values in relation (3.3), the first approximation function $a_1(x)$ has the following mathematical expression:

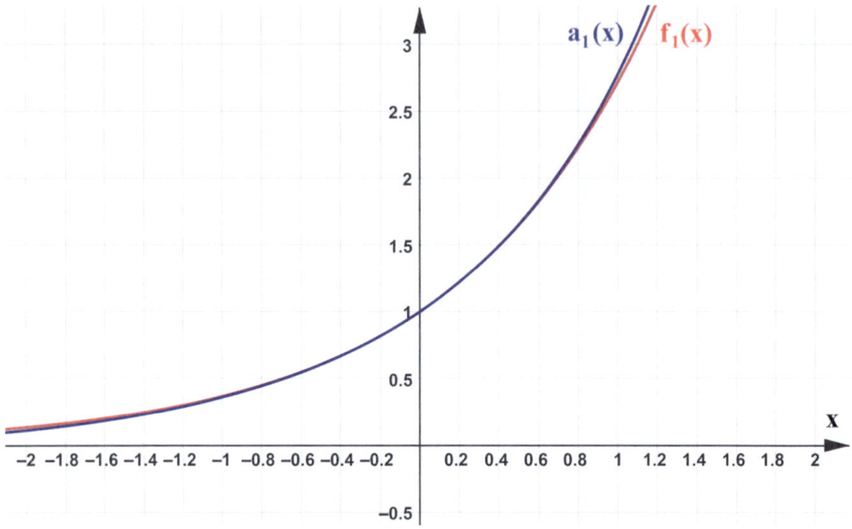

Fig. 3.2 Exponential function $f_1(x)$ and the first approximation function $a_1(x)$

$$a_1(x) = \left(\frac{1 + \frac{x}{4}}{1 - \frac{x}{4}} \right)^2 . \tag{3.4}$$

The graphical representation of the exponential function $f_1(x)$ and of the first approximation function $a_1(x)$ is presented in Fig. 3.2.

The graphical representation of the exponential function $f_1(x)$ and of the first approximation function $a_1(x)$, both of them being expressed in db, is presented in Fig. 3.3.

The error of approximation of the exponential function $f_1(x)$ using the first approximation function $a_1(x)$ (error1, expressed in dB) is illustrated in Fig. 3.4. Considering as acceptable a maximal approximation error of ± 1 dB, the maximal range of the input variable x is $[-1.7\,;\ 1.7]$. This is equivalent with an output dynamic range of the exponential function synthesizer based on the first approximation function $a_1(x)$ equal with 29.55 dB.

3.2.3 Second Approximation Function for Generating the Exponential Function

In order to approximate the exponential function $f_1(x)$, a new approximation function $a_2(x)$ is presented. Its mathematical form is obtained exploiting the following particularity of the exponential function, based on the variable changing $x \to \frac{x}{2}$:

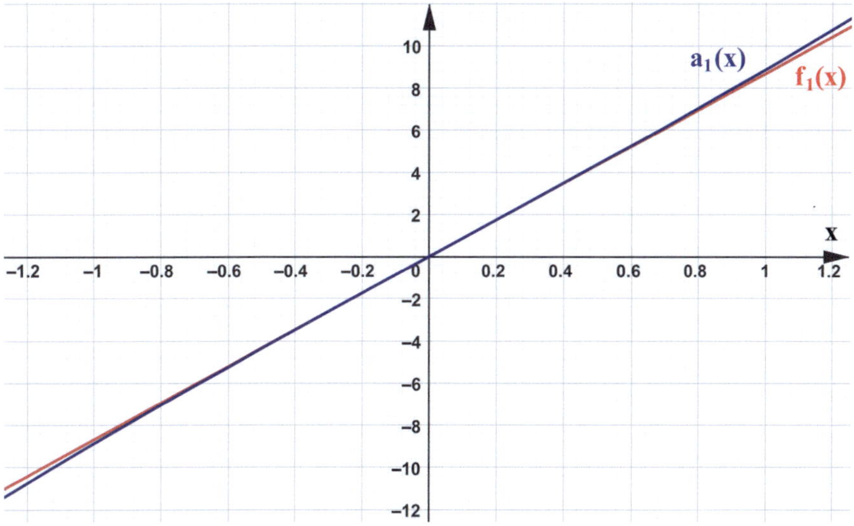

Fig. 3.3 Exponential function $f_1(x)$ and the first approximation function $a_1(x)$ (expressed in dB)

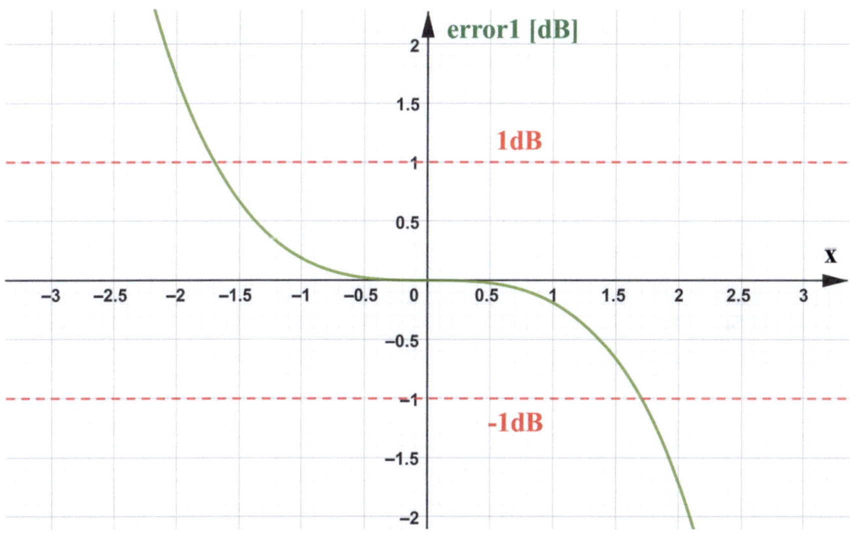

Fig. 3.4 Error of approximation of the exponential function $f_1(x)$ using the first approximation function $a_1(x)$ (expressed in dB)

$$\exp(x) = \exp\left(\frac{x}{2}\right)^2. \tag{3.5}$$

This variable changing strongly increases the accuracy of approximating the exponential function, by decreasing the equivalent variable range of the input variable

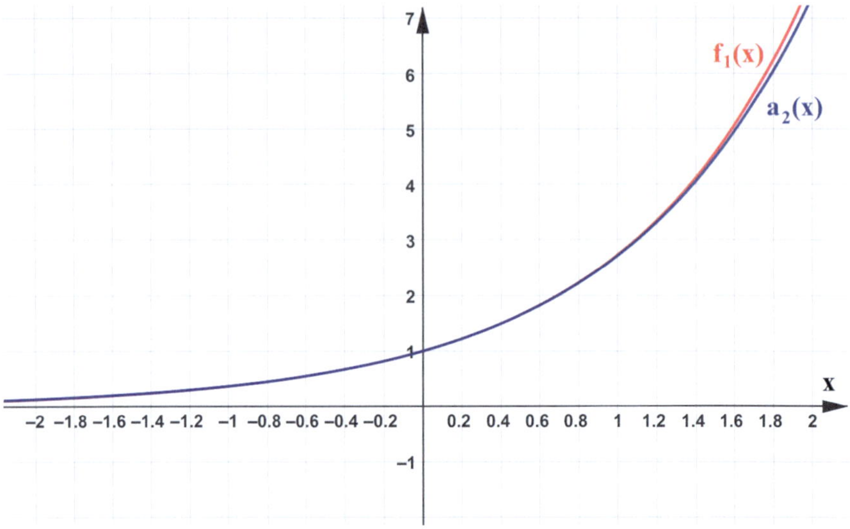

Fig. 3.5 Exponential function $f_1(x)$ and the second approximation function $a_2(x)$

x. Using this variable changing method, the new improved accuracy approximation function $a_2(x)$, based on the previous approximation function $a_1(x)$, has the following expression:

$$a_2(x) = \left(\frac{1 + \frac{x}{8}}{1 - \frac{x}{8}}\right)^4. \tag{3.6}$$

The graphical representation of the exponential function $f_1(x)$ and of the second approximation function $a_2(x)$ is presented in Fig. 3.5.

The graphical representation of the exponential function $f_1(x)$ and of the second approximation function $a_2(x)$, both of them being expressed in db, is presented in Fig. 3.6.

The error of approximation of the exponential function $f_1(x)$ using the second approximation function $a_2(x)$ (error1, expressed in dB) is illustrated in Fig. 3.7. Considering as acceptable a maximal approximation error of ± 1 dB, the maximal range of the input variable x is $[-2.75; 2.75]$. This is equivalent with an output dynamic range of the exponential function synthesizer based on the second approximation function $a_2(x)$ equal with 47.8 dB.

Fig. 3.6 Exponential function $f_1(x)$ and the second approximation function $a_2(x)$ (expressed in dB)

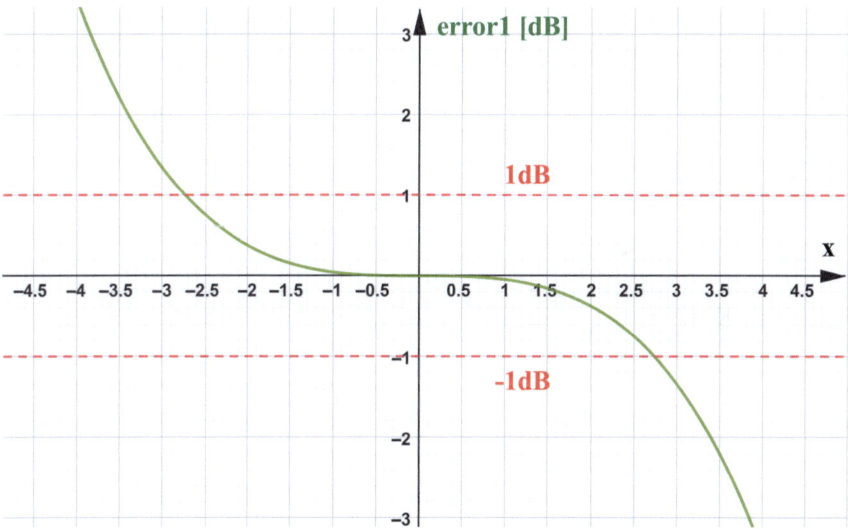

Fig. 3.7 Error of approximation of the exponential function $f_1(x)$ using the second approximation function $a_2(x)$ (expressed in dB)

3.2.4 Third Approximation Function for Generating the Exponential Function

In order to approximate the exponential function $f_1(x)$, a new approximation function $a_3(x)$ is presented. The general form of the proposed third approximation function $a_3(x)$ is the following:

$$a_{3\mathrm{GEN}}(x) = \frac{c_2x^2 + c_1x + c_0}{d_2x^2 + d_1x + d_0}, \tag{3.7}$$

c_0, c_1, c_2, d_0, d_1 and d_2 being constant coefficients having particular values that can be determined by the condition that the superior-order Taylor series expansion of $f_1(x)$ exponential function, expressed by (3.2), to be identical with the superior-order Taylor series expansion of the general form of the approximation function $a_{3\mathrm{GEN}}(x)$, expressed by (3.7).

The values of these coefficients, obtained considering the previous conditions, can be written as: $c_0 = 1, c_1 = \frac{1}{2}, c_2 = \frac{1}{12}, d_0 = 1, d_1 = -\frac{1}{2}$ and $d_2 = \frac{1}{12}$. Replacing these particular values in relation (3.7), the third approximation function $a_3(x)$ has the following mathematical expression:

$$a_3(x) = \frac{x^2 + 6x + 12}{x^2 - 6x + 12}. \tag{3.8}$$

The graphical representation of the exponential function $f_1(x)$ and of the third approximation function $a_3(x)$ is presented in Fig. 3.8.

The graphical representation of the exponential function $f_1(x)$ and of the third approximation function $a_3(x)$, both of them being expressed in db, is presented in Fig. 3.9.

The error of approximation of the exponential function $f_1(x)$ using the third approximation function $a_3(x)$ (error1, expressed in dB) is illustrated in Fig. 3.10. Considering as acceptable a maximal approximation error of ± 1 dB, the maximal range of the input variable x is $[-2.3\,;\ 2.3]$. This is equivalent with an output dynamic range of the exponential function synthesizer based on the third approximation function $a_3(x)$ equal with 39.81 dB.

3.2.5 Fourth Approximation Function for Generating the Exponential Function

Using the previous presented variable changing method $\left(x \to \frac{x}{2}\right)$, the new improved accuracy approximation function $a_4(x)$, based on the previous approximation function $a_3(x)$, has the following expression:

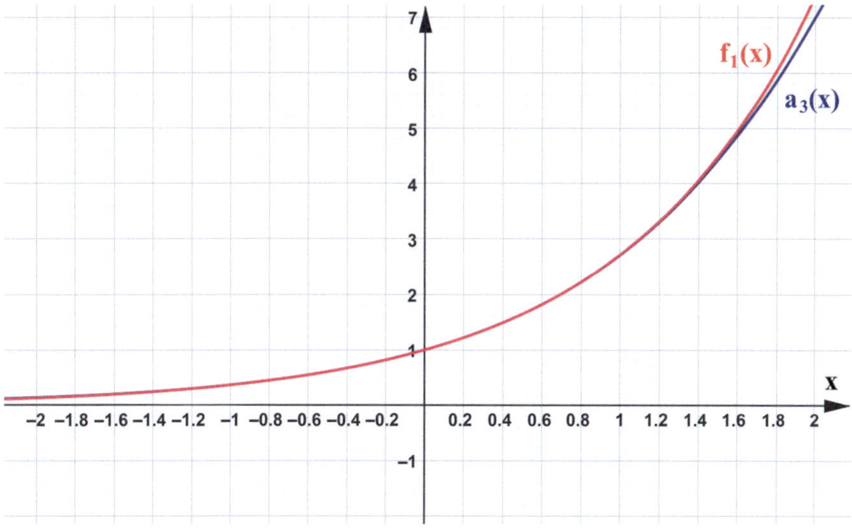

Fig. 3.8 Exponential function $f_1(x)$ and the third approximation function $a_3(x)$

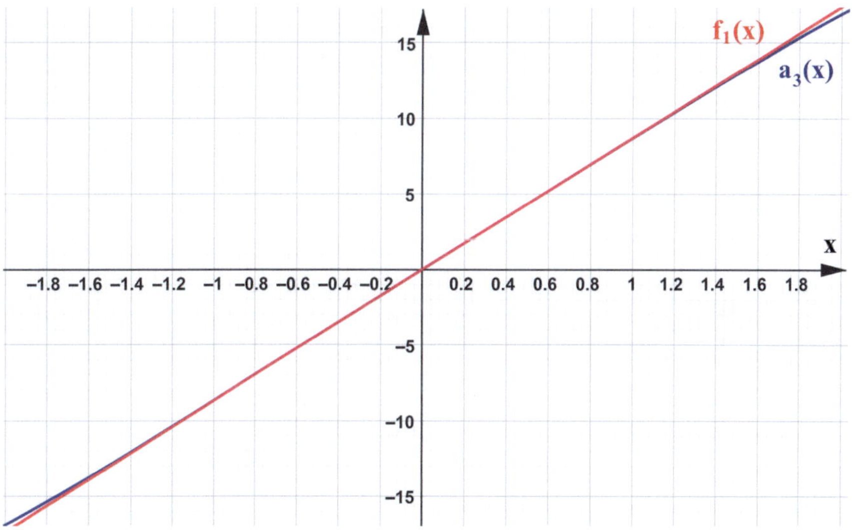

Fig. 3.9 Exponential function $f_1(x)$ and the third approximation function $a_3(x)$ (expressed in dB)

$$a_4(x) = \left(\frac{x^2 + 12x + 48}{x^2 - 12x + 48} \right)^2. \tag{3.9}$$

The graphical representation of the exponential function $f_1(x)$ and of the fourth approximation function $a_4(x)$ is presented in Fig. 3.11.

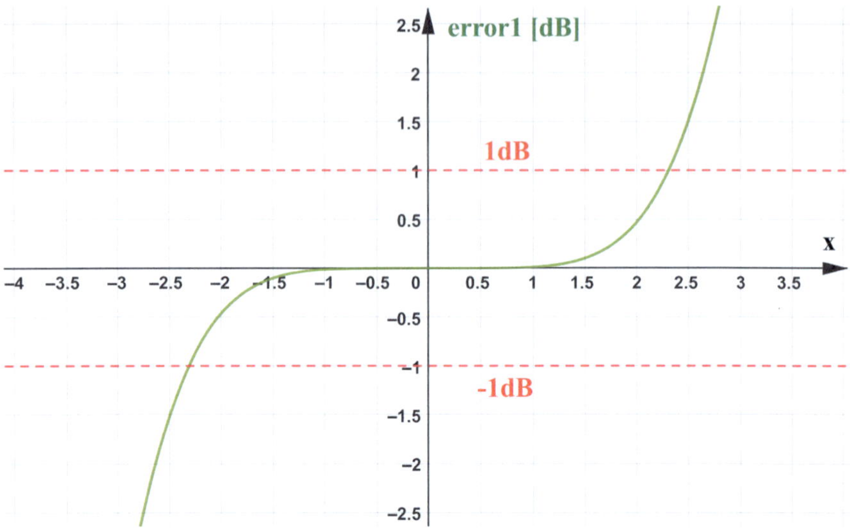

Fig. 3.10 Error of approximation of the exponential function $f_1(x)$ using the third approximation function $a_3(x)$ (expressed in dB)

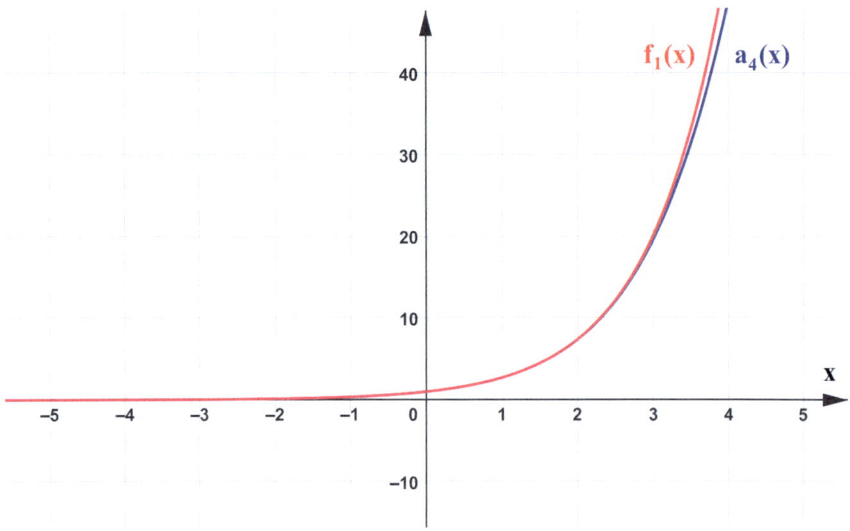

Fig. 3.11 Exponential function $f_1(x)$ and the fourth approximation function $a_4(x)$

The graphical representation of the exponential function $f_1(x)$ and of the fourth approximation function $a_4(x)$, both of them being expressed in db, is presented in Fig. 3.12.

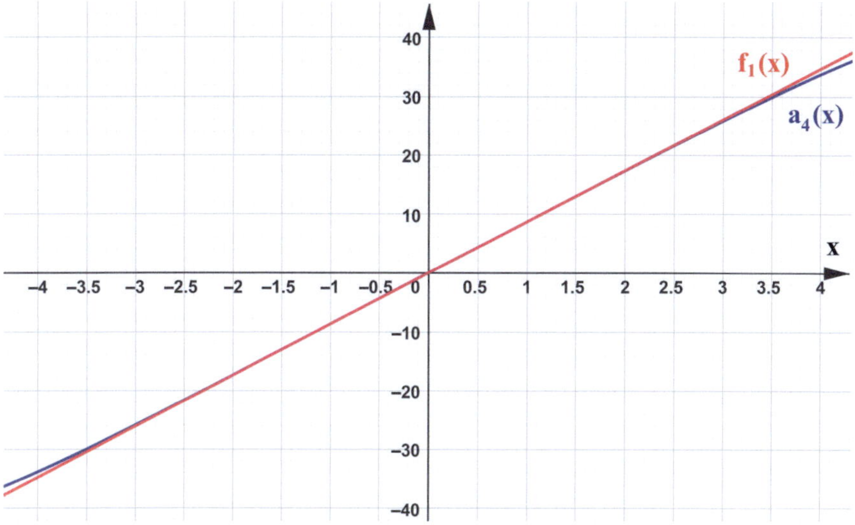

Fig. 3.12 Exponential function $f_1(x)$ and the fourth approximation function $a_4(x)$ (expressed in dB)

The error of approximation of the exponential function $f_1(x)$ using the fourth approximation function $a_4(x)$ (error1, expressed in dB) is illustrated in Fig. 3.13. Considering as acceptable a maximal approximation error of ± 1 dB, the maximal range of the input variable x is $[-4\,;\,4]$. This is equivalent with an output dynamic range of the exponential function synthesizer based on the fourth approximation function $a_4(x)$ equal with 69.53 dB.

3.2.6 *Fifth Approximation Function for Generating the Exponential Function*

Using for the second time the previous presented variable changing method $\left(x \rightarrow \frac{x}{2}\right)$, the new improved accuracy approximation function $a_5(x)$, based on the previous approximation function $a_4(x)$, has the following expression:

$$a_5(x) = \left(\frac{x^2 + 24x + 192}{x^2 - 24x + 192}\right)^4. \tag{3.10}$$

The graphical representation of the exponential function $f_1(x)$ and of the fifth approximation function $a_5(x)$ is presented in Fig. 3.14.

The graphical representation of the exponential function $f_1(x)$ and of the fifth approximation function $a_5(x)$, both of them being expressed in db, is presented in Fig. 3.15.

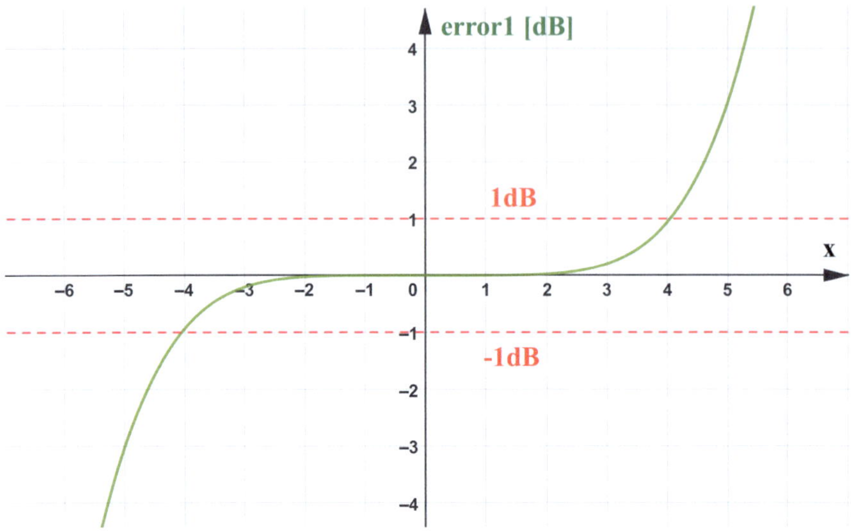

Fig. 3.13 Error of approximation of the exponential function $f_1(x)$ using the fourth approximation function $a_4(x)$ (expressed in dB)

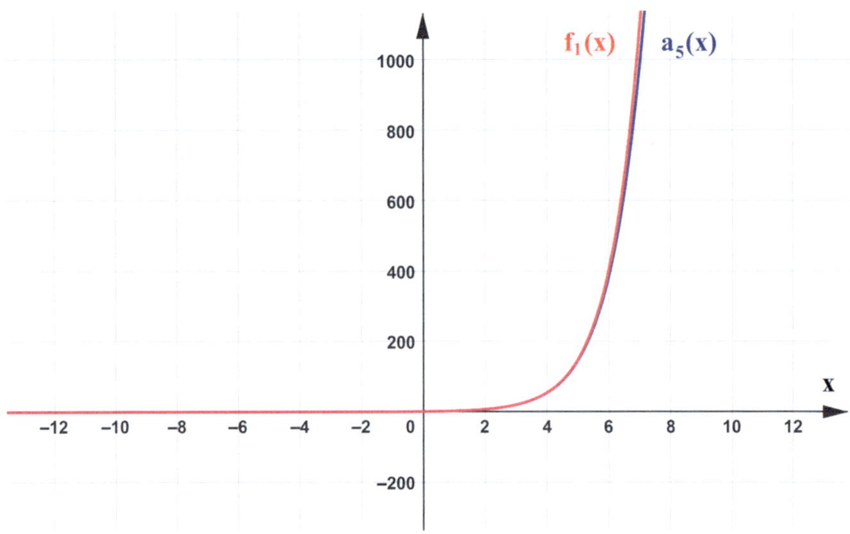

Fig. 3.14 Exponential function $f_1(x)$ and the fifth approximation function $a_5(x)$

The error of approximation of the exponential function $f_1(x)$ using the fifth approximation function $a_5(x)$ (error1, expressed in dB) is illustrated in Fig. 3.16. Considering as acceptable a maximal approximation error of ± 1 dB, the maximal range of the input variable x is $[-7.1\ ;\ 7.1]$. This is equivalent with an output dynamic range

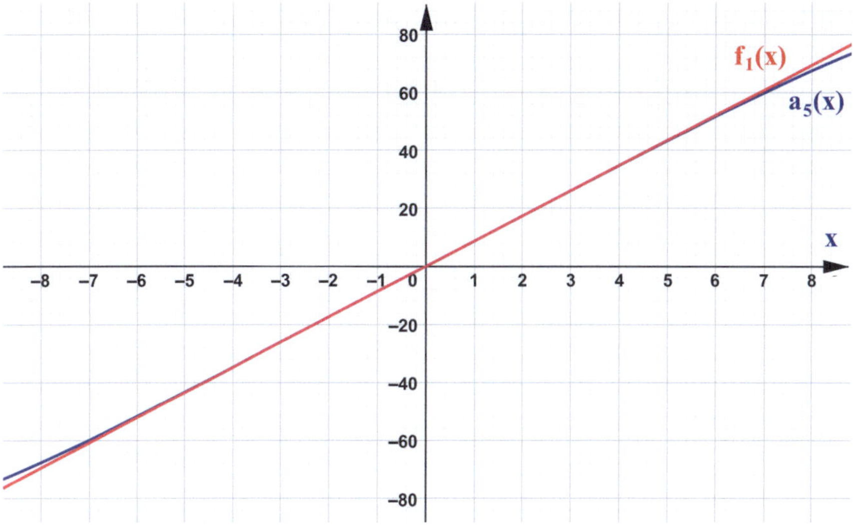

Fig. 3.15 Exponential function $f_1(x)$ and the fifth approximation function $a_5(x)$ (expressed in dB)

of the exponential function synthesizer based on the fifth approximation function $a_5(x)$ equal with 123.42 dB.

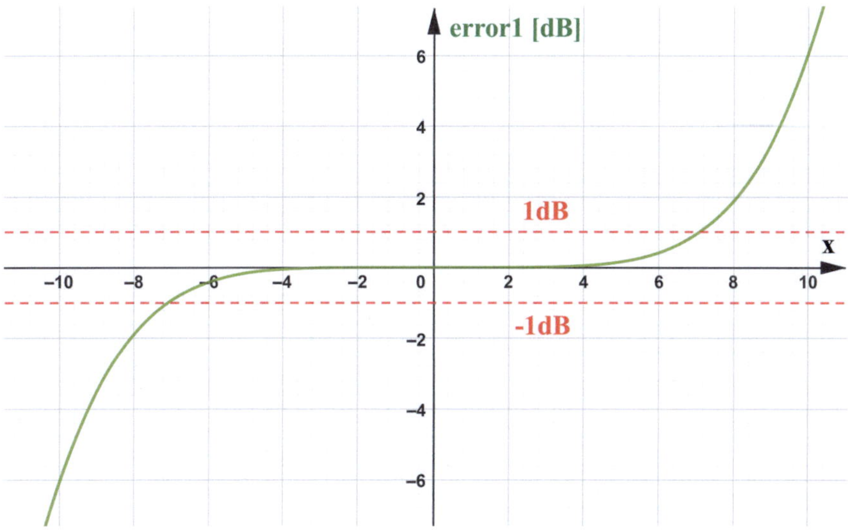

Fig. 3.16 Error of approximation of the exponential function $f_1(x)$ using the fifth approximation function $a_5(x)$ (expressed in dB)

3.2.7 Sixth Approximation Function for Generating the Exponential Function

In order to approximate the exponential function $f_1(x)$, a new approximation function $a_6(x)$ is presented. The general form of the proposed sixth approximation function $a_6(x)$ is the following:

$$a_{6\mathrm{GEN}}(x) = \frac{c_2x^2 + c_1x + c_0}{d_3x^3 + d_2x^2 + d_1x + d_0}, \tag{3.11}$$

c_0, c_1, c_2, d_0, d_1, d_2 and d_3 being constant coefficients having particular values that can be determined by the condition that the superior-order Taylor series expansion of $f_1(x)$ exponential function, expressed by (3.2), must be identical with the superior-order Taylor series expansion of the general form of the approximation function $a_{6\mathrm{GEN}}(x)$, expressed by (3.11).

The values of these coefficients, obtained considering the previous conditions, can be written as: $c_0 = 1$, $c_1 = \frac{2}{5}$, $c_2 = \frac{1}{20}$, $d_0 = 1$, $d_1 = -\frac{3}{5}$, $d_2 = \frac{3}{20}$ and $d_3 = -\frac{1}{60}$. Replacing these particular values in relation (3.11), the sixth approximation function $a_6(x)$ has the following mathematical expression:

$$a_6(x) = \frac{\frac{x^2}{20} + \frac{2x}{5} + 1}{-\frac{x^3}{60} + \frac{3x^2}{20} - \frac{3x}{5} + 1}. \tag{3.12}$$

The graphical representation of the exponential function $f_1(x)$ and of the sixth approximation function $a_6(x)$ is presented in Fig. 3.17.

The graphical representation of the exponential function $f_1(x)$ and of the sixth approximation function $a_6(x)$, both of them being expressed in db, is presented in Fig. 3.18.

The error of approximation of the exponential function $f_1(x)$ using the sixth approximation function $a_6(x)$ (error1, expressed in dB) is illustrated in Fig. 3.19. Considering as acceptable a maximal approximation error of ± 1 dB, the maximal range of the input variable x is $[-3.15;\ 2.7]$. This is equivalent with an output dynamic range of the exponential function synthesizer based on the sixth approximation function $a_6(x)$ equal with 50.85 dB.

3.2.8 Seventh Approximation Function for Generating the Exponential Function

Using the previous presented variable changing method $\left(x \to \frac{x}{2}\right)$, the new improved accuracy approximation function $a_7(x)$, based on the previous approximation function $a_6(x)$, has the following expression:

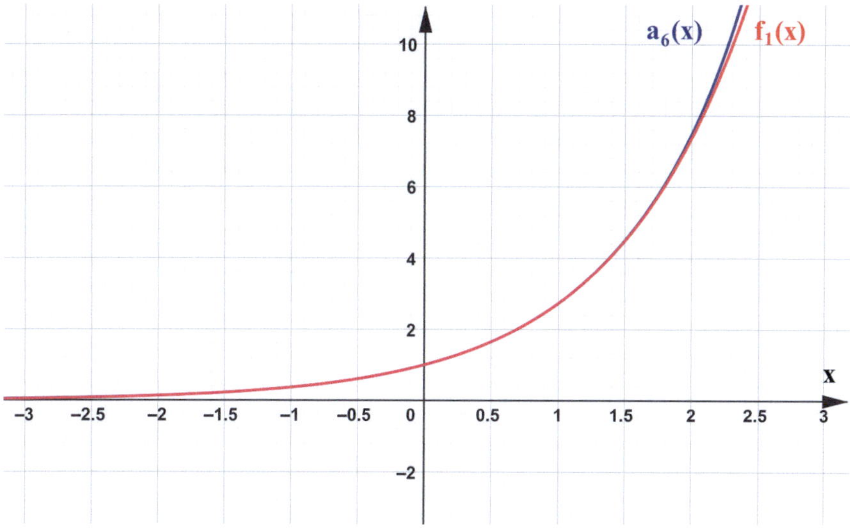

Fig. 3.17 Exponential function $f_1(x)$ and the sixth approximation function $a_6(x)$

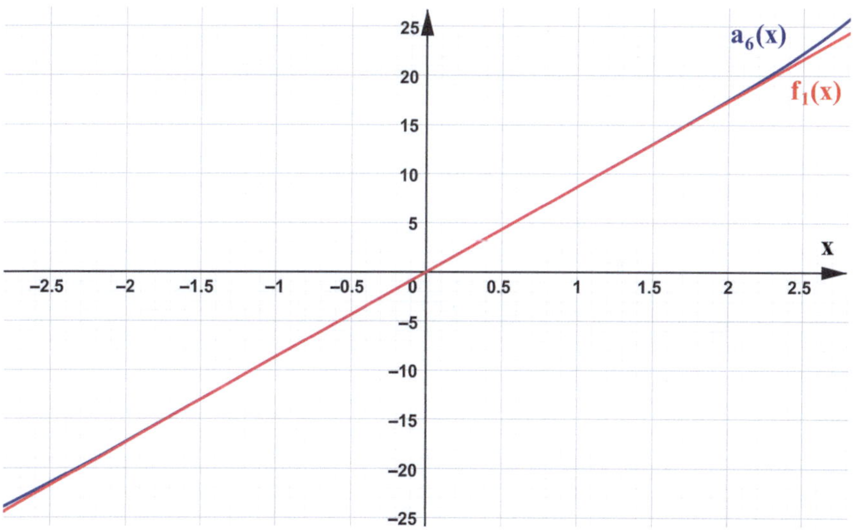

Fig. 3.18 Exponential function $f_1(x)$ and the sixth approximation function $a_6(x)$ (expressed in dB)

$$a_7(x) = \left(\frac{\frac{x^2}{80} + \frac{x}{5} + 1}{-\frac{x^3}{480} + \frac{3x^2}{80} - \frac{3x}{10} + 1} \right)^2. \tag{3.13}$$

The graphical representation of the exponential function $f_1(x)$ and of the seventh approximation function $a_7(x)$ is presented in Fig. 3.20.

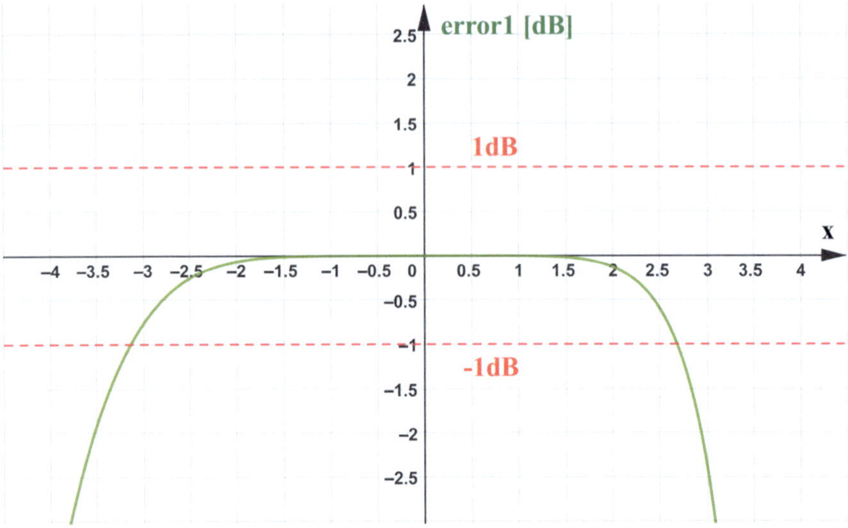

Fig. 3.19 Error of approximation of the exponential function $f_1(x)$ using the sixth approximation function $a_6(x)$ (expressed in dB)

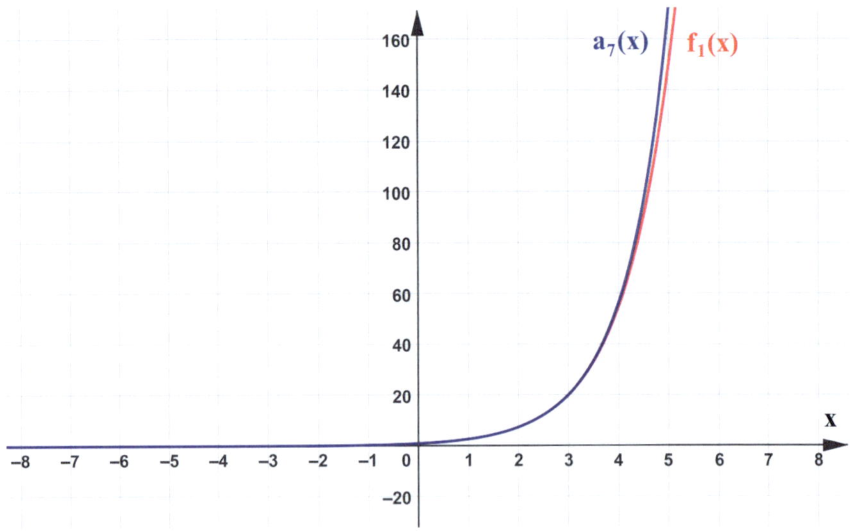

Fig. 3.20 Exponential function $f_1(x)$ and the seventh approximation function $a_7(x)$

The graphical representation of the exponential function $f_1(x)$ and of the seventh approximation function $a_7(x)$, both of them being expressed in db, is presented in Fig. 3.21.

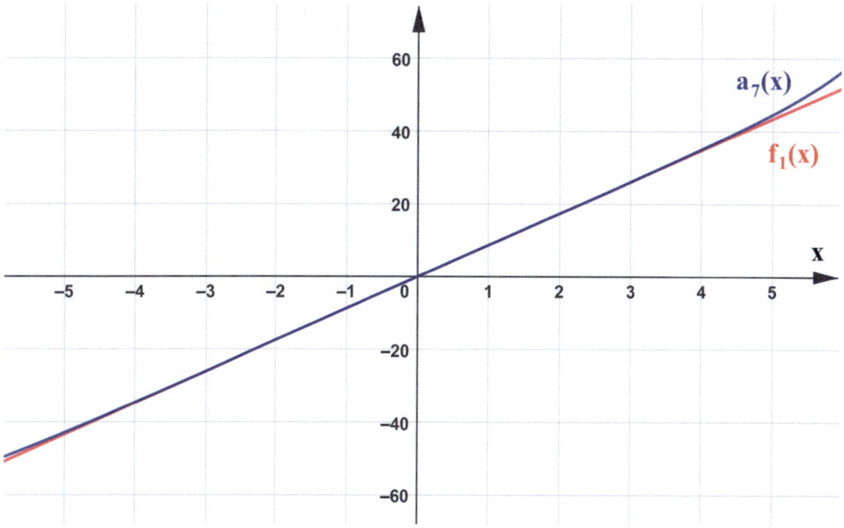

Fig. 3.21 Exponential function $f_1(x)$ and the seventh approximation function $a_7(x)$ (expressed in dB)

The error of approximation of the exponential function $f_1(x)$ using the seventh approximation function $a_7(x)$ (error1, expressed in dB) is illustrated in Fig. 3.22. Considering as acceptable a maximal approximation error of ± 1 dB, the maximal range of the input variable x is $[-5.6\,;\ 4.9]$. This is equivalent with an output dynamic range of the exponential function synthesizer based on the seventh approximation function $a_7(x)$ equal with 91.26 dB.

3.2.9 Eighth Approximation Function for Generating the Exponential Function

In order to approximate the exponential function $f_1(x)$, a new approximation function $a_8(x)$ is presented. The general form of the proposed eighth approximation function $a_8(x)$ is the following:

$$a_{8\mathrm{GEN}}(x) = \frac{c_3x^3 + c_2x^2 + c_1x + c_0}{d_3x^3 + d_2x^2 + d_1x + d_0}, \tag{3.14}$$

$c_0, c_1, c_2, c_3, d_0, d_1, d_2$ and d_3 being constant coefficients having particular values that can be determined by the condition that the superior-order Taylor series expansion of $f_1(x)$ exponential function, expressed by (3.2), to be identical with the superior-order Taylor series expansion of the general form of the approximation function $a_{8\mathrm{GEN}}(x)$, expressed by (3.14).

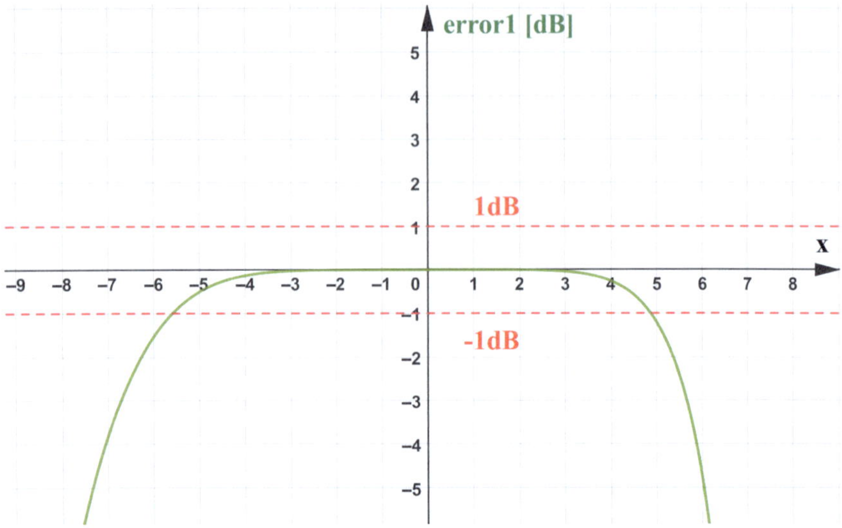

Fig. 3.22 Error of approximation of the exponential function $f_1(x)$ using the seventh approximation function $a_7(x)$ (expressed in dB)

The values of these coefficients, obtained considering the previous conditions, can be written as: $c_0 = 1$, $c_1 = \frac{1}{2}$, $c_2 = \frac{1}{10}$, $c_3 = \frac{1}{120}$, $d_0 = 1$, $d_1 = -\frac{1}{2}$, $d_2 = \frac{1}{10}$ and $d_3 = -\frac{1}{120}$. Replacing these particular values in relation (3.14), the eighth approximation function $a_8(x)$ has the following mathematical expression:

$$a_8(x) = \frac{\frac{x^3}{120} + \frac{x^2}{10} + \frac{x}{2} + 1}{-\frac{x^3}{120} + \frac{x^2}{10} - \frac{x}{2} + 1}. \tag{3.15}$$

The graphical representation of the exponential function $f_1(x)$ and of the eighth approximation function $a_8(x)$ is presented in Fig. 3.23.

The graphical representation of the exponential function $f_1(x)$ and of the eighth approximation function $a_8(x)$, both of them being expressed in db, is presented in Fig. 3.24.

The error of approximation of the exponential function $f_1(x)$ using the eighth approximation function $a_8(x)$ (error1, expressed in dB) is illustrated in Fig. 3.25. Considering as acceptable a maximal approximation error of ± 1 dB, the maximal range of the input variable x is $[-3.55\,;\,3.55]$. This is equivalent with an output dynamic range of the exponential function synthesizer based on the eighth approximation function $a_8(x)$ equal with 61.71 dB.

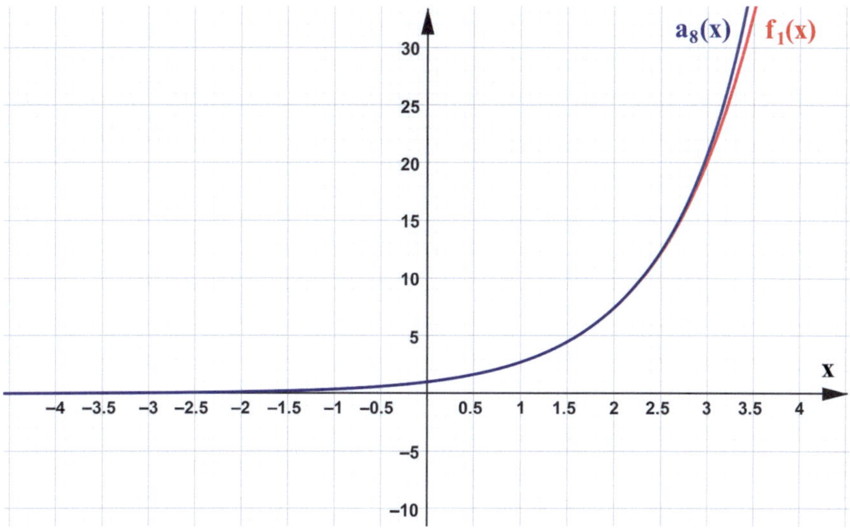

Fig. 3.23 Exponential function $f_1(x)$ and the eighth approximation function $a_8(x)$

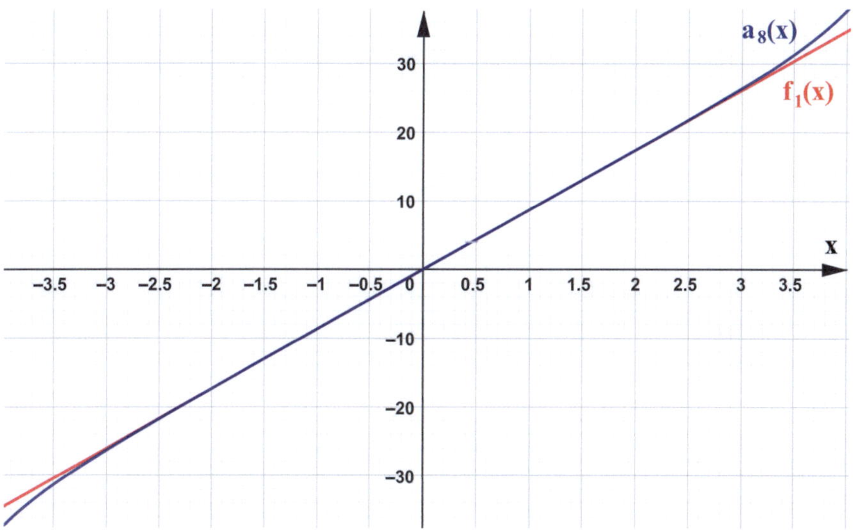

Fig. 3.24 Exponential function $f_1(x)$ and the eighth approximation function $a_8(x)$ (expressed in dB)

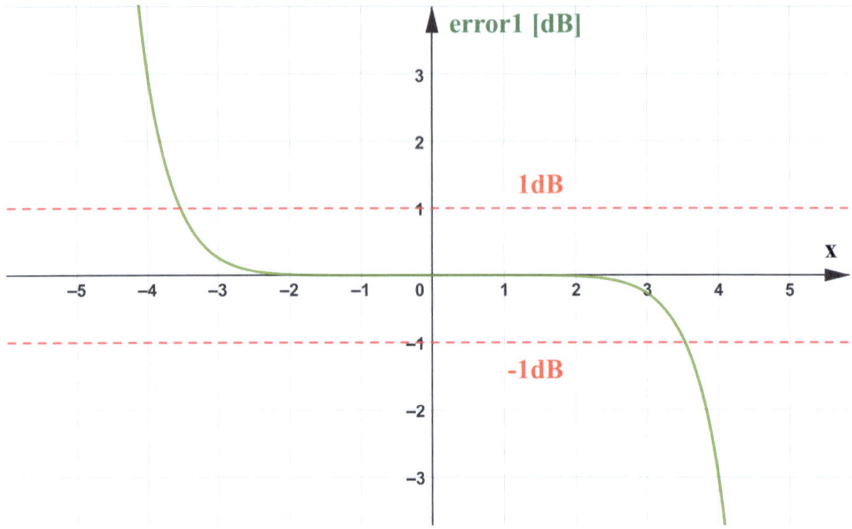

Fig. 3.25 Error of approximation of the exponential function $f_1(x)$ using the eighth approximation function $a_8(x)$ (expressed in dB)

3.2.10 Nineth Approximation Function for Generating the Exponential Function

Using the previous presented variable changing method $\left(x \rightarrow \frac{x}{2}\right)$, the new improved accuracy approximation function $a_9(x)$, based on the previous approximation function $a_8(x)$, has the following expression:

$$a_9(x) = \left(\frac{\frac{x^3}{960} + \frac{x^2}{40} + \frac{x}{4} + 1}{-\frac{x^3}{960} + \frac{x^2}{40} - \frac{x}{4} + 1} \right)^2. \tag{3.16}$$

The graphical representation of the exponential function $f_1(x)$ and of the nineth approximation function $a_9(x)$ is presented in Fig. 3.26.

The graphical representation of the exponential function $f_1(x)$ and of the nineth approximation function $a_9(x)$, both of them being expressed in db, is presented in Fig. 3.27.

The error of approximation of the exponential function $f_1(x)$ using the nineth approximation function $a_9(x)$ (error1, expressed in dB) is illustrated in Fig. 3.28. Considering as acceptable a maximal approximation error of ± 1 dB, the maximal range of the input variable x is $[-6.5\,;\ 6.5]$. This is equivalent with an output dynamic range of the exponential function synthesizer based on the nineth approximation function $a_9(x)$ equal with 112.99 dB.

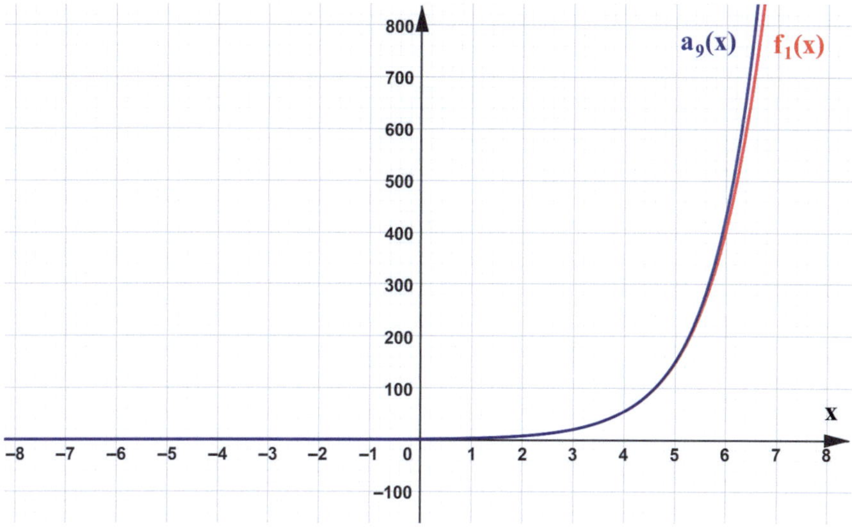

Fig. 3.26 Exponential function $f_1(x)$ and the nineth approximation function $a_9(x)$

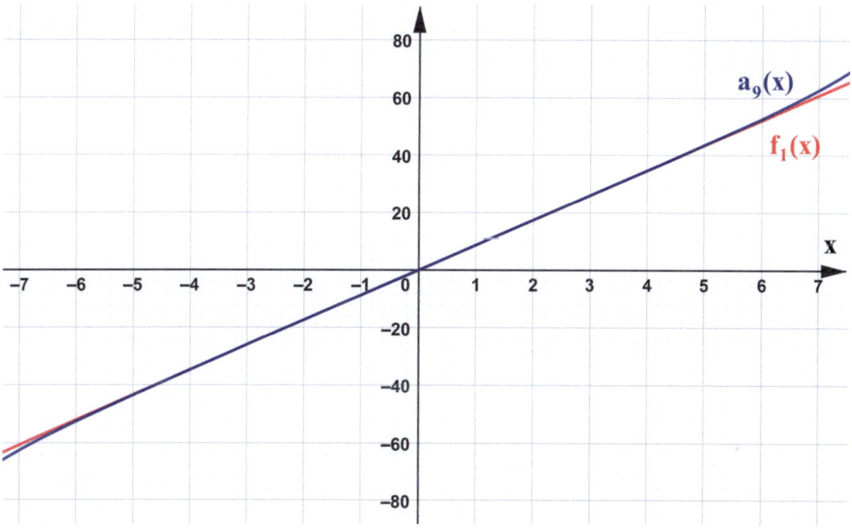

Fig. 3.27 Exponential function $f_1(x)$ and the nineth approximation function $a_9(x)$ (expressed in dB)

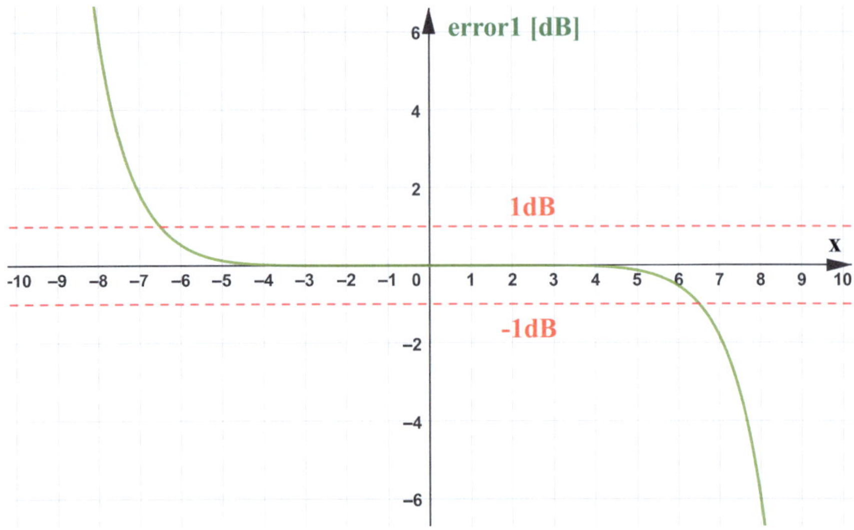

Fig. 3.28 Error of approximation of the exponential function $f_1(x)$ using the nineth approximation function $a_9(x)$ (expressed in dB)

3.2.11 Tenth Approximation Function for Generating the Exponential Function

In order to approximate the exponential function $f_1(x)$, a new approximation function $a_{10}(x)$ is presented. The general form of the proposed tenth approximation function $a_{10}(x)$ is the following:

$$a_{10\mathrm{GEN}}(x) = \frac{c_3x^3 + c_2x^2 + c_1x + c_0}{d_4x^4 + d_3x^3 + d_2x^2 + d_1x + d_0}, \tag{3.17}$$

c_0, c_1, c_2, c_3, d_0, d_1, d_2, d_3 and d_4 being constant coefficients having particular values that can be determined by the condition that the superior-order Taylor series expansion of $f_1(x)$ exponential function, expressed by (3.2), to be identical with the superior-order Taylor series expansion of the general form of the approximation function $a_{10\mathrm{GEN}}(x)$, expressed by (3.17).

The values of these coefficients, obtained considering the previous conditions, can be written as: $c_0 = 1, c_1 = \frac{3}{7}, c_2 = \frac{1}{14}, c_3 = \frac{1}{210}, d_0 = 1, d_1 = -\frac{4}{7}, d_2 = \frac{1}{7}, d_3 = -\frac{2}{105}$ and $d_4 = \frac{1}{840}$. Replacing these particular values in relation (3.17), the tenth approximation function $a_8(x)$ has the following mathematical expression:

$$a_{10}(x) = \frac{\frac{x^3}{210} + \frac{x^2}{14} + \frac{3x}{7} + 1}{\frac{x^4}{840} - \frac{2x^3}{105} + \frac{x^2}{7} - \frac{4x}{7} + 1}. \tag{3.18}$$

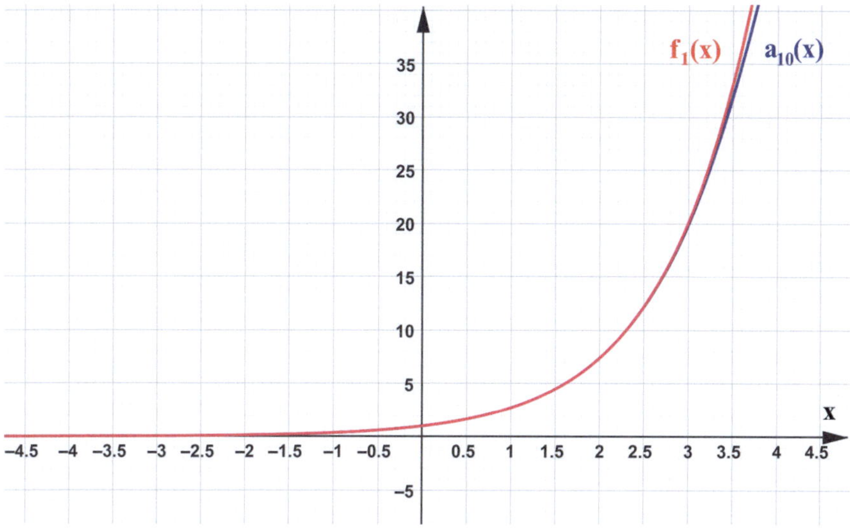

Fig. 3.29 Exponential function $f_1(x)$ and the tenth approximation function $a_{10}(x)$

The graphical representation of the exponential function $f_1(x)$ and of the tenth approximation function $a_{10}(x)$ is presented in Fig. 3.29.

The graphical representation of the exponential function $f_1(x)$ and of the tenth approximation function $a_{10}(x)$, both of them being expressed in db, is presented in Fig. 3.30.

The error of approximation of the exponential function $f_1(x)$ using the tenth approximation function $a_{10}(x)$ (error1, expressed in dB) is illustrated in Fig. 3.31. Considering as acceptable a maximal approximation error of ± 1 dB, the maximal range of the input variable x is $[-4.4\,;\ 4]$. This is equivalent with an output dynamic range of the exponential function synthesizer based on the tenth approximation function $a_{10}(x)$ equal with 73.01 dB.

3.2.12 Eleventh Approximation Function for Generating the Exponential Function

In order to approximate the exponential function $f_1(x)$, a new approximation function $a_{11}(x)$ is presented. The general form of the proposed 11th approximation function $a_{11}(x)$ is the following:

$$a_{11\text{GEN}}(x) = \frac{c_4x^4 + c_3x^3 + c_2x^2 + c_1x + c_0}{d_4x^4 + d_3x^3 + d_2x^2 + d_1x + d_0}, \tag{3.19}$$

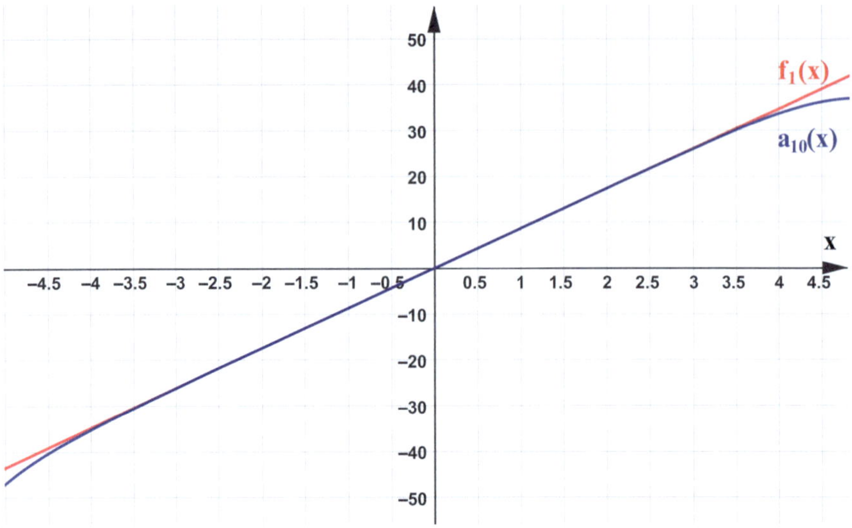

Fig. 3.30 Exponential function $f_1(x)$ and the tenth approximation function $a_{10}(x)$ (expressed in dB)

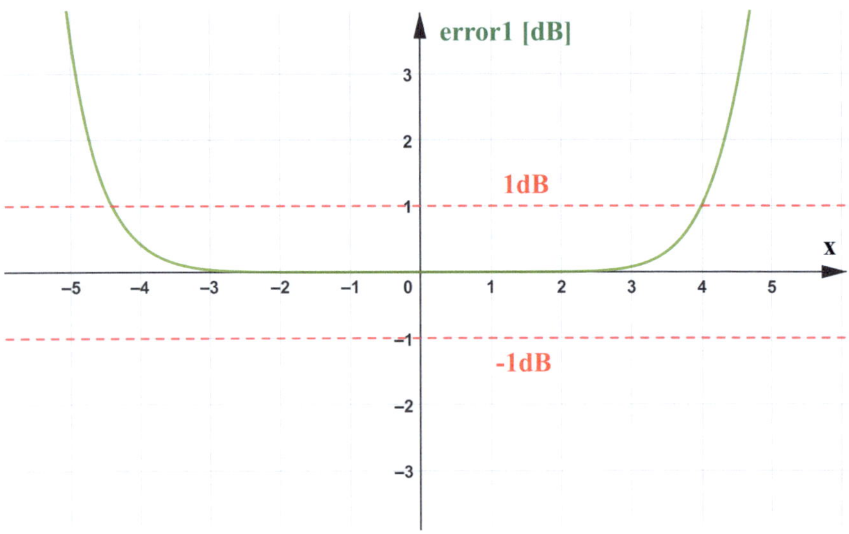

Fig. 3.31 Error of approximation of the exponential function $f_1(x)$ using the tenth approximation function $a_{10}(x)$ (expressed in dB)

c_0, c_1, c_2, c_3, c_4, d_0, d_1, d_2, d_3 and d_4 being constant coefficients having particular values that can be determined by the condition that the superior-order Taylor series expansion of $f_1(x)$ exponential function, expressed by (3.2), to be identical with the superior-order Taylor series expansion of the general form of the approximation function $a_{11GEN}(x)$, expressed by (3.19).

The values of these coefficients, obtained considering the previous conditions, can be written as: $c_0 = 1$, $c_1 = \frac{1}{2}$, $c_2 = \frac{3}{28}$, $c_3 = \frac{1}{84}$, $c_4 = \frac{1}{1680}$, $d_0 = 1$, $d_1 = -\frac{1}{2}$, $d_2 = \frac{3}{28}$, $d_3 = -\frac{1}{84}$ and $d_4 = \frac{1}{1680}$. Replacing these particular values in relation (3.19), the 11th approximation function $a_{11}(x)$ has the following mathematical expression:

$$a_{11}(x) = \frac{x^4 + 20x^3 + 180x^2 + 840x + 1680}{x^4 - 20x^3 + 180x^2 - 840x + 1680}. \quad (3.20)$$

The graphical representation of the exponential function $f_1(x)$ and of the 11th approximation function $a_{11}(x)$ is presented in Fig. 3.32.

The graphical representation of the exponential function $f_1(x)$ and of the 11th approximation function $a_{11}(x)$, both of them being expressed in db, is presented in Fig. 3.33.

The error of approximation of the exponential function $f_1(x)$ using the 11th approximation function $a_{11}(x)$ (error1, expressed in dB) is illustrated in Fig. 3.34. Considering as acceptable a maximal approximation error of ± 1 dB, the maximal range of the input variable x is $[-4.9; 4.9]$. This is equivalent with an output dynamic range of the exponential function synthesizer based on the 11th approximation function $a_{11}(x)$ equal with 85.18 dB.

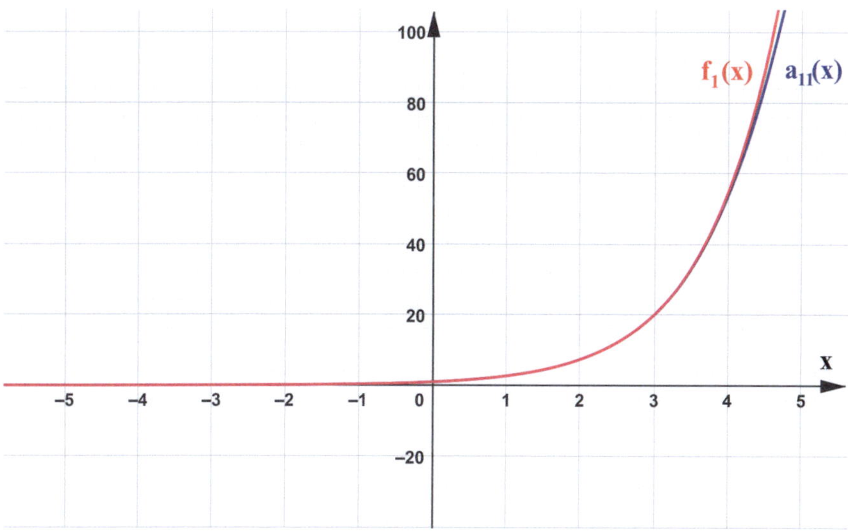

Fig. 3.32 Exponential function $f_1(x)$ and the 11th approximation function $a_{11}(x)$

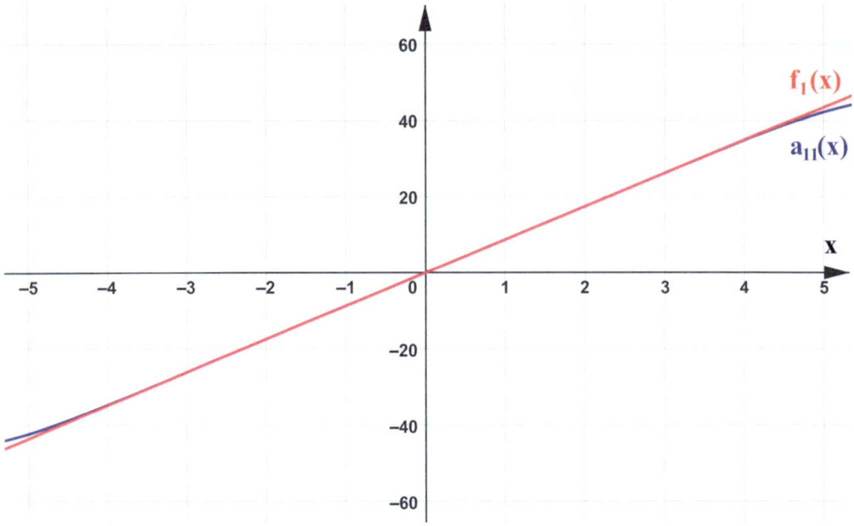

Fig. 3.33 Exponential function $f_1(x)$ and the 11th approximation function $a_{11}(x)$ (expressed in dB)

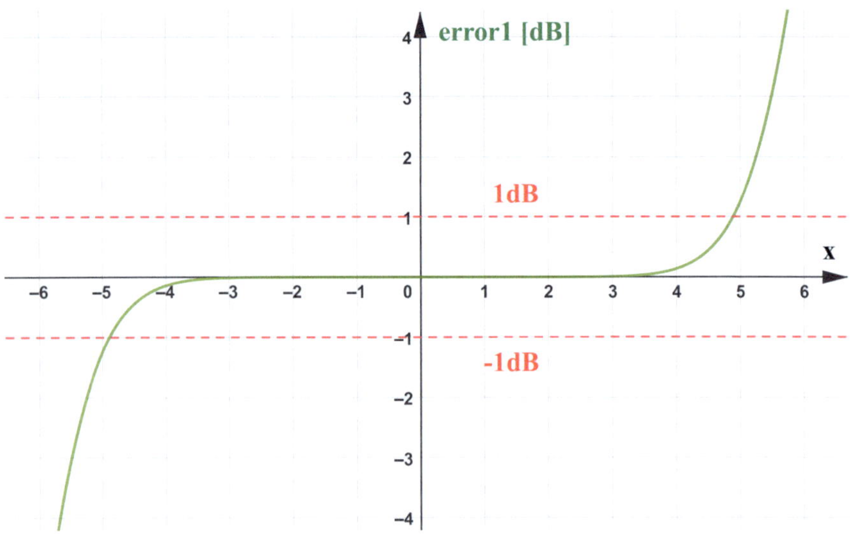

Fig. 3.34 Error of approximation of the exponential function $f_1(x)$ using the 11th approximation function $a_{11}(x)$ (expressed in dB)

3.2.13 Twelfth Approximation Function for Generating the Exponential Function

In order to approximate the exponential function $f_1(x)$, a new approximation function $a_{12}(x)$ is presented. The general form of the proposed 12th approximation function $a_{12}(x)$ is the following:

$$a_{12\text{GEN}}(x) = \left(\frac{c_0}{d_0 + d_1 x} + e_2 x^2 + e_1 x + e_0\right)^4, \tag{3.21}$$

c_1, d_0, d_1, e_0, e_1 and e_2 being constant coefficients having particular values that can be determined by the condition that the superior-order Taylor series expansion of $f_1(x)$ exponential function, expressed by (3.2), to be identical with the superior-order Taylor series expansion of the general form of the approximation function $a_{12\text{GEN}}(x)$, expressed by (3.21).

The values of these coefficients, obtained considering the previous conditions, can be written as: $c_0 = \frac{32}{3}$, $d_0 = 1$, $d_1 = -\frac{1}{16}$, $e_0 = -\frac{29}{3}$, $e_1 = -\frac{5}{12}$ and $e_2 = -\frac{1}{64}$. Replacing these particular values in relation (3.21), the 12th approximation function $a_{12}(x)$ has the following mathematical expression:

$$a_{12}(x) = \left(\frac{32}{3}\frac{1}{1 - \frac{x}{16}} - \frac{x^2}{64} - \frac{5x}{12} - \frac{29}{3}\right)^4. \tag{3.22}$$

The graphical representation of the exponential function $f_1(x)$ and of the 12th approximation function $a_{12}(x)$ is presented in Fig. 3.35.

The graphical representation of the exponential function $f_1(x)$ and of the 12th approximation function $a_{12}(x)$, both of them being expressed in db, is presented in Fig. 3.36.

The error of approximation of the exponential function $f_1(x)$ using the 12th approximation function $a_{12}(x)$ (error1, expressed in dB) is illustrated in Fig. 3.37. Considering as acceptable a maximal approximation error of ± 1 dB, the maximal range of the input variable x is $[-1.85; 3.8] \cup [6.5; 9.2]$. This is equivalent with an output dynamic range of the exponential function synthesizer based on the 12th approximation function $a_{12}(x)$ equal with 72.58 dB.

3.2.14 Thirteenth Approximation Function for Generating the Exponential Function

In order to approximate the exponential function $f_1(x)$, a new approximation function $a_{13}(x)$ is presented. The general form of the proposed 13th approximation function $a_{13}(x)$ is the following:

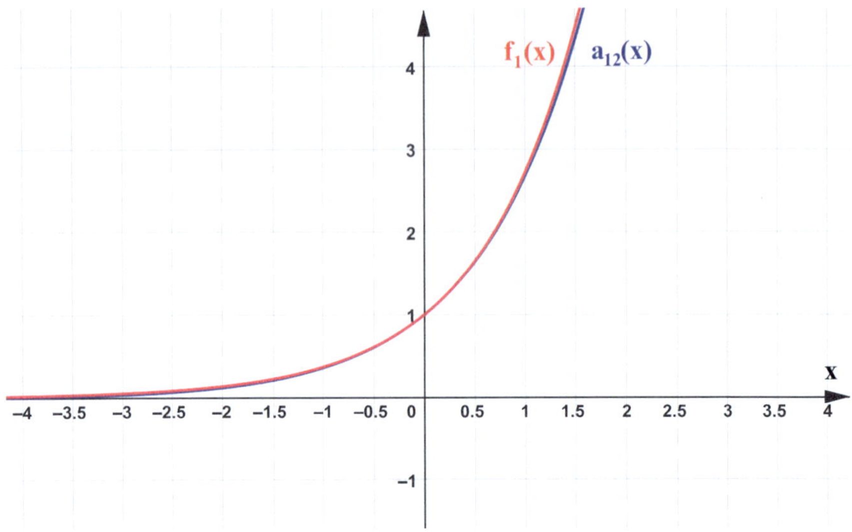

Fig. 3.35 Exponential function $f_1(x)$ and the 12th approximation function $a_{12}(x)$

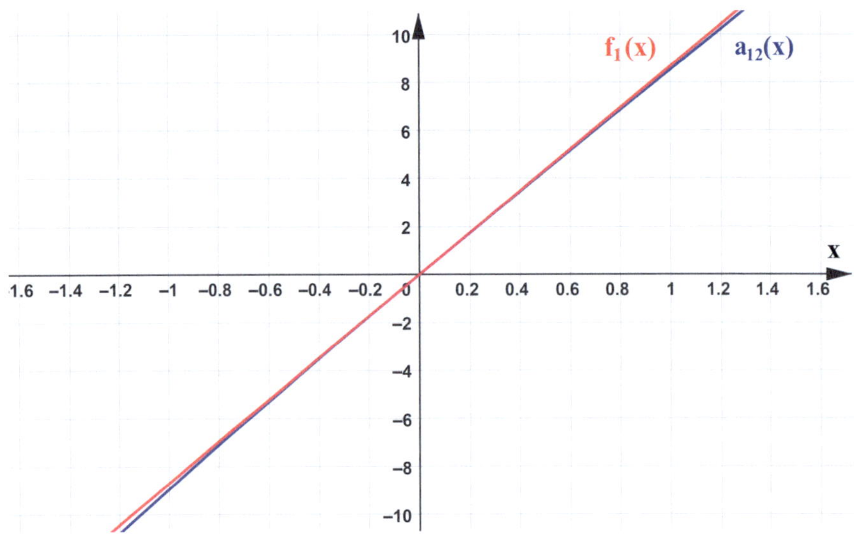

Fig. 3.36 Exponential function $f_1(x)$ and the 12th approximation function $a_{12}(x)$ (expressed in dB)

$$a_{13\mathrm{GEN}}(x) = \frac{c_2x^2 + c_1x + c_0}{d_2x^2 + d_1x + d_0} + \frac{e_2x^2 + e_1x}{f_0}, \tag{3.23}$$

c_0, c_1, c_2, d_0, d_1, d_2, e_0, e_1 and f_0 being constant coefficients having particular values that can be determined by the condition that the superior-order Taylor series

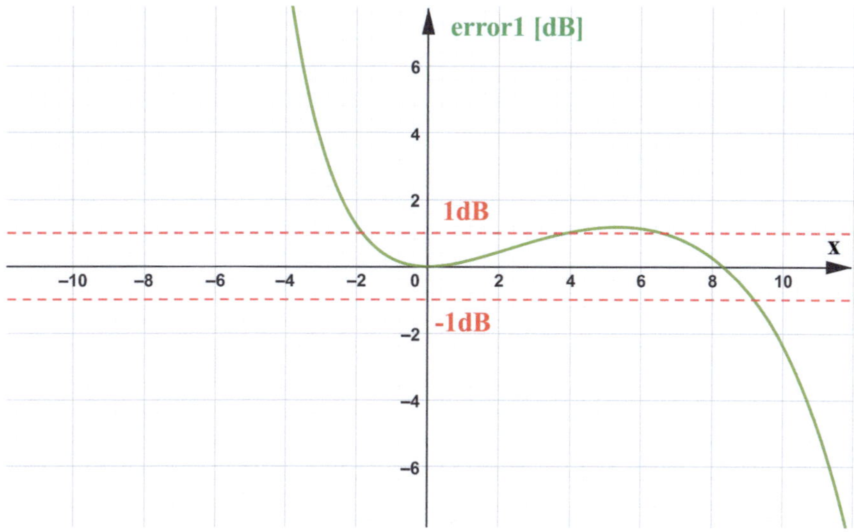

Fig. 3.37 Error of approximation of the exponential function $f_1(x)$ using the 12th approximation function $a_{12}(x)$ (expressed in dB)

expansion of $f_1(x)$ exponential function, expressed by (3.2), to be identical with the superior-order Taylor series expansion of the general form of the approximation function $a_{13GEN}(x)$, expressed by (3.23).

The values of these coefficients, obtained considering the previous conditions, can be written as: $c_0 = 1$, $c_1 = -\frac{7}{6}$, $c_2 = \frac{131}{180}$, $d_0 = 1$, $d_1 = -\frac{1}{3}$, $d_2 = \frac{1}{30}$, $e_1 = 22$, $e_2 = 1$ and $f_0 = 12$. Replacing these particular values in relation (3.23), the 13th approximation function $a_{13}(x)$ has the following mathematical expression:

$$a_{13}(x) = \frac{131x^2 - 210x + 180}{6x^2 - 60x + 180} + \frac{x^2 + 22x}{12}. \tag{3.24}$$

The graphical representation of the exponential function $f_1(x)$ and of the 13th approximation function $a_{13}(x)$ is presented in Fig. 3.38.

The graphical representation of the exponential function $f_1(x)$ and of the 13th approximation function $a_{13}(x)$, both of them being expressed in db, is presented in Fig. 3.39.

The error of approximation of the exponential function $f_1(x)$ using the 13th approximation function $a_{13}(x)$ (error1, expressed in dB) is illustrated in Fig. 3.40. Considering as acceptable a maximal approximation error of ±1 dB, the maximal range of the input variable x is [−3.1 ; 4]. This is equivalent with an output dynamic range of the exponential function synthesizer based on the 13th approximation function $a_{13}(x)$ equal with 61.71 dB.

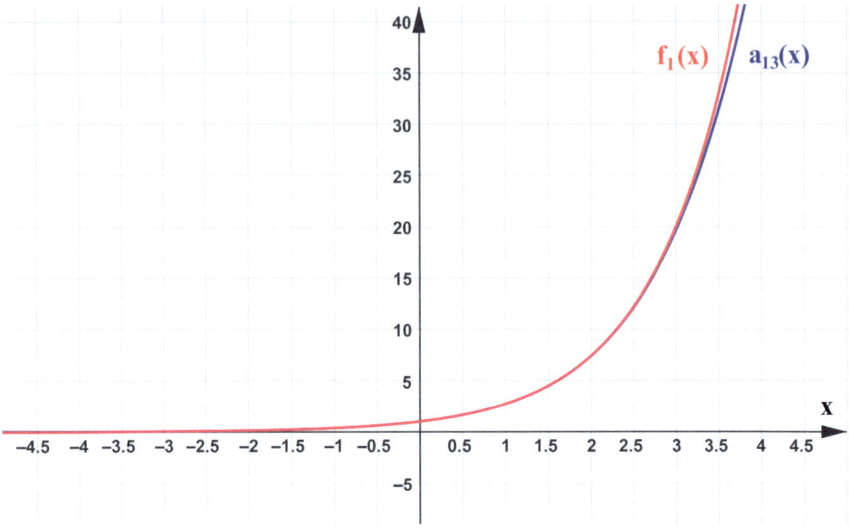

Fig. 3.38 Exponential function $f_1(x)$ and the 13th approximation function $a_{13}(x)$

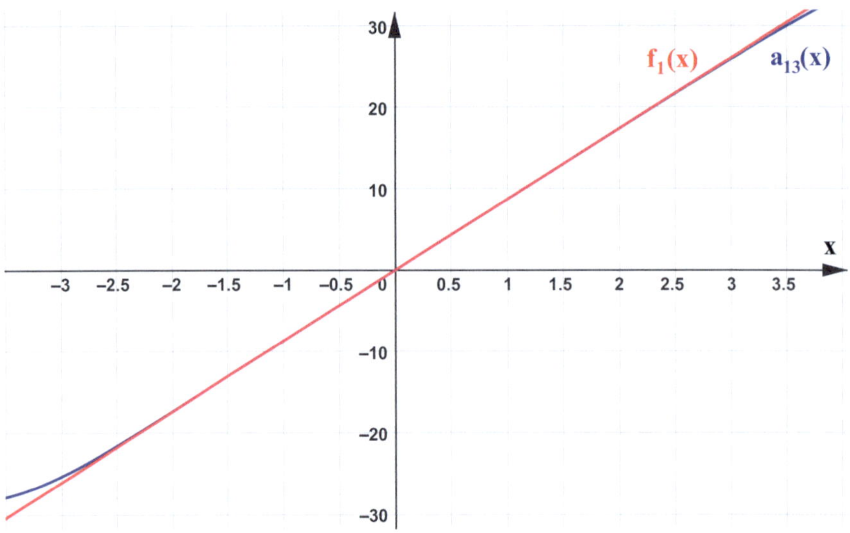

Fig. 3.39 Exponential function $f_1(x)$ and the 13th approximation function $a_{13}(x)$ (expressed in dB)

3.2.15 Fourteenth Approximation Function for Generating the Exponential Function

In order to approximate the exponential function $f_1(x)$, a new approximation function $a_{14}(x)$ is presented. The general form of the proposed 14th approximation function

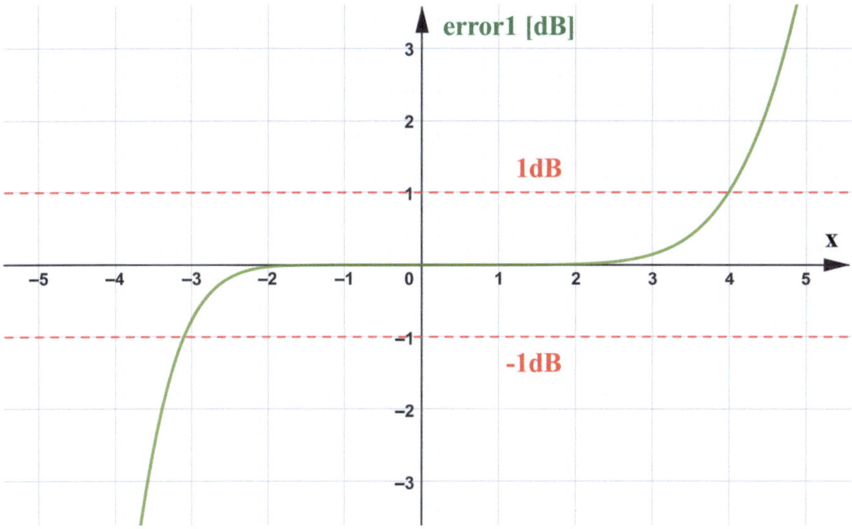

Fig. 3.40 Error of approximation of the exponential function $f_1(x)$ using the 13th approximation function $a_{13}(x)$ (expressed in dB)

$a_{14}(x)$ is the following:

$$a_{14GEN}(x) = \left(\frac{c_0}{d_1x + d_0} + \frac{f_0}{e_1x + e_0} + \frac{g_1x + g_0}{h_0}\right)^4, \tag{3.25}$$

c_0, d_0, d_1, e_0, e_1, f_0, g_0, g_1 and h_0 being constant coefficients having particular values that can be determined by the condition that the superior-order Taylor series expansion of $f_1(x)$ exponential function, expressed by (3.2), to be identical with the superior-order Taylor series expansion of the general form of the approximation function $a_{14GEN}(x)$, expressed by (3.25).

The values of these coefficients, obtained considering the previous conditions, can be written as $c_0 = \frac{26}{5}$, $d_0 = 1$, $d_1 = -\frac{5}{64}$, $e_0 = 1$, $e_1 = \frac{1}{4}$, $f_0 = \frac{1}{128}$, $g_0 = 924$, $g_1 = 35$ and $h_0 = -220$. Replacing these particular values in relation (3.25), the 14th approximation function $a_{14}(x)$ has the following mathematical expression:

$$a_{14}(x) = \left(\frac{1664}{5}\frac{1}{64 - 5x} + \frac{1}{32}\frac{1}{x + 4} - \frac{35x + 924}{220}\right)^4. \tag{3.26}$$

This approximation function could be used for obtaining the exponential function only for positive values of the input variable x. The graphical representation of the exponential function $f_1(x)$ and of the 14th approximation function $a_{14}(x)$ is presented in Fig. 3.41.

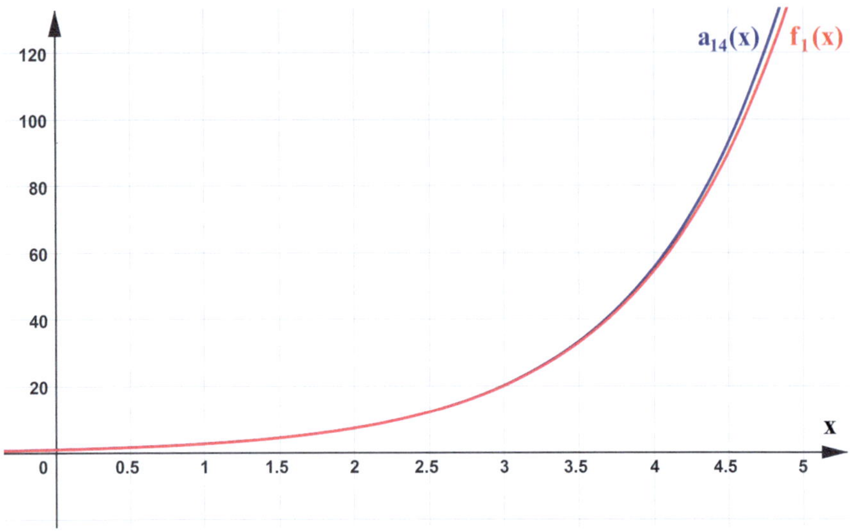

Fig. 3.41 Exponential function $f_1(x)$ and the 14th approximation function $a_{14}(x)$

The graphical representation of the exponential function $f_1(x)$ and of the 14th approximation function $a_{14}(x)$, both of them being expressed in db, is presented in Fig. 3.42.

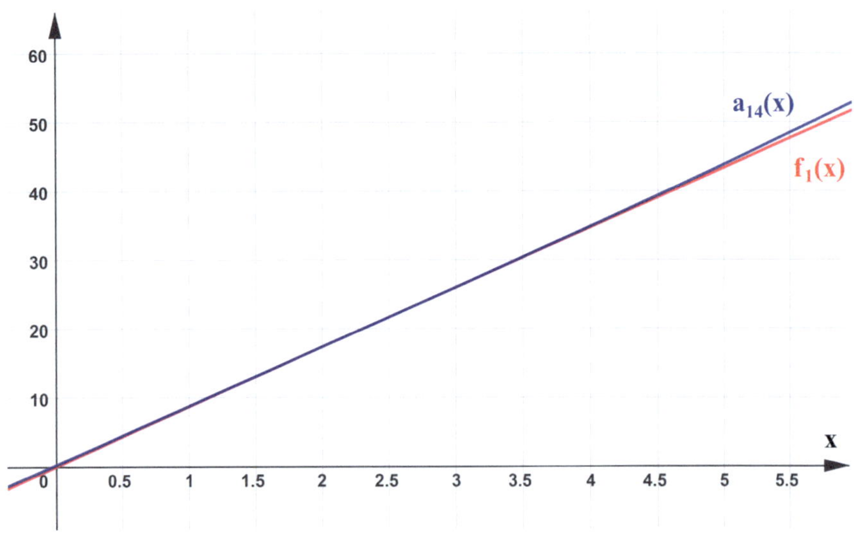

Fig. 3.42 Exponential function $f_1(x)$ and the 14th approximation function $a_{14}(x)$ (expressed in dB)

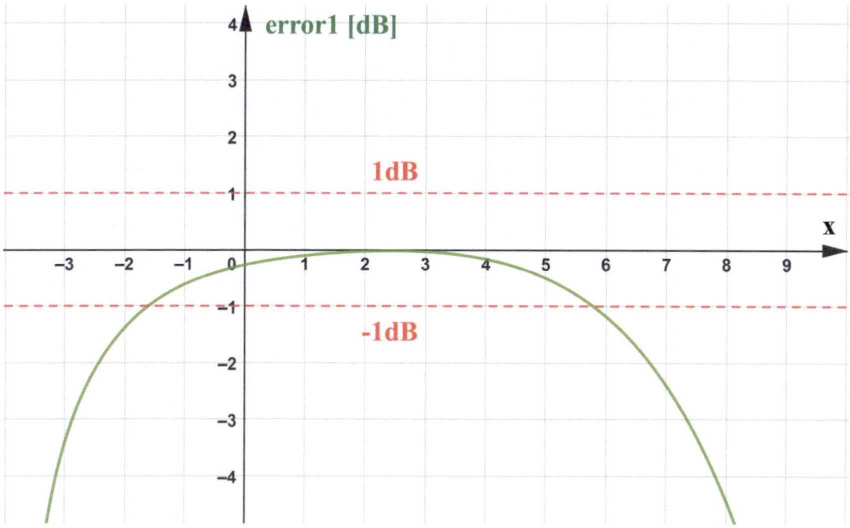

Fig. 3.43 Error of approximation of the exponential function $f_1(x)$ using the 14th approximation function $a_{14}(x)$ (expressed in dB)

The error of approximation of the exponential function $f_1(x)$ using the 14th approximation function $a_{14}(x)$ (error1, expressed in dB) is illustrated in Fig. 3.43. Considering as acceptable a maximal approximation error of ± 1 dB, the maximal range of the input variable x is $[-1.6,\ 5.8]$. This is equivalent with an output dynamic range of the exponential function synthesizer based on the 14th approximation function $a_{14}(x)$ equal with 64.32 dB.

Table 3.1 contains a comparison between the previous approximation functions, developed for obtaining the exponential functions with as good as possible accuracy.

3.3 Approximation Functions for Exponential Ramp Function

The exponential ramp activation function is a custom or less commonly used activation function designed to combine properties of both linear and exponential functions. It is typically used in niche or experimental architectures, often in attempts to improve gradient flow, response smoothness, or biological plausibility. It is actually similar to the exponential linear unit (ELU) activation, which is the most common real-world implementation of an exponential ramp. The areas of utilization of the exponential ramp function are:

a. ***Deep Learning (Hidden Layers)***

Table. 3.1 Comparison between the previous approximation functions

Approximation function	Range of input variable, x	Output dynamic range [dB]
a_1	– 1.7 to 1.7	29.55
a_2	– 2.75 to 2.75	47.80
a_3	– 2.3 to 2.3	39.81
a_4	– 4 to 4	69.53
a_5	– 7.1 to 7.1	123.42
a_6	– 3.15 to 2.7	50.85
a_7	– 5.6 to 4.9	91.26
a_8	– 3.55 to 3.55	61.71
a_9	– 6.5 to 6.5	112.99
a_{10}	– 4.4 to 4	73.01
a_{11}	– 4.9 to 4.9	85.18
a_{12}	– 1.85 to 3.8 ∪ 6.5 –9.2	72.58
a_{13}	– 3.1 to 4	61.71
a_{14}	– 1.6 to 5.8	64.32

- Used as an alternative to ReLU to address its non-differentiability at 0 and zero gradient for negative inputs.
- Provides smooth transitions, helping gradient descent converge faster and more stably.
- Example: ELU and Scaled ELU (SELU) are used in feedforward networks for improved convergence speed and robustness.

b. ***Biologically Inspired Neural Networks***

- Exponential ramps can mimic biological neuron behavior, where negative stimuli decay exponentially rather than being completely cut off (as in ReLU).

c. ***Self-normalizing Networks (SNNs)***

- The SELU function (a scaled exponential ramp) enables internal normalization of outputs, removing the need for batch normalization.
- Helps maintain zero mean and unit variance across layers automatically.

d. ***Reinforcement Learning and Robotics***

- When smooth response curves are needed for continuous control tasks, exponential ramps are sometimes preferred over sharp cutoffs like ReLU.

e. ***Custom Activation Design***

- Exponential ramps are explored in experimental architectures to fine tune learning dynamics, especially in areas like neural ordinary differential equations (ODEs), transformers with continuous depth, or adaptive learning rate networks.

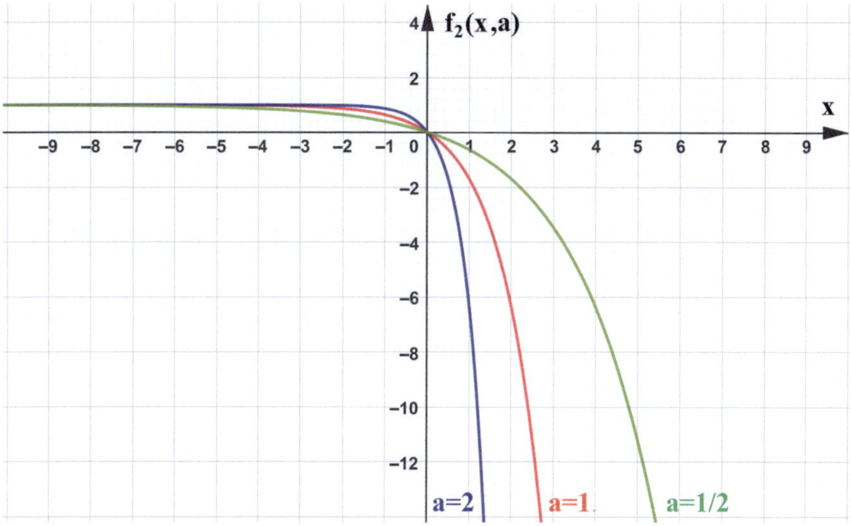

Fig. 3.44 Exponential ramp function $f_2(x, a)$ for $a = 1$, $a = 2$ and $a = 1/2$

3.3.1 General Form of the Exponential Ramp Function

The mathematical form of this function can be expressed as follows:

$$f_2(x, a) = 1 - \exp(ax). \tag{3.27}$$

a is being a parameter. The graphical representation of the exponential ramp function $f_1(x, a)$, considering three values of a parameter ($a = 1$, $a = 2$ and $a = 1/2$), is presented in Fig. 3.44.

In order to obtain an analog implementation of the exponential ramp function, a multitude of superior-order approximation functions is presented. The most important differences between these functions are mainly related to the accuracy of their approximation, as well as to the hardware resources required for their implementations in CMOS technology.

3.3.2 First Approximation Function for Generating the Exponential Ramp Function (for a = 1)

Considering the simplest case $a = 1$, the mathematical form (3.27) of the exponential ramp function becomes:

$$f_3(x) \stackrel{\text{not}}{=} f_2(x, 1) = 1 - \exp(x). \tag{3.28}$$

In order to approximate the exponential ramp function $f_3(x)$, a new approximation function $a_{15}(x)$ is presented. Its particular mathematical form is developed having two very important objectives. The first objective is represented by the accuracy of the approximation that could be obtained using this approximation function. The second objective is related to the possibility of its facile analog implementation in CMOS technology using as reasonable as possible hardware resources (see Chaps. 4 and 7).

The general form of the proposed first approximation function $a_{15}(x)$ is the following:

$$a_{15\mathrm{GEN}}(x) = \frac{c_1 x}{d_2 x^2 + d_1 x + d_0}, \tag{3.29}$$

c_1, d_0, d_1 and d_2 being constant coefficients having particular values that can be determined by the condition that the superior-order Taylor series expansion of $f_3(x)$ exponential ramp function, expressed by (3.28), to be identical with the superior-order Taylor series expansion of the general form of the approximation function $a_{15\mathrm{GEN}}(x)$, expressed by (3.29).

The values of these coefficients, obtained considering the previous conditions, can be written as: $c_1 = -1$, $d_0 = 1$, $d_1 = -\frac{1}{2}$ and $d_2 = \frac{1}{12}$. Replacing these particular values in relation (3.29), the first approximation function $a_{15}(x)$ has the following mathematical expression:

$$a_{15}(x) = -\frac{x}{\frac{x^2}{12} - \frac{x}{2} + 1}. \tag{3.30}$$

The graphical representation of the exponential ramp function $f_3(x)$ and of the first approximation function $a_{15}(x)$ is presented in Fig. 3.45.

The absolute error of approximation of the exponential ramp function $f_3(x)$ using the first approximation function $a_{15}(x)$ is illustrated in Fig. 3.46. It is defined as the difference between the previous functions, $\mathrm{error1} = f_3(x) - a_{15}(x)$.

In order to more accurately evaluate the error of approximation, the relative error of approximation is shown in Fig. 3.47. This error is defined as $\mathrm{error2} = 100\frac{f_3(x) - a_{15}(x)}{f_3(x)}$ (in percents). Considering as acceptable a maximal approximation error of $\pm 1\%$, the maximal range of the input variable x is $[-2.1\,;\ 1.4]$.

3.3.3 *Second Approximation Function for Generating the Exponential Ramp Function (for a = 1)*

In order to approximate the exponential ramp function $f_3(x)$, a new approximation function $a_{16}(x)$ is presented. The general form of the proposed second approximation function $a_{16}(x)$ is the following:

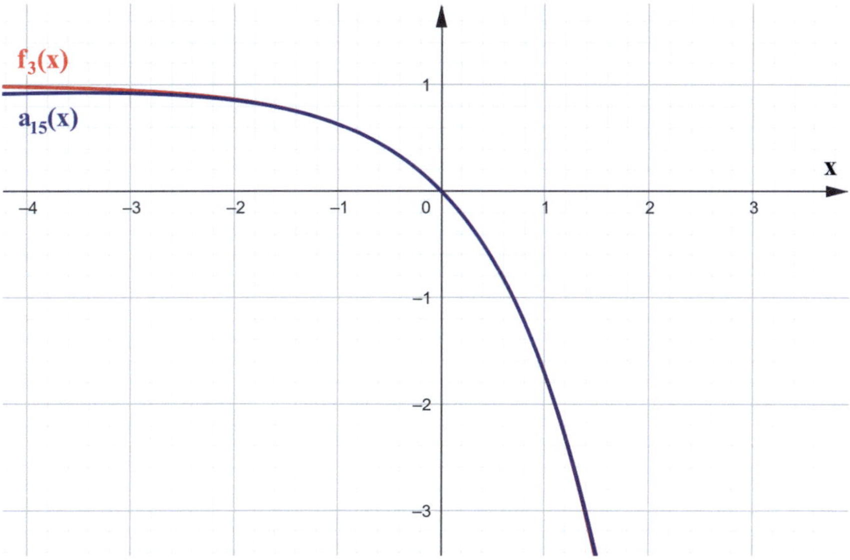

Fig. 3.45 Exponential ramp function $f_3(x)$ and the first approximation function $a_{15}(x)$

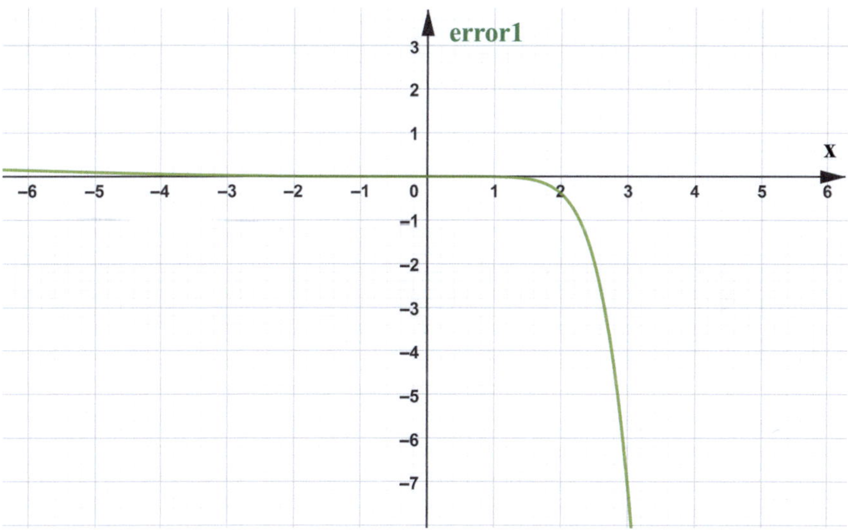

Fig. 3.46 Absolute error of approximation of the exponential ramp function $f_3(x)$ using the first approximation function $a_{15}(x)$

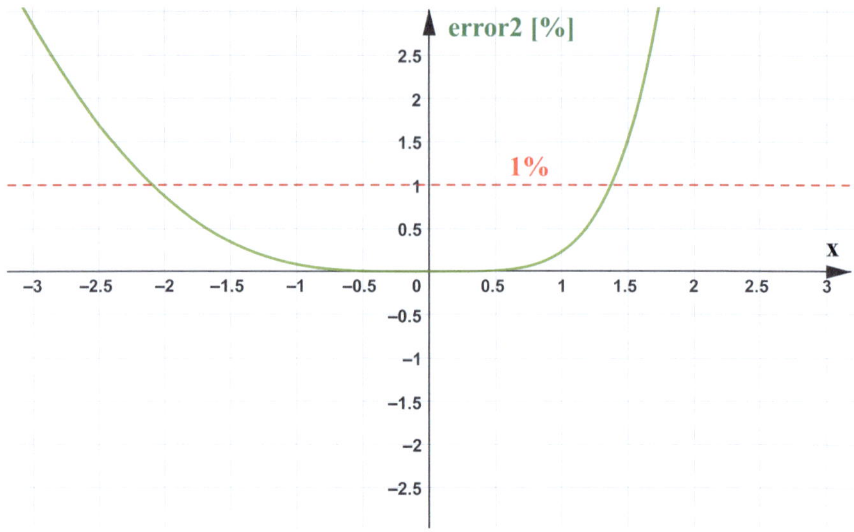

Fig. 3.47 Relative error of approximation of the exponential ramp function $f_3(x)$ using the first approximation function $a_{15}(x)$

$$a_{16\mathrm{GEN}}(x) = \frac{c_3x^3 + c_2x^2 + c_1x}{d_2x^2 + d_1x + d_0}, \tag{3.31}$$

c_1, c_2, c_3, d_0, d_1 and d_2 being constant coefficients having particular values that can be determined by the condition that the superior-order Taylor series expansion of $f_3(x)$ exponential ramp function, expressed by (3.28), to be identical with the superior-order Taylor series expansion of the general form of the approximation function $a_{16\mathrm{GEN}}(x)$, expressed by (3.31).

The values of these coefficients, obtained considering the previous conditions, can be written as: $c_1 = -1$, $c_2 = -\frac{1}{10}$, $c_3 = -\frac{1}{60}$, $d_0 = 1$, $d_1 = -\frac{2}{5}$ and $d_2 = \frac{1}{20}$. Replacing these particular values in relation (3.31), the second approximation function $a_{16}(x)$ has the following mathematical expression:

$$a_{16}(x) = \frac{-\frac{x^3}{60} - \frac{x^2}{10} - x}{\frac{x^2}{20} - \frac{2x}{5} + 1}. \tag{3.32}$$

The graphical representation of the exponential ramp function $f_3(x)$ and of the second approximation function $a_{16}(x)$ is presented in Fig. 3.48.

The absolute error of approximation of the exponential ramp function $f_3(x)$ using the second approximation function $a_{16}(x)$ is illustrated in Fig. 3.49. It is defined as the difference between the previous functions, $\mathrm{error1} = f_3(x) - a_{16}(x)$.

In order to more accurately evaluate the error of approximation, the relative error of approximation is shown in Fig. 3.50. This error is defined as $\mathrm{error2} = 100\frac{f_3(x) - a_{16}(x)}{f_3(x)}$

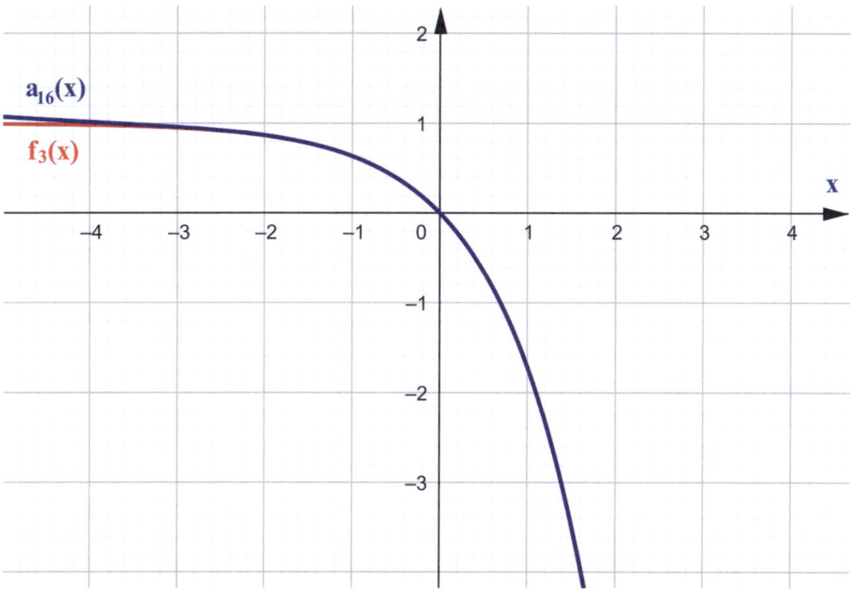

Fig. 3.48 Exponential ramp function $f_3(x)$ and the second approximation function $a_{16}(x)$

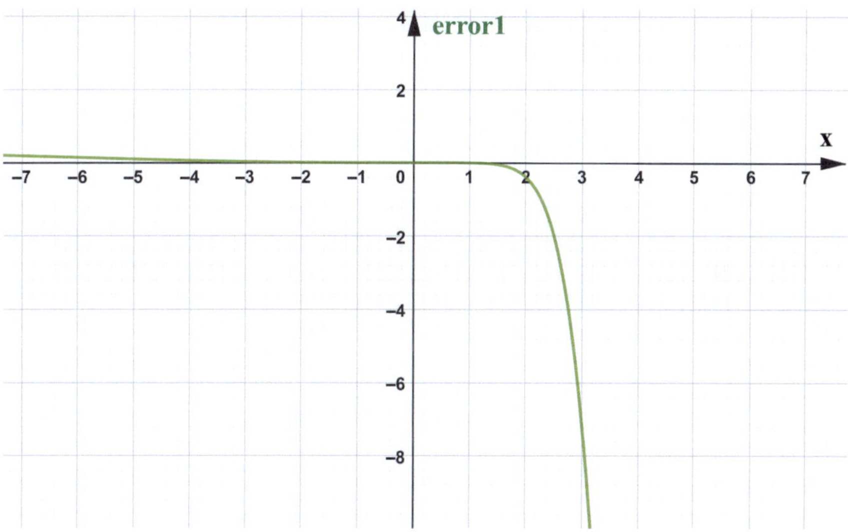

Fig. 3.49 Absolute error of approximation of the exponential ramp function $f_3(x)$ using the second approximation function $a_{16}(x)$

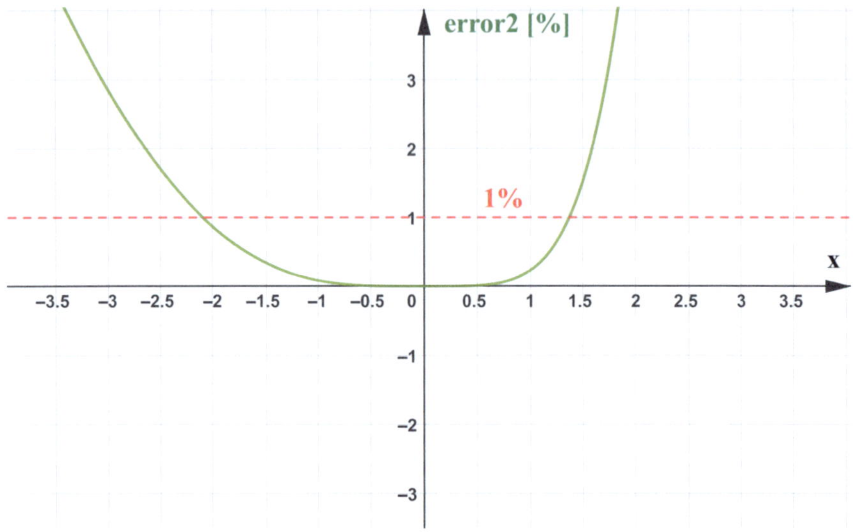

Fig. 3.50 Relative error of approximation of the exponential ramp function $f_3(x)$ using the second approximation function $a_{16}(x)$

(in percents). Considering as acceptable a maximal approximation error of $\pm 1\%$, the maximal range of the input variable x is $[-2.1\ ;\ 1.4]$.

3.3.4 Third Approximation Function for Generating the Exponential Ramp Function (for a = 1)

In order to approximate the exponential ramp function $f_3(x)$, a new approximation function $a_{17}(x)$ is presented. The general form of the proposed third approximation function $a_{17}(x)$ is the following:

$$a_{17\text{GEN}}(x) = \frac{c_3 x^3 + c_1 x}{d_3 x^3 + d_2 x^2 + d_1 x + d_0}, \tag{3.33}$$

c_1, c_3, d_0, d_1, d_2 and d_3 being constant coefficients having particular values that can be determined by the condition that the superior-order Taylor series expansion of $f_3(x)$ exponential ramp function, expressed by (3.28), to be identical with the superior-order Taylor series expansion of the general form of the approximation function $a_{17\text{GEN}}(x)$, expressed by (3.33).

The values of these coefficients, obtained considering the previous conditions, can be written as: $c_1 = -1$, $c_3 = -\frac{1}{60}$, $d_0 = 1$, $d_1 = -\frac{1}{2}$, $d_2 = \frac{1}{10}$ and $d_3 = -\frac{1}{120}$. Replacing these particular values in relation (3.33), the third approximation function $a_{17}(x)$ has the following mathematical expression:

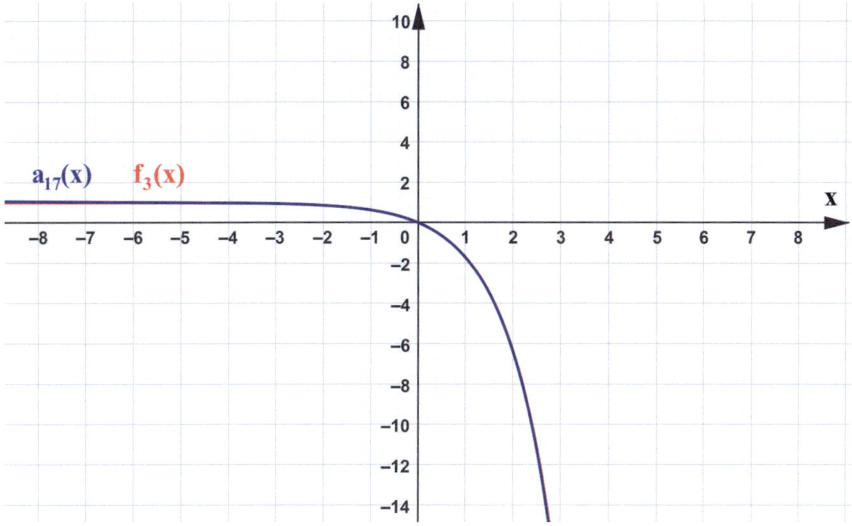

Fig. 3.51 Exponential ramp function $f_3(x)$ and the third approximation function $a_{17}(x)$

$$a_{17}(x) = \frac{-\frac{x^3}{60} - x}{-\frac{x^3}{120} + \frac{x^2}{10} - \frac{x}{2} + 1}. \tag{3.34}$$

The graphical representation of the exponential ramp function $f_3(x)$ and of the third approximation function $a_{17}(x)$ is presented in Fig. 3.51.

The absolute error of approximation of the exponential ramp function $f_3(x)$ using the third approximation function $a_{17}(x)$ is illustrated in Fig. 3.52. It is defined as the difference between the previous functions, error1 $= f_3(x) - a_{17}(x)$.

In order to more accurately evaluate the error of approximation, the relative error of approximation is shown in Fig. 3.53. This error is defined as error2 $= 100\frac{f_3(x) - a_{17}(x)}{f_3(x)}$ (in percents). Considering as acceptable a maximal approximation error of $\pm 1\%$, the maximal range of the input variable x is $[-4.7\,;\ 2.6]$.

3.3.5 *Fourth Approximation Function for Generating the Exponential Ramp Function (for a = 1)*

In order to approximate the exponential ramp function $f_3(x)$, a new approximation function $a_{18}(x)$ is presented. The general form of the proposed fourth approximation function $a_{18}(x)$ is the following:

$$a_{18\mathrm{GEN}}(x) = \frac{c_3 x^3 + c_1 x}{d_4 x^4 + d_3 x^3 + d_2 x^2 + d_1 x + d_0}, \tag{3.35}$$

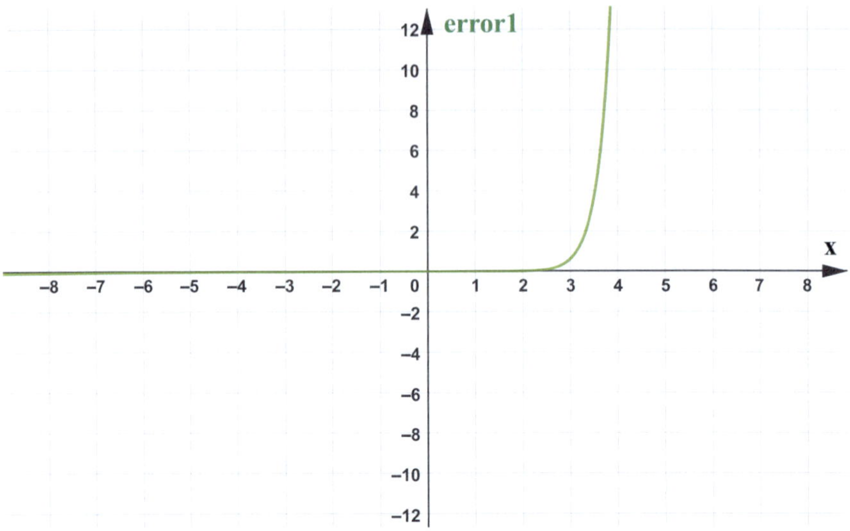

Fig. 3.52 Absolute error of approximation of the exponential ramp function $f_3(x)$ using the third approximation function $a_{17}(x)$

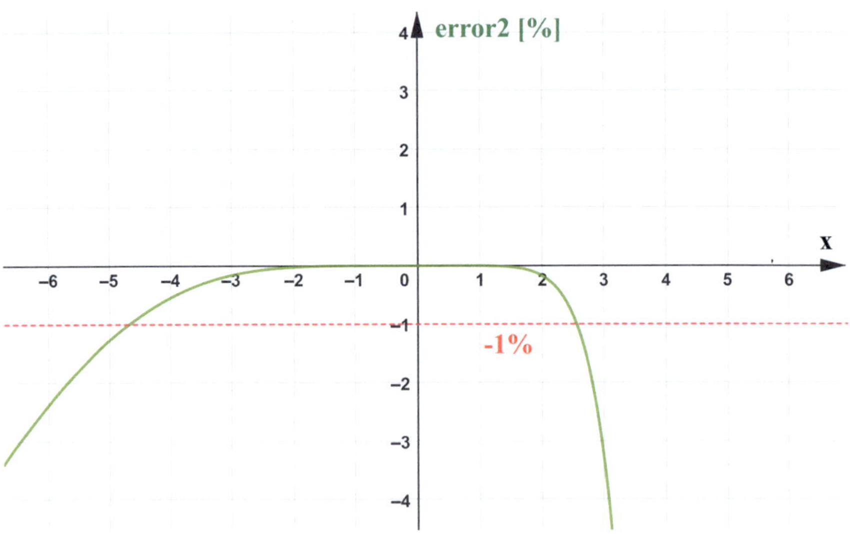

Fig. 3.53 Relative error of approximation of the exponential ramp function $f_3(x)$ using the third approximation function $a_{17}(x)$

c_1, c_3, d_0, d_1, d_2, d_3 and d_4 being constant coefficients having particular values that can be determined by the condition that the superior-order Taylor series expansion of $f_3(x)$ exponential ramp function, expressed by (3.28), to be identical with the superior-order Taylor series expansion of the general form of the approximation function $a_{18GEN}(x)$, expressed by (3.35).

The values of these coefficients, obtained considering the previous conditions, can be written as: $c_1 = -1$, $c_3 = -\frac{1}{42}$, $d_0 = 1$, $d_1 = -\frac{1}{2}$, $d_2 = \frac{3}{28}$, $d_3 = -\frac{1}{84}$ and $d_4 = \frac{1}{1680}$. Replacing these particular values in relation (3.35), the fourth approximation function $a_{18}(x)$ has the following mathematical expression:

$$a_{18}(x) = \frac{-\frac{x^3}{42} - x}{\frac{x^4}{1680} - \frac{x^3}{84} + \frac{3x^2}{28} - \frac{x}{2} + 1}. \tag{3.36}$$

The graphical representation of the exponential ramp function $f_3(x)$ and of the fourth approximation function $a_{18}(x)$ is presented in Fig. 3.54.

The absolute error of approximation of the exponential ramp function $f_3(x)$ using the fourth approximation function $a_{18}(x)$ is illustrated in Fig. 3.55. It is defined as the difference between the previous functions, error1 $= f_3(x) - a_{18}(x)$.

In order to more accurately evaluate the error of approximation, the relative error of approximation is shown in Fig. 3.56. This error is defined as error2 $= 100\frac{f_3(x)-a_{18}(x)}{f_3(x)}$ (in percents). Considering as acceptable a maximal approximation error of $\pm 1\%$, the maximal range of the input variable x is $[-8.2\,;\; 3.8]$.

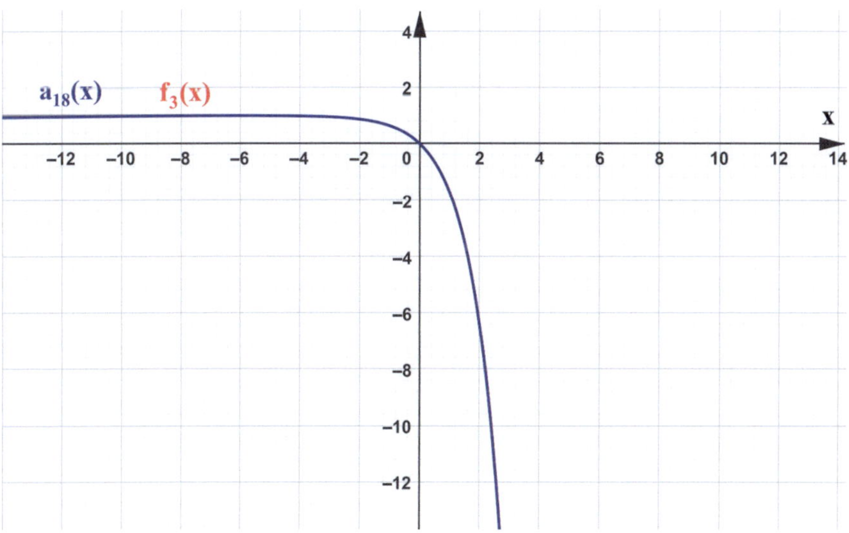

Fig. 3.54 Exponential ramp function $f_3(x)$ and the fourth approximation function $a_{18}(x)$

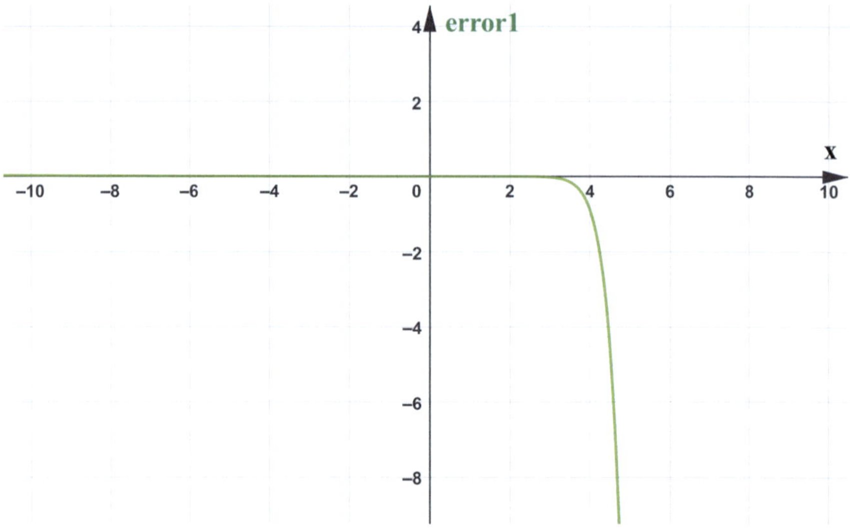

Fig. 3.55 Absolute error of approximation of the exponential ramp function $f_3(x)$ using the fourth approximation function $a_{18}(x)$

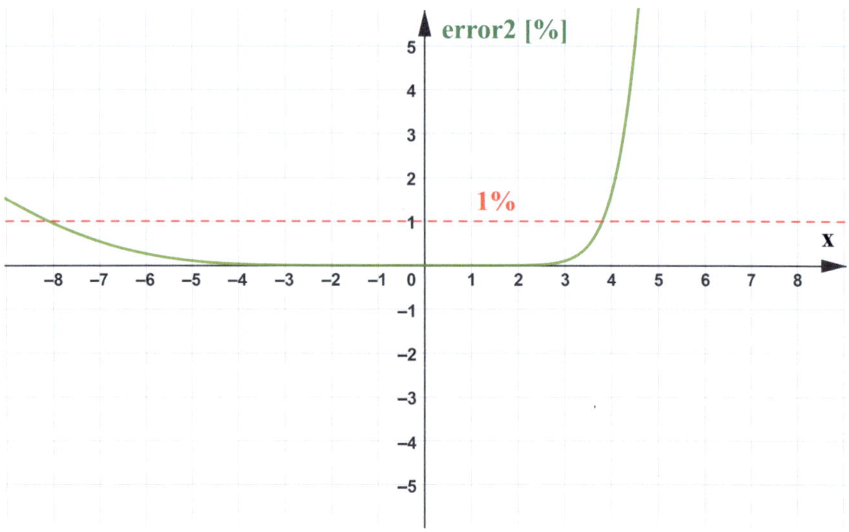

Fig. 3.56 Relative error of approximation of the exponential ramp function $f_3(x)$ using the fourth approximation function $a_{18}(x)$

3.3.6 First Approximation Function for Generating the Exponential Ramp Function (for a = 2)

Considering the case $a = 2$, the mathematical form (3.27) of the exponential ramp function becomes:

$$f_4(x) \overset{\text{not}}{=} f_2(x, 2) = 1 - \exp(2x). \tag{3.37}$$

In order to approximate the exponential ramp function $f_4(x)$, a new approximation function $a_{19}(x)$ is presented. The general form of the proposed first approximation function $a_{19}(x)$ is the following:

$$a_{19\text{GEN}}(x) = \frac{c_3x^3 + c_1x}{d_4x^4 + d_3x^3 + d_2x^2 + d_1x + d_0}, \tag{3.38}$$

c_1, c_3, d_0, d_1, d_2, d_3 and d_4 being constant coefficients having particular values that can be determined by the condition that the superior-order Taylor series expansion of $f_4(x)$ exponential ramp function, expressed by (3.37), to be identical with the superior-order Taylor series expansion of the general form of the approximation function $a_{19\text{GEN}}(x)$, expressed by (3.38).

The values of these coefficients, obtained considering the previous conditions, can be written as: $c_1 = -2$, $c_3 = -\frac{4}{21}$, $d_0 = 1$, $d_1 = -1$, $d_2 = \frac{3}{7}$, $d_3 = -\frac{2}{21}$ and $d_4 = \frac{1}{105}$. Replacing these particular values in relation (3.38), the first approximation function $a_{19}(x)$ has the following mathematical expression:

$$a_{19}(x) = \frac{-\frac{4x^3}{21} - 2x}{\frac{x^4}{105} - \frac{2x^3}{21} + \frac{3x^2}{7} - x + 1}. \tag{3.39}$$

The graphical representation of the exponential ramp function $f_4(x)$ and of the first approximation function $a_{19}(x)$ is presented in Fig. 3.57.

The absolute error of approximation of the exponential ramp function $f_4(x)$ using the first approximation function $a_{19}(x)$ is illustrated in Fig. 3.58. It is defined as the difference between the previous functions, error1 $= f_1(x) - a_{19}(x)$.

In order to more accurately evaluate the error of approximation, the relative error of approximation is shown in Fig. 3.59. This error is defined as error2 $= 100\frac{f_4(x) - a_{19}(x)}{f_1(x)}$ (in percents). Considering as acceptable a maximal approximation error of $\pm 1\%$, the maximal range of the input variable x is $[-4.1\,;\ 1.9]$.

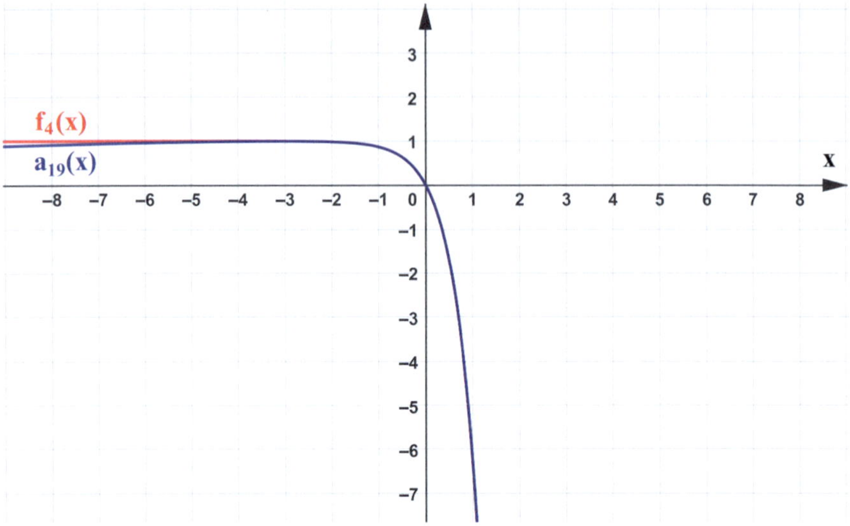

Fig. 3.57 Exponential ramp function $f_4(x)$ and the first approximation function $a_{19}(x)$

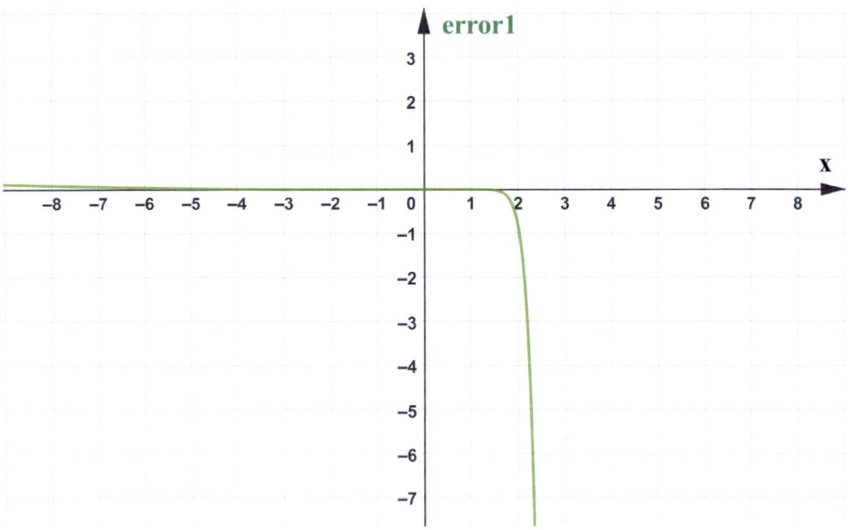

Fig. 3.58 Absolute error of approximation of the exponential ramp function $f_4(x)$ using the first approximation function $a_{19}(x)$

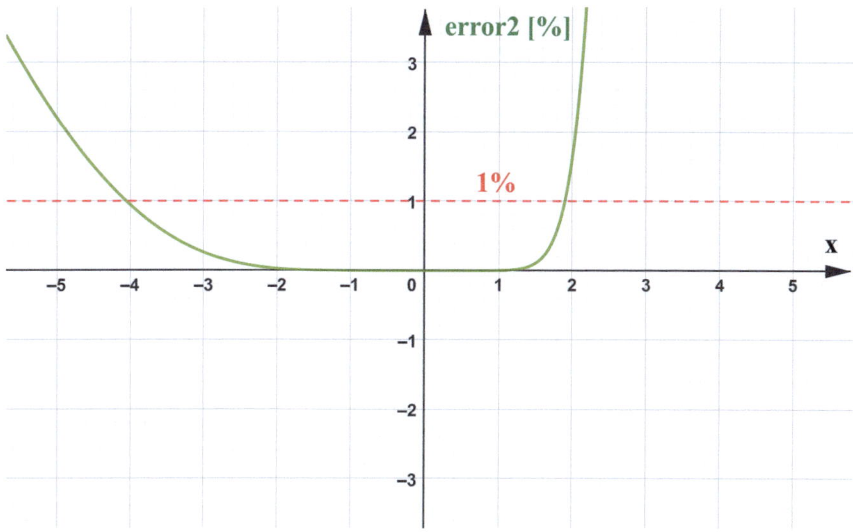

Fig. 3.59 Relative error of approximation of the exponential ramp function $f_4(x)$ using the first approximation function $a_{19}(x)$

3.3.7 *First Approximation Function for Generating the Exponential Ramp Function (for a = 1/2)*

Considering the case $a = \frac{1}{2}$, the mathematical form (3.27) of the exponential ramp function becomes:

$$f_5(x) \overset{\text{not}}{=} f_2(x, 1) = 1 - \exp\left(\frac{x}{2}\right). \tag{3.40}$$

In order to approximate the exponential ramp function $f_5(x)$, a new approximation function $a_{20}(x)$ is presented. The general form of the proposed first approximation function $a_{20}(x)$ is the following:

$$a_{20\text{GEN}}(x) = \frac{c_1 x}{d_2 x^2 + d_1 x + d_0}. \tag{3.41}$$

c_1, d_0, d_1 and d_2 being constant coefficients having particular values that can be determined by the condition that the superior-order Taylor series expansion of $f_5(x)$ exponential ramp function, expressed by (3.40), to be identical with the superior-order Taylor series expansion of the general form of the approximation function $a_{20\text{GEN}}(x)$, expressed by (3.41).

The values of these coefficients, obtained considering the previous conditions, can be written as: $c_1 = -1$, $d_0 = 2$, $d_1 = -\frac{1}{2}$ and $d_2 = \frac{1}{24}$. Replacing these particular values in relation (3.41), the first approximation function $a_{20}(x)$ has the following mathematical expression:

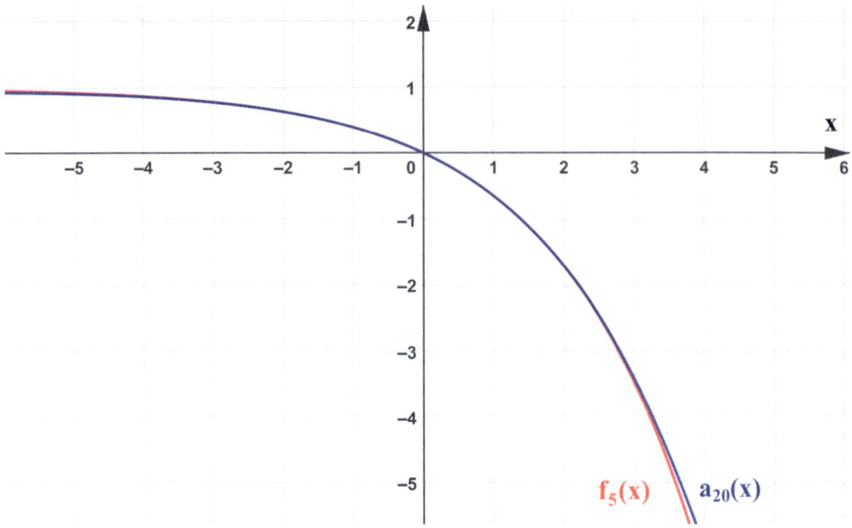

Fig. 3.60 Exponential ramp function $f_5(x)$ and the first approximation function $a_{20}(x)$

$$a_{20}(x) = \frac{-x}{\frac{x^2}{24} - \frac{x}{2} + 2}. \tag{3.42}$$

The graphical representation of the exponential ramp function $f_5(x)$ and of the first approximation function $a_{20}(x)$ is presented in Fig. 3.60.

The absolute error of approximation of the exponential ramp function $f_5(x)$ using the first approximation function $a_{20}(x)$ is illustrated in Fig. 3.61. It is defined as the difference between the previous functions, error1 $= f_5(x) - a_{20}(x)$.

In order to more accurately evaluate the error of approximation, the relative error of approximation is shown in Fig. 3.62. This error is defined as error2 $= 100\frac{f_5(x) - a_{20}(x)}{f_5(x)}$ (in percents). Considering as acceptable a maximal approximation error of $\pm 1\%$, the maximal range of the input variable x is $[-\infty\,;\ 2.2]$.

3.3.8 *Second Approximation Function for Generating the Exponential Ramp Function (for a = 1/2)*

In order to approximate the exponential ramp function $f_5(x)$, a new approximation function $a_{21}(x)$ is presented. The general form of the proposed second approximation function $a_{21}(x)$ is the following:

$$a_{21\mathrm{GEN}}(x) = \frac{c_3x^3 + c_2x^2 + c_1x}{d_2x^2 + d_1x + d_0}, \tag{3.43}$$

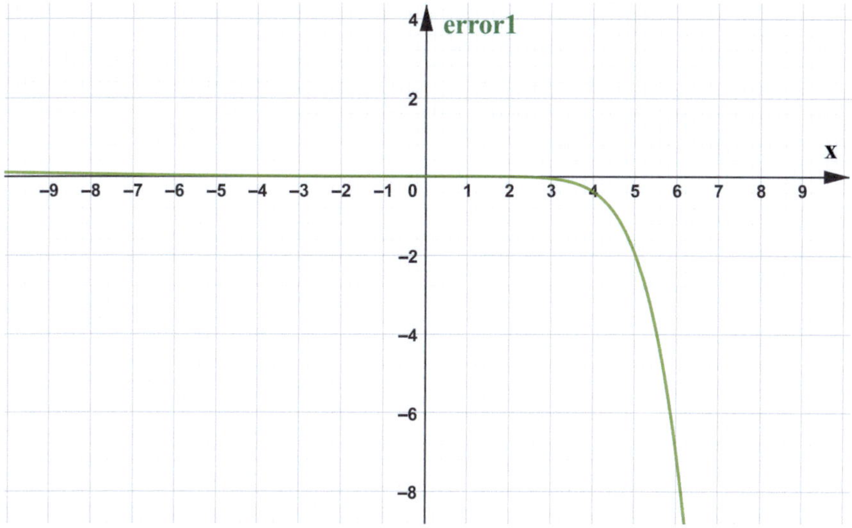

Fig. 3.61 Absolute error of approximation of the exponential ramp function $f_5(x)$ using the first approximation function $a_{20}(x)$

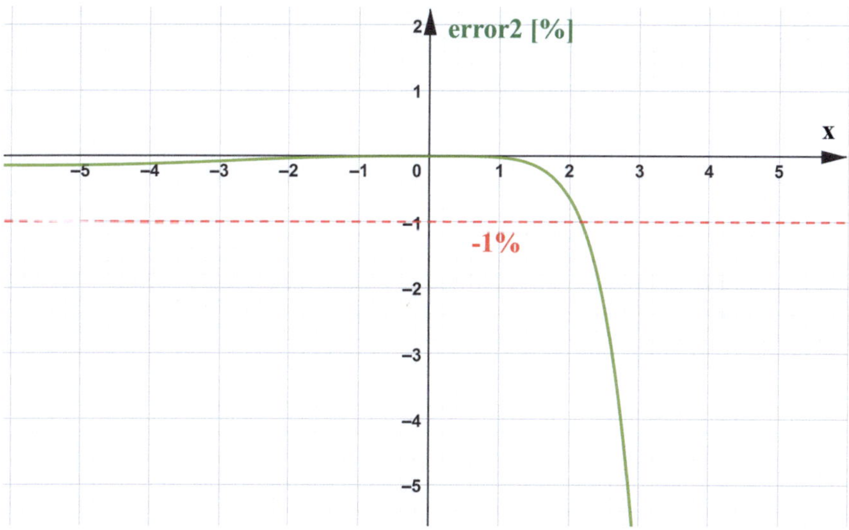

Fig. 3.62 Relative error of approximation of the exponential ramp function $f_5(x)$ using the first approximation function $a_{20}(x)$

c_1, c_2, c_3, d_0, d_1 and d_2 being constant coefficients having particular values that can be determined by the condition that the superior-order Taylor series expansion of $f_5(x)$ exponential ramp function, expressed by (3.40), to be identical with the superior-order Taylor series expansion of the general form of the approximation function $a_{21\text{GEN}}(x)$, expressed by (3.43).

The values of these coefficients, obtained considering the previous conditions, can be written as: $c_1 = -\frac{1}{2}$, $c_2 = -\frac{1}{40}$, $c_3 = -\frac{1}{480}$, $d_0 = 1$, $d_1 = -\frac{1}{5}$ and $d_2 = \frac{1}{80}$. Replacing these particular values in relation (3.43), the second approximation function $a_{21}(x)$ has the following mathematical expression:

$$a_{21}(x) = \frac{-\frac{x^3}{480} - \frac{x^2}{40} - \frac{x}{2}}{\frac{x^2}{80} - \frac{x}{5} + 1}. \tag{3.44}$$

The graphical representation of the exponential ramp function $f_5(x)$ and of the second approximation function $a_{21}(x)$ is presented in Fig. 3.63.

The absolute error of approximation of the exponential ramp function $f_5(x)$ using the second approximation function $a_{21}(x)$ is illustrated in Fig. 3.64. It is defined as the difference between the previous functions, error1 $= f_5(x) - a_{21}(x)$.

In order to more accurately evaluate the error of approximation, the relative error of approximation is shown in Fig. 3.65. This error is defined as error2 $= 100\frac{f_5(x) - a_{21}(x)}{f_5(x)}$ (in percents). Considering as acceptable a maximal approximation error of $\pm 1\%$, the maximal range of the input variable x is $[-5.7\,;\ 4.1]$.

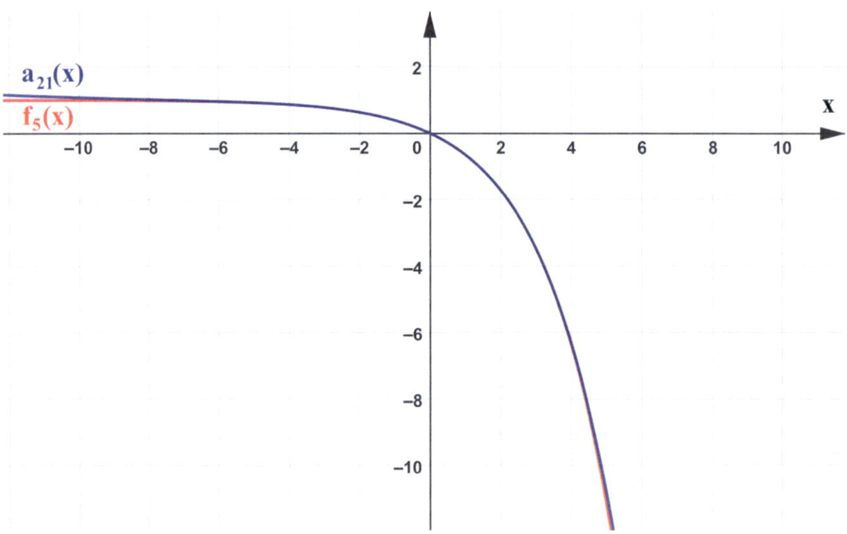

Fig. 3.63 Exponential ramp function $f_5(x)$ and the second approximation function $a_{21}(x)$

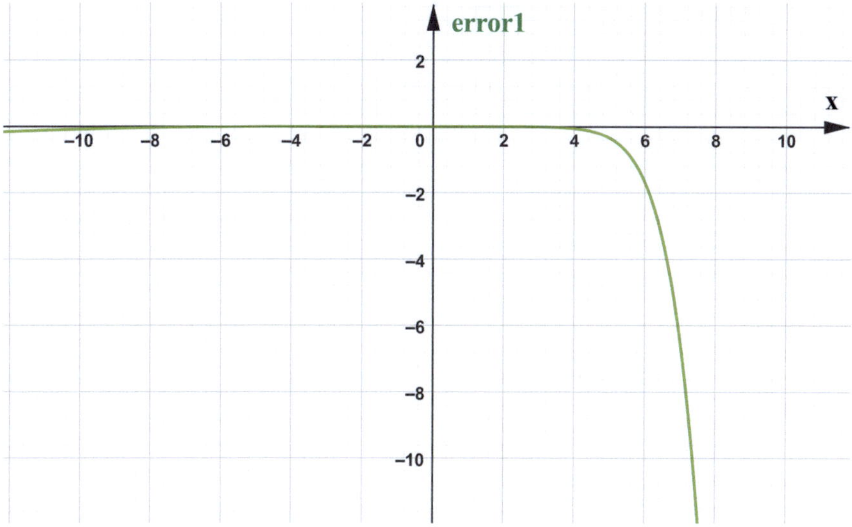

Fig. 3.64 Absolute error of approximation of the exponential ramp function $f_5(x)$ using the second approximation function $a_{21}(x)$

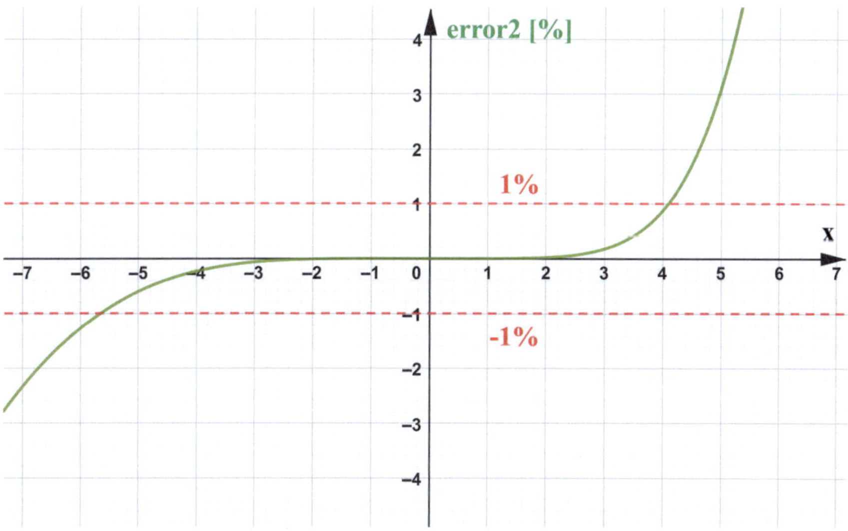

Fig. 3.65 Relative error of approximation of the exponential ramp function $f_5(x)$ using the second approximation function $a_{21}(x)$

3.3.9 Third Approximation Function for Generating the Exponential Ramp Function (for a = 1/2)

In order to approximate the exponential ramp function $f_5(x)$, a new approximation function $a_{22}(x)$ is presented. The general form of the proposed third approximation function $a_{22}(x)$ is the following:

$$a_{22\mathrm{GEN}}(x) = \frac{c_3x^3 + c_1x}{d_3x^3 + d_2x^2 + d_1x + d_0}, \tag{3.45}$$

c_1, c_3, d_0, d_1, d_2 and d_3 being constant coefficients having particular values that can be determined by the condition that the superior-order Taylor series expansion of $f_5(x)$ exponential ramp function, expressed by (3.40), to be identical with the superior-order Taylor series expansion of the general form of the approximation function $a_{22\mathrm{GEN}}(x)$, expressed by (3.45).

The values of these coefficients, obtained considering the previous conditions, can be written as: $c_1 = -\frac{1}{2}$, $c_3 = -\frac{1}{480}$, $d_0 = 1$, $d_1 = -\frac{1}{4}$, $d_2 = \frac{1}{40}$ and $d_3 = \frac{1}{960}$. Replacing these particular values in relation (3.45), the third approximation function $a_{22}(x)$ has the following mathematical expression:

$$a_{22}(x) = \frac{-\frac{x^3}{480} - \frac{x}{2}}{-\frac{x^3}{960} + \frac{x^2}{40} - \frac{x}{4} + 1}. \tag{3.46}$$

The graphical representation of the exponential ramp function $f_5(x)$ and of the third approximation function $a_{22}(x)$ is presented in Fig. 3.66.

The absolute error of approximation of the exponential ramp function $f_5(x)$ using the third approximation function $a_{22}(x)$ is illustrated in Fig. 3.67. It is defined as the difference between the previous functions, error1 $= f_5(x) - a_{22}(x)$.

In order to more accurately evaluate the error of approximation, the relative error of approximation is shown in Fig. 3.68. This error is defined as error2 $= 100\frac{f_5(x) - a_{22}(x)}{f_5(x)}$ (in percents). Considering as acceptable a maximal approximation error of $\pm 1\%$, the maximal range of the input variable x is $[-9.4\,;\ 5.2]$.

3.4 Approximation Functions for Hyperbolic Sinusoidal Function

The hyperbolic sinusoidal activation function is not commonly used in mainstream deep learning, but it does appear in specialized contexts due to its smooth, unbounded and symmetric properties. The areas of utilization of the hyperbolic sinusoidal functions are:

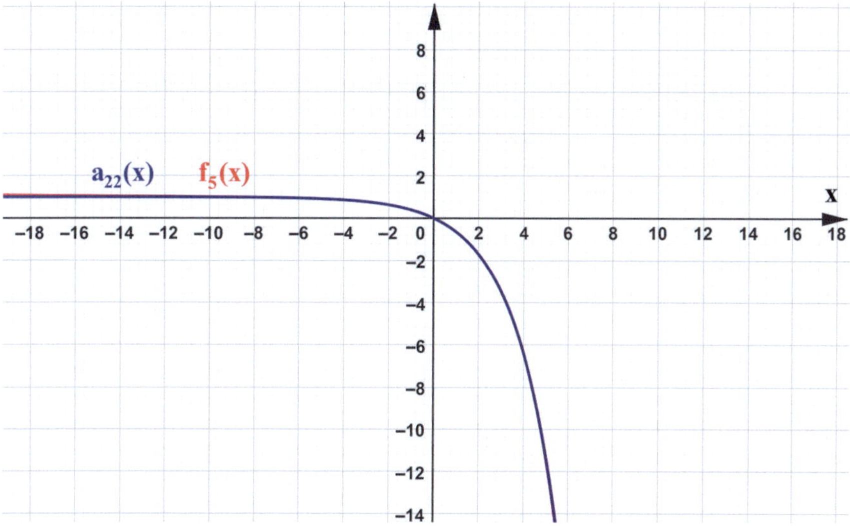

Fig. 3.66 Exponential ramp function $f_5(x)$ and the third approximation function $a_{22}(x)$

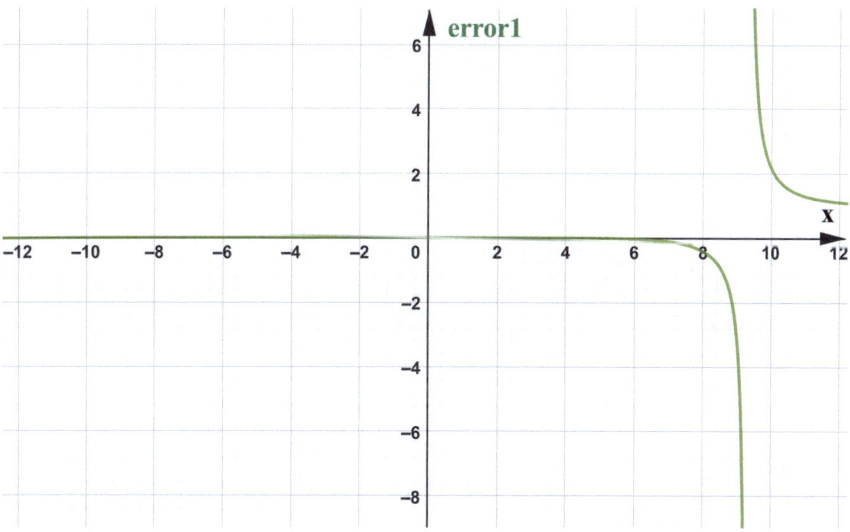

Fig. 3.67 Absolute error of approximation of the exponential ramp function $f_5(x)$ using the third approximation function $a_{22}(x)$

a. ***Mathematical and Physics-Inspired Neural Models***

- Used in neural networks that model physical systems (for example, quantum systems, wave propagation, heat transfer), where hyperbolic functions appear naturally in differential equations.

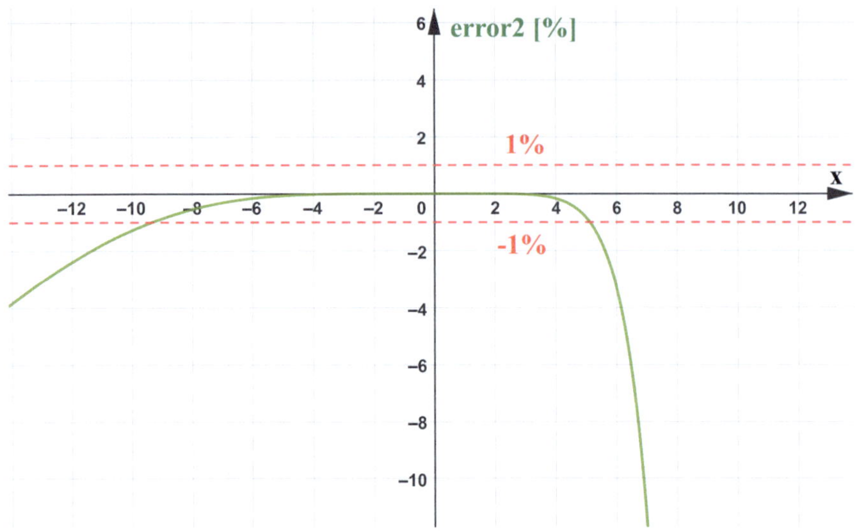

Fig. 3.68 Relative error of approximation of the exponential ramp function $f_5(x)$ using the third approximation function $a_{22}(x)$

- Particularly useful in physics-informed neural networks (PINNs) that solve partial differential equations (PDEs).

b. ***Custom Activation Design***

- Applied in experimental activations, where unbounded, antisymmetric functions are needed.
- Offers strong gradients for large inputs (due to exponential growth), which can help or hurt depending on context.

c. ***Symbolic and Analytical Neural Models***

- In models aimed at learning symbolic representations or solving symbolic regression, hyperbolic sinusoidal activation function may be part of the function set used in expression trees (for example, genetic programming, neural-symbolic solvers).

d. ***Bio-inspired Systems***

- Occasionally used in bio-inspired neural circuits to simulate certain types of excitation/inhibition curves.

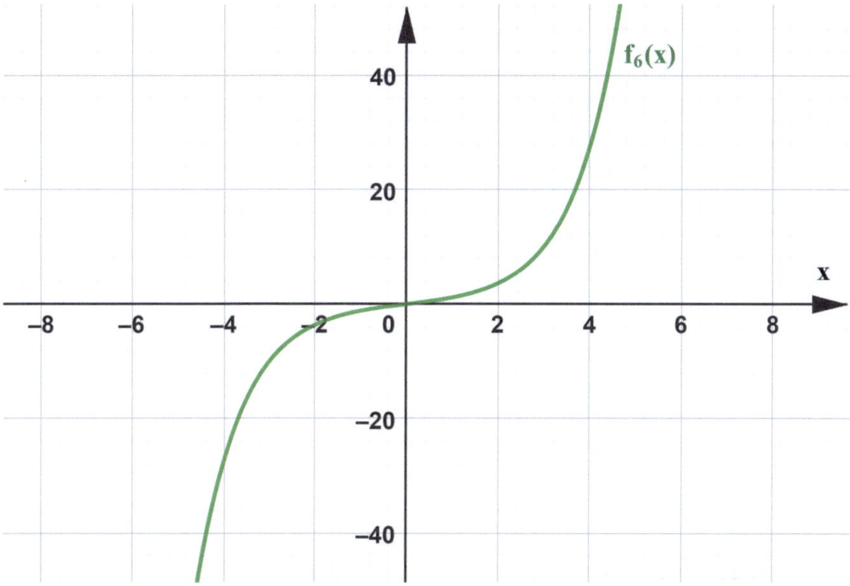

Fig. 3.69 Hyperbolic sinusoidal function $f_6(x)$

3.4.1 General Form of the Hyperbolic Sinusoidal Function

The mathematical form of this function can be expressed as follows:

$$f_6(x) = \frac{\exp(x) - \exp(-x)}{2}. \tag{3.47}$$

The graphical representation of the hyperbolic sinusoidal function $f_6(x)$ is presented in Fig. 3.69.

In order to obtain an analog implementation of the hyperbolic sinusoidal function, a multitude of superior-order approximation functions is presented. The most important differences between these functions are mainly related to the accuracy of their approximation, as well as to the hardware resources required for their implementations in CMOS technology.

3.4.2 First Approximation Function for Generating the Hyperbolic Sinusoidal Function

In order to approximate the hyperbolic sinusoidal function $f_6(x)$, a new approximation function $a_{23}(x)$ is presented. The general form of the proposed first approximation function $a_{23}(x)$ is the following:

$$a_{23GEN}(x) = \frac{c_1 x + c_3 x^3}{d_2 x^2 + d_0}, \tag{3.48}$$

c_1, c_3, d_0 and d_2 being constant coefficients having particular values that can be determined by the condition that the superior-order Taylor series expansion of $f_6(x)$ hyperbolic sinusoidal function, expressed by (3.47), to be identical with the superior-order Taylor series expansion of the general form of the approximation function $a_{23GEN}(x)$, expressed by (3.48).

The values of these coefficients, obtained considering the previous conditions, can be written as: $c_1 = 1$, $c_3 = \frac{7}{60}$, $d_0 = 1$ and $d_2 = -\frac{1}{20}$. Replacing these particular values in relation (3.48), the first approximation function $a_{23}(x)$ has the following mathematical expression:

$$a_{23}(x) = \frac{\frac{7x^3}{60} + x}{1 - \frac{x^2}{20}}. \tag{3.49}$$

The graphical representation of the hyperbolic sinusoidal function $f_6(x)$ and of the first approximation function $a_{23}(x)$ is presented in Fig. 3.70.

The absolute error of approximation of the hyperbolic sinusoidal function $f_6(x)$ using the first approximation function $a_{23}(x)$ is illustrated in Fig. 3.71. It is defined as the difference between the previous functions, error1 $= f_6(x) - a_{23}(x)$.

In order to more accurately evaluate the error of approximation, the relative error of approximation is shown in Fig. 3.72. This error is defined as error2 $= 100\frac{f_6(x) - a_{23}(x)}{f_6(x)}$ (in percents). Considering as acceptable a maximal approximation error of $\pm 1\%$, the maximal range of the input variable x is $[-1.95\,;\ 1.95]$.

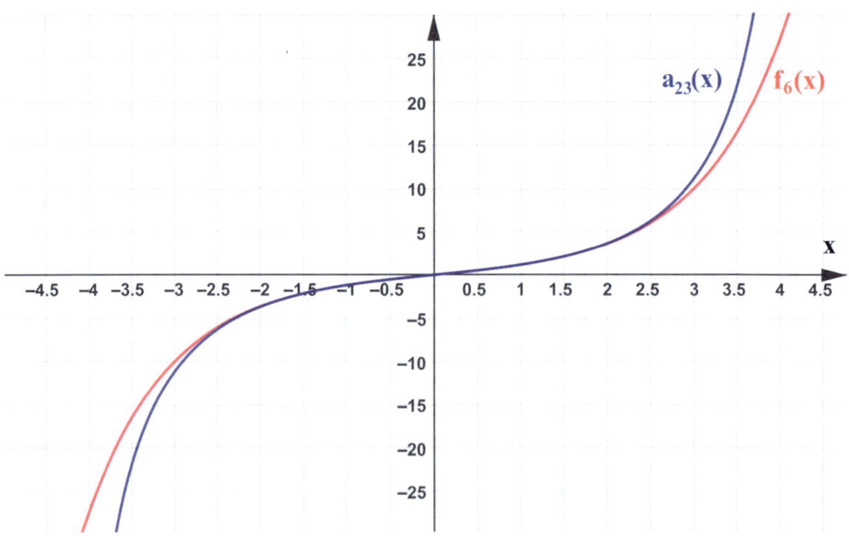

Fig. 3.70 Hyperbolic sinusoidal function $f_6(x)$ and the first approximation function $a_{23}(x)$

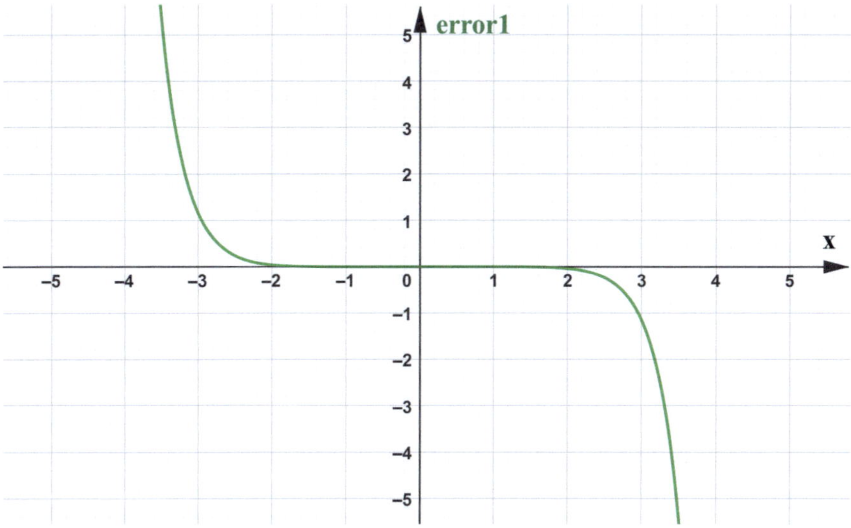

Fig. 3.71 Absolute error of approximation of the hyperbolic sinusoidal function $f_6(x)$ using the first approximation function $a_{23}(x)$

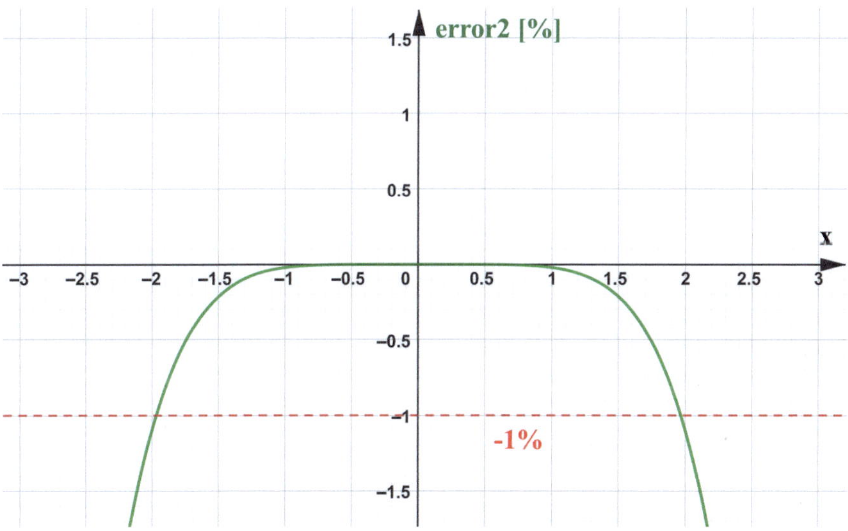

Fig. 3.72 Relative error of approximation of the hyperbolic sinusoidal function $f_6(x)$ using the first approximation function $a_{23}(x)$

3.4.3 Second Approximation Function for Generating the Hyperbolic Sinusoidal Function

In order to approximate the hyperbolic sinusoidal function $f_6(x)$, a new approximation function $a_{24}(x)$ is presented. The general form of the proposed second approximation function $a_{24}(x)$ is the following:

$$a_{24\mathrm{GEN}}(x) = \frac{c_1 x + c_3 x^3}{d_4 x^4 + d_2 x^2 + d_0}, \tag{3.50}$$

c_1, c_3, d_0, d_2 and d_2 being constant coefficients having particular values that can be determined by the condition that the superior-order Taylor series expansion of $f_6(x)$ hyperbolic sinusoidal function, expressed by (3.47), to be identical with the superior-order Taylor series expansion of the general form of the approximation function $a_{24\mathrm{GEN}}(x)$, expressed by (3.50).

The values of these coefficients, obtained considering the previous conditions, can be written as: $c_1 = 1$, $c_3 = \frac{31}{294}$, $d_0 = 1$, $d_2 = -\frac{3}{49}$ and $d_4 = \frac{11}{5880}$. Replacing these particular values in relation (3.50), the second approximation function $a_{24}(x)$ has the following mathematical expression:

$$a_{24}(x) = \frac{\frac{31x^3}{294} + x}{1 - \frac{3x^2}{49} + \frac{11x^4}{5880}}. \tag{3.51}$$

The graphical representation of the hyperbolic sinusoidal function $f_6(x)$ and of the second approximation function $a_{24}(x)$ is presented in Fig. 3.73.

The absolute error of approximation of the hyperbolic sinusoidal function $f_6(x)$ using the second approximation function $a_{24}(x)$ is illustrated in Fig. 3.74. It is defined as the difference between the previous functions, error1 $= f_6(x) - a_{24}(x)$.

In order to more accurately evaluate the error of approximation, the relative error of approximation is shown in Fig. 3.75. This error is defined as error2 $= 100\frac{f_6(x) - a_{24}(x)}{f_6(x)}$ (in percents). Considering as acceptable a maximal approximation error of $\pm 1\%$, the maximal range of the input variable x is $[-2.6\,;\ 2.6]$.

3.4.4 Third Approximation Function for Generating the Hyperbolic Sinusoidal Function

In order to approximate the hyperbolic sinusoidal function $f_6(x)$, a new approximation function $a_{25}(x)$ is presented. The general form of the proposed third approximation function $a_{25}(x)$ is the following:

$$a_{25\mathrm{GEN}}(x) = x\left(\frac{c_0}{d_0 + d_2 x^2} + e_0 + e_2 x^2\right), \tag{3.52}$$

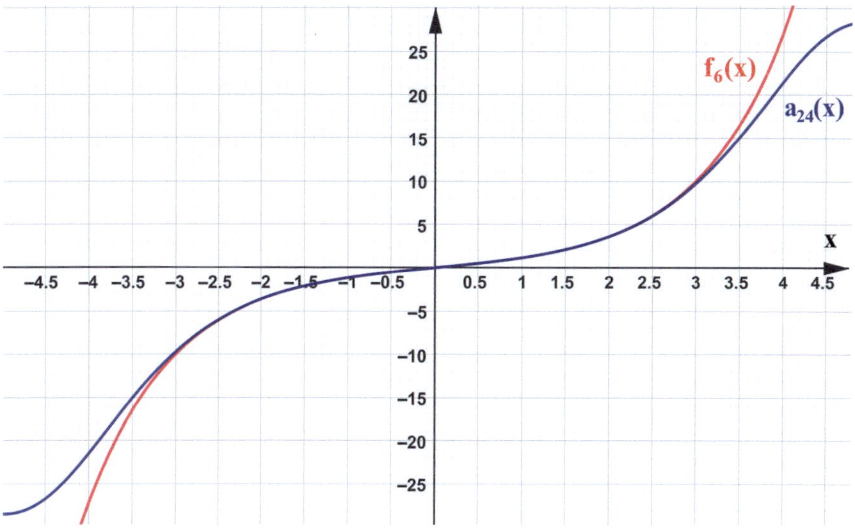

Fig. 3.73 hyperbolic sinusoidal function $f_6(x)$ and the second approximation function $a_{24}(x)$

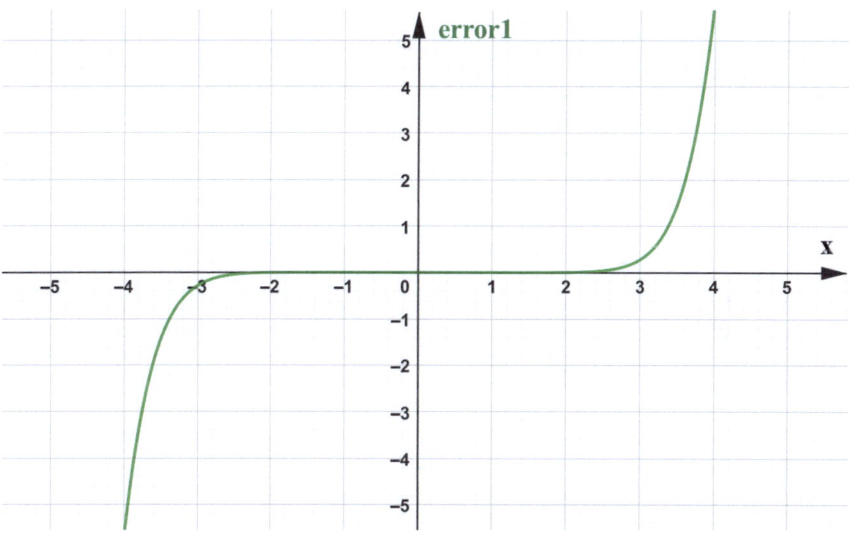

Fig. 3.74 Absolute error of approximation of the hyperbolic sinusoidal function $f_6(x)$ using the second approximation function $a_{24}(x)$

c_1, d_0, d_2, e_0 and $e2$ being constant coefficients having particular values that can be determined by the condition that the superior-order Taylor series expansion of $f_6(x)$ hyperbolic sinusoidal function, expressed by (3.47), to be identical with the

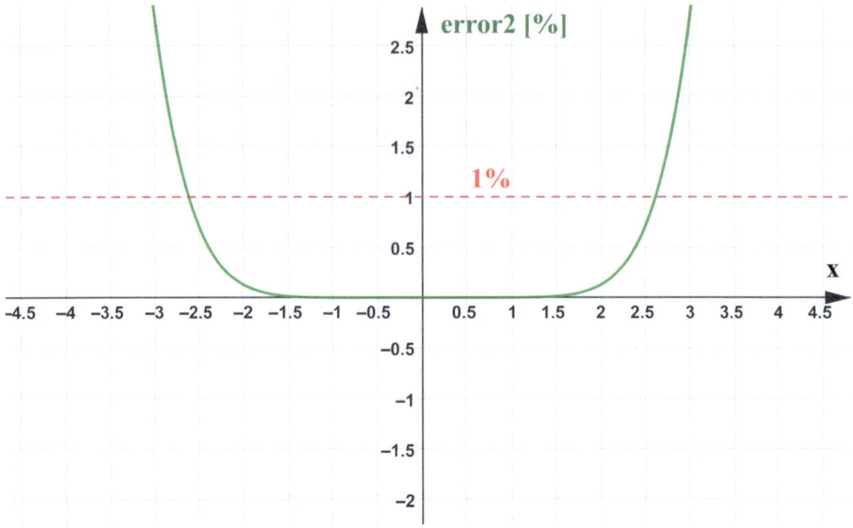

Fig. 3.75 Relative error of approximation of the hyperbolic sinusoidal function $f_6(x)$ using the second approximation function $a_{24}(x)$

superior-order Taylor series expansion of the general form of the approximation function $a_{25\text{GEN}}(x)$, expressed by (3.52).

The values of these coefficients, obtained considering the previous conditions, can be written as: $c_0 = \frac{147}{10}$, $d_0 = 1$, $d_2 = -\frac{1}{42}$, $e_0 = -\frac{137}{10}$ and $e_2 = -\frac{11}{60}$. Replacing these particular values in relation (3.52), the third approximation function $a_{25}(x)$ has the following mathematical expression:

$$a_{25}(x) = \frac{x}{10}\left(\frac{6174}{42 - x^2} - \frac{822 + 11x^2}{6}\right). \tag{3.53}$$

The graphical representation of the hyperbolic sinusoidal function $f_6(x)$ and of the third approximation function $a_{25}(x)$ is presented in Fig. 3.76.

The absolute error of approximation of the hyperbolic sinusoidal function $f_6(x)$ using the third approximation function $a_{25}(x)$ is illustrated in Fig. 3.77. It is defined as the difference between the previous functions, error1 $= f_6(x) - a_{25}(x)$.

In order to more accurately evaluate the error of approximation, the relative error of approximation is shown in Fig. 3.78. This error is defined as error2 $= 100\frac{f_6(x) - a_{25}(x)}{f_6(x)}$ (in percents). Considering as acceptable a maximal approximation error of $\pm 1\%$, the maximal range of the input variable x is $[-3.25\,;\;3.25]$.

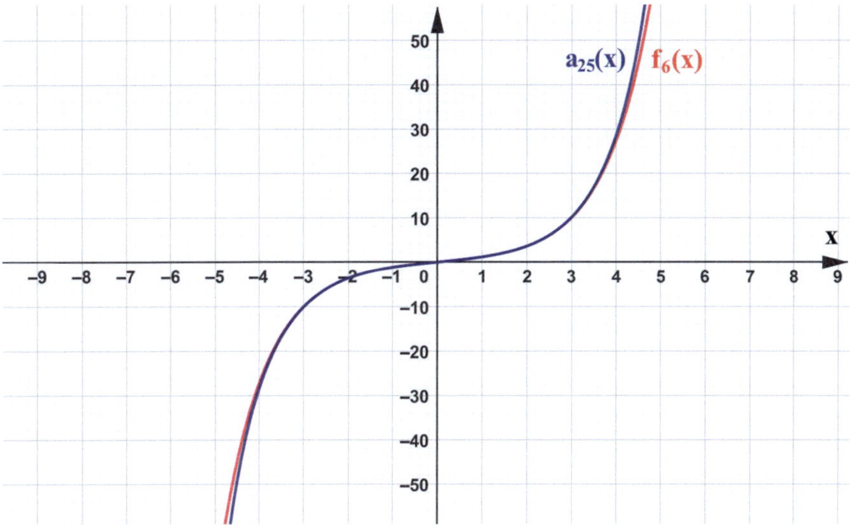

Fig. 3.76 Hyperbolic sinusoidal function $f_6(x)$ and the third approximation function $a_{25}(x)$

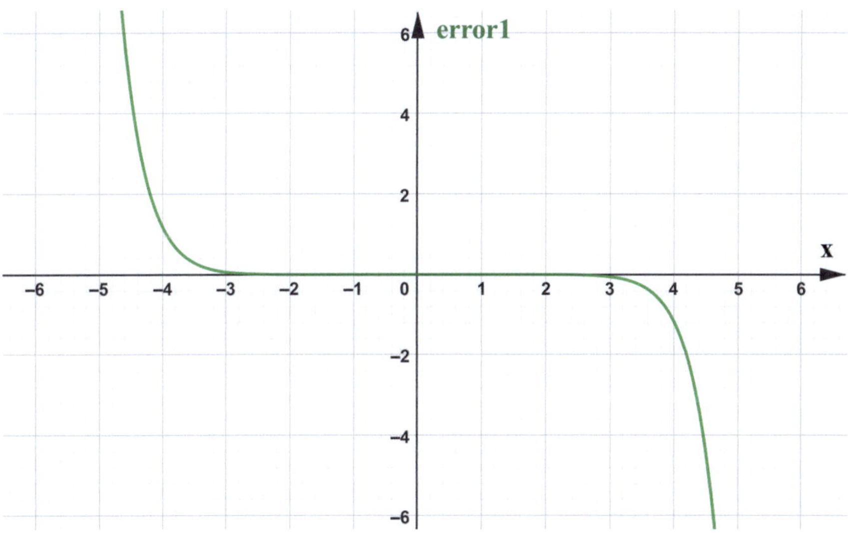

Fig. 3.77 Absolute error of approximation of the hyperbolic sinusoidal function $f_6(x)$ using the third approximation function $a_{25}(x)$

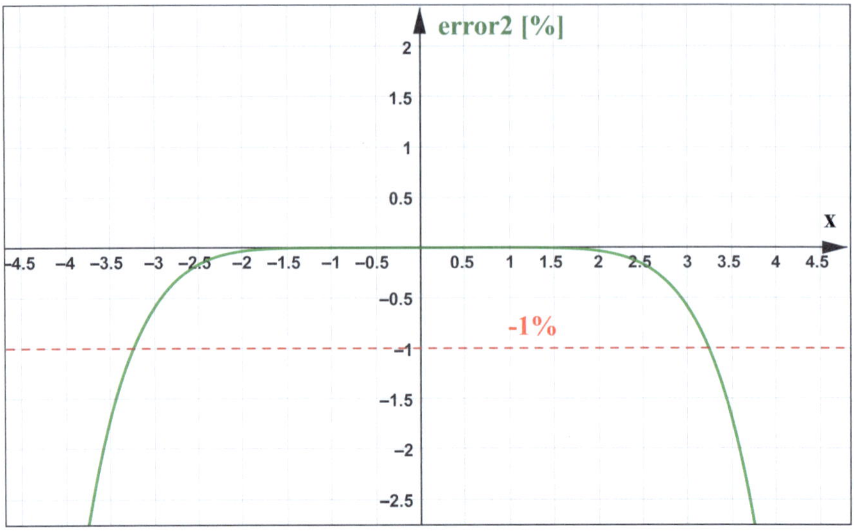

Fig. 3.78 Relative error of approximation of the hyperbolic sinusoidal function $f_6(x)$ using the third approximation function $a_{25}(x)$

3.4.5 *Fourth Approximation Function for Generating the Hyperbolic Sinusoidal Function*

In order to approximate the hyperbolic sinusoidal function $f_6(x)$, a new approximation function $a_{26}(x)$ is presented. The general form of the proposed fourth approximation function $a_{26}(x)$ is the following:

$$a_{26\text{GEN}}(x) = x\left(\frac{c_0}{d_0 + d_2x^2} + e_0\right), \tag{3.54}$$

c_0, d_0, d_2 and e_0 being constant coefficients having particular values that can be determined by the condition that the superior-order Taylor series expansion of $f_6(x)$ hyperbolic sinusoidal function, expressed by (3.47), to be identical with the superior-order Taylor series expansion of the general form of the approximation function $a_{26\text{GEN}}(x)$, expressed by (3.54).

The values of these coefficients, obtained considering the previous conditions, can be written as: $c_0 = \frac{10}{3}$, $d_0 = 1$, $d_2 = -\frac{1}{20}$ and $e_0 = -\frac{7}{3}$. Replacing these particular values in relation (3.54), the fourth approximation function $a_{26}(x)$ has the following mathematical expression:

$$a_{26}(x) = x\left(\frac{10}{3}\frac{1}{1 - \frac{x^2}{20}} - \frac{7}{3}\right). \tag{3.55}$$

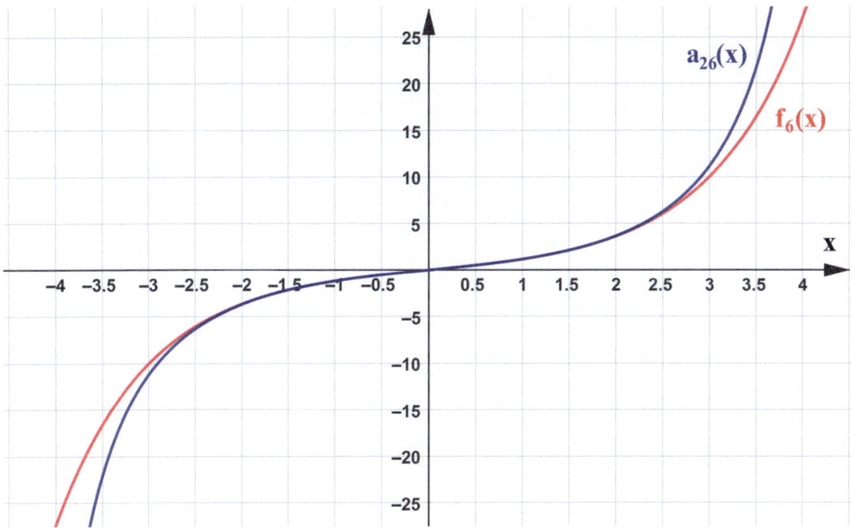

Fig. 3.79 Hyperbolic sinusoidal function $f_6(x)$ and the fourth approximation function $a_{26}(x)$

The graphical representation of the hyperbolic sinusoidal function $f_6(x)$ and of the fourth approximation function $a_{26}(x)$ is presented in Fig. 3.79.

The absolute error of approximation of the hyperbolic sinusoidal function $f_6(x)$ using the fourth approximation function $a_{26}(x)$ is illustrated in Fig. 3.80. It is defined as the difference between the previous functions, error1 $= f_6(x) - a_{26}(x)$.

In order to more accurately evaluate the error of approximation, the relative error of approximation is shown in Fig. 3.81. This error is defined as error2 $= 100\frac{f_6(x)-a_{26}(x)}{f_6(x)}$ (in percents). Considering as acceptable a maximal approximation error of $\perp 1\%$, the maximal range of the input variable x is $[-1.95\,;\ 1.95]$.

3.5 Approximation Functions for Hyperbolic Cosinusoidal Function

The hyperbolic cosinusoidal activation function is a smooth, symmetric and non-negative function. Although it is not widely used in standard neural network architectures, it has applications in specialized or theoretical domains due to its mathematical properties. The areas of utilization of the hyperbolic cosinusoidal activation function are:

a. ***Physics-Informed and Engineering Neural Networks***

- Common in physics-informed neural networks (PINNs), where hyperbolic functions naturally appear in solutions to partial differential equations (PDEs), especially in heat transfer, wave propagation and quantum mechanics.

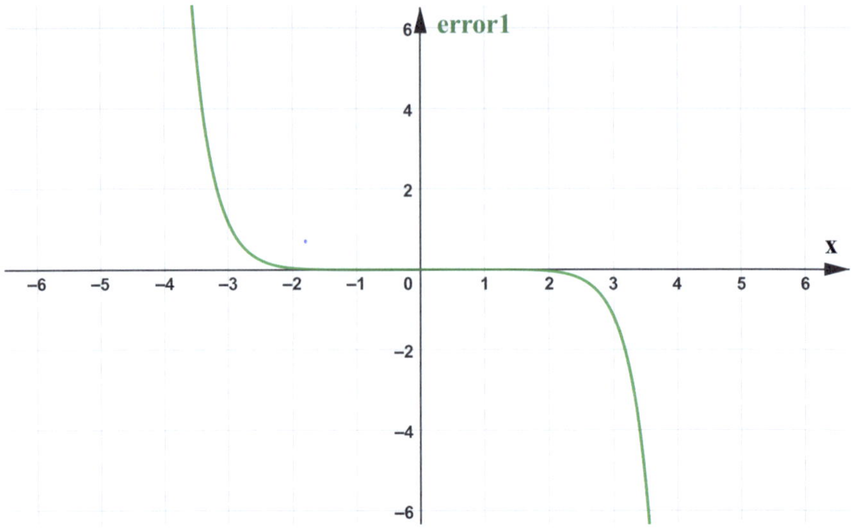

Fig. 3.80 Absolute error of approximation of the hyperbolic sinusoidal function $f_6(x)$ using the fourth approximation function $a_{26}(x)$

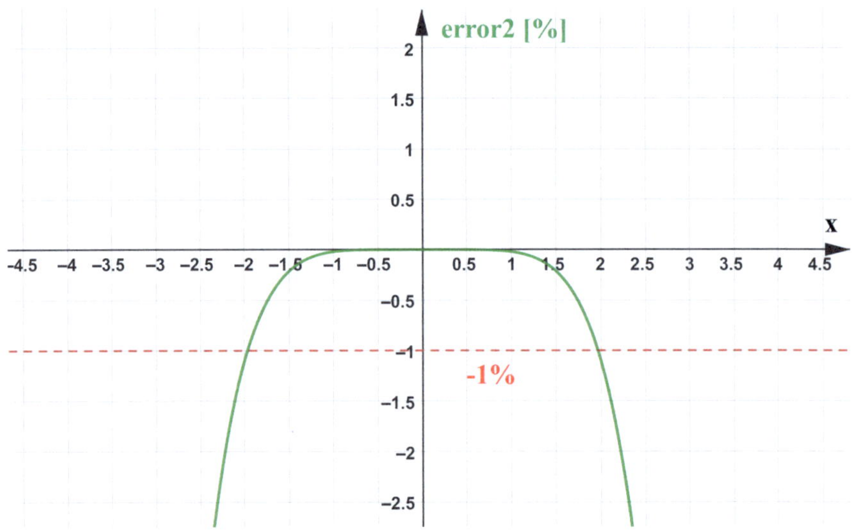

Fig. 3.81 Relative error of approximation of the hyperbolic sinusoidal function $f_6(x)$ using the fourth approximation function $a_{26}(x)$

- In such networks, hyperbolic cosinusoidal activation function may be used to embed known physical behavior into the model.

b. ***Symbolic Regression/Genetic Programming***

- The hyperbolic cosinusoidal activation function is often included in the primitive function set in symbolic regression tasks, where the goal is to discover mathematical expressions that describe data.
- Used in neural-symbolic systems and function composition models.

c. ***Custom and Experimental Activation Functions***

- Sometimes used in experimental architectures where a non-negative, smooth and strongly convex function is desirable.
- Can serve as a component in compound activation functions.

d. ***Robust Loss Function Derivatives***

- log(cosh(x)) is used as a robust alternative to mean square error or mean absolute error, especially in regression tasks with outliers.

3.5.1 General Form of the Hyperbolic Cosinusoidal Function

The mathematical form of this function can be expressed as follows:

$$f_7(x) = \frac{\exp(x) + \exp(-x)}{2}. \tag{3.56}$$

The graphical representation of the hyperbolic cosinusoidal function $f_7(x)$ is presented in Fig. 3.82.

In order to obtain an analog implementation of the hyperbolic cosinusoidal function, a many superior-order approximation functions is presented. The most important differences between these functions are mainly related to the accuracy of their approximation, as well as to the hardware resources required for their implementations in CMOS technology.

3.5.2 First Approximation Function for Generating the Hyperbolic Cosinusoidal Function

In order to approximate the hyperbolic cosinusoidal function $f_7(x)$, a new approximation function $a_{27}(x)$ is presented. The general form of the proposed first approximation function $a_{27}(x)$ is the following:

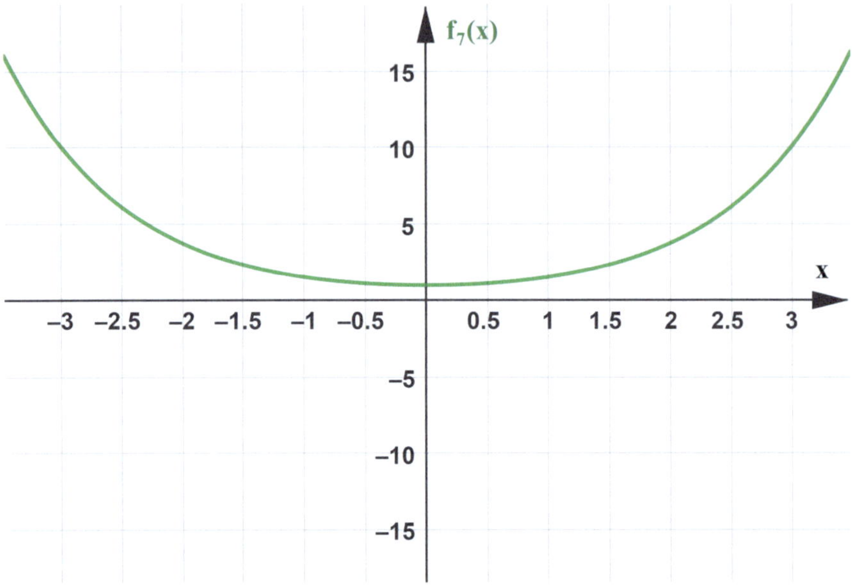

Fig. 3.82 Hyperbolic cosinusoidal function $f_7(x)$

$$a_{27\mathrm{GEN}}(x) = \frac{c_0 + c_2x^2}{d_4x^4 + d_2x^2 + d_0}, \tag{3.57}$$

c_0, c_3, d_0, d_2 and d_4 being constant coefficients having particular values that can be determined by the condition that the superior-order Taylor series expansion of $f_7(x)$ hyperbolic cosinusoidal function, expressed by (3.56), to be identical with the superior-order Taylor series expansion of the general form of the approximation function $a_{27\mathrm{GEN}}(x)$, expressed by (3.57).

The values of these coefficients, obtained considering the previous conditions, can be written as: $c_0 = 1$, $c_2 = \frac{61}{150}$, $d_0 = 1$, $d_2 = -\frac{7}{75}$ and $d_4 = \frac{1}{200}$. Replacing these particular values in relation (3.57), the first approximation function $a_{27}(x)$ has the following mathematical expression:

$$a_{27}(x) = \frac{\frac{61x^2}{150} + 1}{1 - \frac{7x^2}{75} + \frac{x^4}{200}}. \tag{3.58}$$

The graphical representation of the hyperbolic cosinusoidal function $f_7(x)$ and of the first approximation function $a_{27}(x)$ is presented in Fig. 3.83.

The absolute error of approximation of the hyperbolic cosinusoidal function $f_7(x)$ using the first approximation function $a_{27}(x)$ is illustrated in Fig. 3.84. It is defined as the difference between the previous functions, error1 $= f_7(x) - a_{27}(x)$.

In order to more accurately evaluate the error of approximation, the relative error of approximation is shown in Fig. 3.85. This error is defined as error2 $= 100\frac{f_7(x)-a_{27}(x)}{f_7(x)}$

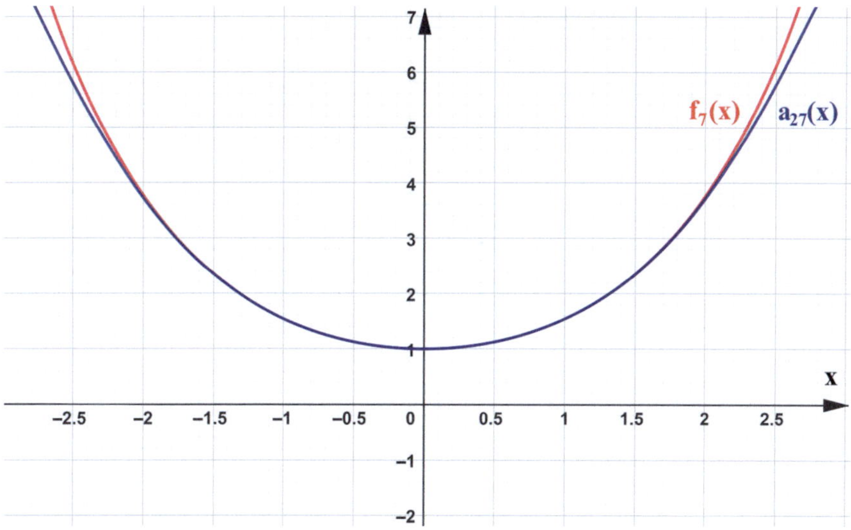

Fig. 3.83 Hyperbolic cosinusoidal function $f_7(x)$ and the first approximation function $a_{27}(x)$

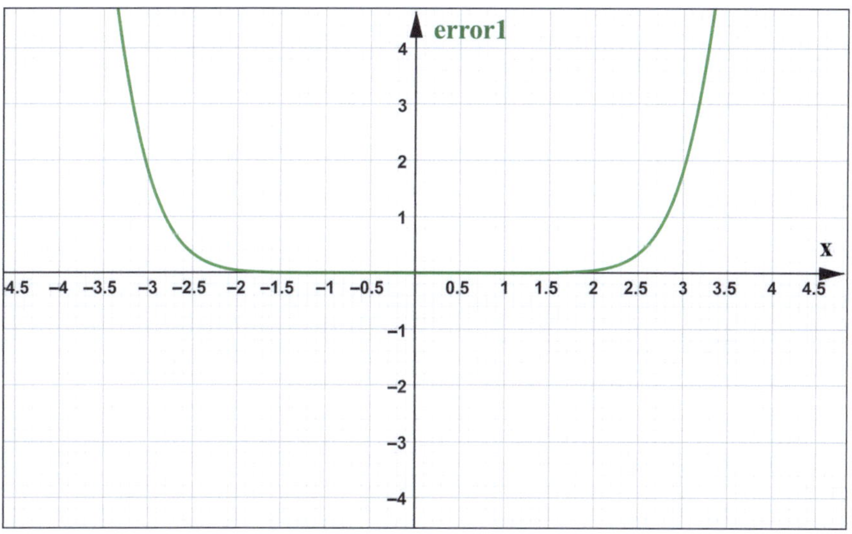

Fig. 3.84 Absolute error of approximation of the hyperbolic cosinusoidal function $f_7(x)$ using the first approximation function $a_{27}(x)$

(in percents). Considering as acceptable a maximal approximation error of $\pm 1\%$, the maximal range of the input variable x is $[-1.95\,;\ 1.95]$.

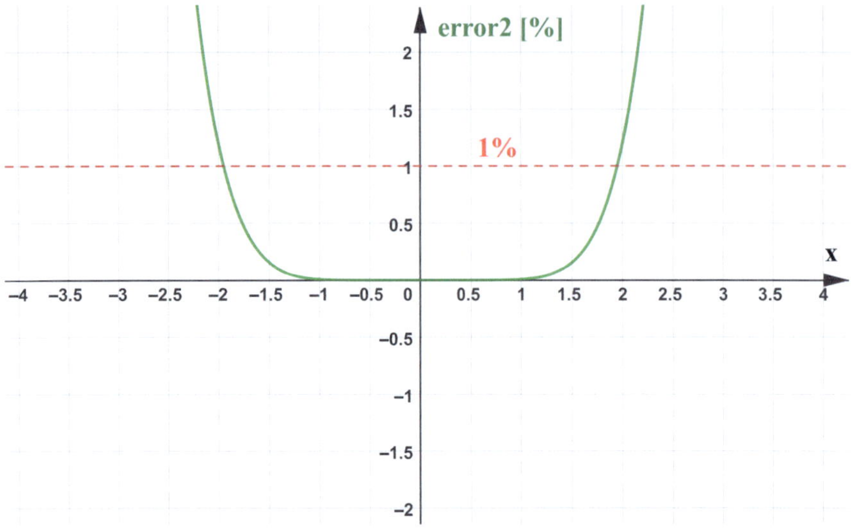

Fig. 3.85 Relative error of approximation of the hyperbolic cosinusoidal function $f_7(x)$ using the first approximation function $a_{27}(x)$

3.5.3 *Second Approximation Function for Generating the Hyperbolic Cosinusoidal Function*

In order to approximate the hyperbolic cosinusoidal function $f_7(x)$, a new approximation function $a_{28}(x)$ is presented. The general form of the proposed second approximation function $a_{28}(x)$ is the following:

$$a_{28\mathrm{GEN}}(x) = \frac{c_4x^4 + c_2x^2 + c_0}{d_2x^2 + d_0}, \tag{3.59}$$

c_1, c_3, c_4, d_0 and d_2 being constant coefficients having particular values that can be determined by the condition that the superior-order Taylor series expansion of $f_7(x)$ hyperbolic cosinusoidal function, expressed by (3.56), to be identical with the superior-order Taylor series expansion of the general form of the approximation function $a_{28\mathrm{GEN}}(x)$, expressed by (3.59).

The values of these coefficients, obtained considering the previous conditions, can be written as: $c_0 = 1$, $c_2 = \frac{7}{15}$, $c_4 = \frac{1}{40}$, $d_0 = 1$ and $d_2 = -\frac{1}{30}$. Replacing these particular values in relation (3.59), the second approximation function $a_{28}(x)$ has the following mathematical expression:

$$a_{28}(x) = \frac{\frac{x^4}{40} + \frac{7x^2}{15} + 1}{1 - \frac{x^2}{30}}. \tag{3.60}$$

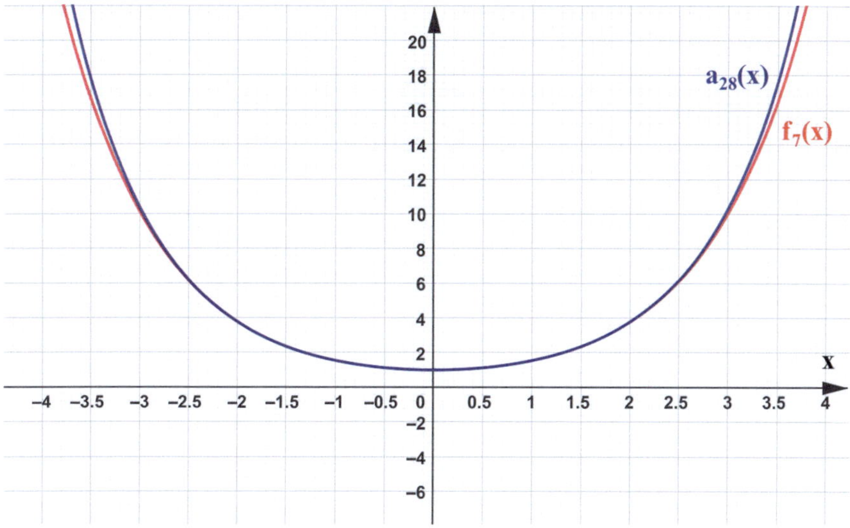

Fig. 3.86 Hyperbolic cosinusoidal function $f_7(x)$ and the second approximation function $a_{28}(x)$

The graphical representation of the hyperbolic cosinusoidal function $f_7(x)$ and of the second approximation function $a_{28}(x)$ is presented in Fig. 3.86.

The absolute error of approximation of the hyperbolic cosinusoidal function $f_7(x)$ using the second approximation function $a_{28}(x)$ is illustrated in Fig. 3.87. It is defined as the difference between the previous functions, error1 $= f_7(x) - a_{28}(x)$.

In order to more accurately evaluate the error of approximation, the relative error of approximation is shown in Fig. 3.88. This error is defined as error2 $= 100\frac{f_7(x)-a_{28}(x)}{f_7(x)}$ (in percents). Considering as acceptable a maximal approximation error of $\pm 1\%$, the maximal range of the input variable x is $[-2.6\,;\ 2.6]$.

3.6 Approximation Functions for General Analog Function Synthesizers

In order to obtain an analog implementation of any continuous mathematical function, three superior-order approximation functions are presented. The most important differences between these functions are mainly related to the accuracy of their approximation, as well as to the hardware resources required for their implementations in CMOS technology.

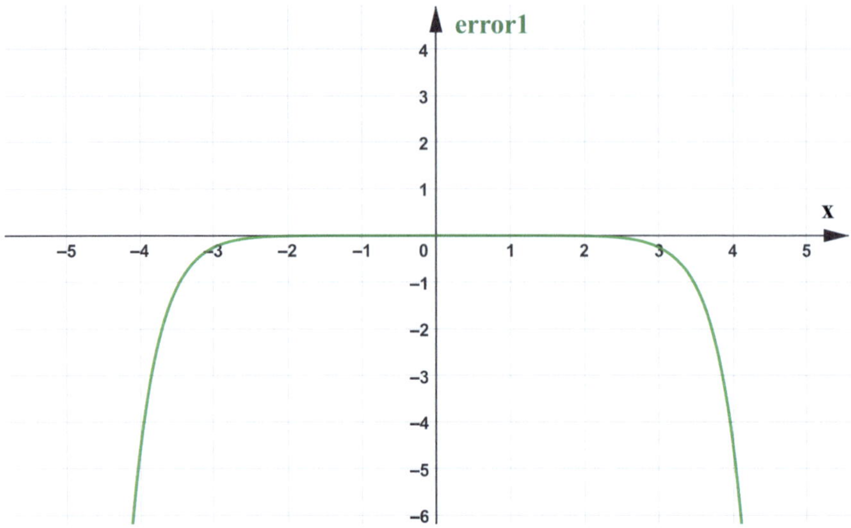

Fig. 3.87 Absolute error of approximation of the hyperbolic cosinusoidal function $f_7(x)$ using the second approximation function $a_{28}(x)$

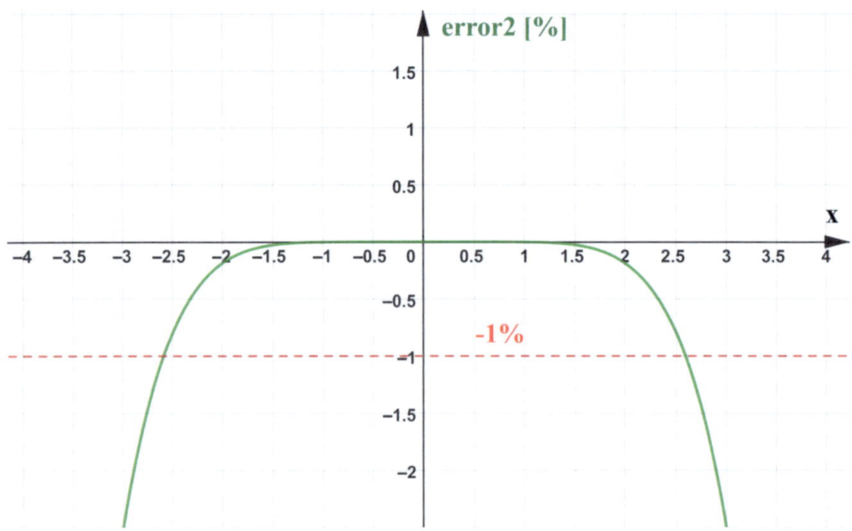

Fig. 3.88 Relative error of approximation of the hyperbolic cosinusoidal function $f_7(x)$ using the second approximation function $a_{28}(x)$

3.6.1 First Approximation Function for Generating Continuous Mathematical Functions

In order to approximate any continuous mathematical function $f(x)$, a new approximation function $a_{29}(x)$ is presented. Its particular mathematical form is developed having two very important objectives. The first objective is represented by the accuracy of the approximation that could be obtained using this approximation function. The second objective is related to the possibility of its facile analog implementation in CMOS technology using as reasonable as possible hardware resources (see Chaps. 4 and 7).

The proposed method for approximating the general $f(x)$ mathematical function is to develop an approximation function $a_{29}(x)$ that must have the same Taylor series expansion with the approximated function $f(x)$ (in a superior-order approximation).

Considering that the Taylor series expansion of $f(x)$ function can be expressed as follows:

$$f(x) = a + bx + cx^2 + dx^3 + ex^4 + fx^5 + gx^6 + \cdots, \tag{3.61}$$

a, b, c, d, e, f and g being constant coefficients that depend on the particular mathematical form of the function $f(x)$, and the first proposed third-order approximation function $a_{29}(x)$ has the following expression:

$$a_{29}(x) = \frac{c^3}{d^2}\frac{1}{1-\frac{dx}{c}} + \left(b - \frac{c^2}{d}\right)x + \left(a - \frac{c^3}{d^2}\right). \tag{3.62}$$

3.6.2 Second Approximation Function for Generating Continuous Mathematical Functions

Considering the Taylor series expansion of $f(x)$ function expressed by relation (3.61), a, b, c, d, e, f and g being constant coefficients that depend on the particular mathematical form of the function $f(x)$, the second proposed fourth-order approximation function $a_{30}(x)$ has the following expression:

$$a_{30}(x) = \frac{d^4}{e^3}\frac{1}{1-\frac{ex}{d}} + \left(c - \frac{d^2}{e}\right)x^2 + \left(b - \frac{d^3}{e^2}\right)x + \left(a - \frac{d^4}{e^3}\right). \tag{3.63}$$

3.6.3 Third Approximation Function for Generating Continuous Mathematical Functions

Considering the Taylor series expansion of $f(x)$ function expressed by relation using relation (3.61), a, b, c, d, e, f and g being constant coefficients that depend on the particular mathematical form of the function $f(x)$, the third proposed fourth-order approximation function $a_{31}(x)$ has the following expression:

$$a_{31}(x) = \frac{1 + a_1 x}{1 + a_2 x} + \frac{a_3 x}{1 - x} + a_4 x^2 + a_5 x + a_6. \tag{3.64}$$

a_1–a_6 being constant coefficients having the following expressions:

$$a_2 = \frac{e^2 + df + ef + ed}{(e - d)^2}, \tag{3.65}$$

$$a_3 = \frac{e^2 - df}{2e - d - f}, \tag{3.66}$$

$$a_1 = a_2 + \frac{d - a_3}{a_2^2}, \tag{3.67}$$

$$a_4 = c - a_3 - a_2(a_2 - a_1), \tag{3.68}$$

$$a_5 = b - a_3 - a_1 + a_2 \tag{3.69}$$

and

$$a_6 = a - 1. \tag{3.70}$$

References

Popa C (2012) High-accuracy function synthesizer circuit with applications in signal processing. EURASIP J Adv Signal Process 2012:129. https://doi.org/10.1186/1687-6180-2012-129

Popa CR (2023) High-accuracy Gaussian function generator for neural networks. Electronics 12(1):24. https://doi.org/10.3390/electronics12010024

Chapter 4
Low-Voltage Low-Power Current-Mode CMOS Computational Circuits for Implementing Activation Functions

4.1 Introduction

In order to implement in CMOS technology all the previous approximation functions, detailed analyzed in Chap. 1, Chap. 2 and Chap. 3, the most convenient fundamental CMOS computational circuits must be designed. A multitude of very important objectives must be taken into account for a proper choice of these fundamental computational circuits (Popa 2012, 2023):

- the accuracy of implementing the required mathematical function;
- the immunity of the computation on technological errors and, also, on temperature variations;
- the low-voltage operation of the computational circuits;
- the low-power operation of the computational circuits;
- the hardware resources required for implementing in CMOS technology of the computational circuits.

4.2 CMOS Increased Accuracy Current-Mode Computational Circuits

Evaluating all previous conditions for designing the fundamental computational circuits, the best choice is represented by the following two elementary circuits:

- current-mode CMOS squaring circuits (SQ);
- current-mode CMOS multiplier/divider circuits (MD).

C. R. Popa, *Analog Current-Mode Computational Circuits for Artificial Neural Networks*, Analog Circuits and Signal Processing,
https://doi.org/10.1007/978-3-032-03989-7_4

4.3 Low-Voltage Low-Power Techniques for Current-Mode Computational Circuits

In order to design low-voltage computational structures with increased accuracy, the following methods are used:

- the current-mode operation of the circuits; comparing with the alternative of using a voltage-mode operation, the current-mode operated circuit permits an important reduction of the minimal supply voltage of the designed CMOS computational circuits;
- particular design techniques for developing specific architectures of CMOS computational circuits, that allow an additional reduction of the minimal supply voltage of the designed CMOS computational circuits.

The low-power operation of the developed CMOS computational circuits is achieved by a proper biasing in saturation of all MOS transistors from the proposed architectures (having an order of magnitude between nanoamperes and hundreds of nanoamperes).

4.4 Improved Accuracy CMOS Current-Mode Squaring Circuits

The first class of fundamental CMOS computational structures, designed for implementing all previous approximation functions, is represented by current-mode squaring circuits (SQ).

4.4.1 First CMOS Implementation of the Current-Mode Squaring Circuit

The proposed implementation in CMOS technology of the first architecture of the current-mode squaring circuit (SQ) is presented in Fig. 4.1.

I_{IN} represents the input current, I_{REF} is the reference constant current and I_{OUT} represents the output current, that must be proportional with the square of the input current. The characteristic equation of the translinear loop containing T1–T4 transistors can be written as follows:

$$V_{GS1} + V_{GS2} = V_{GS3} + V_{GS4}. \tag{4.1}$$

Imposing by design identical transistors (except T8, which has an aspect ratio double than the aspect ratios of T6 and T7 transistors) and considering the squaring low of the MOS transistors biased in saturation region:

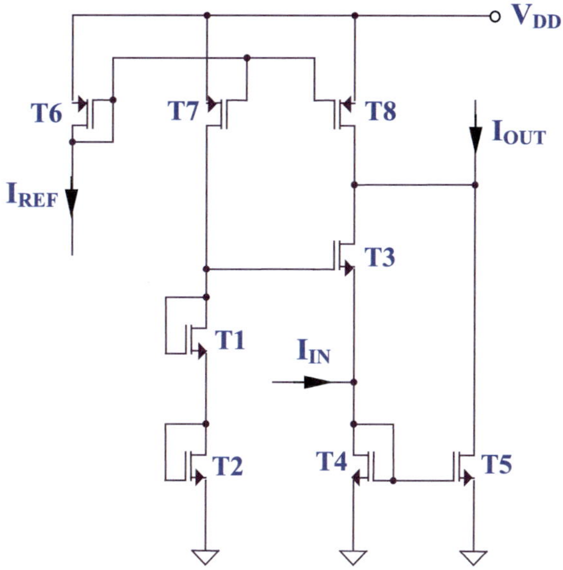

Fig. 4.1 CMOS implementation of the first current-mode squaring circuit (SQ)

$$I_D = \frac{K}{2}(V_{GS} - V_T)^2, \tag{4.2}$$

it results that:

$$2\left(V_T + \sqrt{\frac{2I_{REF}}{K}}\right) = \left(V_T + \sqrt{\frac{2I_{D3}}{K}}\right) + \left[V_T + \sqrt{\frac{2(D_3 + I_{IN})}{K}}\right], \tag{4.3}$$

equivalent with:

$$I_{D3} = I_{REF} - \frac{I_{IN}}{2} + \frac{I_{IN}^2}{16 I_{REF}}. \tag{4.4}$$

Because:

$$I_{OUT} = I_{D3} + (I_{D3} + I_{IN}) - 2I_{REF}, \tag{4.5}$$

it results that:

$$I_{OUT} = \frac{I_{IN}^2}{8 I_{REF}}. \tag{4.6}$$

The first proposed CMOS squaring circuit can generate an expression of the output current independent on any technological parameter. So, temperature variations and technological errors do not affect the operation of the CMOS squaring circuit from

Fig. 4.1 (in the hypothesis of the squaring characteristic (4.2) of MOS transistor biased in saturation region).

The simulation of $I_{OUT}(I_{IN})$ dependence for the first squaring circuit presented in Fig. 4.1 is shown in Fig. 4.2. The variation range for I_{IN} current is set between – 200 and 200 nA and the reference current I_{REF} is equal with 100 nA.

The parametric simulation of $I_{OUT}(I_{IN})$ dependence for the first squaring circuit presented in Fig. 4.1 is shown in Fig. 4.3. The variation range for I_{IN} current is set between – 50 nA and 50 nA and the reference current I_{REF} is considered to be a parameter, having the following values: 25 nA, 50 nA, 75 nA and 100 nA.

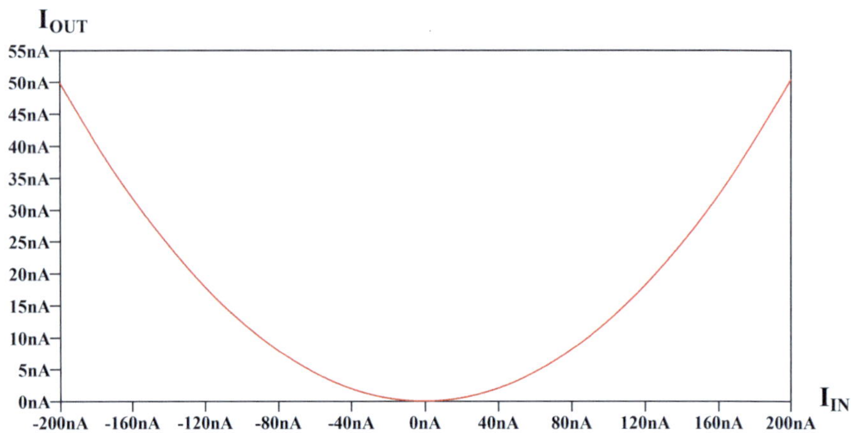

Fig. 4.2 Simulation of $I_{OUT}(I_{IN})$ dependence for the first squaring circuit presented in Fig. 4.1

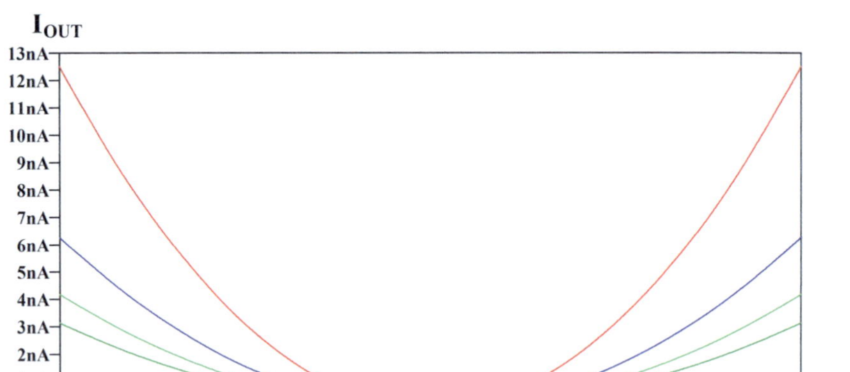

Fig. 4.3 Parametric simulation of $I_{OUT}(I_{IN})$ dependence for the first squaring circuit presented in Fig. 4.1

4.4.2 *Second CMOS Implementation of the Current-Mode Squaring Circuit*

The proposed implementation in CMOS technology of the second architecture of the current-mode squaring circuit (SQ) is presented in Fig. 4.4. All transistors are identical, except T12, which has an aspect ratio double comparing with all other transistors from the circuit.

I_{IN} represents the input current, I_{REF} is the reference constant current and I_{OUT} represents the output current, that must be proportional with the square of the input current. The characteristic equation of the translinear loop containing T1–T4 transistors can be written as follows:

$$V_{GS1} + V_{GS4} = V_{GS2} + V_{GS3}. \tag{4.7}$$

Imposing by design identical transistors and considering the squaring low of the MOS transistors biased in saturation region, expressed by relation (4.2), it results that:

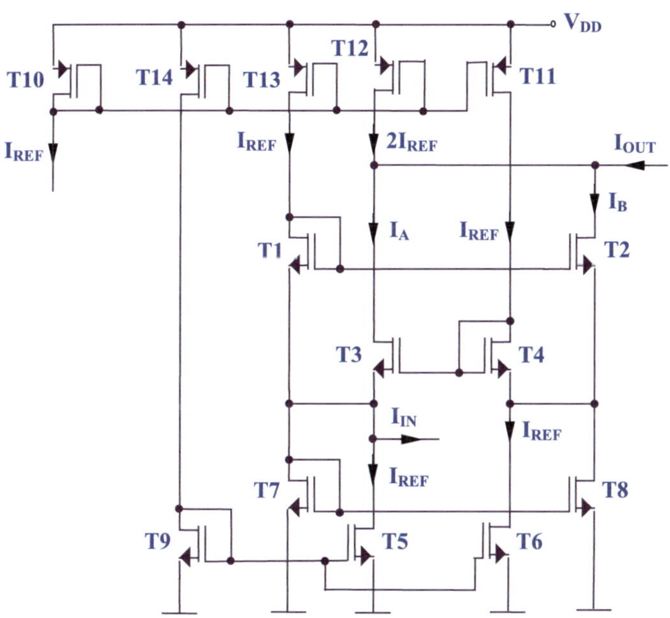

Fig. 4.4 CMOS implementation of the second current-mode squaring circuit (SQ). Reproduced with permission from: Popa (2023). High-Accuracy Gaussian Function Generator for Neural Networks. Electronics, 12(1), 24. https://doi.org/. Licensed under CC BY 4.0

$$2\left(V_T + \sqrt{\frac{2I_{\mathrm{REF}}}{K}}\right) = \left(V_T + \sqrt{\frac{2I_A}{K}}\right) + \left(V_T + \sqrt{\frac{2I_B}{K}}\right), \quad (4.8)$$

equivalent with:

$$2\sqrt{I_{\mathrm{REF}}} = \sqrt{I_A} + \sqrt{I_B}. \quad (4.9)$$

Because:

$$I_A + I_B = I_{\mathrm{OUT}} + 2I_{\mathrm{REF}} \quad (4.10)$$

and

$$I_A - I_B = I_{\mathrm{IN}}, \quad (4.11)$$

it results:

$$I_A = I_{\mathrm{REF}} + \frac{I_{\mathrm{OUT}} + I_{\mathrm{IN}}}{2} \quad (4.12)$$

and

$$I_B = I_{\mathrm{REF}} + \frac{I_{\mathrm{OUT}} - I_{\mathrm{IN}}}{2} \quad (4.13)$$

From (4.9), (4.12) and (4.13), the expression of the output current as function on the input current and the reference current is:

$$I_{\mathrm{OUT}} = \frac{I_{\mathrm{IN}}^2}{8I_{\mathrm{REF}}}. \quad (4.14)$$

The second proposed CMOS squaring circuit can generate an expression of the output current independent on any technological parameter. So, temperature variations and technological errors do not affect the operation of the CMOS squaring circuit from Fig. 4.4 (in the hypothesis of the squaring characteristic (4.2) of MOS transistor biased in saturation region).

The simulation of $I_{\mathrm{OUT}}(I_{\mathrm{IN}})$ dependence for the second squaring circuit presented in Fig. 4.4 is shown in Fig. 4.5. The variation range for I_{IN} current is set between − 100 nA and 100 nA and the reference current I_{REF} is equal with 100 nA.

The parametric simulation of $I_{\mathrm{OUT}}(I_{\mathrm{IN}})$ dependence for the second squaring circuit presented in Fig. 4.4 is shown in Fig. 4.6. The variation range for I_{IN} current is set between − 50 nA and 50 nA and the reference current I_{REF} is considered to be a parameter, having the following values: 50 nA, 100 nA, 150 nA and 200 nA.

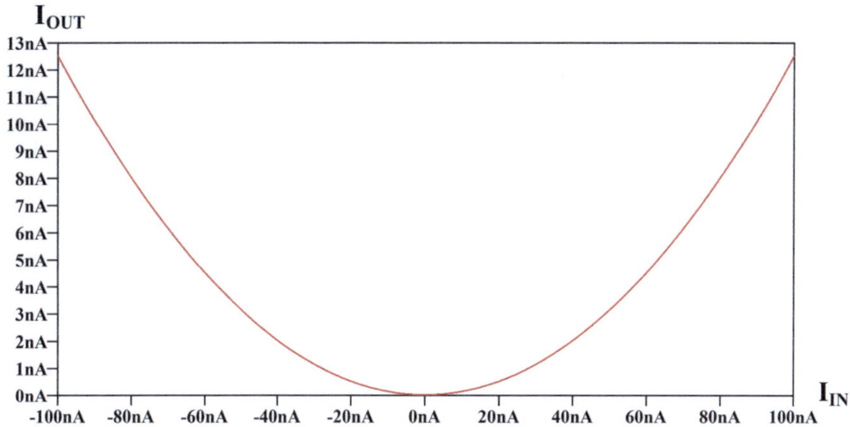

Fig. 4.5 Simulation of $I_{OUT}(I_{IN})$ dependence for the second squaring circuit presented in Fig. 4.4

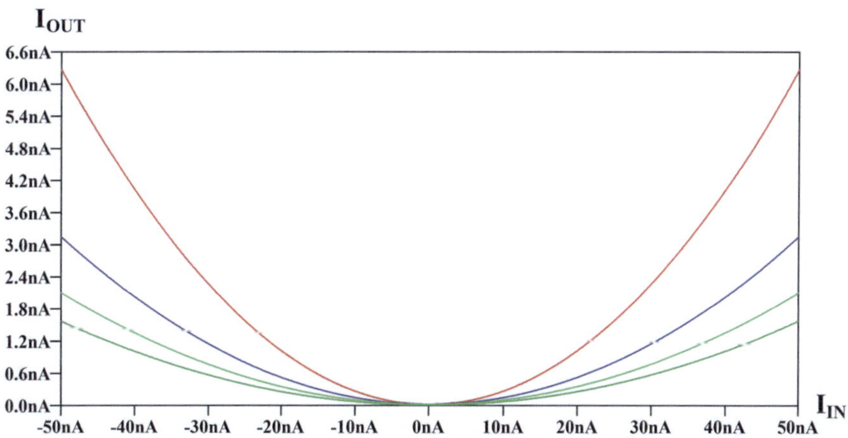

Fig. 4.6 Parametric simulation of $I_{OUT}(I_{IN})$ dependence for the second squaring circuit presented in Fig. 4.4

4.4.3 Third CMOS Implementation of the Current-Mode Squaring Circuit

The proposed implementation in CMOS technology of the third architecture of the current-mode squaring circuit (SQ) is presented in Fig. 4.7.

I_{IN} represents the input current, I_{REF} is the reference constant current and I_{OUT} represents the output current, that must be proportional with the square of the input current. The characteristic equation of the translinear loop containing T1, T2, T3 and T5 transistors can be written as follows:

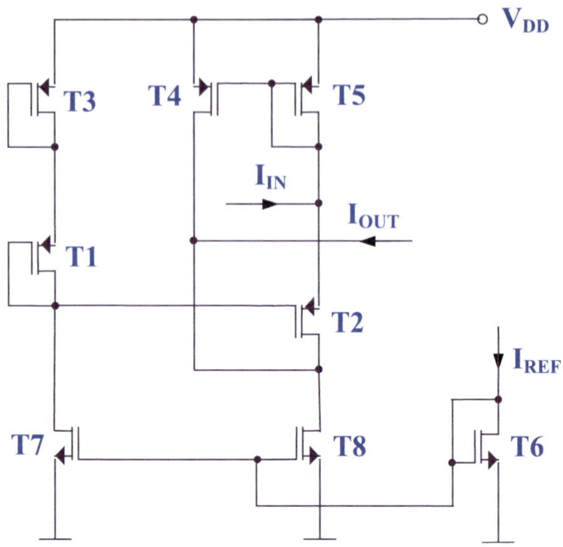

Fig. 4.7 CMOS implementation of the third current-mode squaring circuit (SQ)

$$V_{SG1} + V_{SG3} = V_{SG2} + V_{SG5}. \tag{4.15}$$

Imposing by design identical transistors (except T8, which has an aspect ratio equal with double the aspect ratios of T6 and T7 transistors) and considering the squaring low of the MOS transistors biased in saturation region—relation (4.2), it results that:

$$2\left(V_T + \sqrt{\frac{2I_{REF}}{K}}\right) = \left(V_T + \sqrt{\frac{2I_{D4}}{K}}\right) + \left[V_T + \sqrt{\frac{2(I_{D4} + I_{IN})}{K}}\right], \tag{4.16}$$

equivalent with:

$$I_{D4} = I_{REF} - \frac{I_{IN}}{2} + \frac{I_{IN}^2}{16 I_{REF}}. \tag{4.17}$$

Because:

$$I_{OUT} = I_{D4} + (I_{D4} + I_{IN}) - 2I_{REF}, \tag{4.18}$$

it results that:

$$I_{OUT} = \frac{I_{IN}^2}{8 I_{REF}}. \tag{4.19}$$

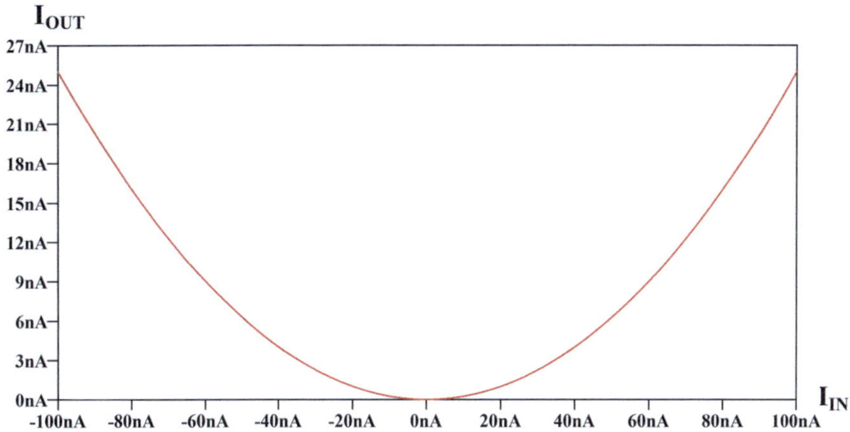

Fig. 4.8 Simulation of $I_{OUT}(I_{IN})$ dependence for the third squaring circuit presented in Fig. 4.7

The original proposed CMOS squaring circuit can generate an expression of the output current independent on any technological parameter. So, temperature variations and technological errors do not affect the operation of the third CMOS squaring circuit from Fig. 4.7 (in the hypothesis of the squaring characteristic (4.2) of MOS transistor biased in saturation region).

The simulation of $I_{OUT}(I_{IN})$ dependence for the third squaring circuit presented in Fig. 4.7 is shown in Fig. 4.8. The variation range for I_{IN} current is set between − 100 and 100 nA and the reference current I_{REF} is equal with 50 nA.

The parametric simulation of $I_{OUT}(I_{IN})$ dependence for the third squaring circuit presented in Fig. 4.7 is shown in Fig. 4.9. The variation range for I_{IN} current is set between − 100 nA and 100 nA and the reference current I_{REF} is considered to be a parameter, having the following values: 25 nA, 50 nA, 75 nA and 100 nA.

4.4.4 *Fourth CMOS Implementation of the Current-Mode Squaring Circuit*

The new proposed method for designing an accurate current-mode squaring circuit (presented in Fig. 4.10) uses two translinear loops (T1–T4 and T1, T3, T5 and T6, respectively), all transistors from the circuit being identical.

The characteristic equation for the first translinear loop can be expressed as follows:

$$V_{GS3} + V_{SG1} = V_{GS4} + V_{SG2}. \tag{4.20}$$

Biasing in saturation all MOS transistors from Fig. 4.10, the previous relation becomes:

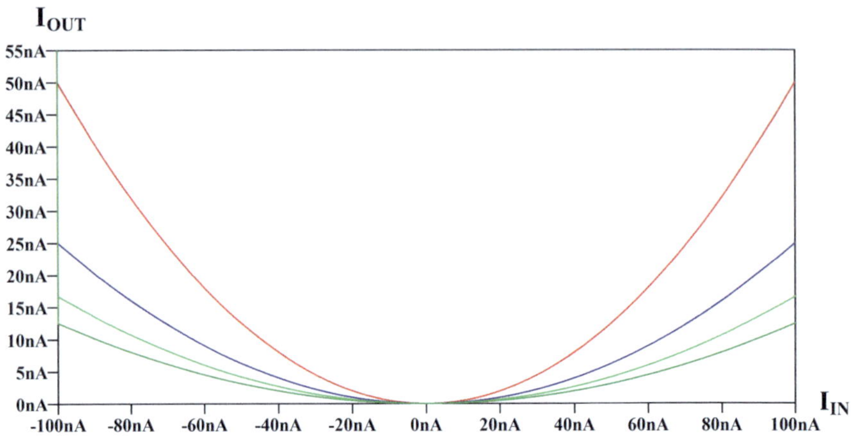

Fig. 4.9 Parametric simulation of $I_{OUT}(I_{IN})$ dependence for the third squaring circuit presented in Fig. 4.7

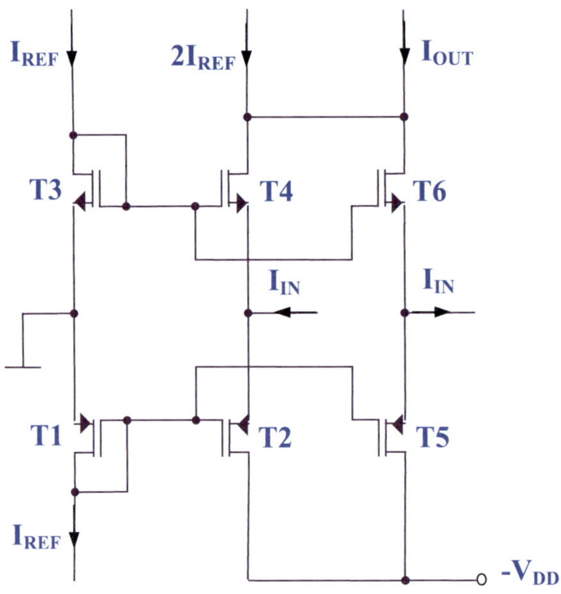

Fig. 4.10 CMOS implementation of the fourth current-mode squaring circuit (SQ)

$$2\sqrt{I_{REF}} = \sqrt{I_{D4}} + \sqrt{I_{D4} + I_{IN}}, \tag{4.21}$$

resulting that:

$$I_{D4} = I_{REF} - \frac{I_{IN}}{2} + \frac{I_{IN}^2}{16 I_{REF}}. \tag{4.22}$$

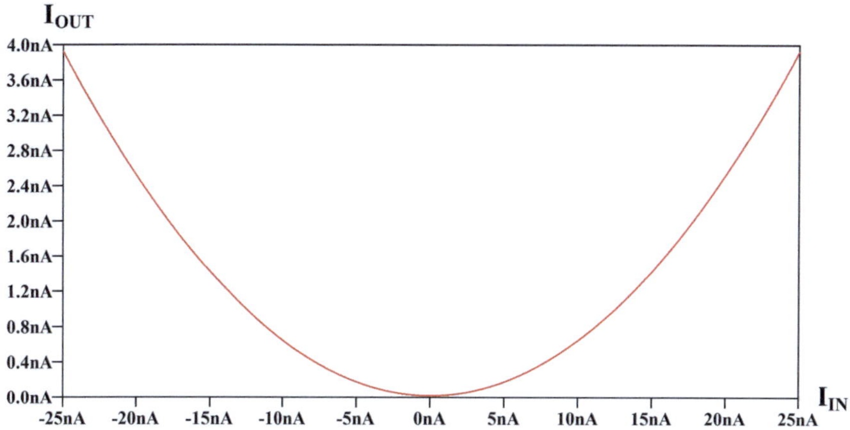

Fig. 4.11 Simulation of $I_{\text{OUT}}(I_{\text{IN}})$ dependence for the fourth squaring circuit presented in Fig. 4.10

Similarly, analyzing the second translinear loop, it results:

$$I_{D6} = I_{\text{REF}} + \frac{I_{\text{IN}}}{2} + \frac{I_{\text{IN}}^2}{16 I_{\text{REF}}}. \tag{4.23}$$

The expression of the output current for the squaring circuit is:

$$I_{\text{OUT}} = I_{D4} + I_{D6} - 2 I_{\text{REF}} = \frac{I_{\text{IN}}^2}{8 I_{\text{REF}}}. \tag{4.24}$$

The proposed fourth squaring circuit from Fig. 4.10, based on a current-mode operation, allows to obtain an independence of the circuit behavior on technological errors and, also, on temperature variations.

The simulation of $I_{\text{OUT}}(I_{\text{IN}})$ dependence for the fourth squaring circuit presented in Fig. 4.10 is shown in Fig. 4.11. The variation range for I_{IN} current is set between – 25 nA and 25 nA and the reference current I_{REF} is equal with 20 nA.

The parametric simulation of $I_{\text{OUT}}(I_{\text{IN}})$ dependence for the squaring circuit presented in Fig. 4.10 is shown in Fig. 4.12. The variation range for I_{IN} current is set between – 5 nA and 5 nA and the reference current I_{REF} is considered to be a parameter, having the following values: 2.5 nA, 5 nA, 7.5 nA and 10 nA.

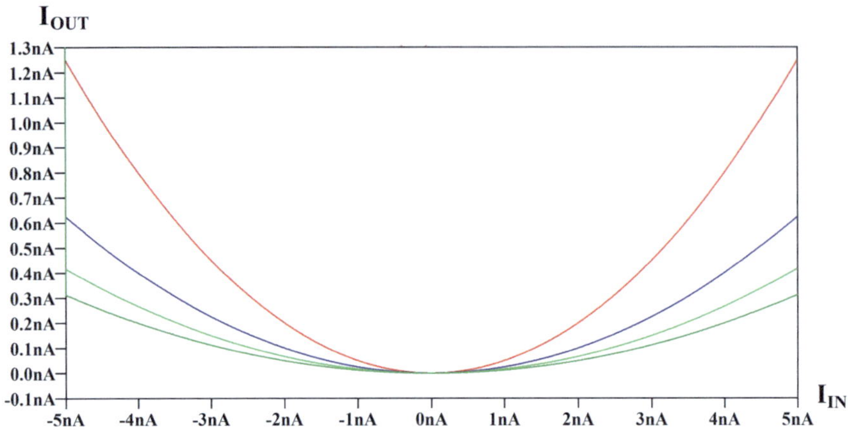

Fig. 4.12 Parametric simulation of $I_{OUT}(I_{IN})$ dependence for the fourth squaring circuit presented in Fig. 4.10

4.4.5 Graphical Representation of the Current-Mode Squaring Circuits (SQ)

All previous presented squaring circuits could be represented as block diagrams as follows:

I_{IN} represents the input current, I_{REF} is the reference current and I_{OUT} represents the output current, that is proportional with the square of the input current, $I_{OUT} = \frac{I_{IN}^2}{8I_{REF}}$. The squaring circuits present the important advantage of an independence of their operation on temperature variation and on technological caused errors (Fig. 4.13).

This circuits are used in the following chapters (Chaps. 5, 6 and 7) for implementing with excellent accuracy the approximation functions from previous chapters (Chaps. 1, 2 and 3).

Fig. 4.13 Graphical represented of the current-mode squaring circuits (SQ)

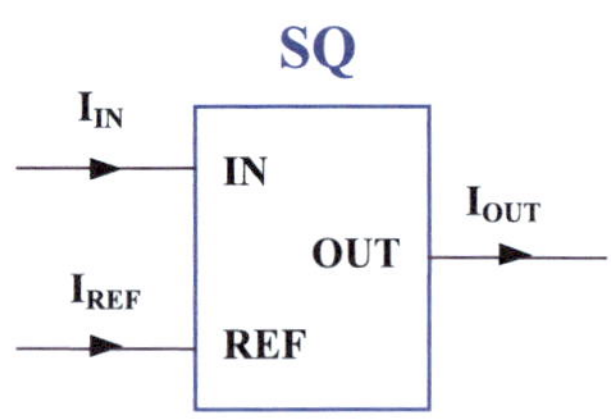

4.5 Improved Accuracy CMOS Current-Mode Multiplier/ Divider Circuits

The second class of fundamental CMOS computational structures, designed for implementing all previous approximation functions, is represented by current-mode multiplier/divider circuits (MD).

4.5.1 First CMOS Implementation of the Current-Mode Multiplier/Divider Circuit

The proposed implementation in CMOS technology of the first architecture of the current-mode multiplier/divider (MD) is presented in Fig. 4.14.

Considering that all transistors are identical, the characteristic equation for the first translinear loop T1, T2, T9 and T10 can be expressed as follows:

$$V_{\mathrm{GS1}} + V_{\mathrm{GS2}} = V_{\mathrm{GS9}} + V_{\mathrm{GS10}}. \tag{4.25}$$

Biasing in saturation all MOS transistors from Fig. 4.14, the previous relation becomes:

$$2\sqrt{I_C} = \sqrt{I_{D1}} + \sqrt{I_{D1} + (I_A + I_B)}, \tag{4.26}$$

resulting that

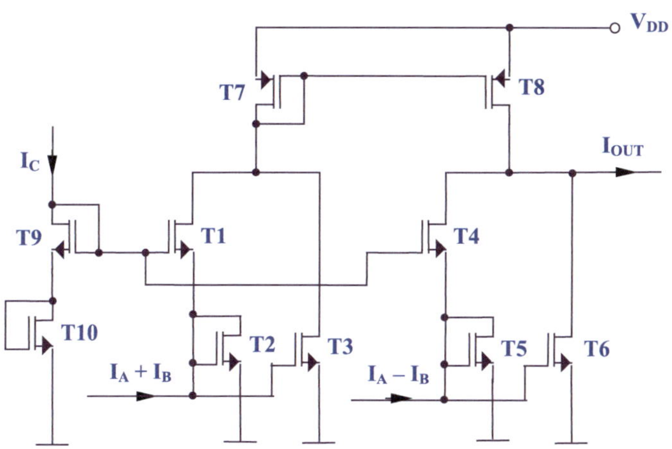

Fig. 4.14 CMOS implementation of the first current-mode multiplier/divider circuit (MD)

$$I_{D1} = I_C - \frac{I_A + I_B}{2} + \frac{(I_A + I_B)^2}{16I_C}. \quad (4.27)$$

Similarly, analyzing the second translinear loop T4, T5, T9 and T10, it results:

$$I_{D4} = I_C - \frac{I_A - I_B}{2} + \frac{(I_A - I_B)^2}{16I_C}. \quad (4.28)$$

The expression of the output current for the multiplier/divider circuit is:

$$I_{OUT} = (2I_{D1} + I_A + I_B) - (2I_{D4} + I_A - I_B). \quad (4.29)$$

In conclusion:

$$I_{OUT} = \frac{I_A I_B}{2I_C}. \quad (4.30)$$

The proposed first multiplier/divider circuit from Fig. 4.14, based on a current-mode operation, allows to obtain an independence of the circuit behavior on technological errors and, also, on temperature variations.

The parametric simulation of $I_{OUT}(I_A)$ dependence for the first multiplier/divider circuit presented in Fig. 4.14 is shown in Fig. 4.15. The variation range for I_A current is set between 0 and 10 nA, the I_C current is equal with 10 nA, while I_B current is considered to be a parameter, having the following values: 10 nA, 15 nA, 17.5 nA and 20 nA.

The parametric simulation of $I_{OUT}(I_B)$ dependence for the first multiplier/divider circuit presented in Fig. 4.14 is shown in Fig. 4.16. The variation range for I_B current

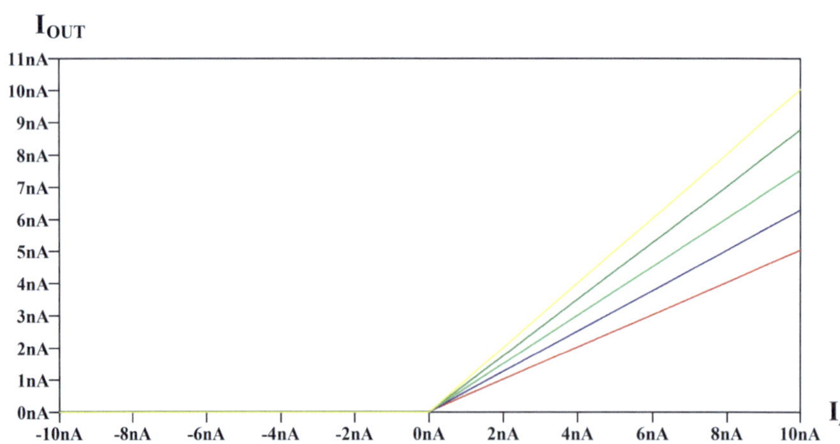

Fig. 4.15 Parametric simulation of $I_{OUT}(I_A)$ dependence for the first multiplier/divider circuit presented in Fig. 4.14

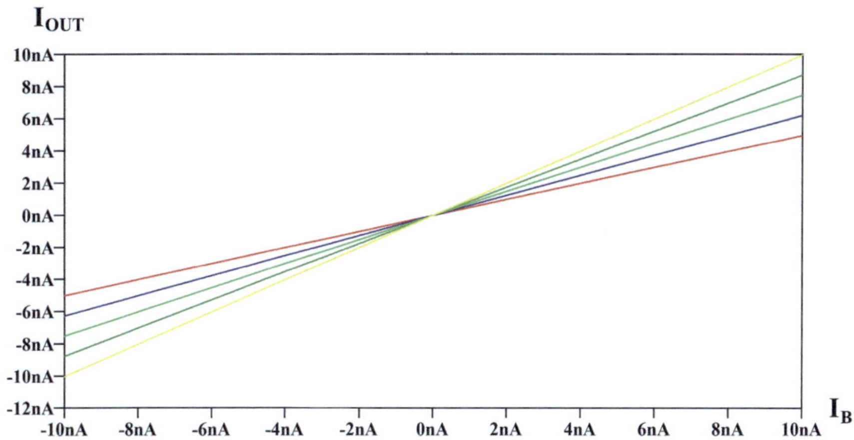

Fig. 4.16 Parametric simulation of $I_{\mathrm{OUT}}(I_B)$ dependence for the first multiplier/divider circuit presented in Fig. 4.14

is set between − 10 nA and 10 nA, the I_C current is equal with 10 nA, while I_A current is considered to be a parameter, having the following values: 10 nA, 12.5 nA, 15 nA, 17.5 nA and 20 nA.

4.5.2 Second CMOS Implementation of the Current-Mode Multiplier/Divider Circuit

The proposed implementation in CMOS technology of the second architecture of the current-mode multiplier/divider (MD) is presented in Fig. 4.17.

The characteristic equation of the translinear loop containing T1–T4 transistors can be written as follows:

$$V_{\mathrm{GS1}} + V_{\mathrm{GS4}} = V_{\mathrm{GS2}} + V_{\mathrm{GS3}}. \tag{4.31}$$

Imposing by design identical transistors and considering the squaring low of the MOS transistors biased in saturation region, expressed by relation (4.2), it results that:

$$2\left(V_T + \sqrt{\frac{2I_C}{K}}\right) = \left(V_T + \sqrt{\frac{2I_{D1}}{K}}\right) + \left(V_T + \sqrt{\frac{2I_{D4}}{K}}\right), \tag{4.32}$$

equivalent with:

$$2\sqrt{I_C} = \sqrt{I_{D1}} + \sqrt{I_{D4}}. \tag{4.33}$$

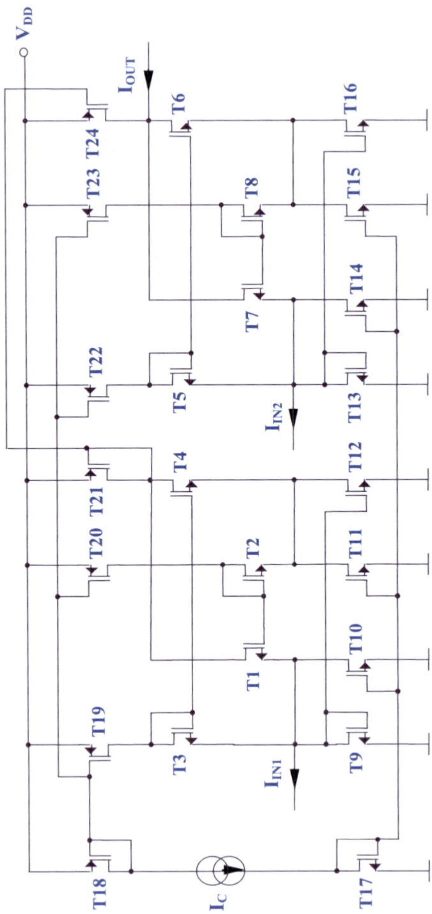

Fig. 4.17 CMOS implementation of the second current-mode multiplier/divider circuit (MD)

Because:

$$I_{D1} + (I_C - I_{IN1}) = I_{D9} + I_C, \tag{4.34}$$

$$I_{D4} + I_C = I_C + I_{D12} \tag{4.35}$$

and

$$I_{D9} = I_{D12}, \tag{4.36}$$

it results that:

$$I_{D1} - I_{D4} = I_{IN1}. \tag{4.37}$$

Because $I_{D21} = I_{D1} + I_{D4}$, it is possible to write that:

$$I_{D1} = \frac{I_{D21} + I_{IN1}}{2} \tag{4.38}$$

and

$$I_{D4} = \frac{I_{D21} - I_{IN1}}{2}. \tag{4.39}$$

From (4.33), (4.38) and (4.39) it results the following relation:

$$I_{D21} = I_{D1} + I_{D4} = 2I_C + \frac{I_{IN1}^2}{8I_C}. \tag{4.40}$$

Similarly, analyzing the characteristic equation of the translinear loop containing T5-T8 transistors, it is possible to write that:

$$I_{D6} + I_{D7} = 2I_C + \frac{I_{IN2}^2}{8I_C}. \tag{4.41}$$

Imposing the following expressions of I_{IN1} and I_{IN2} currents (using a double current mirror, not shown in Fig. 4.17):

$$I_{IN1} = I_B - I_A \tag{4.42}$$

and

$$I_{IN2} = I_B + I_A, \tag{4.43}$$

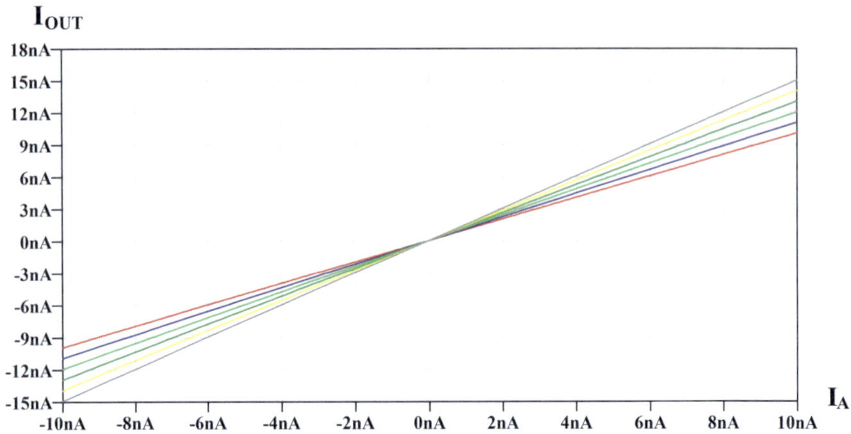

Fig. 4.18 Parametric simulation of $I_{OUT}(I_A)$ dependence for the second multiplier/divider circuit presented in Fig. 4.17

and replacing their values in (4.40) and (4.41), it results that the output current of the second multiplier/divider circuit presented in Fig. 4.17 can be expressed as follows:

$$I_{OUT} = I_{D6} + I_{D7} - I_{21} = \frac{I_A I_B}{2I_C}. \tag{4.44}$$

The proposed second multiplier/divider circuit from Fig. 4.17, based on a current-mode operation, allows to obtain an independence of the circuit behavior on technological errors and, also, on temperature variations.

The parametric simulation of $I_{OUT}(I_A)$ dependence for the second multiplier/divider circuit presented in Fig. 4.17 is shown in Fig. 4.18. The variation range for I_A current is set between − 10 nA and 10 nA, the I_C current is equal with 10 nA, while I_B current is considered to be a parameter, having the following values: 20 nA, 22.5 nA, 25 nA, 27.5 nA and 30 nA.

The parametric simulation of $I_{OUT}(I_B)$ dependence for the second multiplier/divider circuit presented in Fig. 4.17 is shown in Fig. 4.19. The variation range for I_B current is set between 0 nA and 10 nA, the I_C current is equal with 10 nA, while I_A current is considered to be a parameter, having the following values: 20 nA, 22.5 nA, 25 nA, 27.5 nA and 30 nA.

4.5.3 Third CMOS Implementation of the Current-Mode Multiplier/Divider Circuit

The third multiplier/divider circuit presented in Fig. 4.20 uses two identical previously presented squaring circuits, having the architecture proposed in Fig. 4.7.

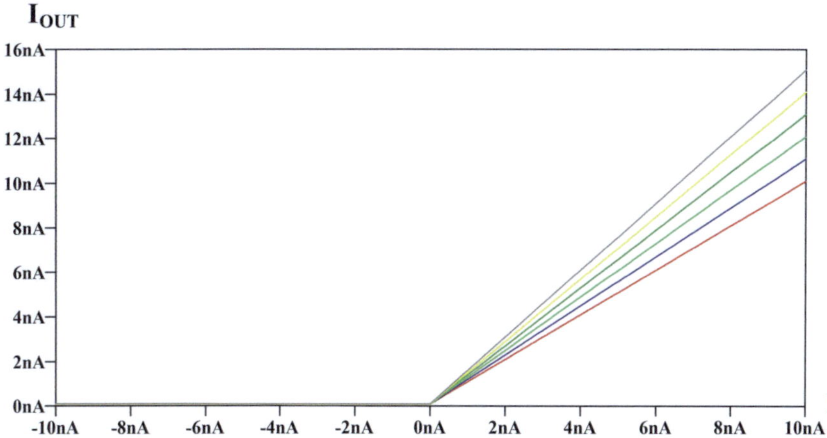

Fig. 4.19 Parametric simulation of $I_{\mathrm{OUT}}(I_B)$ dependence for the second multiplier/divider circuit presented in Fig. 4.17

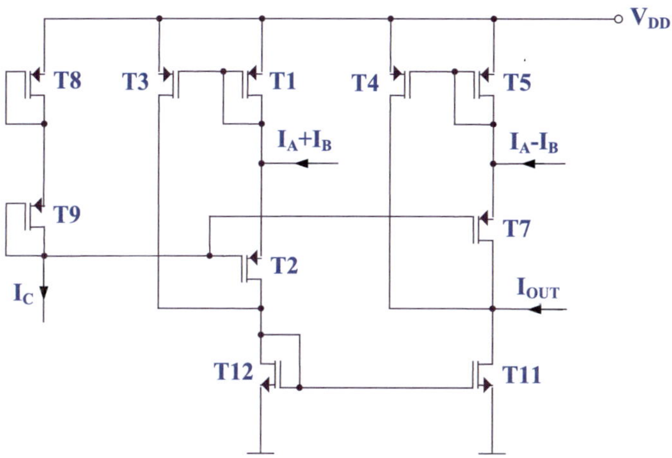

Fig. 4.20 CMOS implementation of the third current-mode multiplier/divider circuit (MD)

To minimize the multiplier/divider complexity, redundant transistors have been eliminated from Fig. 4.20 (the double utilization of diode-connected T1, T3 transistors from Fig. 4.7). Using the relation (4.17) obtained for the squaring circuit presented in Fig. 4.7, it is possible to write that:

$$I_{D3} = I_C - \frac{I_A + I_B}{2} + \frac{(I_A + I_B)^2}{16 I_C} \tag{4.45}$$

and

$$I_{D4} = I_C - \frac{I_A - I_B}{2} + \frac{(I_A - I_B)^2}{16I_C}. \tag{4.46}$$

Because

$$I_{OUT} = (2I_{D3} + I_A + I_B) - (2I_{D4} + I_A - I_B). \tag{4.47}$$

It results that:

$$I_{OUT} = \frac{(I_A + I_B)^2}{8I_C} - \frac{(I_A - I_B)^2}{8I_C}, \tag{4.48}$$

equivalent with

$$I_{OUT} = \frac{I_A I_B}{2I_C}. \tag{4.49}$$

The proposed second multiplier/divider circuit from Fig. 4.20, based on a current-mode operation, allows to obtain an independence of the circuit behavior on technological errors and, also, on temperature variations.

The parametric simulation of $I_{OUT}(I_A)$ dependence for the third multiplier/divider circuit presented in Fig. 4.20 is shown in Fig. 4.21. The variation range for I_A current is set between 0 nA and 20 nA, the I_C current is equal with 10 nA, while I_B current is considered to be a parameter, having the following values: 10 nA, 12.5 nA, 15 nA, 17.5 nA and 20 nA.

The parametric simulation of $I_{OUT}(I_B)$ dependence for the third multiplier/divider circuit presented in Fig. 4.20 is shown in Fig. 4.22. The variation range for I_B current

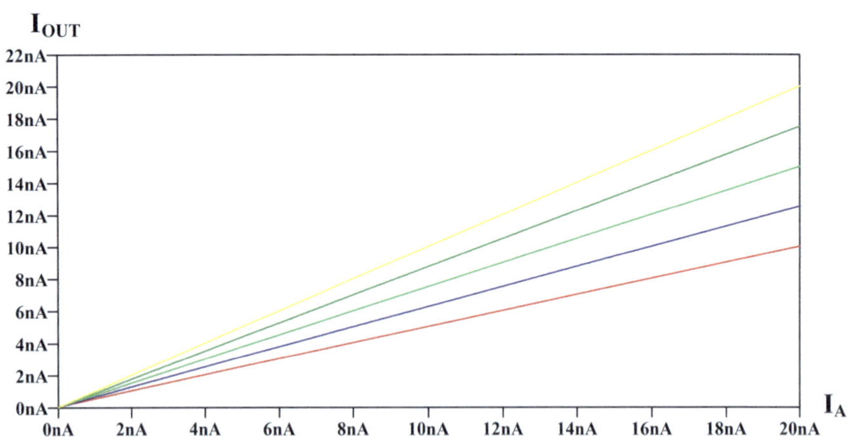

Fig. 4.21 Parametric simulation of $I_{OUT}(I_A)$ dependence for the third multiplier/divider circuit presented in Fig. 4.20

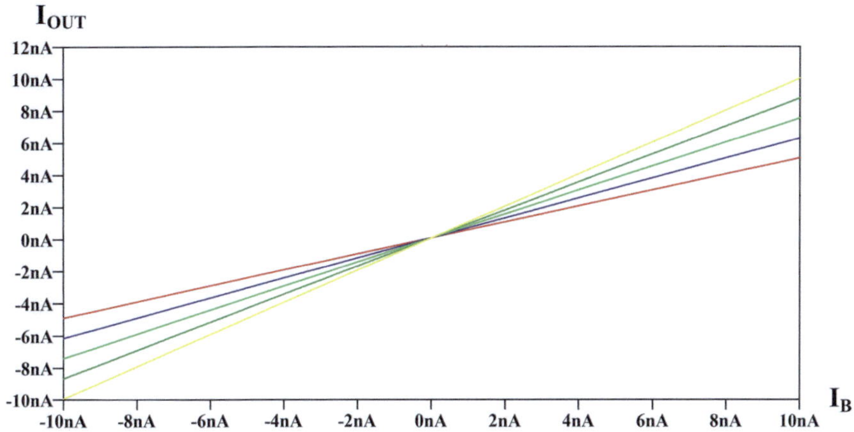

Fig. 4.22 Parametric simulation of $I_{OUT}(I_B)$ dependence for the third multiplier/divider circuit presented in Fig. 4.20

is set between − 10 nA and 10 nA, the I_C current is equal with 10 nA, while I_A current is considered to be a parameter, having the following values: 10 nA, 12.5 nA, 15 nA, 17.5 nA and 20 nA.

4.5.4 Fourth CMOS Implementation of the Current-Mode Multiplier/Divider Circuit

The proposed fourth architecture of the multiplier/divider circuit from Fig. 4.23 uses four translinear loops (T9–T10–T1–T2, T9–T10–T3–T4, T9–T10–T5–T6 and T9–T10–T7–T8, respectively).

The characteristic equations for these translinear loops are the following:

$$V_{GS9} + V_{SG10} = V_{GS1} + V_{SG2}, \tag{4.50}$$

$$V_{GS9} + V_{SG10} = V_{GS3} + V_{SG4}, \tag{4.51}$$

$$V_{GS9} + V_{SG10} = V_{GS5} + V_{SG6}, \tag{4.52}$$

and

$$V_{GS9} + V_{SG10} = V_{GS7} + V_{SG8}. \tag{4.53}$$

It is possible to write the following relations using I_A and I_B currents:

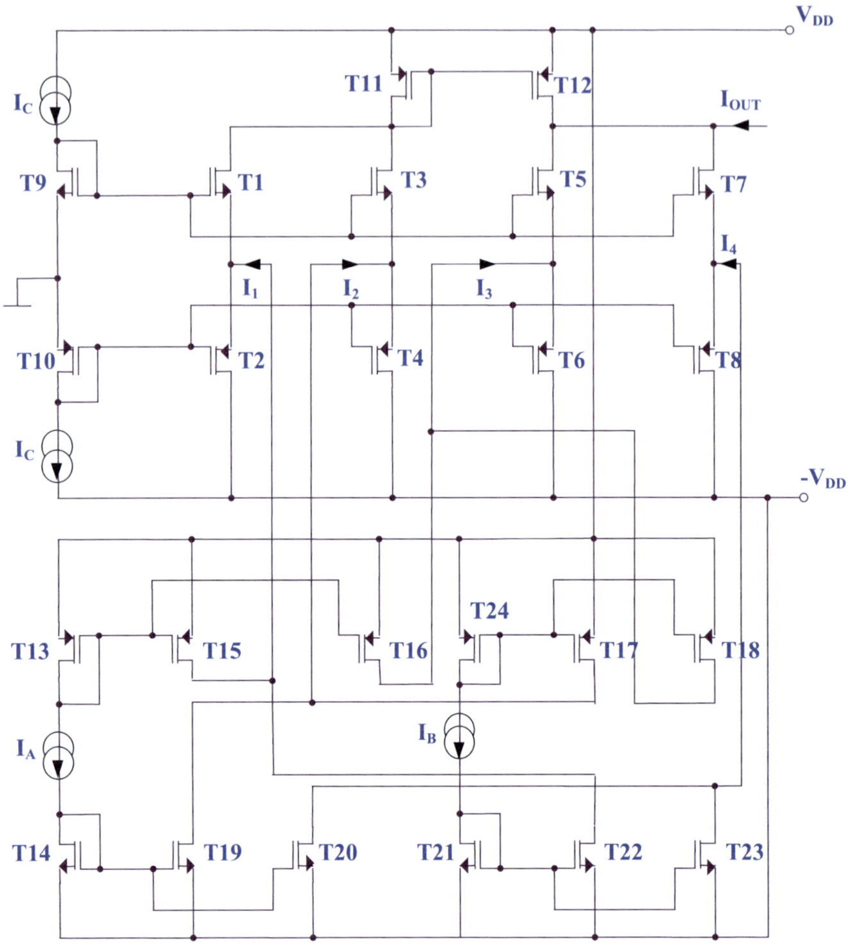

Fig. 4.23 CMOS implementation of the fourth current-mode multiplier/divider circuit (MD)

$$I_1 = I_A - I_B, \tag{4.54}$$

$$I_2 = -I_A + I_B, \tag{4.55}$$

$$I_C = I_1 + I_2 \tag{4.56}$$

and

$$I_4 = -I_A - I_B. \tag{4.57}$$

Biasing in saturation region of all MOS transistors from Fig. 4.23, relation (4.50) becomes:

$$2\sqrt{I_C} = \sqrt{I_{D1}} + \sqrt{I_{D1} + I_1}, \tag{4.58}$$

resulting

$$I_{D1} = I_C - \frac{I_1}{2} + \frac{I_1^2}{16I_C}. \tag{4.59}$$

Similarly, analyzing relations (4.51), (4.52) and (4.53), it is possible to write that:

$$I_{D3} = I_C - \frac{I_2}{2} + \frac{I_2^2}{16I_C}, \tag{4.60}$$

$$I_{D5} = I_C - \frac{I_3}{2} + \frac{I_3^2}{16I_C} \tag{4.61}$$

and

$$I_{D7} = I_C - \frac{I_4}{2} + \frac{I_4^2}{16I_C}. \tag{4.62}$$

Replacing relations (4.54)–(4.57) in expressions (4.59)–(4.62) of drain currents, the output current of the fourth multiplier/divider circuit presented in Fig. 4.23 can be expressed as follows:

$$I_{OUT} = I_{D7} + I_{D5} - I_{D1} - I_{D3}, \tag{4.63}$$

resulting

$$I_{OUT} = \frac{I_A I_B}{2I_C}. \tag{4.64}$$

The parametric simulation of $I_{OUT}(I_A)$ dependence for the fourth multiplier/divider circuit presented in Fig. 4.23 is shown in Fig. 4.24. The variation range for I_A current is set between 0 nA and 10 nA, the I_C current is equal with 10 nA, while I_B current is considered to be a parameter, having the following values: 5 nA, 10 nA, 15 nA and 20 nA.

The parametric simulation of $I_{OUT}(I_B)$ dependence for the fourth multiplier/divider circuit presented in Fig. 4.23 is shown in Fig. 4.25. The variation range for I_B current is set between 0 nA and 10 nA, the I_C current is equal with 10 nA, while I_A current is considered to be a parameter, having the following values: 10 nA, 20 nA, 30 nA and 40 nA.

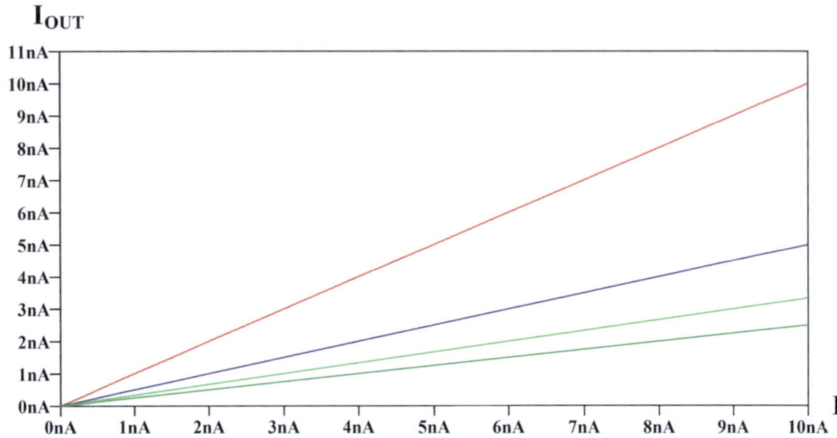

Fig. 4.24 Parametric simulation of $I_{OUT}(I_A)$ dependence for the fourth multiplier/divider circuit presented in Fig. 4.23

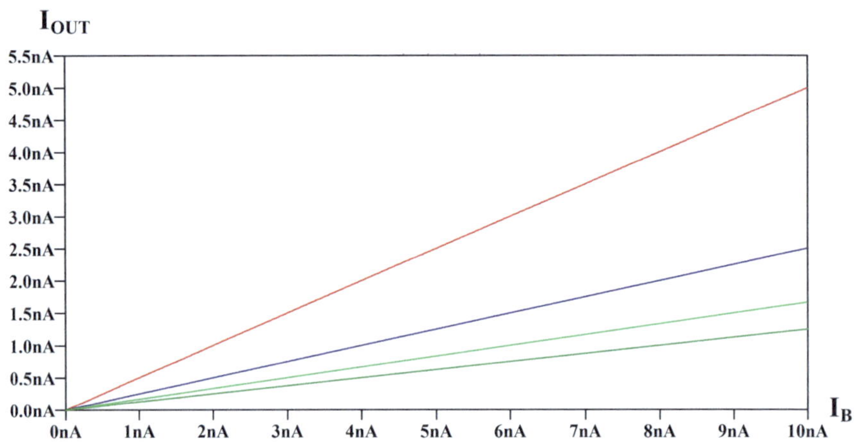

Fig. 4.25 Parametric simulation of $I_{\mathrm{OUT}}(I_B)$ dependence for the fourth multiplier/divider circuit presented in Fig. 4.23

4.5.5 Graphical Representation of the Current-Mode Multiplier/Divider Circuits (MD)

All previous presented multiplier/divider circuits could be represented as block diagrams as follows:

I_A, I_B and I_C represent input currents and I_{OUT} is the output current, $I_{\mathrm{OUT}} = \frac{I_A I_B}{2I_C}$. The multiplier/divider circuits present the important advantage of an independence

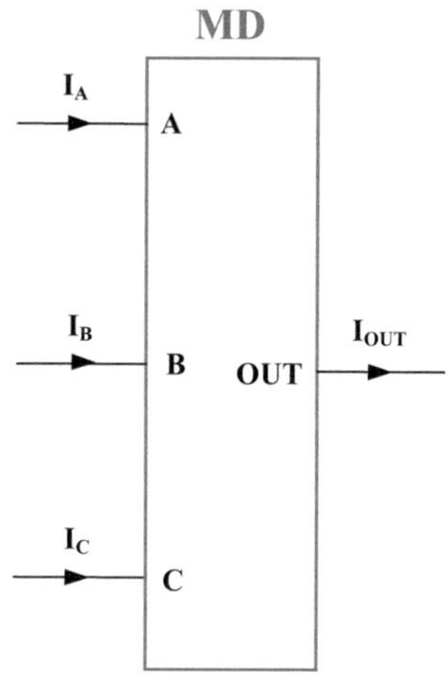

Fig. 4.26 Graphical representation of the current-mode multiplier/divider circuits (MD)

of their operation on temperature variation and on technological caused errors (Fig. 4.26).

These circuits are used in the following chapters (Chaps. 5, 6 and 7) for implementing with excellent accuracy the approximation functions from previous chapters (Chaps. 1, 2 and 3).

References

Popa C (2012) High-accuracy function synthesizer circuit with applications in signal processing. EURASIP J Adv Signal Process 2012:129. https://doi.org/10.1186/1687-6180-2012-129

Popa CR (2023) High-accuracy Gaussian function generator for neural networks. Electronics 12(1):24. https://doi.org/10.3390/electronics12010024

Chapter 5
Analysis and Design of Analog Function Synthesizers for Generating Sigmoidal Activation Functions

5.1 Introduction

This chapter presents a multitude of analog function synthesizers (Popa 2012, 2023), developed for implementing in CMOS technology the sigmoidal activation functions, detailed and analyzed in Chap. 1 (unipolar and bipolar sigmoidal activation function, Einstein activation functions, hyperbolic tangent activation function and continuous-log sigmoid activation functions). The superior-order approximation functions described in Chap. 1 represent the basis for elaborating the block diagrams from this chapter, while the concrete implementations in CMOS technology based on reasonable hardware resources use the fundamental CMOS computational circuits (current-mode squaring and multiplier/divider circuits), analyzed in detail in Chap. 4.

5.2 Analog Function Synthesizers for Unipolar Sigmoidal Activation Function

In order to implement in CMOS technology the unipolar sigmoidal activation function, the approximation functions described in Chap. 1 and the CMOS squaring and multiplies/divider circuits from Chap. 4 are used.

5.2.1 First Analog Function Synthesizer for Generating the Unipolar Sigmoidal Activation Function

The first analog function synthesizer, designed for generating the unipolar sigmoidal activation function, uses $a_1(x)$ approximation function described in Chap. 1. The

C. R. Popa, *Analog Current-Mode Computational Circuits for Artificial Neural Networks*, Analog Circuits and Signal Processing,
https://doi.org/10.1007/978-3-032-03989-7_5

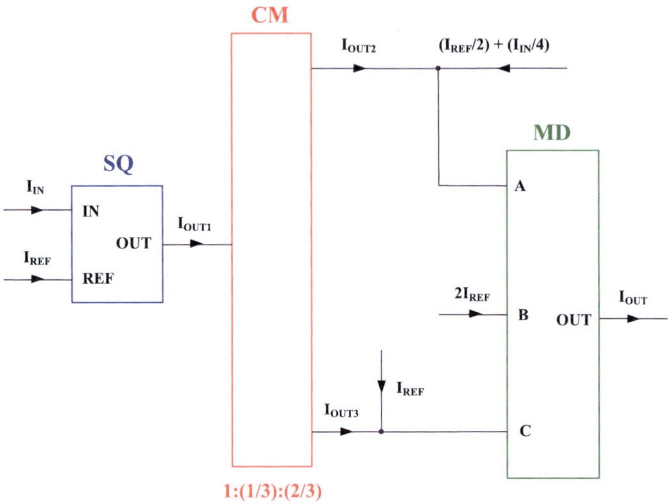

Fig. 5.1 Block diagram of the first analog function synthesizer for generating the unipolar sigmoidal activation function

utilization of the current-mode squaring circuits (SQ) and multiplier/divider circuits (MD) analyzed in Chap. 4 allows to obtain CMOS implementations of the analog function synthesizer, having very good accuracy and reasonable complexity.

The block diagram of the first analog function synthesizer for generating the unipolar sigmoidal activation function is presented in Fig. 5.1.

The expressions of the currents from the circuit are:

$$I_{\text{OUT1}} = \frac{I_{\text{IN}}^2}{8I_{\text{REF}}}, \tag{5.1}$$

$$I_{\text{OUT2}} = \frac{1}{3}I_{\text{OUT1}} = \frac{I_{\text{IN}}^2}{24I_{\text{REF}}} \tag{5.2}$$

and

$$I_{\text{OUT3}} = \frac{2}{3}I_{\text{OUT1}} = \frac{I_{\text{IN}}^2}{12I_{\text{REF}}}. \tag{5.3}$$

The output current of the circuit from Fig. 5.1 can be expressed as follows:

$$I_{\text{OUT}} = \frac{I_A I_B}{2I_C} = I_{\text{REF}}\frac{\frac{I_{\text{IN}}^2}{24I_{\text{REF}}} + \frac{I_{\text{IN}}}{4} + \frac{I_{\text{REF}}}{2}}{\frac{I_{\text{IN}}^2}{12I_{\text{REF}}} + I_{\text{REF}}}, \tag{5.4}$$

equivalent with

$$I_{OUT} = I_{REF} \frac{\frac{1}{24}\left(\frac{I_{IN}}{I_{REF}}\right)^2 + \frac{1}{4}\left(\frac{I_{IN}}{I_{REF}}\right) + \frac{1}{2}}{\frac{1}{12}\left(\frac{I_{IN}}{I_{REF}}\right)^2 + 1}. \tag{5.5}$$

Defining the undimensional variable x as the ratio between the input variable current I_{IN} and the constant reference current I_{REF}, $x = \frac{I_{IN}}{I_{REF}}$, the previous relation becomes:

$$I_{OUT} = I_{REF} \frac{\frac{x^2}{24} + \frac{x}{4} + \frac{1}{2}}{\frac{x^2}{12} + 1}. \tag{5.6}$$

In conclusion, the analog function synthesizer shown in Fig. 5.1 approximates the unipolar sigmoidal activation function:

$$I_{OUT} = I_{REF} a_1(x) \cong I_{REF} f_1(x). \tag{5.7}$$

5.2.2 *Second Analog Function Synthesizer for Generating the Unipolar Sigmoidal Activation Function*

The second analog function synthesizer, designed for generating the unipolar sigmoidal activation function, uses $a_2(x)$ approximation function described in Chap. 1.

The block diagram of the second analog function synthesizer for generating the unipolar sigmoidal activation function is presented in Fig. 5.2.

The expressions of the currents from the circuit are:

$$I_{OUT1} = \frac{I_{IN}^2}{8 I_{REF}}, \tag{5.8}$$

$$I_{OUT2} = \frac{1}{30} I_{OUT1} = \frac{I_{IN}^2}{240 I_{REF}}, \tag{5.9}$$

$$I_{OUT3} = \frac{2}{5} I_{OUT1} = \frac{I_{IN}^2}{20 I_{REF}}, \tag{5.10}$$

$$I_{OUT4} = \frac{4}{5} I_{OUT1} = \frac{I_{IN}^2}{10 I_{REF}} \tag{5.11}$$

and

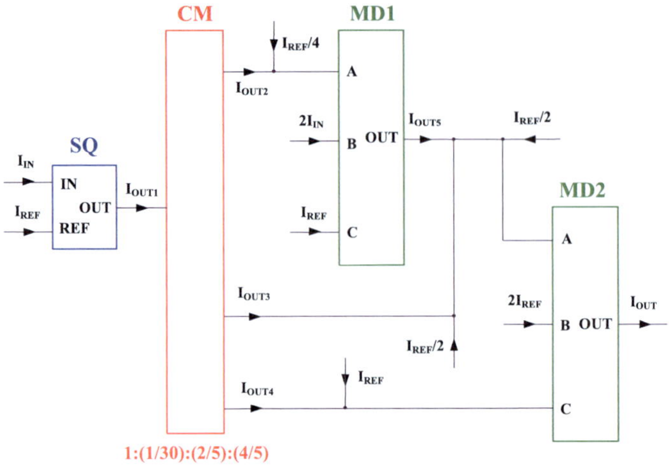

Fig. 5.2 Block diagram of the second analog function synthesizer for generating the unipolar sigmoidal activation function

$$I_{\mathrm{OUT5}} = \frac{\left(\frac{I_{\mathrm{IN}}^2}{240 I_{\mathrm{REF}}} + \frac{I_{\mathrm{REF}}}{4}\right) I_{\mathrm{IN}}}{I_{\mathrm{REF}}}. \tag{5.12}$$

So, the output current of the circuit from Fig. 5.2 can be expressed as follows:

$$I_{\mathrm{OUT}} = I_{\mathrm{REF}} \frac{\frac{I_{\mathrm{IN}}^3}{240 I_{\mathrm{REF}}^2} + \frac{I_{\mathrm{IN}}^2}{20 I_{\mathrm{REF}}} + \frac{I_{\mathrm{IN}}}{4} + \frac{I_{\mathrm{REF}}}{2}}{\frac{I_{\mathrm{IN}}^2}{10 I_{\mathrm{REF}}} + I_{\mathrm{REF}}}, \tag{5.13}$$

equivalent with

$$I_{\mathrm{OUT}} = I_{\mathrm{REF}} \frac{\frac{1}{240}\left(\frac{I_{\mathrm{IN}}}{I_{\mathrm{REF}}}\right)^3 + \frac{1}{20}\left(\frac{I_{\mathrm{IN}}}{I_{\mathrm{REF}}}\right)^2 + \frac{1}{4}\left(\frac{I_{\mathrm{IN}}}{I_{\mathrm{REF}}}\right) + \frac{1}{2}}{\frac{1}{10}\left(\frac{I_{\mathrm{IN}}}{I_{\mathrm{REF}}}\right)^2 + 1}. \tag{5.14}$$

Using the definition of the x variable, the previous relation becomes:

$$I_{\mathrm{OUT}} = I_{\mathrm{REF}} \frac{\frac{x^3}{240} + \frac{x^2}{20} + \frac{x}{4} + \frac{1}{2}}{\frac{x^2}{10} + 1}. \tag{5.15}$$

In conclusion, the analog function synthesizer shown in Fig. 5.2 approximates the unipolar sigmoidal activation function:

$$I_{\mathrm{OUT}} = I_{\mathrm{REF}} a_2(x) \cong I_{\mathrm{REF}} f_1(x). \tag{5.16}$$

5.2.3 Third Analog Function Synthesizer for Generating the Unipolar Sigmoidal Activation Function

The third analog function synthesizer, designed for generating the unipolar sigmoidal activation function, uses $a_3(x)$ approximation function described in Chap. 1.

The block diagram of the third analog function synthesizer for generating the unipolar sigmoidal activation function is presented in Fig. 5.3.

The expressions of the currents from the circuit are:

$$I_{\mathrm{OUT1}} = \frac{I_{\mathrm{IN}}^2}{8I_{\mathrm{REF}}}, \tag{5.17}$$

$$I_{\mathrm{OUT2}} = \frac{1}{21} I_{\mathrm{OUT1}} = \frac{I_{\mathrm{IN}}^2}{168I_{\mathrm{REF}}}, \tag{5.18}$$

$$I_{\mathrm{OUT3}} = \frac{\left(I_{\mathrm{OUT2}} + \frac{I_{\mathrm{REF}}}{4}\right) 2I_{\mathrm{IN}}}{2I_{\mathrm{REF}}} = \frac{\left(\frac{I_{\mathrm{IN}}^2}{168I_{\mathrm{REF}}} + \frac{I_{\mathrm{REF}}}{4}\right) I_{\mathrm{IN}}}{I_{\mathrm{REF}}}, \tag{5.19}$$

$$I_{\mathrm{OUT4}} = I_{\mathrm{OUT1}} = \frac{I_{\mathrm{IN}}^2}{8I_{\mathrm{REF}}}, \tag{5.20}$$

$$I_{\mathrm{OUT5}} = \frac{16 I_{\mathrm{OUT4}}^2}{8 \times 105 \times I_{\mathrm{REF}}^3} = \frac{I_{\mathrm{IN}}^4}{3360 I_{\mathrm{REF}}^3}, \tag{5.21}$$

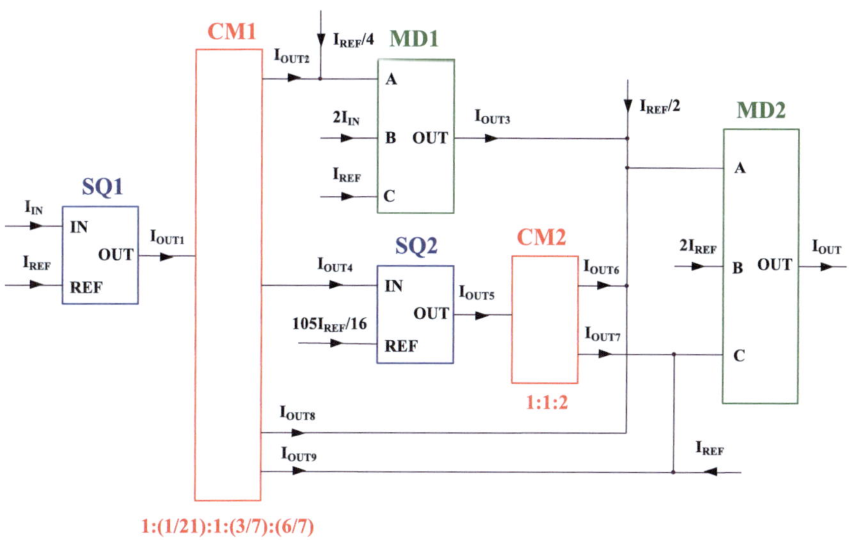

Fig. 5.3 Block diagram of the third analog function synthesizer for generating the unipolar sigmoidal activation function

$$I_{OUT6} = I_{OUT5} = \frac{I_{IN}^4}{3360 I_{REF}^3}, \tag{5.22}$$

$$I_{OUT7} = 2I_{OUT5} = \frac{I_{IN}^4}{1680 I_{REF}^3}, \tag{5.23}$$

$$I_{OUT8} = \frac{3}{7} I_{OUT1} = \frac{3I_{IN}^2}{56 I_{REF}} \tag{5.24}$$

and

$$I_{OUT9} = \frac{6}{7} I_{OUT1} = \frac{3I_{IN}^2}{28 I_{REF}}. \tag{5.25}$$

So, the output current of the circuit from Fig. 5.3 can be expressed as follows:

$$I_{OUT} = I_{REF} \frac{\frac{I_{IN}^4}{3360 I_{REF}^3} + \frac{I_{IN}^3}{168 I_{REF}^2} + \frac{3I_{IN}^2}{56 I_{REF}} + \frac{I_{IN}}{4} + \frac{I_{REF}}{2}}{\frac{I_{IN}^4}{1680 I_{REF}^3} + \frac{3I_{IN}^2}{28 I_{REF}} + I_{REF}}. \tag{5.26}$$

Using the definition of the x variable, the previous relation becomes:

$$I_{OUT} = I_{REF} \frac{\frac{x^4}{3360} + \frac{x^3}{168} + \frac{3x^2}{56} + \frac{x}{4} + \frac{1}{2}}{\frac{x^4}{1680} + \frac{3x^2}{28} + 1}. \tag{5.27}$$

In conclusion, the analog function synthesizer shown in Fig. 5.3 approximates the unipolar sigmoidal activation function:

$$I_{OUT} = I_{REF} a_3(x) \cong I_{REF} f_1(x). \tag{5.28}$$

5.3 Analog Function Synthesizers for Bipolar Sigmoidal Activation Function

In order to implement in CMOS technology the bipolar sigmoidal activation function, the approximation functions described in Chap. 1 and the CMOS squaring and multiplies/divider circuits from Chap. 4 are used.

5.3.1 *First Analog Function Synthesizer for Generating the Bipolar Sigmoidal Activation Function*

The first analog function synthesizer, designed for generating the bipolar sigmoidal activation function, uses $a_4(x)$ approximation function described in Chap. 1. The utilization of the current-mode squaring circuits (SQ) and multiplier/divider circuits (MD) analyzed in Chap. 4 allows to obtain CMOS implementations of the analog function synthesizer, having very good accuracy and reasonable complexity.

The block diagram of the first analog function synthesizer for generating the bipolar sigmoidal activation function is presented in Fig. 5.4.

The expressions of the currents from the circuit are:

$$I_{\mathrm{OUT1}} = \frac{I_{\mathrm{IN}}^2}{8I_{\mathrm{REF}}}, \tag{5.29}$$

$$I_{\mathrm{OUT2}} = \frac{1}{15}I_{\mathrm{OUT1}} = \frac{I_{\mathrm{IN}}^2}{120I_{\mathrm{REF}}}, \tag{5.30}$$

and

$$I_{\mathrm{OUT3}} = \frac{4}{5}I_{\mathrm{OUT1}} = \frac{I_{\mathrm{IN}}^2}{10I_{\mathrm{REF}}}. \tag{5.31}$$

The output current of the circuit from Fig. 5.4 can be expressed as follows:

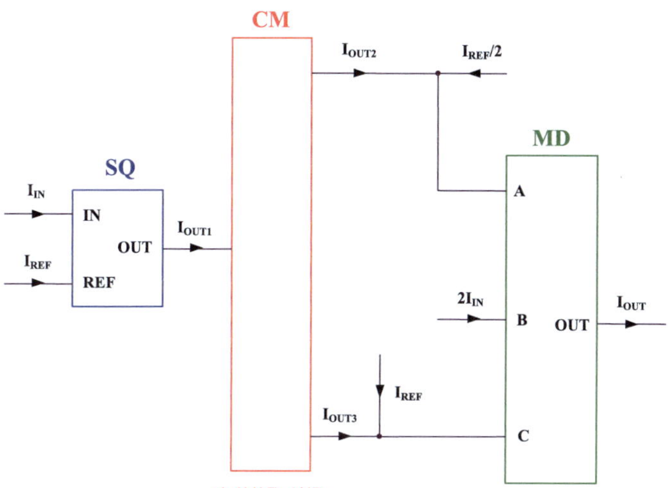

Fig. 5.4 Block diagram of the first analog function synthesizer for generating the bipolar sigmoidal activation function

$$I_{\mathrm{OUT}} = I_{\mathrm{REF}} \frac{\frac{I_{\mathrm{IN}}^3}{120 I_{\mathrm{REF}}^2} + \frac{I_{\mathrm{IN}}}{2}}{\frac{I_{\mathrm{IN}}^2}{10 I_{\mathrm{REF}}} + I_{\mathrm{REF}}}, \tag{5.32}$$

equivalent with

$$I_{\mathrm{OUT1}} = I_{\mathrm{REF}} \frac{\frac{1}{120}\left(\frac{I_{\mathrm{IN}}}{I_{\mathrm{REF}}}\right)^3 + \frac{1}{2}\left(\frac{I_{\mathrm{IN}}}{I_{\mathrm{REF}}}\right)}{\frac{1}{10}\left(\frac{I_{\mathrm{IN}}}{I_{\mathrm{REF}}}\right)^2 + 1}. \tag{5.33}$$

Using the definition of the x variable, the previous relation becomes:

$$I_{\mathrm{OUT}} = I_{\mathrm{REF}} \frac{\frac{x^3}{120} + \frac{x}{2}}{\frac{x^2}{10} + 1}. \tag{5.34}$$

In conclusion, the analog function synthesizer shown in Fig. 5.4 approximates the bipolar sigmoidal activation function:

$$I_{\mathrm{OUT}} = I_{\mathrm{REF}} a_4(x) \cong I_{\mathrm{REF}} f_2(x). \tag{5.35}$$

5.3.2 *Second Analog Function Synthesizer for Generating the Bipolar Sigmoidal Activation Function*

The second analog function synthesizer, designed for generating the bipolar sigmoidal activation function, uses $a_5(x)$ approximation function described in Chap. 1.

The block diagram of the second analog function synthesizer for generating the bipolar sigmoidal activation function is presented in Fig. 5.5.

The expressions of the currents from the circuit are:

$$I_{\mathrm{OUT1}} = \frac{I_{\mathrm{IN}}^2}{8 I_{\mathrm{REF}}}, \tag{5.36}$$

$$I_{\mathrm{OUT2}} = \frac{2}{21} I_{\mathrm{OUT1}} = \frac{I_{\mathrm{IN}}^2}{84 I_{\mathrm{REF}}}, \tag{5.37}$$

$$I_{\mathrm{OUT3}} = \frac{6}{7} I_{\mathrm{OUT1}} = \frac{3 I_{\mathrm{IN}}^2}{28 I_{\mathrm{REF}}}, \tag{5.38}$$

$$I_{\mathrm{OUT4}} = I_{\mathrm{OUT1}} = \frac{I_{\mathrm{IN}}^2}{8 I_{\mathrm{REF}}} \tag{5.39}$$

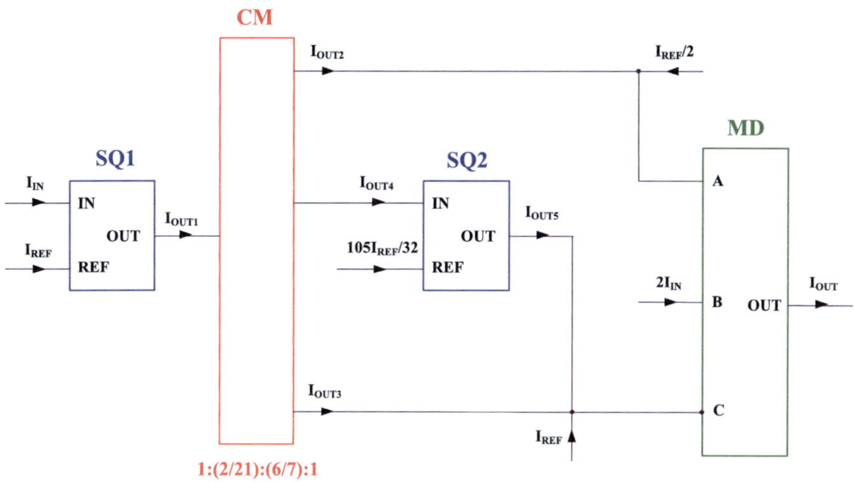

Fig. 5.5 Block diagram of the second analog function synthesizer for generating the bipolar sigmoidal activation function

and

$$I_{OUT5} = \frac{32 I_{OUT4}^2}{8 \times 105 \times I_{REF}} = \frac{I_{IN}^4}{1680 I_{REF}^3}. \tag{5.40}$$

So, the output current of the circuit from Fig. 5.5 can be expressed as follows:

$$I_{OUT} = I_{REF} \frac{\left(\frac{I_{IN}^2}{84 I_{REF}} + \frac{I_{REF}}{2}\right) I_{IN}}{\frac{I_{IN}^4}{1680 I_{REF}^3} + \frac{3 I_{IN}^2}{28 I_{REF}} + I_{REF}}, \tag{5.41}$$

equivalent with

$$I_{OUT} = I_{REF} \frac{\frac{1}{84}\left(\frac{I_{IN}}{I_{REF}}\right)^3 + \frac{1}{2}\left(\frac{I_{IN}}{I_{REF}}\right)}{\frac{1}{1680}\left(\frac{I_{IN}}{I_{REF}}\right)^4 + \frac{3}{28}\left(\frac{I_{IN}}{I_{REF}}\right)^2 + 1}. \tag{5.42}$$

Using the definition of the x variable, the previous relation becomes:

$$I_{OUT} = I_{REF} \frac{\frac{x^3}{84} + \frac{x}{2}}{\frac{x^4}{1680} + \frac{3x^2}{28} + 1}. \tag{5.43}$$

In conclusion, the analog function synthesizer shown in Fig. 5.5 approximates the bipolar sigmoidal activation function:

$$I_{OUT} = I_{REF} a_5(x) \cong I_{REF} f_2(x). \quad (5.44)$$

5.4 Analog Function Synthesizers for Einstein Activation Functions

In order to implement in CMOS technology the Einstein activation functions, the approximation functions described in Chap. 1 and the CMOS squaring and multiplies/divider circuits from Chap. 4 are used.

5.4.1 First Analog Function Synthesizer for Generating the First Einstein Activation Function

The first analog function synthesizer, designed for generating the first Einstein activation function, uses $a_6(x)$ approximation function described in Chap. 1. The utilization of the current-mode squaring circuits (SQ) and multiplier/divider circuits (MD) analyzed in Chap. 4 allows to obtain CMOS implementations of the analog function synthesizer, having very good accuracy and reasonable complexity.

The block diagram of the first analog function synthesizer for generating the first Einstein activation function is presented in Fig. 5.6.

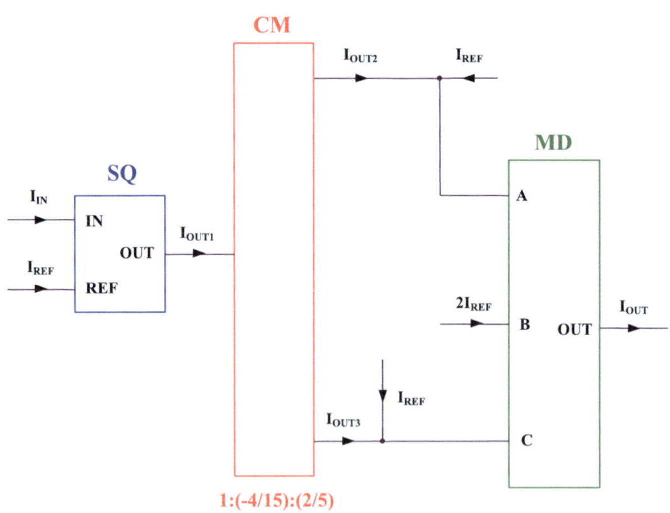

Fig. 5.6 Block diagram of the first analog function synthesizer for generating the first Einstein activation function

The expressions of the currents from the circuit are:

$$I_{\mathrm{OUT1}} = \frac{I_{\mathrm{IN}}^2}{8I_{\mathrm{REF}}}, \tag{5.45}$$

$$I_{\mathrm{OUT2}} = -\frac{4}{15}I_{\mathrm{OUT1}} = -\frac{I_{\mathrm{IN}}^2}{30I_{\mathrm{REF}}}, \tag{5.46}$$

and

$$I_{\mathrm{OUT3}} = \frac{2}{5}I_{\mathrm{OUT1}} = \frac{I_{\mathrm{IN}}^2}{20I_{\mathrm{REF}}}. \tag{5.47}$$

The output current of the circuit from Fig. 5.6 can be expressed as follows:

$$I_{\mathrm{OUT}} = I_{\mathrm{REF}}\frac{I_{\mathrm{REF}} - \frac{I_{\mathrm{IN}}^2}{30I_{\mathrm{REF}}}}{\frac{I_{\mathrm{IN}}^2}{20I_{\mathrm{REF}}} + I_{\mathrm{REF}}}, \tag{5.48}$$

equivalent with

$$I_{\mathrm{OUT}} = I_{\mathrm{REF}}\frac{1 - \frac{1}{30}\left(\frac{I_{\mathrm{IN}}}{I_{\mathrm{REF}}}\right)^2}{\frac{1}{20}\left(\frac{I_{\mathrm{IN}}}{I_{\mathrm{REF}}}\right)^2 + 1}. \tag{5.49}$$

Using the definition of the x variable, the previous relation becomes:

$$I_{\mathrm{OUT}} = I_{\mathrm{REF}}\frac{1 - \frac{x^2}{30}}{\frac{x^2}{20} + 1}. \tag{5.50}$$

In conclusion, the analog function synthesizer shown in Fig. 5.6 approximates the bipolar sigmoidal activation function:

$$I_{\mathrm{OUT}} = I_{\mathrm{REF}}a_6(x) \cong I_{\mathrm{REF}}f_3(x). \tag{5.51}$$

5.4.2 *Second Analog Function Synthesizer for Generating the First Einstein Activation Function*

The second analog function synthesizer, designed for generating the first Einstein activation function, uses $a_7(x)$ approximation function described in Chap. 1.

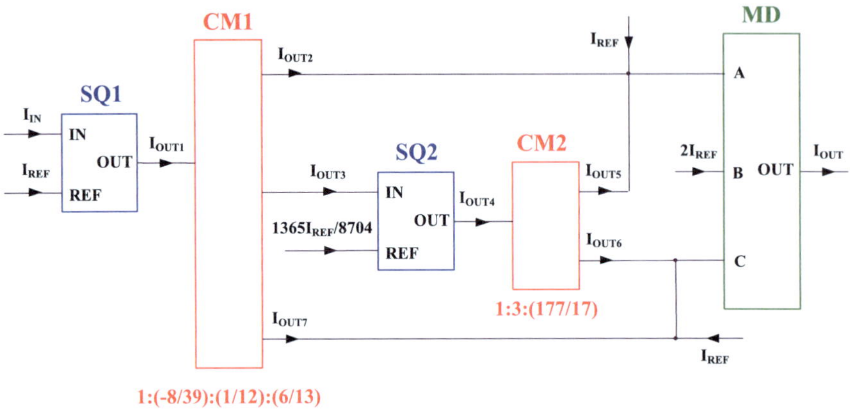

Fig. 5.7 Block diagram of the second analog function synthesizer for generating the first Einstein activation function

The block diagram of the second analog function synthesizer for generating the first Einstein activation function is presented in Fig. 5.7.

The expressions of the currents from the circuit are:

$$I_{OUT1} = \frac{I_{IN}^2}{8I_{REF}}, \tag{5.52}$$

$$I_{out2} = -\frac{8}{39}I_{out1} = -\frac{I_{IN}^2}{I_{REF}}, \tag{5.53}$$

$$I_{OUT3} = \frac{1}{12}I_{OUT1} = \frac{I_{IN}^2}{96I_{REF}}, \tag{5.54}$$

$$I_{OUT4} = \frac{I_{OUT3}^2}{8 \times 1365I_{REF}}8704 = \frac{17I_{IN}^4}{196560I_{REF}^3} \tag{5.55}$$

$$I_{OUT5} = 3I_{OUT4} = \frac{17}{65520}\frac{I_{IN}^2}{I_{REF}^3}, \tag{5.56}$$

$$I_{OUT6} = \frac{177}{17}I_{OUT4} = \frac{59}{65520}\frac{I_{IN}^4}{I_{REF}^3} \tag{5.57}$$

and

$$I_{OUT7} = \frac{6}{13}I_{OUT1} = \frac{3}{52}\frac{I_{IN}^2}{I_{REF}}. \tag{5.58}$$

So, the output current of the circuit from Fig. 5.7 can be expressed as follows:

$$I_{OUT1} = I_{REF} \frac{\frac{17}{65520}\frac{I_{IN}^4}{I_{REF}^3} - \frac{I_{IN}^2}{39 I_{REF}} + I_{REF}}{\frac{59}{65520}\frac{I_{IN}^4}{I_{REF}^3} + \frac{3}{52}\frac{I_{IN}^2}{I_{REF}} + I_{REF}}. \tag{5.59}$$

Using the definition of the x variable, the previous relation becomes:

$$I_{OUT} = I_{REF} \frac{\frac{17x^4}{65520} - \frac{x^2}{39} + 1}{\frac{59x^4}{65520} + \frac{3x^2}{52} + 1}. \tag{5.60}$$

In conclusion, the analog function synthesizer shown in Fig. 5.7 approximates the first Einstein activation function:

$$I_{OUT} = I_{REF} a_7(x) \cong I_{REF} f_3(x). \tag{5.61}$$

5.4.3 Third Analog Function Synthesizer for Generating the First Einstein Activation Function

The third analog function synthesizer, designed for generating the first Einstein activation function, uses $a_8(x)$ approximation function described in Chap. 1.

The block diagram of the third analog function synthesizer for generating the first Einstein activation function is presented in Fig. 5.8.

The expressions of the currents from the circuit are:

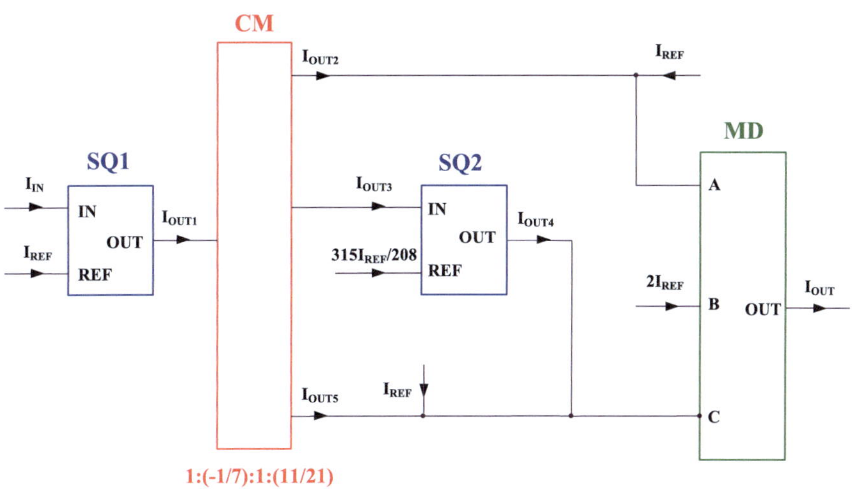

Fig. 5.8 Block diagram of the third analog function synthesizer for generating the first Einstein activation function

$$I_{\mathrm{OUT1}} = \frac{I_{\mathrm{IN}}^2}{8 I_{\mathrm{REF}}}, \tag{5.62}$$

$$I_{\mathrm{OUT2}} = -\frac{1}{7} I_{\mathrm{OUT1}} = -\frac{I_{\mathrm{IN}}^2}{56 I_{\mathrm{REF}}}, \tag{5.63}$$

$$I_{\mathrm{OUT3}} = I_{\mathrm{OUT1}} = \frac{I_{\mathrm{IN}}^2}{8 I_{\mathrm{REF}}}, \tag{5.64}$$

$$I_{\mathrm{OUT4}} = \frac{I_{\mathrm{OUT3}}^2}{8 \times 315 I_{\mathrm{REF}}} 208 = \frac{13 I_{\mathrm{IN}}^4}{10080 I_{\mathrm{REF}}^3}, \tag{5.65}$$

and

$$I_{\mathrm{OUT5}} = \frac{11}{21} I_{\mathrm{OUT1}} = \frac{11}{168} \frac{I_{\mathrm{IN}}^2}{I_{\mathrm{REF}}}. \tag{5.66}$$

So, the output current of the circuit from Fig. 5.8 can be expressed as follows:

$$I_{\mathrm{OUT}} = I_{\mathrm{REF}} \frac{I_{\mathrm{REF}} - \frac{I_{\mathrm{IN}}^2}{56 I_{\mathrm{REF}}}}{\frac{13 I_{\mathrm{IN}}^4}{10080 I_{\mathrm{REF}}^3} + \frac{11 I_{\mathrm{IN}}^2}{168 I_{\mathrm{REF}}} + I_{\mathrm{REF}}}. \tag{5.67}$$

Using the definition of the x variable, the previous relation becomes:

$$I_{\mathrm{OUT}} = I_{\mathrm{REF}} \frac{1 - \frac{x^2}{56}}{\frac{13x^4}{10080} + \frac{11x^2}{168} + 1}. \tag{5.68}$$

In conclusion, the analog function synthesizer shown in Fig. 5.8 approximates the first Einstein activation function:

$$I_{\mathrm{OUT}} = I_{\mathrm{REF}} a_8(x) \cong I_{\mathrm{REF}} f_3(x). \tag{5.69}$$

5.4.4 *First Analog Function Synthesizer for Generating the Second Einstein Activation Function*

The first analog function synthesizer, designed for generating the second Einstein activation function, uses $a_9(x)$ approximation function described in Chap. 1.

The block diagram of the first analog function synthesizer for generating the second Einstein activation function is presented in Fig. 5.9.

The expressions of the currents from the circuit are:

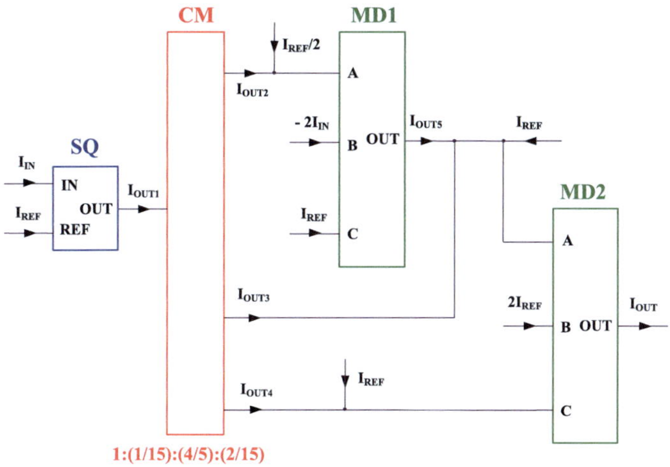

Fig. 5.9 Block diagram of the first analog function synthesizer for generating the second Einstein activation function

$$I_{\text{OUT1}} = \frac{I_{\text{IN}}^2}{8I_{\text{REF}}}, \tag{5.70}$$

$$I_{\text{OUT2}} = \frac{1}{15}I_{\text{OUT1}} = \frac{I_{\text{IN}}^2}{120I_{\text{REF}}}, \tag{5.71}$$

$$I_{\text{OUT3}} = \frac{4}{5}I_{\text{OUT1}} = \frac{I_{\text{IN}}^2}{10I_{\text{REF}}}, \tag{5.72}$$

$$I_{\text{OUT4}} = \frac{2}{15}I_{\text{OUT1}} = \frac{I_{\text{IN}}^2}{60I_{\text{REF}}} \tag{5.73}$$

and

$$I_{\text{OUT5}} = \frac{-\left(\frac{I_{\text{IN}}^2}{120I_{\text{REF}}} + \frac{I_{\text{REF}}}{2}\right)I_{\text{IN}}}{I_{\text{REF}}}. \tag{5.74}$$

So, the output current of the circuit from Fig. 5.9 can be expressed as follows:

$$I_{\text{OUT}} = I_{\text{REF}}\frac{-\frac{1}{120}\frac{I_{\text{IN}}^3}{I_{\text{REF}}^2} + \frac{I_{\text{IN}}^2}{10I_{\text{REF}}} - \frac{I_{\text{IN}}}{2} + I_{\text{REF}}}{\frac{I_{\text{IN}}^2}{60I_{\text{REF}}} + I_{\text{REF}}}. \tag{5.75}$$

Using the definition of the x variable, the previous relation becomes:

$$I_{OUT} = I_{REF} \frac{-\frac{x^3}{120} + \frac{x^2}{10} - \frac{x}{2} + 1}{\frac{x^2}{60} + 1}. \tag{5.76}$$

In conclusion, the analog function synthesizer shown in Fig. 5.9 approximates the second Einstein activation function:

$$I_{OUT} = I_{REF} a_9(x) \cong I_{REF} f_4(x). \tag{5.77}$$

5.4.5 *Second Analog Function Synthesizer for Generating the Second Einstein Activation Function*

The second analog function synthesizer, designed for generating the second Einstein activation function, uses $a_{10}(x)$ approximation function described in Chap. 1.

The block diagram of the second analog function synthesizer for generating the second Einstein activation function is presented in Fig. 5.10.

The expressions of the currents from the circuit are:

$$I_{OUT1} = \frac{I_{IN}^2}{8I_{REF}}, \tag{5.78}$$

$$I_{OUT2} = \frac{4}{7} I_{OUT1} = \frac{I_{IN}^2}{14I_{REF}}, \tag{5.79}$$

$$I_{OUT3} = \frac{4}{15} I_{OUT1} = \frac{I_{IN}^2}{30I_{REF}}, \tag{5.80}$$

$$I_{OUT4} = -\frac{I_{IN}}{7I_{REF}} \left(\frac{I_{IN}^2}{30I_{REF}} + 3I_{REF} \right), \tag{5.81}$$

$$I_{OUT5} = \frac{2}{15} I_{OUT1} = \frac{I_{IN}^2}{60I_{REF}}, \tag{5.82}$$

$$I_{OUT6} = \frac{I_{IN}}{14I_{REF}} \left(\frac{I_{IN}^2}{60I_{REF}} + I_{REF} \right) \tag{5.83}$$

and

$$I_{OUT7} = \frac{4}{21} I_{OUT1} = \frac{I_{IN}^2}{42I_{REF}}. \tag{5.84}$$

The output current of the circuit from Fig. 5.10 can be expressed as follows:

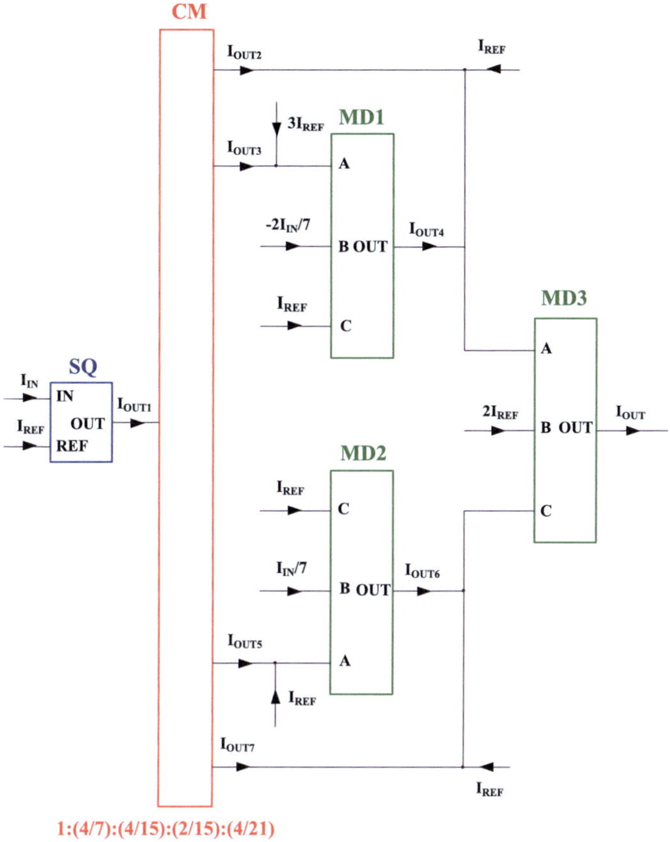

Fig. 5.10 Block diagram of the second analog function synthesizer for generating the second Einstein activation function

$$I_{\mathrm{OUT}} = I_{\mathrm{REF}} \frac{-\frac{I_{\mathrm{IN}}^3}{210 I_{\mathrm{REF}}^2} + \frac{I_{\mathrm{IN}}^2}{14 I_{\mathrm{REF}}} - \frac{3 I_{\mathrm{IN}}}{7} + I_{\mathrm{REF}}}{\frac{I_{\mathrm{IN}}^3}{840 I_{\mathrm{REF}}^2} + \frac{I_{\mathrm{IN}}^2}{42 I_{\mathrm{REF}}} + \frac{I_{\mathrm{IN}}}{14} + I_{\mathrm{REF}}}. \tag{5.85}$$

Using the definition of the x variable, the previous relation becomes:

$$I_{\mathrm{OUT}} = I_{\mathrm{REF}} \frac{-\frac{x^3}{210} + \frac{x^2}{14} - \frac{3x}{7} + 1}{\frac{x^3}{840} + \frac{x^2}{42} + \frac{x}{14} + 1}. \tag{5.86}$$

In conclusion, the analog function synthesizer shown in Fig. 5.10 approximates the second Einstein activation function:

$$I_{\mathrm{OUT}} = I_{\mathrm{REF}} a_{10}(x) \cong I_{\mathrm{REF}} f_4(x). \tag{5.87}$$

5.5 Analog Function Synthesizers for Hyperbolic Tangent (Tan-Sigmoid) Activation Function

In order to implement in CMOS technology the hyperbolic tangent activation functions, the approximation functions described in Chap. 1 and the CMOS squaring and multiplies/divider circuits from Chap. 4 are used.

5.5.1 *First Analog Function Synthesizer for Generating the Hyperbolic Tangent Activation Function*

The first analog function synthesizer, designed for generating the hyperbolic tangent activation function, uses $a_{11}(x)$ approximation function described in Chap. 1.

The block diagram of the second analog function synthesizer for generating the hyperbolic tangent activation function is presented in Fig. 5.11.

The expressions of the currents from the circuit are:

$$I_{\mathrm{OUT1}} = \frac{I_{\mathrm{IN}}^2}{8I_{\mathrm{REF}}}, \tag{5.88}$$

$$I_{\mathrm{OUT2}} = \frac{8}{15}I_{\mathrm{OUT1}} = \frac{I_{\mathrm{IN}}^2}{15I_{\mathrm{REF}}}, \tag{5.89}$$

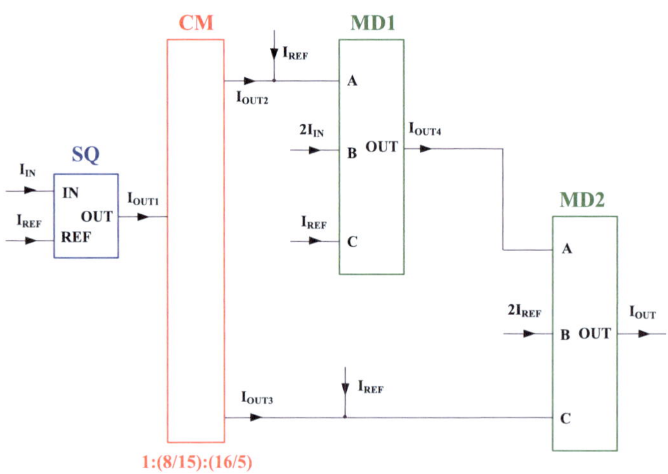

Fig. 5.11 Block diagram of the first analog function synthesizer for generating the hyperbolic tangent activation function

$$I_{OUT3} = \frac{16}{5} I_{OUT1} = \frac{2I_{IN}^2}{5I_{REF}} \tag{5.90}$$

and

$$I_{OUT4} = \frac{\left(\frac{I_{IN}^2}{15I_{REF}} + I_{REF}\right) I_{IN}}{I_{REF}}. \tag{5.91}$$

So, the output current of the circuit from Fig. 5.11 can be expressed as follows:

$$I_{OUT} = I_{REF} \frac{\frac{I_{IN}^3}{15I_{REF}^2} + I_{IN}}{\frac{2I_{IN}^2}{5I_{REF}} + I_{REF}}, \tag{5.92}$$

equivalent with

$$I_{OUT} = I_{REF} \frac{\frac{1}{15}\left(\frac{I_{IN}}{I_{REF}}\right)^3 + \left(\frac{I_{IN}}{I_{REF}}\right)}{\frac{2}{5}\left(\frac{I_{IN}}{I_{REF}}\right)^2 + 1}. \tag{5.93}$$

Using the definition of the x variable, the previous relation becomes:

$$I_{OUT} = I_{REF} \frac{\frac{x^3}{15} + x}{\frac{2x^2}{5} + 1}. \tag{5.94}$$

In conclusion, the analog function synthesizer shown in Fig. 5.11 approximates the hyperbolic tangent activation function:

$$I_{OUT} = I_{REF} a_{11}(x) \cong I_{REF} f_5(x). \tag{5.95}$$

5.5.2 *Second Analog Function Synthesizer for Generating the Hyperbolic Tangent Activation Function*

The second analog function synthesizer, designed for generating the hyperbolic tangent activation function, uses $a_{12}(x)$ approximation function described in Chap. 1.

The block diagram of the second analog function synthesizer for generating the hyperbolic tangent activation function is presented in Fig. 5.12.

The expressions of the currents from the circuit are:

$$I_{OUT1} = \frac{I_{IN}^2}{8I_{REF}}, \tag{5.96}$$

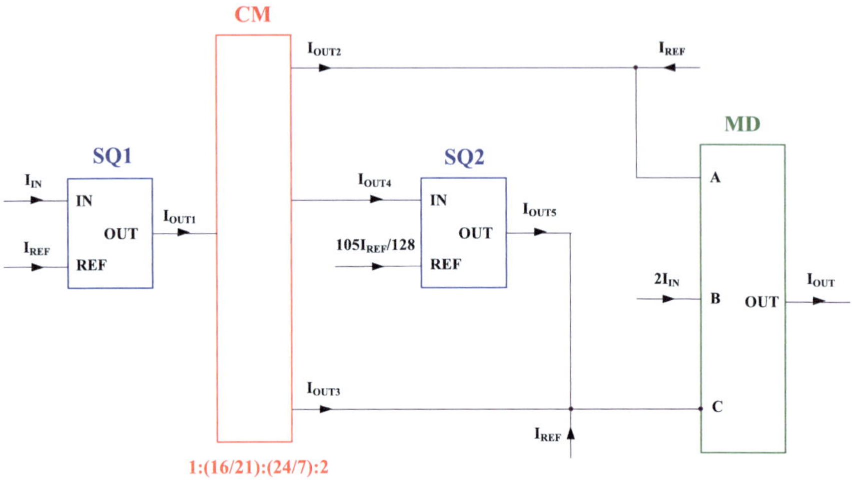

Fig. 5.12 Block diagram of the second analog function synthesizer for generating the hyperbolic tangent activation function

$$I_{OUT2} = \frac{16}{21} I_{OUT1} = \frac{2I_{IN}^2}{21I_{REF}}, \tag{5.97}$$

$$I_{OUT3} = \frac{24}{7} I_{OUT1} = \frac{3I_{IN}^2}{7I_{REF}}, \tag{5.98}$$

$$I_{OUT4} = 2I_{OUT1} = \frac{I_{IN}^2}{4I_{REF}} \tag{5.99}$$

and

$$I_{OUT5} = \frac{I_{OUT4}^2}{8 \times (105I_{REF}/128)} = \frac{I_{IN}^4}{105I_{REF}^3}. \tag{5.100}$$

So, the output current of the circuit from Fig. 5.12 can be expressed as follows:

$$I_{OUT} = I_{REF} \frac{\left(\frac{2I_{IN}^2}{21I_{REF}} + I_{REF} \right) I_{IN}}{\frac{I_{IN}^4}{105I_{REF}^3} + \frac{3I_{IN}^2}{7I_{REF}} + I_{REF}}, \tag{5.101}$$

equivalent with

$$I_{OUT} = I_{REF} \frac{\frac{2}{21}\left(\frac{I_{IN}}{I_{REF}}\right)^3 + \left(\frac{I_{IN}}{I_{REF}}\right)}{\frac{1}{105}\left(\frac{I_{IN}}{I_{REF}}\right)^4 + \frac{3}{7}\left(\frac{I_{IN}}{I_{REF}}\right)^2 + 1}. \tag{5.102}$$

Using the definition of the x variable, the previous relation becomes:

$$I_{\mathrm{OUT}} = I_{\mathrm{REF}} \frac{10x^3 + 105x}{x^4 + 45x^2 + 105}. \tag{5.103}$$

In conclusion, the analog function synthesizer shown in Fig. 5.12 approximates the hyperbolic tangent activation function:

$$I_{\mathrm{OUT}} = I_{\mathrm{REF}} a_{12}(x) \cong I_{\mathrm{REF}} f_5(x). \tag{5.104}$$

5.5.3 *Third Analog Function Synthesizer for Generating the Hyperbolic Tangent Activation Function*

The third analog function synthesizer, designed for generating the hyperbolic tangent activation function, uses $a_{13}(x)$ approximation function described in Chap. 1.

The block diagram of the third analog function synthesizer for generating the hyperbolic tangent activation function is presented in Fig. 5.13.

The expressions of the currents from the circuit are:

$$I_{\mathrm{OUT1}} = \frac{I_{\mathrm{IN}}^2}{8 I_{\mathrm{REF}}}, \tag{5.105}$$

$$I_{\mathrm{OUT2}} = \frac{8}{9} I_{\mathrm{OUT1}} = \frac{I_{\mathrm{IN}}^2}{9 I_{\mathrm{REF}}}, \tag{5.106}$$

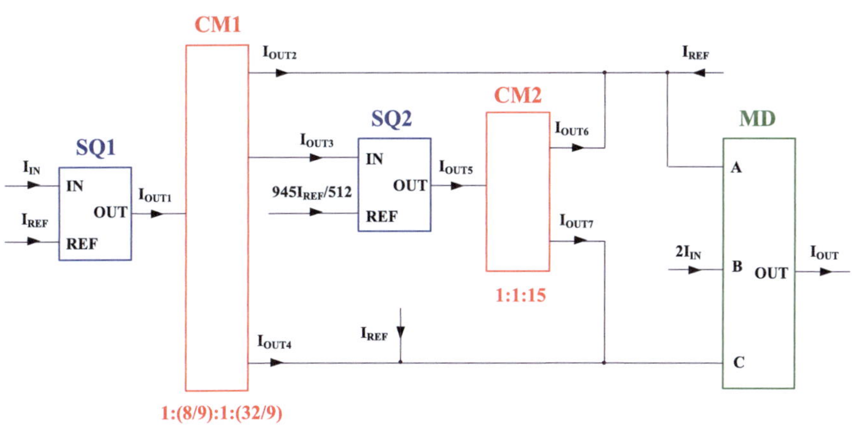

Fig. 5.13 Block diagram of the third analog function synthesizer for generating the hyperbolic tangent activation function

$$I_{OUT3} = I_{OUT1} = \frac{I_{IN}^2}{8I_{REF}}, \tag{5.107}$$

$$I_{OUT4} = \frac{32}{9} I_{OUT1} = \frac{4I_{IN}^2}{9I_{REF}}, \tag{5.108}$$

$$I_{OUT5} = \frac{I_{OUT3}^2}{8(945I_{REF}/512)} = \frac{I_{IN}^4}{945I_{REF}^3}, \tag{5.109}$$

$$I_{OUT6} = I_{OUT5} = \frac{I_{IN}^4}{945I_{REF}^3} \tag{5.110}$$

and

$$I_{OUT7} = 15I_{OUT5} = \frac{I_{IN}^4}{63I_{REF}^3}. \tag{5.111}$$

The output current of the circuit from Fig. 5.13 can be expressed as follows:

$$I_{OUT} = I_{IN} \frac{\frac{I_{IN}^4}{945I_{REF}^3} + \frac{I_{IN}^2}{9I_{REF}} + I_{REF}}{\frac{I_{IN}^4}{63I_{REF}^3} + \frac{4I_{IN}^2}{9I_{REF}} + I_{REF}}. \tag{5.112}$$

Using the definition of the x variable, the previous relation becomes:

$$I_{OUT} = I_{REF} \frac{x}{15} \frac{x^4 + 105x^2 + 945}{x^4 + 28x^2 + 63}. \tag{5.113}$$

In conclusion, the analog function synthesizer shown in Fig. 5.13 approximates the hyperbolic tangent activation function:

$$I_{OUT} = I_{REF} a_{13}(x) \cong I_{REF} f_5(x). \tag{5.114}$$

5.5.4 *Fourth Analog Function Synthesizer for Generating the Hyperbolic Tangent Activation Function*

The fourth analog function synthesizer, designed for generating the hyperbolic tangent activation function, uses $a_{14}(x)$ approximation function described in Chap. 1.

The block diagram of the fourth analog function synthesizer for generating the hyperbolic tangent activation function is presented in Fig. 5.14.

The expressions of the currents from the circuit are:

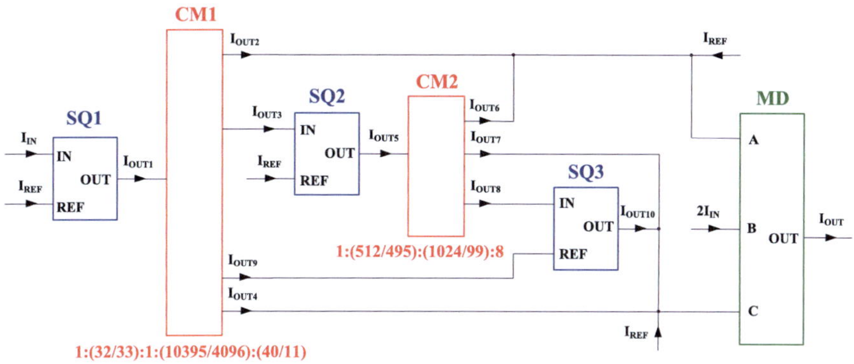

Fig. 5.14 Block diagram of the fourth analog function synthesizer for generating the hyperbolic tangent activation function

$$I_{\text{OUT1}} = \frac{I_{\text{IN}}^2}{8I_{\text{REF}}}, \tag{5.115}$$

$$I_{\text{OUT2}} = \frac{32}{33} I_{\text{OUT1}} = \frac{4I_{\text{IN}}^2}{33I_{\text{REF}}}, \tag{5.116}$$

$$I_{\text{OUT3}} = I_{\text{OUT1}} = \frac{I_{\text{IN}}^2}{8I_{\text{REF}}}, \tag{5.117}$$

$$I_{\text{OUT4}} = \frac{40}{11} I_{\text{OUT1}} = \frac{5I_{\text{IN}}^2}{11I_{\text{REF}}}, \tag{5.118}$$

$$I_{\text{OUT5}} = \frac{I_{\text{OUT3}}^2}{8I_{\text{REF}}} = \frac{I_{\text{IN}}^4}{512I_{\text{REF}}^3}, \tag{5.119}$$

$$I_{\text{OUT6}} = \frac{512}{495} I_{\text{OUT5}} = \frac{I_{\text{IN}}^4}{495I_{\text{REF}}^3}, \tag{5.120}$$

$$I_{\text{OUT7}} = \frac{1024}{99} I_{\text{OUT5}} = \frac{2I_{\text{IN}}^4}{99I_{\text{REF}}^3}, \tag{5.121}$$

$$I_{\text{OUT8}} = 8I_{\text{OUT5}} = \frac{I_{\text{IN}}^4}{64I_{\text{REF}}^3}, \tag{5.122}$$

$$I_{\text{OUT9}} = \frac{10395}{4096} I_{\text{OUT1}} = \frac{10395}{32768} \frac{I_{\text{IN}}^2}{I_{\text{REF}}}, \tag{5.123}$$

and

$$I_{\mathrm{OUT10}} = \frac{I_{\mathrm{OUT8}}^2}{8I_{\mathrm{OUT9}}} = \frac{1}{10395}\frac{I_{\mathrm{IN}}^6}{I_{\mathrm{REF}}^5}. \tag{5.124}$$

The output current of the circuit from Fig. 5.14 can be expressed as follows:

$$I_{\mathrm{OUT}} = I_{\mathrm{IN}} \frac{\frac{I_{\mathrm{IN}}^4}{495I_{\mathrm{REF}}^3} + \frac{4I_{\mathrm{IN}}^2}{33I_{\mathrm{REF}}} + I_{\mathrm{REF}}}{\frac{I_{\mathrm{IN}}^6}{10395I_{\mathrm{REF}}^5} + \frac{2I_{\mathrm{IN}}^4}{99I_{\mathrm{REF}}^3} + \frac{5I_{\mathrm{IN}}^2}{11I_{\mathrm{REF}}} + I_{\mathrm{REF}}}. \tag{5.125}$$

Using the definition of the x variable, the previous relation becomes:

$$I_{\mathrm{OUT}} = I_{\mathrm{REF}} \frac{\frac{x^5}{495} + \frac{4x^3}{33} + x}{\frac{x^6}{10395} + \frac{2x^4}{99} + \frac{5x^2}{11} + 1}. \tag{5.126}$$

In conclusion, the analog function synthesizer shown in Fig. 5.14 approximates the hyperbolic tangent activation function:

$$I_{\mathrm{OUT}} = I_{\mathrm{REF}} a_{14}(x) \cong I_{\mathrm{REF}} f_5(x). \tag{5.127}$$

5.6 Analog Function Synthesizers for Continuous-Log Sigmoid Activation Function

In order to implement in CMOS technology the continuous-log activation functions, the approximation functions described in Chap. 1 and the CMOS squaring and multiplies/divider circuits from Chap. 4 are used.

5.6.1 First Analog Function Synthesizer for Generating the Continuous-Log Sigmoid Activation Function (for b = 2)

The first analog function synthesizer, designed for generating the continuous-log activation function (for $b = 2$), uses $a_{15}(x)$ approximation function described in Chap. 1.

The block diagram of the second analog function synthesizer for generating the continuous-log activation function (for $b = 2$) is presented in Fig. 5.15.

The expressions of the currents from the circuit are:

$$I_{\mathrm{OUT1}} = \frac{I_{\mathrm{IN}}^2}{8I_{\mathrm{REF}}}, \tag{5.128}$$

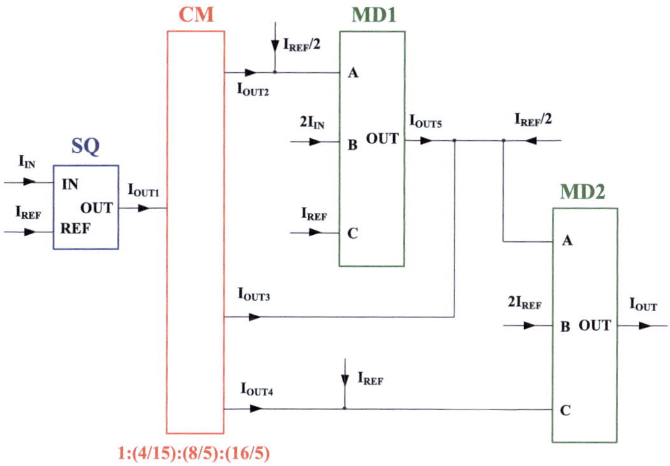

Fig. 5.15 Block diagram of the first analog function synthesizer for generating the continuous-log activation function (for $b = 2$)

$$I_{\text{OUT2}} = \frac{4}{15} I_{\text{OUT1}} = \frac{I_{\text{IN}}^2}{30 I_{\text{REF}}}, \tag{5.129}$$

$$I_{\text{OUT3}} = \frac{8}{5} I_{\text{OUT1}} = \frac{I_{\text{IN}}^2}{5 I_{\text{REF}}}, \tag{5.130}$$

$$I_{\text{OUT4}} = \frac{16}{5} I_{\text{OUT1}} = \frac{2 I_{\text{IN}}^2}{5 I_{\text{REF}}} \tag{5.131}$$

and

$$I_{\text{OUT5}} = \frac{\left(\frac{I_{\text{IN}}^2}{30 I_{\text{REF}}} + \frac{I_{\text{REF}}}{2} \right) I_{\text{IN}}}{I_{\text{REF}}}. \tag{5.132}$$

So, the output current of the circuit from Fig. 5.15 can be expressed as follows:

$$I_{\text{OUT}} = I_{\text{REF}} \frac{\frac{I_{\text{IN}}^3}{30 I_{\text{REF}}^2} + \frac{I_{\text{IN}}^2}{5 I_{\text{REF}}} + \frac{I_{\text{IN}}}{2} + \frac{I_{\text{REF}}}{2}}{\frac{2 I_{\text{IN}}^2}{5 I_{\text{REF}}} + I_{\text{REF}}}, \tag{5.133}$$

equivalent with

$$I_{\text{OUT}} = I_{\text{REF}} \frac{\frac{1}{30} \left(\frac{I_{\text{IN}}}{I_{\text{REF}}} \right)^3 + \frac{1}{5} \left(\frac{I_{\text{IN}}}{I_{\text{REF}}} \right)^2 + \frac{1}{2} \left(\frac{I_{\text{IN}}}{I_{\text{REF}}} \right) + \frac{1}{2}}{\frac{2}{5} \left(\frac{I_{\text{IN}}}{I_{\text{REF}}} \right)^2 + 1}. \tag{5.134}$$

Using the definition of the x variable, the previous relation becomes:

$$I_{\mathrm{OUT}} = I_{\mathrm{REF}} \frac{\frac{x^3}{30} + \frac{x^2}{5} + \frac{x}{2} + \frac{1}{2}}{\frac{2x^2}{5} + 1}. \tag{5.135}$$

In conclusion, the analog function synthesizer shown in Fig. 5.15 approximates the continuous-log activation function (for $b = 2$):

$$I_{\mathrm{OUT}} = I_{\mathrm{REF}} a_{15}(x) \cong I_{\mathrm{REF}} f_8(x). \tag{5.136}$$

5.6.2 *Second Analog Function Synthesizer for Generating the Continuous-Log Sigmoid Activation Function (for b = 2)*

The second analog function synthesizer, designed for generating the continuous-log sigmoid activation function (for $b = 2$), uses $a_{16}(x)$ approximation function described in Chap. 1.

The block diagram of the second analog function synthesizer for generating the continuous-log sigmoid activation function (for $b = 2$) is presented in Fig. 5.16.

The expressions of the currents from the circuit are:

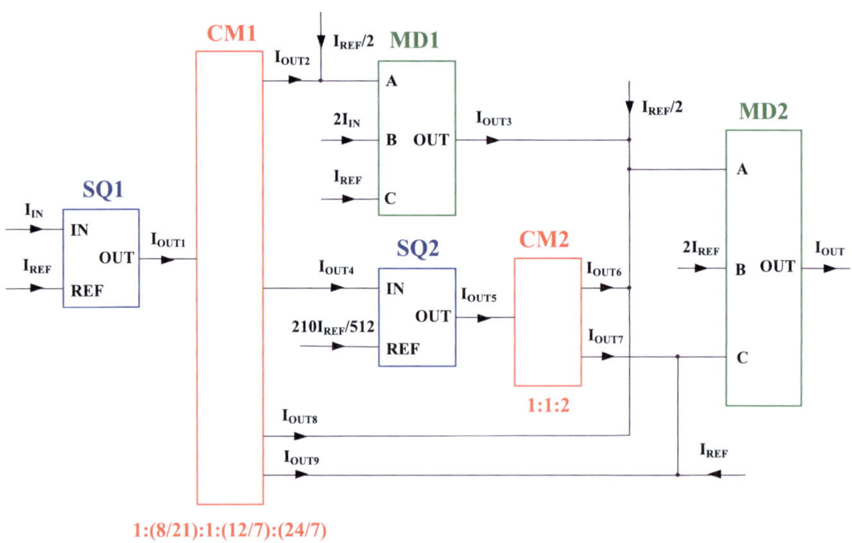

Fig. 5.16 Block diagram of the second analog function synthesizer for generating the continuous-log sigmoid activation function (for $b = 2$)

$$I_{OUT1} = \frac{I_{IN}^2}{8I_{REF}}, \tag{5.137}$$

$$I_{OUT2} = \frac{8}{21} I_{OUT1} = \frac{I_{IN}^2}{21I_{REF}}, \tag{5.138}$$

$$I_{OUT3} = \frac{\left(I_{OUT2} + \frac{I_{REF}}{2}\right) 2I_{IN}}{2I_{REF}} = \frac{\left(\frac{I_{IN}^2}{21I_{REF}} + \frac{I_{REF}}{2}\right) I_{IN}}{I_{REF}}, \tag{5.139}$$

$$I_{OUT4} = I_{OUT1} = \frac{I_{IN}^2}{8I_{REF}}, \tag{5.140}$$

$$I_{OUT5} = \frac{I_{OUT4}^2}{8(210I_{REF}/512)} = \frac{I_{IN}^4}{210I_{REF}^3}, \tag{5.141}$$

$$I_{OUT6} = I_{OUT5} = \frac{I_{IN}^4}{210I_{REF}^3}, \tag{5.142}$$

$$I_{OUT7} = 2I_{OUT5} = \frac{I_{IN}^4}{105I_{REF}^3}, \tag{5.143}$$

$$I_{OUT8} = \frac{12}{7} I_{OUT1} = \frac{3I_{IN}^2}{14I_{REF}} \tag{5.144}$$

and

$$I_{OUT9} = \frac{24}{7} I_{OUT1} = \frac{3I_{IN}^2}{7I_{REF}}. \tag{5.145}$$

So, the output current of the circuit from Fig. 5.16 can be expressed as follows:

$$I_{OUT} = I_{REF} \frac{\frac{I_{IN}^4}{210I_{REF}^3} + \frac{I_{IN}^3}{21I_{REF}^2} + \frac{3I_{IN}^2}{14I_{REF}} + \frac{I_{IN}}{2} + \frac{I_{REF}}{2}}{\frac{I_{IN}^4}{105I_{REF}^3} + \frac{3I_{IN}^2}{7I_{REF}} + I_{REF}}. \tag{5.146}$$

Using the definition of the x variable, the previous relation becomes:

$$I_{OUT} = I_{REF} \frac{\frac{x^4}{210} + \frac{x^3}{21} + \frac{3x^2}{14} + \frac{x}{2} + \frac{1}{2}}{\frac{x^4}{105} + \frac{3x^2}{7} + 1}. \tag{5.147}$$

In conclusion, the analog function synthesizer shown in Fig. 5.16 approximates the continuous-log sigmoid activation function (for $b = 2$):

$$I_{OUT} = I_{REF} a_{16}(x) \cong I_{REF} f_8(x). \tag{5.148}$$

5.6.3 *First Analog Function Synthesizer for Generating the Continuous-Log Sigmoid Activation Function (for b = 3)*

The first analog function synthesizer, designed for generating the continuous-log sigmoid activation function (for $b = 3$), uses $a_{17}(x)$ approximation function described in Chap. 1.

The block diagram of the second analog function synthesizer for generating the continuous-log sigmoid activation function (for $b = 3$) is presented in Fig. 5.17.

The expressions of the currents from the circuit are:

$$I_{\mathrm{OUT1}} = \frac{I_{\mathrm{IN}}^2}{8I_{\mathrm{REF}}}, \tag{5.149}$$

$$I_{\mathrm{OUT2}} = \frac{18}{5} I_{\mathrm{OUT1}} = \frac{9I_{\mathrm{IN}}^2}{20I_{\mathrm{REF}}}, \tag{5.150}$$

$$I_{\mathrm{OUT3}} = \frac{18}{5} I_{\mathrm{OUT1}} = \frac{9I_{\mathrm{IN}}^2}{20I_{\mathrm{REF}}}, \tag{5.151}$$

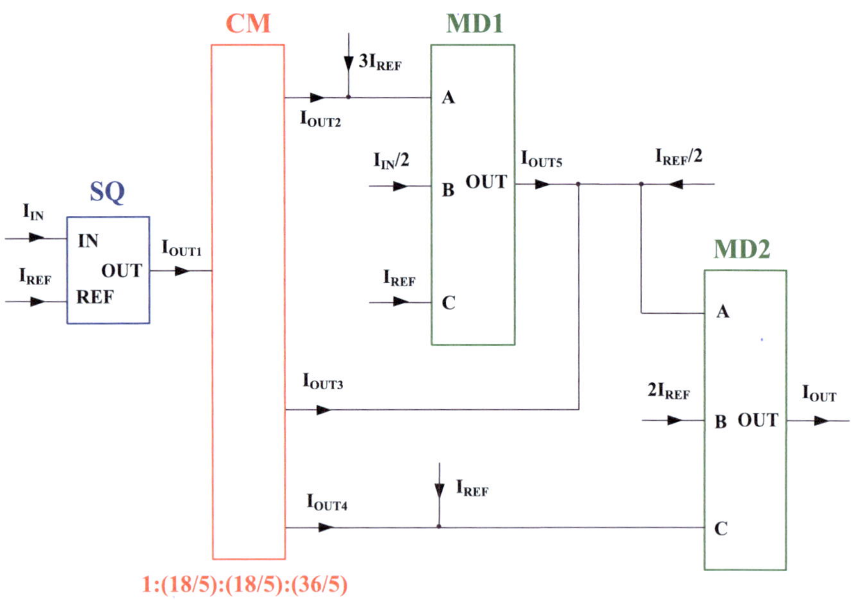

Fig. 5.17 Block diagram of the second analog function synthesizer for generating the continuous-log sigmoid activation function (for $b = 3$)

$$I_{\text{OUT4}} = \frac{36}{5} I_{\text{OUT1}} = \frac{9I_{\text{IN}}^2}{10I_{\text{REF}}} \tag{5.152}$$

and

$$I_{\text{OUT5}} = \frac{\left(\frac{9I_{\text{IN}}^2}{20I_{\text{REF}}} + 3I_{\text{REF}}\right)\frac{I_{\text{IN}}}{2}}{2I_{\text{REF}}}. \tag{5.153}$$

So, the output current of the circuit from Fig. 5.17 can be expressed as follows:

$$I_{\text{OUT}} = I_{\text{REF}} \frac{\frac{9I_{\text{IN}}^3}{80I_{\text{REF}}^2} + \frac{9I_{\text{IN}}^2}{20I_{\text{REF}}} + \frac{3I_{\text{IN}}}{4} + \frac{I_{\text{REF}}}{2}}{\frac{9I_{\text{IN}}^2}{10I_{\text{REF}}} + I_{\text{REF}}}, \tag{5.154}$$

equivalent with

$$I_{\text{OUT}} = I_{\text{REF}} \frac{\frac{9}{80}\left(\frac{I_{\text{IN}}}{I_{\text{REF}}}\right)^3 + \frac{9}{20}\left(\frac{I_{\text{IN}}}{I_{\text{REF}}}\right)^2 + \frac{3}{4}\left(\frac{I_{\text{IN}}}{I_{\text{REF}}}\right) + \frac{1}{2}}{\frac{9}{10}\left(\frac{I_{\text{IN}}}{I_{\text{REF}}}\right)^2 + 1}. \tag{5.155}$$

Using the definition of the x variable, the previous relation becomes:

$$I_{\text{OUT}} = I_{\text{REF}} \frac{\frac{9x^3}{80} + \frac{9x^2}{20} + \frac{3x}{4} + \frac{1}{2}}{\frac{9x^2}{10} + 1}. \tag{5.156}$$

In conclusion, the analog function synthesizer shown in Fig. 5.17 approximates the continuous-log sigmoid activation function (for $b = 3$):

$$I_{\text{OUT}} = I_{\text{REF}} a_{17}(x) \cong I_{\text{REF}} f_8(x). \tag{5.157}$$

5.6.4 *Second Analog Function Synthesizer for Generating the Continuous-Log Sigmoid Activation Function (for b = 3)*

The second analog function synthesizer, designed for generating the continuous-log sigmoid activation function (for $b = 3$), uses $a_{18}(x)$ approximation function described in Chap. 1.

The block diagram of the second analog function synthesizer for generating the continuous-log sigmoid activation function (for $b = 3$) is presented in Fig. 5.18.

The expressions of the currents from the circuit are:

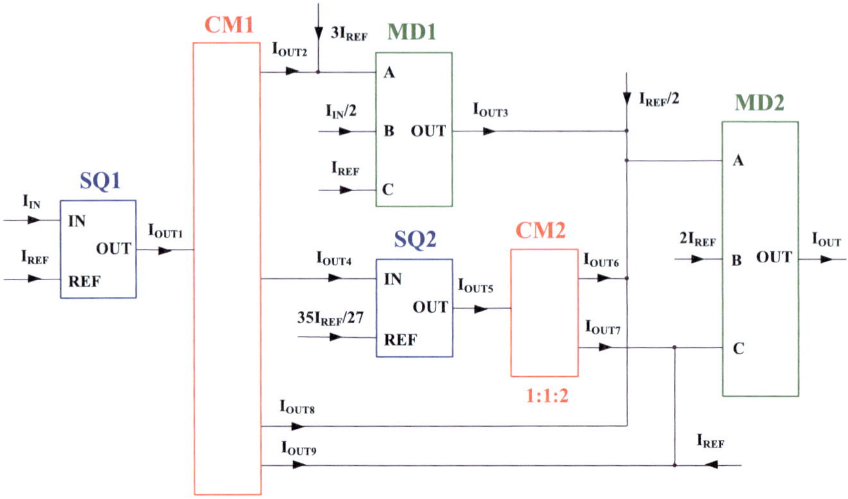

Fig. 5.18 Block diagram of the third analog function synthesizer for generating the continuous-log sigmoid activation function (for $b = 3$)

$$I_{\mathrm{OUT1}} = \frac{I_{\mathrm{IN}}^2}{8I_{\mathrm{REF}}}, \tag{5.158}$$

$$I_{\mathrm{OUT2}} = \frac{36}{7} I_{\mathrm{OUT1}} = \frac{9I_{\mathrm{IN}}^2}{14I_{\mathrm{REF}}}, \tag{5.159}$$

$$I_{\mathrm{OUT3}} = \frac{(I_{\mathrm{OUT2}} + 3I_{\mathrm{REF}})\frac{I_{\mathrm{IN}}}{2}}{2I_{\mathrm{REF}}} = \frac{\left(\frac{9I_{\mathrm{IN}}^2}{14I_{\mathrm{REF}}} + 3I_{\mathrm{REF}}\right)\frac{I_{\mathrm{IN}}}{2}}{2I_{\mathrm{REF}}}, \tag{5.160}$$

$$I_{\mathrm{OUT4}} = 4I_{\mathrm{OUT1}} = \frac{I_{\mathrm{IN}}^2}{2I_{\mathrm{REF}}}, \tag{5.161}$$

$$I_{\mathrm{OUT5}} = \frac{I_{\mathrm{OUT4}}^2}{8(35I_{\mathrm{REF}}/27)} = \frac{27I_{\mathrm{IN}}^4}{1120I_{\mathrm{REF}}^3}, \tag{5.162}$$

$$I_{\mathrm{OUT6}} = I_{\mathrm{OUT5}} = \frac{27I_{\mathrm{IN}}^4}{1120I_{\mathrm{REF}}^3}, \tag{5.163}$$

$$I_{\mathrm{OUT7}} = 2I_{\mathrm{OUT5}} = \frac{27I_{\mathrm{IN}}^4}{560I_{\mathrm{REF}}^3}, \tag{5.164}$$

$$I_{\mathrm{OUT8}} = \frac{27}{7} I_{\mathrm{OUT1}} = \frac{27I_{\mathrm{IN}}^2}{56I_{\mathrm{REF}}} \tag{5.165}$$

and

$$I_{OUT9} = \frac{54}{7} I_{OUT1} = \frac{27 I_{IN}^2}{28 I_{REF}}. \tag{5.166}$$

So, the output current of the circuit from Fig. 5.18 can be expressed as follows:

$$I_{OUT} = I_{REF} \frac{\frac{27 I_{IN}^4}{1120 I_{REF}^3} + \frac{9 I_{IN}^3}{56 I_{REF}^2} + \frac{27 I_{IN}^2}{56 I_{REF}} + \frac{3 I_{IN}}{4} + \frac{I_{REF}}{2}}{\frac{27 I_{IN}^4}{560 I_{REF}^3} + \frac{27 I_{IN}^2}{28 I_{REF}} + I_{REF}}. \tag{5.167}$$

Using the definition of the x variable, the previous relation becomes:

$$I_{OUT} = I_{REF} \frac{\frac{27x^4}{1120} + \frac{9x^3}{56} + \frac{27x^2}{56} + \frac{3x}{4} + \frac{1}{2}}{\frac{27x^4}{560} + \frac{27x^2}{28} + 1}. \tag{5.168}$$

In conclusion, the analog function synthesizer shown in Fig. 5.18 approximates the continuous-log sigmoid activation function (for $b = 3$):

$$I_{OUT} = I_{REF} a_{18}(x) \cong I_{REF} f_9(x). \tag{5.169}$$

References

Popa C (2012) High-accuracy function synthesizer circuit with applications in signal processing. EURASIP J Adv Signal Process 2012:129. https://doi.org/10.1186/1687-6180-2012-129

Popa CR (2023) High-accuracy Gaussian function generator for neural networks. Electronics 12(1):24. https://doi.org/10.3390/electronics12010024

Chapter 6
Analysis and Design of Analog Function Synthesizers for Generating Radial Basis Activation Functions

6.1 Introduction

This chapter presents a multitude of analog function synthesizers (Popa 2012, 2023), developed for implementing in CMOS technology the radial basis activation functions, detailed analyzed in Chap. 2 (cubic activation function, Gaussian activation function, quadratic, inverse quadratic activation functions, generalized multiquadratic and generalized inverse multiquadratic activation functions). The superior-order approximation functions described in Chap. 2 represent the basis for elaborating the block diagrams from this chapter, while the concrete implementations in CMOS technology using reasonable hardware resources use the fundamental CMOS computational circuits (current-mode squaring and multiplier/divider circuits), analyzed in details in Chap. 4.

6.2 Analog Function Synthesizers for Cubic Activation Function

In order to implement in CMOS technology the cubic activation function, the following simple circuit is used, based on two current-mode squaring circuits (SQ) analyzed in Chap. 4. The block diagram of the analog function synthesizer for generating the cubic activation function is shown in Fig. 6.1

The expression of the current I_{OUT1} is:

$$I_{OUT1} = \frac{I_{IN}^2}{8(I_{REF}/8)} = \frac{I_{IN}^2}{I_{REF}}. \tag{6.1}$$

The output current of the circuit from Fig. 6.1 can be expressed as follows:

C. R. Popa, *Analog Current-Mode Computational Circuits for Artificial Neural Networks*, Analog Circuits and Signal Processing,
https://doi.org/10.1007/978-3-032-03989-7_6

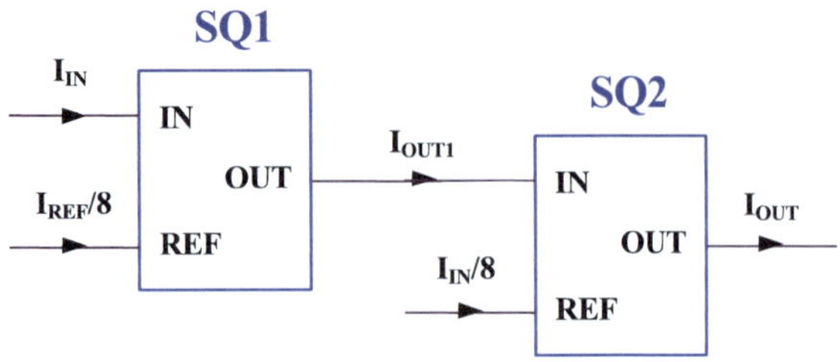

Fig. 6.1 Block diagram of the analog function synthesizer for generating the cubic activation function

$$I_{OUT} = \frac{I_{OUT1}^2}{8(I_{IN}/8)} = I_{REF}\left(\frac{I_{IN}}{I_{REF}}\right)^3. \tag{6.2}$$

Defining the undimensional variable x as the ratio between the input variable current I_{IN} and the constant reference current I_{REF}, $x = \frac{I_{IN}}{I_{REF}}$, the previous relation becomes:

$$I_{OUT} = I_{REF}x^3. \tag{6.3}$$

In conclusion, the analog function synthesizer shown in Fig. 6.1 generates the cubic activation function.

6.3 Analog Function Synthesizers for Gaussian Activation Function

In order to implement in CMOS technology the Gaussian activation function, the approximation functions described in Chap. 2 and the CMOS squaring and multiplies/divider circuits from Chap. 4 are used.

6.3.1 First Analog Function Synthesizer for Generating the Gaussian Activation Function

The first analog function synthesizer, designed for generating the Gaussian activation function, uses $a_3(x)$ approximation function described in Chap. 2. The utilization of the current-mode squaring circuits (SQ) and multiplier/divider circuits (MD) analyzed in Chap. 4 allows to obtain CMOS implementations of the analog function synthesizer, having very good accuracy and reasonable complexity.

The block diagram of the first analog function synthesizer for generating the Gaussian activation function is presented in Fig. 6.2.

The expressions of the currents from the circuit are:

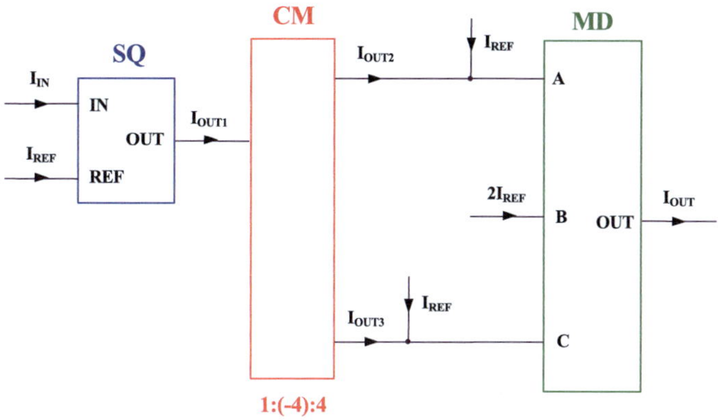

Fig. 6.2 Block diagram of the first analog function synthesizer for generating the Gaussian activation function

$$I_{\mathrm{OUT1}} = \frac{I_{\mathrm{IN}}^2}{8I_{\mathrm{REF}}}, \tag{6.4}$$

$$I_{\mathrm{OUT2}} = (-4)I_{\mathrm{OUT1}} = -\frac{I_{\mathrm{IN}}^2}{2I_{\mathrm{REF}}} \tag{6.5}$$

and

$$I_{\mathrm{OUT3}} = 4I_{\mathrm{OUT1}} = \frac{I_{\mathrm{IN}}^2}{2I_{\mathrm{REF}}}. \tag{6.6}$$

The output current of the circuit from Fig. 6.2 can be expressed as follows:

$$I_{\mathrm{OUT}} = \frac{I_{\mathrm{REF}} - \frac{I_{\mathrm{IN}}^2}{2I_{\mathrm{REF}}}}{I_{\mathrm{REF}} + \frac{I_{\mathrm{IN}}^2}{2I_{\mathrm{REF}}}}. \tag{6.7}$$

Using the definition of x variable, the previous relation becomes:

$$I_{\mathrm{OUT}} = I_{\mathrm{REF}} \frac{1 - \frac{x^2}{2}}{1 + \frac{x^2}{2}}. \tag{6.8}$$

In conclusion, the analog function synthesizer shown in Fig. 6.2 approximates the Gaussian activation function:

$$I_{\mathrm{OUT}} = I_{\mathrm{REF}} a_3(x) \cong I_{\mathrm{REF}} f_3(x). \tag{6.9}$$

6.3.2 Second Analog Function Synthesizer for Generating the Gaussian Activation Function

The second analog function synthesizer, designed for generating the Gaussian activation function, uses $a_4(x)$ approximation function described in Chap. 2.

The block diagram of the second analog function synthesizer for generating the Gaussian activation function is presented in Fig. 6.3.

The expressions of the currents from the circuit are:

$$I_{\text{OUT1}} = \frac{I_{\text{IN}}^2}{8I_{\text{REF}}}, \tag{6.10}$$

$$I_{\text{OUT2}} = -I_{\text{OUT1}} = -\frac{I_{\text{IN}}^2}{8I_{\text{REF}}}, \tag{6.11}$$

$$I_{\text{OUT3}} = I_{\text{OUT1}} = \frac{I_{\text{IN}}^2}{8I_{\text{REF}}}, \tag{6.12}$$

$$I_{\text{OUT4}} = I_{\text{REF}} \frac{I_{\text{REF}} - \frac{I_{\text{IN}}^2}{8I_{\text{REF}}}}{I_{\text{REF}} + \frac{I_{\text{IN}}^2}{8I_{\text{REF}}}} \tag{6.13}$$

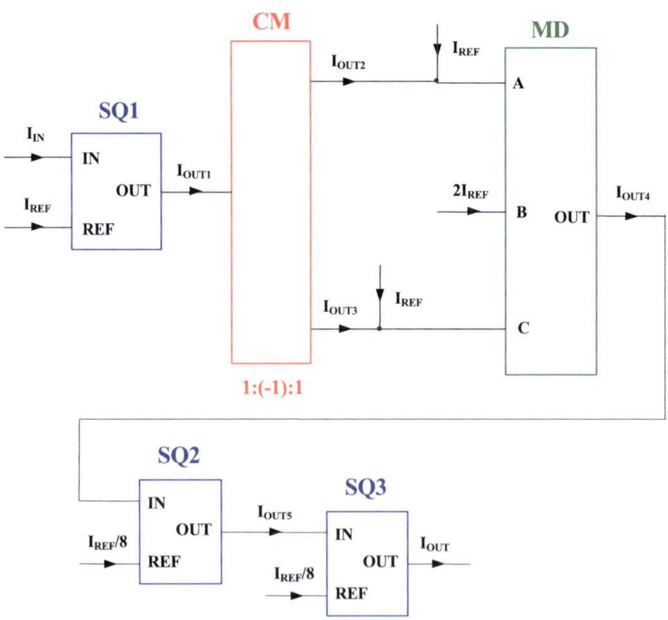

Fig. 6.3 Block diagram of the second analog function synthesizer for generating the Gaussian activation function

and

$$I_{OUT5} = \frac{I_{OUT4}^2}{8(I_{REF}/8)} = I_{REF}\left(\frac{I_{REF} - \frac{I_{IN}^2}{8I_{REF}}}{I_{REF} + \frac{I_{IN}^2}{8I_{REF}}}\right)^2. \tag{6.14}$$

The output current of the circuit from Fig. 6.3 can be expressed as follows:

$$I_{OUT} = \frac{I_{OUT5}^2}{8(I_{REF}/8)} = I_{REF}\left(\frac{I_{REF} - \frac{I_{IN}^2}{8I_{REF}}}{I_{REF} + \frac{I_{IN}^2}{8I_{REF}}}\right)^4. \tag{6.15}$$

Using the definition of x variable, the previous relation becomes:

$$I_{OUT} = I_{REF}\left(\frac{1 - \frac{x^2}{8}}{1 + \frac{x^2}{8}}\right)^4 \tag{6.16}$$

In conclusion, the analog function synthesizer shown in Fig. 6.3 approximates the Gaussian activation function:

$$I_{OUT} = I_{REF}a_4(x) \cong I_{REF}f_3(x). \tag{6.17}$$

6.3.3 *Third Analog Function Synthesizer for Generating the Gaussian Activation Function*

The third analog function synthesizer, designed for generating the Gaussian activation function, uses $a_5(x)$ approximation function described in Chap. 2.

The block diagram of the third analog function synthesizer for generating the Gaussian activation function is presented in Fig. 6.4.

The expressions of the currents from the circuit are:

$$I_{OUT1} = \frac{I_{IN}^2}{8I_{REF}}, \tag{6.18}$$

$$I_{OUT2} = -\frac{16}{3}I_{OUT1} = -\frac{2I_{IN}^2}{3I_{REF}}, \tag{6.19}$$

$$I_{OUT3} = 8I_{OUT1} = \frac{I_{IN}^2}{I_{REF}}, \tag{6.20}$$

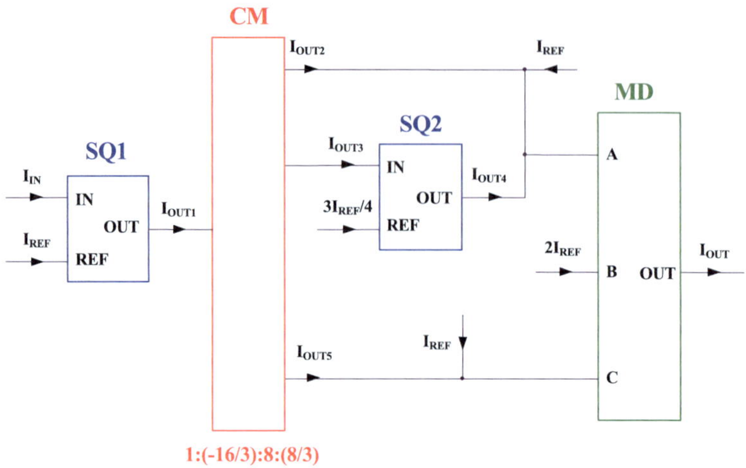

Fig. 6.4 Block diagram of the third analog function synthesizer for generating the Gaussian activation function

$$I_{\mathrm{OUT4}} = \frac{I_{\mathrm{OUT3}}^2}{8(3I_{\mathrm{REF}}/4)} = \frac{I_{\mathrm{IN}}^4}{6I_{\mathrm{REF}}^3} \tag{6.21}$$

and

$$I_{\mathrm{OUT5}} = \frac{8}{3}I_{\mathrm{OUT1}} = \frac{I_{\mathrm{IN}}^2}{3I_{\mathrm{REF}}}. \tag{6.22}$$

The output current of the circuit from Fig. 6.4 can be expressed as follows:

$$I_{\mathrm{OUT}} = I_{\mathrm{REF}} \frac{\frac{I_{\mathrm{IN}}^4}{6I_{\mathrm{REF}}^3} - \frac{2I_{\mathrm{IN}}^2}{3I_{\mathrm{REF}}} + I_{\mathrm{REF}}}{\frac{I_{\mathrm{IN}}^2}{3I_{\mathrm{REF}}} + I_{\mathrm{REF}}}. \tag{6.23}$$

Using the definition of x variable, the previous relation becomes:

$$I_{\mathrm{OUT}} = I_{\mathrm{REF}} \frac{\frac{x^4}{6} - \frac{2x^2}{3} + 1}{\frac{x^2}{3} + 1}. \tag{6.24}$$

In conclusion, the analog function synthesizer shown in Fig. 6.4 approximates the Gaussian activation function:

$$I_{\mathrm{OUT}} = I_{\mathrm{REF}} a_5(x) \cong I_{\mathrm{REF}} f_3(x). \tag{6.25}$$

6.3.4 Fourth Analog Function Synthesizer for Generating the Gaussian Activation Function

The fourth analog function synthesizer, designed for generating the Gaussian activation function, uses $a_6(x)$ approximation function described in Chap. 2.

The block diagram of the fourth analog function synthesizer for generating the Gaussian activation function is presented in Fig. 6.5.

The expressions of the currents from the circuit are:

$$I_{\mathrm{OUT1}} = \frac{I_{\mathrm{IN}}^2}{8I_{\mathrm{REF}}}, \tag{6.26}$$

$$I_{\mathrm{OUT2}} = -\frac{4}{3}I_{\mathrm{OUT1}} = -\frac{I_{\mathrm{IN}}^2}{6I_{\mathrm{REF}}}, \tag{6.27}$$

$$I_{\mathrm{OUT3}} = 2I_{\mathrm{OUT1}} = \frac{I_{\mathrm{IN}}^2}{4I_{\mathrm{REF}}}, \tag{6.28}$$

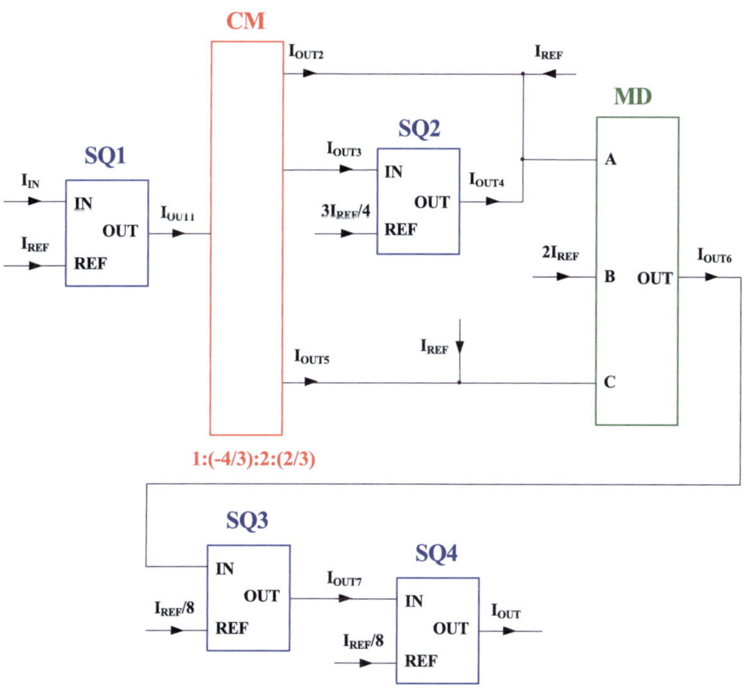

Fig. 6.5 Block diagram of the fourth analog function synthesizer for generating the Gaussian activation function

$$I_{\mathrm{OUT4}} = \frac{I_{\mathrm{OUT3}}^2}{8(3I_{\mathrm{REF}}/4)} = \frac{I_{\mathrm{IN}}^4}{96I_{\mathrm{REF}}^3}, \tag{6.29}$$

$$I_{\mathrm{OUT5}} = \frac{2}{3}I_{\mathrm{OUT1}} = \frac{I_{\mathrm{IN}}^2}{12I_{\mathrm{REF}}}, \tag{6.30}$$

$$I_{\mathrm{OUT6}} = I_{\mathrm{REF}}\frac{\frac{I_{\mathrm{IN}}^4}{96I_{\mathrm{REF}}^3} - \frac{I_{\mathrm{IN}}^2}{6I_{\mathrm{REF}}} + I_{\mathrm{REF}}}{\frac{I_{\mathrm{IN}}^2}{12I_{\mathrm{REF}}} + I_{REREFF}} \tag{6.31}$$

and

$$I_{\mathrm{OUT7}} = \frac{I_{\mathrm{OUT6}}^2}{8(I_{\mathrm{REF}}/8)} = I_{\mathrm{REF}}\left(\frac{\frac{I_{\mathrm{IN}}^4}{96I_{\mathrm{REF}}^3} - \frac{I_{\mathrm{IN}}^2}{6I_{\mathrm{REF}}} + I_{\mathrm{REF}}}{\frac{I_{IN}^2}{12I_{\mathrm{REF}}} + I_{\mathrm{REF}}}\right)^2. \tag{6.32}$$

The output current of the circuit from Fig. 6.5 can be expressed as follows:

$$I_{\mathrm{OUT}} = \frac{I_{\mathrm{OUT7}}^2}{8(I_{\mathrm{REF}}/8)} = I_{\mathrm{REF}}\left(\frac{\frac{I_{\mathrm{IN}}^4}{96I_{\mathrm{REF}}^3} - \frac{I_{\mathrm{IN}}^2}{6I_{\mathrm{REF}}} + I_{\mathrm{REF}}}{\frac{I_{\mathrm{IN}}^2}{12I_{\mathrm{REF}}} + I_{\mathrm{REF}}}\right)^4. \tag{6.33}$$

Using the definition of x variable, the previous relation becomes:

$$I_{\mathrm{OUT}} = I_{\mathrm{REF}}\left(\frac{\frac{x^4}{96} - \frac{x^2}{6} + 1}{\frac{x^2}{12} + 1}\right)^4. \tag{6.34}$$

In conclusion, the analog function synthesizer shown in Fig. 6.5 approximates the Gaussian activation function:

$$I_{\mathrm{OUT}} = I_{\mathrm{REF}}a_6(x) \cong I_{\mathrm{REF}}f_3(x). \tag{6.35}$$

6.3.5 Fifth Analog Function Synthesizer for Generating the Gaussian Activation Function

The fifth analog function synthesizer, designed for generating the Gaussian activation function, uses $a_7(x)$ approximation function described in Chap. 2.

The block diagram of the fifth analog function synthesizer for generating the Gaussian activation function is presented in Fig. 6.6.

The expressions of the currents from the circuit are:

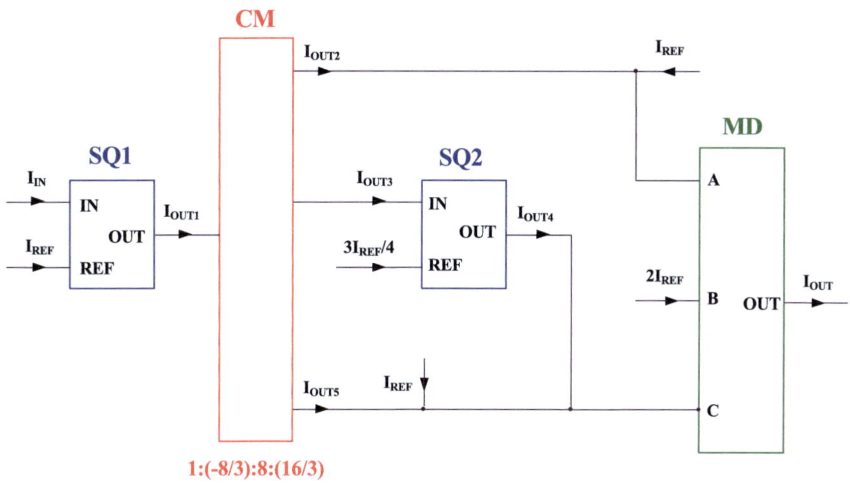

Fig. 6.6 Block diagram of the fifth analog function synthesizer for generating the Guassian activation function

$$I_{\mathrm{OUT1}} = \frac{I_{\mathrm{IN}}^2}{8I_{\mathrm{REF}}}, \tag{6.36}$$

$$I_{\mathrm{OUT2}} = -\frac{8}{3}I_{\mathrm{OUT1}} = -\frac{I_{\mathrm{IN}}^2}{3I_{\mathrm{REF}}}, \tag{6.37}$$

$$I_{\mathrm{OUT3}} = 8I_{\mathrm{OUT1}} = \frac{I_{\mathrm{IN}}^2}{I_{\mathrm{REF}}}, \tag{6.38}$$

$$I_{\mathrm{OUT4}} = \frac{I_{\mathrm{OUT3}}^2}{8(3I_{\mathrm{REF}}/4)} = \frac{I_{\mathrm{IN}}^4}{6I_{\mathrm{REF}}^3} \tag{6.39}$$

and

$$I_{\mathrm{OUT5}} = \frac{16}{3}I_{\mathrm{OUT1}} = \frac{2}{3}\frac{I_{\mathrm{IN}}^2}{I_{\mathrm{REF}}}. \tag{6.40}$$

So, the output current of the circuit from Fig. 6.6 can be expressed as follows:

$$I_{\mathrm{OUT}} = I_{\mathrm{REF}}\frac{I_{\mathrm{REF}} - \frac{I_{\mathrm{IN}}^2}{3I_{\mathrm{REF}}}}{\frac{I_{\mathrm{IN}}^4}{6I_{\mathrm{REF}}^3} + \frac{2I_{\mathrm{IN}}^2}{3I_{\mathrm{REF}}} + I_{\mathrm{REF}}}. \tag{6.41}$$

Using the definition of the x variable, the previous relation becomes:

$$I_{\mathrm{OUT}} = I_{\mathrm{REF}} \frac{1 - \frac{x^2}{3}}{1 + \frac{2x^2}{3} + \frac{x^4}{6}}. \tag{6.42}$$

In conclusion, the analog function synthesizer shown in Fig. 6.6 approximates the Gaussian activation function:

$$I_{\mathrm{OUT}} = I_{\mathrm{REF}} a_7(x) \cong I_{\mathrm{REF}} f_3(x). \tag{6.43}$$

6.3.6 Sixth Analog Function Synthesizer for Generating the Gaussian Activation Function

The sixth analog function synthesizer, designed for generating the Gaussian activation function, uses $a_8(x)$ approximation function described in Chap. 2.

The block diagram of the sixth analog function synthesizer for generating the Gaussian activation function is presented in Fig. 6.7.

The expressions of the currents from the circuit are:

$$I_{\mathrm{OUT1}} = \frac{I_{\mathrm{IN}}^2}{8 I_{\mathrm{REF}}}, \tag{6.44}$$

$$I_{\mathrm{OUT2}} = -\frac{2}{3} I_{\mathrm{OUT1}} = -\frac{I_{\mathrm{IN}}^2}{12 I_{\mathrm{REF}}}, \tag{6.45}$$

$$I_{\mathrm{OUT3}} = 4 I_{\mathrm{OUT1}} = \frac{I_{\mathrm{IN}}^2}{2 I_{\mathrm{REF}}}, \tag{6.46}$$

$$I_{OUT4} = \frac{I_{OUT3}^2}{8(3 I_{REF})} = \frac{I_{IN}^4}{96 I_{REF}^3}, \tag{6.47}$$

$$I_{\mathrm{OUT5}} = \frac{16}{3} I_{\mathrm{OUT1}} = \frac{1}{6} \frac{I_{\mathrm{IN}}^2}{I_{\mathrm{REF}}}, \tag{6.48}$$

$$I_{\mathrm{OUT6}} = I_{\mathrm{REF}} \frac{I_{\mathrm{REF}} - \frac{I_{\mathrm{IN}}^2}{12 I_{\mathrm{REF}}}}{\frac{I_{\mathrm{IN}}^4}{96 I_{\mathrm{REF}}^3} + \frac{I_{\mathrm{IN}}^2}{6 I_{\mathrm{REF}}} + I_{\mathrm{REF}}}, \tag{6.49}$$

$$I_{\mathrm{OUT7}} = \frac{I_{\mathrm{OUT6}}^2}{8(I_{\mathrm{REF}}/8)} = I_{\mathrm{REF}} \left(\frac{I_{\mathrm{REF}} - \frac{I_{\mathrm{IN}}^2}{12 I_{\mathrm{REF}}}}{\frac{I_{\mathrm{IN}}^4}{96 I_{\mathrm{REF}}^3} + \frac{I_{\mathrm{IN}}^2}{6 I_{\mathrm{REF}}} + I_{\mathrm{REF}}} \right)^2. \tag{6.50}$$

So, the output current of the circuit from Fig. 6.7 can be expressed as follows:

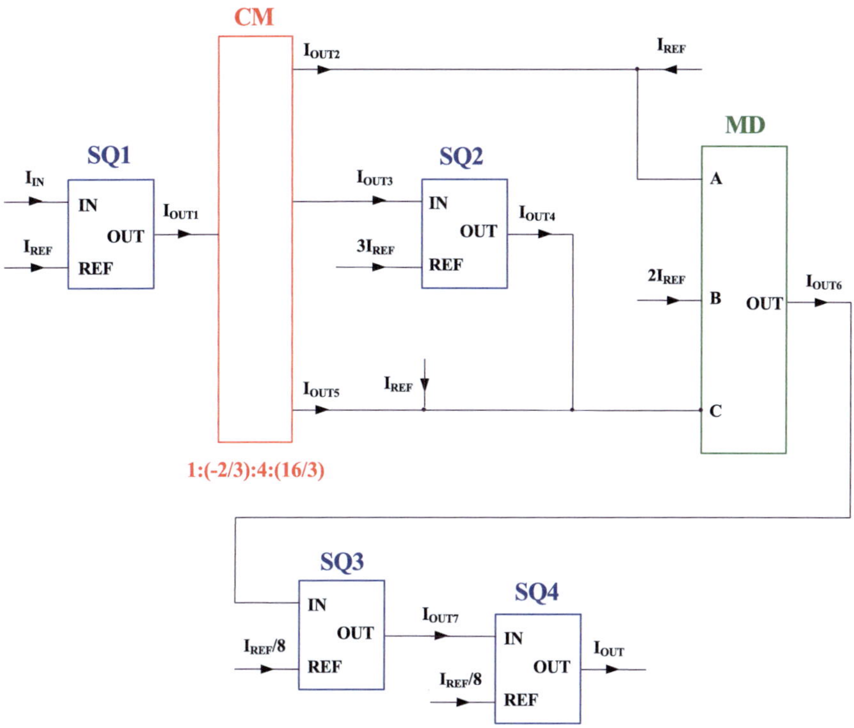

Fig. 6.7 Block diagram of the sixth analog function synthesizer for generating the Gaussian activation function

$$I_{\text{OUT}} = \frac{I_{\text{OUT7}}^2}{8(I_{\text{REF}}/8)} = I_{\text{REF}} \left(\frac{I_{\text{REF}} - \frac{I_{\text{IN}}^2}{12 I_{\text{REF}}}}{\frac{I_{\text{IN}}^4}{96 I_{\text{REF}}^3} + \frac{I_{\text{IN}}^2}{6 I_{\text{REF}}} + I_{\text{REF}}} \right)^4. \tag{6.51}$$

Using the definition of the x variable, the previous relation becomes:

$$I_{\text{OUT}} = I_{\text{REF}} \left(\frac{1 - \frac{x^2}{12}}{1 + \frac{x^2}{6} + \frac{x^4}{96}} \right)^4. \tag{6.52}$$

In conclusion, the analog function synthesizer shown in Fig. 6.7 approximates the Gaussian activation function:

$$I_{\text{OUT}} = I_{\text{REF}} a_8(x) \cong I_{\text{REF}} f_3(x). \tag{6.53}$$

6.3.7 Seventh Analog Function Synthesizer for Generating the Gaussian Activation Function

The seventh analog function synthesizer, designed for generating the Gaussian activation function, uses $a_9(x)$ approximation function described in Chap. 2.

The block diagram of the seventh analog function synthesizer for generating the Gaussian activation function is presented in Fig. 6.8.

The expressions of the currents from the circuit are:

$$I_{\mathrm{OUT1}} = \frac{I_{\mathrm{IN}}^2}{8I_{\mathrm{REF}}}, \tag{6.54}$$

$$I_{\mathrm{OUT2}} = (-4)I_{\mathrm{OUT1}} = -\frac{I_{\mathrm{IN}}^2}{2I_{\mathrm{REF}}}, \tag{6.55}$$

$$I_{\mathrm{OUT3}} = 4I_{\mathrm{OUT1}} = \frac{I_{\mathrm{IN}}^2}{2I_{\mathrm{REF}}}, \tag{6.56}$$

$$I_{\mathrm{OUT4}} = \frac{I_{\mathrm{OUT3}}^2}{8(3I_{\mathrm{REF}}/8)} = \frac{I_{\mathrm{IN}}^4}{12I_{\mathrm{REF}}^3}, \tag{6.57}$$

$$I_{\mathrm{OUT5}} = I_{\mathrm{OUT4}} = \frac{I_{\mathrm{IN}}^4}{12I_{\mathrm{REF}}^3}, \tag{6.58}$$

$$I_{\mathrm{OUT6}} = I_{\mathrm{OUT4}} = \frac{I_{\mathrm{IN}}^4}{12I_{\mathrm{REF}}^3} \tag{6.59}$$

and

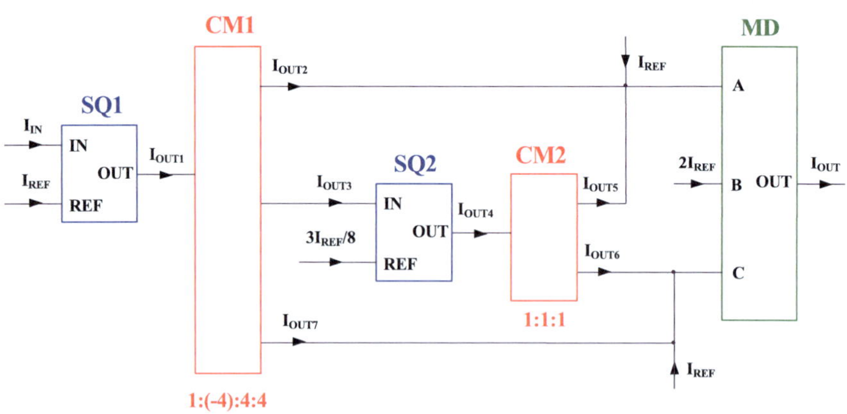

Fig. 6.8 Block diagram of the seventh analog function synthesizer for generating the Gaussian activation function

$$I_{OUT7} = 4I_{OUT1} = \frac{I_{IN}^2}{2I_{REF}}. \tag{6.60}$$

So, the output current of the circuit from Fig. 6.8 can be expressed as follows:

$$I_{OUT} = I_{REF}\frac{I_{REF} - \frac{I_{IN}^2}{2I_{REF}} + \frac{I_{IN}^4}{12I_{REF}^3}}{I_{REF} + \frac{I_{IN}^2}{2I_{REF}} + \frac{I_{IN}^4}{12I_{REF}^3}}. \tag{6.61}$$

Using the definition of the x variable, the previous relation becomes:

$$I_{OUT} = I_{REF}\frac{1 - \frac{x^2}{2} + \frac{x^4}{12}}{1 + \frac{x^2}{2} + \frac{x^4}{12}}. \tag{6.62}$$

In conclusion, the analog function synthesizer shown in Fig. 6.8 approximates the Gaussian activation function:

$$I_{OUT} = I_{REF}a_9(x) \cong I_{REF}f_3(x). \tag{6.63}$$

6.3.8 *Eighth Analog Function Synthesizer for Generating the Gaussian Activation Function*

The eighth analog function synthesizer, designed for generating the Gaussian activation function, uses $a_{10}(x)$ approximation function described in Chap. 2.

The block diagram of the seventh analog function synthesizer for generating the Gaussian activation function is presented in Fig. 6.9.

The expressions of the currents from the circuit are:

$$I_{OUT1} = \frac{I_{IN}^2}{8I_{REF}}, \tag{6.64}$$

$$I_{OUT2} = -I_{OUT1} = -\frac{I_{IN}^2}{8I_{REF}}, \tag{6.65}$$

$$I_{OUT3} = 2I_{OUT1} = \frac{I_{IN}^2}{4I_{REF}}, \tag{6.66}$$

$$I_{OUT4} = \frac{I_{OUT3}^2}{8(3I_{REF}/2)} = \frac{I_{IN}^4}{192I_{REF}^3}, \tag{6.67}$$

$$I_{OUT5} = I_{OUT4} = \frac{I_{IN}^4}{192I_{REF}^3}, \tag{6.68}$$

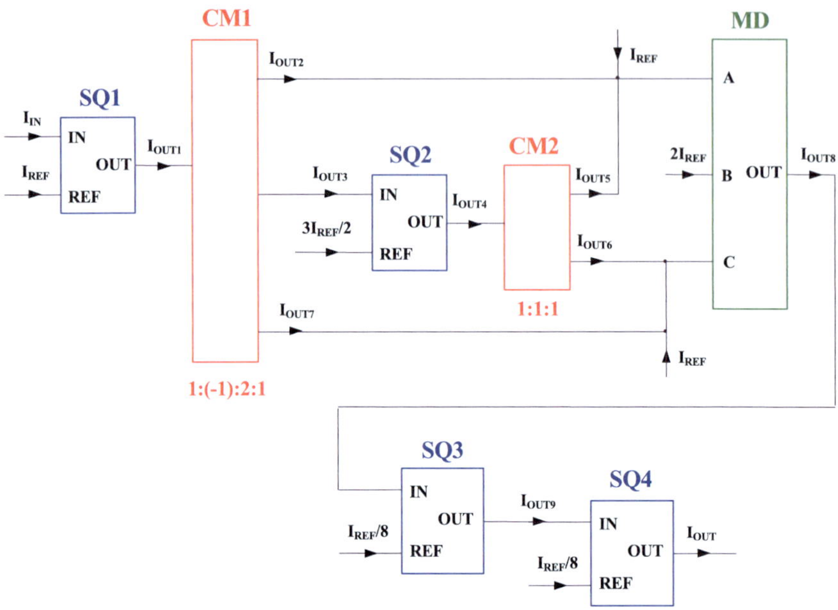

Fig. 6.9 Block diagram of the eighth analog function synthesizer for generating the Gaussian activation function

$$I_{\mathrm{OUT6}} = I_{\mathrm{OUT4}} = \frac{I_{\mathrm{IN}}^4}{192 I_{\mathrm{REF}}^3}, \tag{6.69}$$

$$I_{\mathrm{OUT7}} = I_{\mathrm{OUT1}} = \frac{I_{\mathrm{IN}}^2}{8 I_{\mathrm{REF}}}, \tag{6.70}$$

$$I_{\mathrm{OUT8}} = I_{\mathrm{REF}} \frac{I_{\mathrm{REF}} - \frac{I_{\mathrm{IN}}^2}{8 I_{\mathrm{REF}}} + \frac{I_{\mathrm{IN}}^4}{192 I_{\mathrm{REF}}^3}}{I_{\mathrm{REF}} + \frac{I_{\mathrm{IN}}^2}{8 I_{\mathrm{REF}}} + \frac{I_{\mathrm{IN}}^4}{192 I_{\mathrm{REF}}^3}} \tag{6.71}$$

and

$$I_{\mathrm{OUT9}} = \frac{I_{\mathrm{OUT8}}^2}{8(I_{\mathrm{REF}}/8)} = I_{\mathrm{REF}} \left(\frac{I_{\mathrm{REF}} - \frac{I_{\mathrm{IN}}^2}{8 I_{\mathrm{REF}}} + \frac{I_{\mathrm{IN}}^4}{192 I_{\mathrm{REF}}^3}}{I_{\mathrm{REF}} + \frac{I_{\mathrm{IN}}^2}{8 I_{\mathrm{REF}}} + \frac{I_{\mathrm{IN}}^4}{192 I_{\mathrm{REF}}^3}} \right)^2. \tag{6.72}$$

So, the output current of the circuit from Fig. 6.9 can be expressed as follows:

$$I_{OUT} = \frac{I_{OUT9}^2}{8(I_{REF}/8)} = I_{REF}\left(\frac{I_{REF} - \frac{I_{IN}^2}{8I_{REF}} + \frac{I_{IN}^4}{192I_{REF}^3}}{I_{REF} + \frac{I_{IN}^2}{8I_{REF}} + \frac{I_{IN}^4}{192I_{REF}^3}}\right)^4. \tag{6.73}$$

Using the definition of the x variable, the previous relation becomes:

$$I_{OUT} = I_{REF}\left(\frac{1 - \frac{x^2}{8} + \frac{x^4}{192}}{1 + \frac{x^2}{8} + \frac{x^4}{192}}\right)^4. \tag{6.74}$$

In conclusion, the analog function synthesizer shown in Fig. 6.9 approximates the Gaussian activation function:

$$I_{OUT} = I_{REF}a_{10}(x) \cong I_{REF}f_3(x). \tag{6.75}$$

6.3.9 Ninth Analog Function Synthesizer for Generating the Gaussian Activation Function

The ninth analog function synthesizer, designed for generating the Gaussian activation function, uses $a_{11}(x)$ approximation function described in Chap. 2.

The block diagram of the ninth analog function synthesizer for generating the Gaussian activation function is presented in Fig. 6.10.

The expressions of the currents from the circuit are:

$$I_{OUT1} = \frac{I_{IN}^2}{8I_{REF}}, \tag{6.76}$$

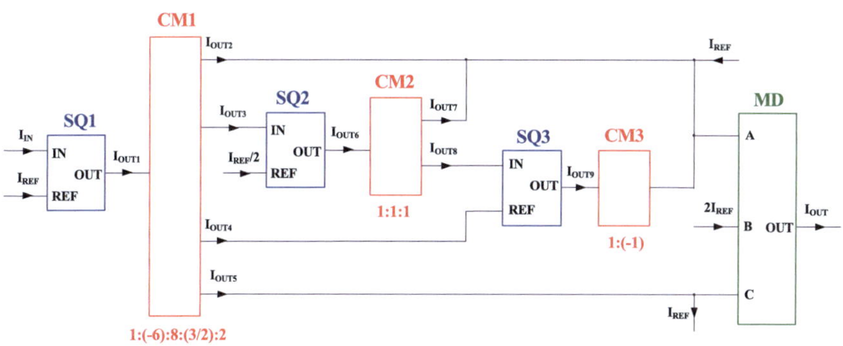

Fig. 6.10 Block diagram of the ninth analog function synthesizer for generating the Gaussian activation function

$$I_{OUT2} = -6I_{OUT1} = -\frac{3I_{IN}^2}{4I_{REF}}, \tag{6.77}$$

$$I_{OUT3} = 8I_{OUT1} = \frac{I_{IN}^2}{I_{REF}}, \tag{6.78}$$

$$I_{OUT4} = \frac{3}{2}I_{OUT1} = \frac{3I_{IN}^2}{16I_{REF}}, \tag{6.79}$$

$$I_{OUT5} = 2I_{OUT1} = \frac{I_{IN}^2}{4I_{REF}}, \tag{6.80}$$

$$I_{OUT6} = \frac{I_{OUT3}^2}{8(I_{REF}/2)} = \frac{I_{IN}^4}{4I_{REF}^3}, \tag{6.81}$$

$$I_{OUT7} = I_{OUT6} = \frac{I_{IN}^4}{4I_{REF}^3}, \tag{6.82}$$

$$I_{OUT8} = I_{OUT6} = \frac{I_{IN}^4}{4I_{REF}^3} \tag{6.83}$$

and

$$I_{OUT9} = \frac{I_{OUT8}^2}{8I_{OUT4}} = \frac{I_{IN}^6}{24I_{REF}^5}. \tag{6.84}$$

So, the output current of the circuit from Fig. 6.10 can be expressed as follows:

$$I_{OUT} = I_{REF}\frac{-\frac{I_{IN}^6}{24I_{TRF}^5} + \frac{I_{IN}^4}{4I_{REF}^3} - \frac{3I_{IN}^2}{4I_{REF}} + I_{REF}}{\frac{I_{IN}^2}{4I_{REF}} + I_{REF}}. \tag{6.85}$$

Using the definition of the x variable, the previous relation becomes:

$$I_{OUT} = I_{REF}\frac{1 - \frac{3x^2}{4} + \frac{x^4}{4} - \frac{x^6}{24}}{1 + \frac{x^2}{4}}. \tag{6.86}$$

In conclusion, the analog function synthesizer shown in Fig. 6.10 approximates the Gaussian activation function:

$$I_{OUT} = I_{REF}a_{11}(x) \cong I_{REF}f_3(x). \tag{6.87}$$

6.3.10 Tenth Analog Function Synthesizer for Generating the Gaussian Activation Function

The tenth analog function synthesizer, designed for generating the Gaussian activation function, uses $a_{12}(x)$ approximation function described in Chap. 2.

The block diagram of the tenth analog function synthesizer for generating the Gaussian activation function is presented in Fig. 6.11.

The expressions of the currents from the circuit are:

$$I_{\text{OUT1}} = \frac{I_{\text{IN}}^2}{8I_{\text{REF}}}, \tag{6.88}$$

$$I_{\text{OUT2}} = \frac{3}{2}I_{\text{OUT1}} = \frac{3I_{\text{IN}}^2}{16I_{\text{REF}}}, \tag{6.89}$$

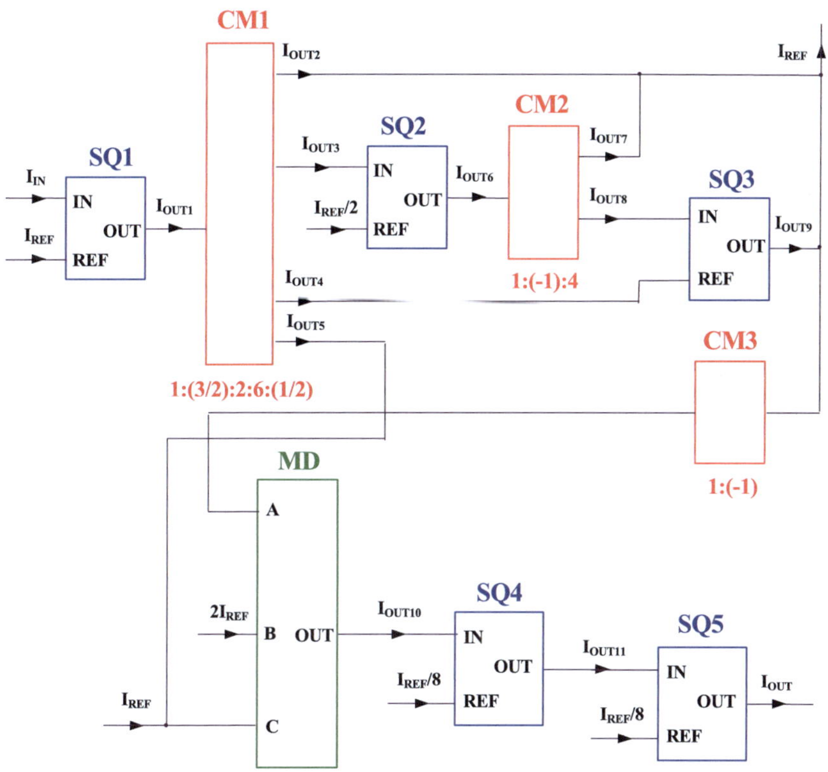

Fig. 6.11 Block diagram of the tenth analog function synthesizer for generating the Gaussian activation function

$$I_{OUT3} = 2I_{OUT1} = \frac{I_{IN}^2}{4I_{REF}}, \tag{6.90}$$

$$I_{OUT4} = 6I_{OUT1} = \frac{3I_{IN}^2}{4I_{REF}}, \tag{6.91}$$

$$I_{OUT5} = \frac{1}{2}I_{OUT1} = \frac{I_{IN}^2}{16I_{REF}}, \tag{6.92}$$

$$I_{OUT6} = \frac{I_{OUT3}^2}{8(I_{REF}/2)} = \frac{I_{IN}^4}{64I_{REF}^3}, \tag{6.93}$$

$$I_{OUT7} = -I_{OUT6} = -\frac{I_{IN}^4}{64I_{REF}^3}, \tag{6.94}$$

$$I_{OUT8} = 4I_{OUT6} = \frac{I_{IN}^4}{16I_{REF}^3}, \tag{6.95}$$

$$I_{OUT9} = \frac{I_{OUT8}^2}{8I_{OUT4}} = \frac{I_{IN}^6}{1536I_{REF}^5}, \tag{6.96}$$

$$I_{OUT10} = I_{REF}\frac{I_{REF} - \frac{3I_{IN}^2}{16I_{REF}} + \frac{I_{IN}^4}{64I_{REF}^3} - \frac{I_{IN}^6}{1536I_{REF}^5}}{I_{REF} + \frac{I_{IN}^2}{16I_{REF}}} \tag{6.97}$$

and

$$I_{OUT11} = \frac{I_{OUT10}^2}{8(I_{REF}/8)} = I_{REF}\left(\frac{I_{REF} - \frac{3I_{IN}^2}{16I_{REF}} + \frac{I_{IN}^4}{64I_{REF}^3} - \frac{I_{IN}^6}{1536I_{REF}^5}}{I_{REF} + \frac{I_{IN}^2}{16I_{REF}}}\right)^2. \tag{6.98}$$

So, the output current of the circuit from Fig. 6.11 can be expressed as follows:

$$I_{OUT} = \frac{I_{OUT11}^2}{8(I_{REF}/8)} = I_{REF}\left(\frac{I_{REF} - \frac{3I_{IN}^2}{16I_{REF}} + \frac{I_{IN}^4}{64I_{REF}^3} - \frac{I_{IN}^6}{1536I_{REF}^5}}{I_{REF} + \frac{I_{IN}^2}{16I_{REF}}}\right)^4. \tag{6.99}$$

Using the definition of the x variable, the previous relation becomes:

$$I_{OUT} = I_{REF}\left(\frac{1 - \frac{3x^2}{16} + \frac{x^4}{64} - \frac{x^6}{1536}}{1 + \frac{x^2}{16}}\right)^4. \tag{6.100}$$

In conclusion, the analog function synthesizer shown in Fig. 6.11 approximates the Gaussian activation function:

$$I_{\mathrm{OUT}} = I_{\mathrm{REF}}a_{12}(x) \cong I_{\mathrm{REF}}f_3(x). \tag{6.101}$$

6.3.11 *Eleventh Analog Function Synthesizer for Generating the Gaussian Activation Function*

The 11th analog function synthesizer, designed for generating the Gaussian activation function, uses $a_{13}(x)$ approximation function described in Chap. 2.

The block diagram of the 11th analog function synthesizer for generating the Gaussian activation function is presented in Fig. 6.12.

The expressions of the currents from the circuit are:

$$I_{\mathrm{OUT1}} = \frac{I_{\mathrm{IN}}^2}{8I_{\mathrm{REF}}}, \tag{6.102}$$

$$I_{\mathrm{OUT2}} = -4I_{\mathrm{OUT1}} = -\frac{I_{\mathrm{IN}}^2}{2I_{\mathrm{REF}}}, \tag{6.103}$$

$$I_{\mathrm{OUT3}} = 8I_{\mathrm{OUT1}} = \frac{I_{\mathrm{IN}}^2}{I_{\mathrm{REF}}}, \tag{6.104}$$

$$I_{\mathrm{OUT4}} = \frac{I_{\mathrm{OUT3}}^2}{8(5I_{\mathrm{REF}}/4)} = \frac{I_{\mathrm{IN}}^4}{10I_{\mathrm{REF}}^3}, \tag{6.105}$$

$$I_{\mathrm{OUT5}} = I_{\mathrm{OUT4}} = \frac{I_{\mathrm{IN}}^4}{10I_{\mathrm{REF}}^3}, \tag{6.106}$$

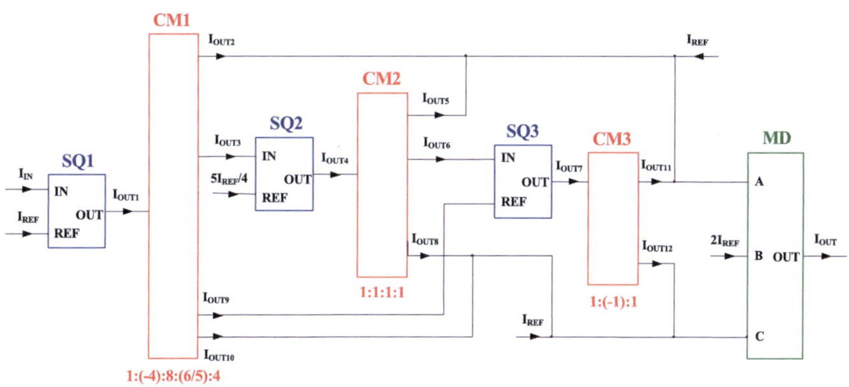

Fig. 6.12 Block diagram of the 11th analog function synthesizer for generating the Gaussian activation function

$$I_{OUT6} = I_{OUT4} = \frac{I_{IN}^4}{10I_{REF}^3}, \quad (6.107)$$

$$I_{OUT7} = \frac{I_{OUT6}^2}{8I_{OUT9}} = \frac{I_{IN}^6}{120I_{REF}^5}, \quad (6.108)$$

$$I_{OUT8} = I_{OUT4} = \frac{I_{IN}^4}{10I_{REF}^3}, \quad (6.109)$$

$$I_{OUT9} = \frac{6}{5}I_{OUT1} = \frac{3}{20}\frac{I_{IN}^2}{I_{REF}}, \quad (6.110)$$

$$I_{OUT10} = 4I_{OUT1} = \frac{I_{IN}^2}{2I_{REF}}, \quad (6.111)$$

$$I_{OUT11} = -I_{OUT7} = -\frac{I_{IN}^6}{120I_{REF}^5} \quad (6.112)$$

and

$$I_{OUT12} = I_{OUT7} = \frac{I_{IN}^6}{120I_{REF}^5}. \quad (6.113)$$

So, the output current of the circuit from Fig. 6.12 can be expressed as follows:

$$I_{OUT} = I_{REF}\frac{-\frac{I_{IN}^6}{120I_{REF}^5} + \frac{I_{IN}^4}{10I_{REF}^3} - \frac{I_{IN}^2}{2I_{REF}} + I_{REF}}{\frac{I_{IN}^6}{120I_{REF}^5} + \frac{I_{IN}^4}{10I_{REF}^3} + \frac{I_{IN}^2}{2I_{REF}} + I_{REF}}. \quad (6.114)$$

Using the definition of the x variable, the previous relation becomes:

$$I_{OUT} = I_{REF}\frac{1 - \frac{x^2}{2} + \frac{x^4}{10} - \frac{x^6}{120}}{1 + \frac{x^2}{2} + \frac{x^4}{10} + \frac{x^6}{120}}. \quad (6.115)$$

In conclusion, the analog function synthesizer shown in Fig. 6.12 approximates the Gaussian activation function:

$$I_{OUT} = I_{REF}a_{13}(x) \cong I_{REF}f_3(x). \quad (6.116)$$

6.3.12 12th Analog Function Synthesizer for Generating the Gaussian Activation Function

The 12th analog function synthesizer, designed for generating the Gaussian activation function, uses $a_{14}(x)$ approximation function described in Chap. 2.

The block diagram of the 12th analog function synthesizer for generating the Gaussian activation function is presented in Fig. 6.13.

The expressions of the currents from the circuit are:

$$I_{\text{OUT1}} = \frac{I_{\text{IN}}^2}{8I_{\text{REF}}}, \tag{6.117}$$

$$I_{\text{OUT2}} = -I_{\text{OUT1}} = -\frac{I_{\text{IN}}^2}{8I_{\text{REF}}}, \tag{6.118}$$

$$I_{\text{OUT3}} = 2I_{\text{OUT1}} = \frac{I_{\text{IN}}^2}{4I_{\text{REF}}}, \tag{6.119}$$

$$I_{\text{OUT4}} = \frac{I_{\text{OUT3}}^2}{8(5I_{\text{REF}}/4)} = \frac{I_{\text{IN}}^4}{160I_{\text{REF}}^3}, \tag{6.120}$$

$$I_{\text{OUT5}} = I_{\text{OUT4}} = \frac{I_{\text{IN}}^4}{160I_{\text{REF}}^3}, \tag{6.121}$$

$$I_{\text{OUT6}} = I_{\text{OUT4}} = \frac{I_{\text{IN}}^4}{160I_{\text{REF}}^3}, \tag{6.122}$$

$$I_{\text{OUT7}} = \frac{I_{\text{OUT6}}^2}{8I_{\text{OUT9}}} = \frac{1}{7680}\frac{I_{\text{IN}}^6}{I_{\text{REF}}^5}, \tag{6.123}$$

$$I_{\text{OUT8}} = I_{\text{OUT4}} = \frac{I_{\text{IN}}^4}{160I_{\text{REF}}^3}, \tag{6.124}$$

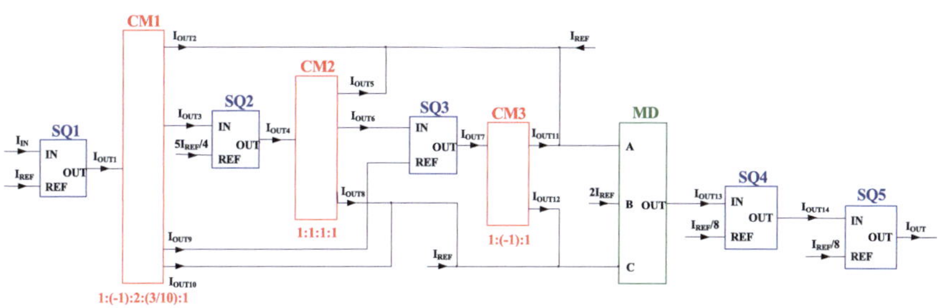

Fig. 6.13 Block diagram of the 12th analog function synthesizer for generating the Gaussian activation function

$$I_{\mathrm{OUT9}} = \frac{3}{10} I_{\mathrm{OUT1}} = \frac{3}{80} \frac{I_{\mathrm{IN}}^2}{I_{\mathrm{REF}}}, \tag{6.125}$$

$$I_{\mathrm{OUT10}} = I_{\mathrm{OUT1}} = \frac{I_{\mathrm{IN}}^2}{8 I_{\mathrm{REF}}}, \tag{6.126}$$

$$I_{\mathrm{OUT11}} = -I_{\mathrm{OUT7}} = -\frac{I_{\mathrm{IN}}^6}{7680 I_{\mathrm{REF}}^5}, \tag{6.127}$$

$$I_{\mathrm{OUT12}} = I_{\mathrm{OUT7}} = \frac{I_{\mathrm{IN}}^6}{7680 I_{\mathrm{REF}}^5}, \tag{6.128}$$

$$I_{\mathrm{OUT13}} = I_{\mathrm{REF}} \frac{-\frac{I_{\mathrm{IN}}^6}{7680 I_{\mathrm{REF}}^5} + \frac{I_{IN}^4}{160 I_{\mathrm{REF}}^3} - \frac{I_{\mathrm{IN}}^2}{8 I_{\mathrm{REF}}} + I_{\mathrm{REF}}}{\frac{I_{\mathrm{IN}}^6}{7680 I_{\mathrm{REF}}^5} + \frac{I_{\mathrm{IN}}^4}{160 I_{\mathrm{REF}}^3} + \frac{I_{\mathrm{IN}}^2}{8 I_{\mathrm{REF}}} + I_{\mathrm{REF}}} \tag{6.129}$$

and

$$I_{\mathrm{OUT14}} = \frac{I_{\mathrm{OUT13}}^2}{8(I_{\mathrm{REF}}/8)} = I_{\mathrm{REF}} \left(\frac{-\frac{I_{\mathrm{IN}}^6}{7680 I_{\mathrm{REF}}^5} + \frac{I_{\mathrm{IN}}^4}{160 I_{\mathrm{REF}}^3} - \frac{I_{\mathrm{IN}}^2}{8 I_{\mathrm{REF}}} + I_{\mathrm{REF}}}{\frac{I_{\mathrm{IN}}^6}{7680 I_{\mathrm{REF}}^5} + \frac{I_{\mathrm{IN}}^4}{160 I_{\mathrm{REF}}^3} + \frac{I_{\mathrm{IN}}^2}{8 I_{\mathrm{REF}}} + I_{\mathrm{REF}}} \right)^2. \tag{6.130}$$

So, the output current of the circuit from Fig. 6.12 can be expressed as follows:

$$I_{\mathrm{OUT}} = \frac{I_{\mathrm{OUT14}}^2}{8(I_{\mathrm{REF}}/8)} = I_{\mathrm{REF}} \left(\frac{-\frac{I_{\mathrm{IN}}^6}{7680 I_{\mathrm{REF}}^5} + \frac{I_{\mathrm{IN}}^4}{160 I_{\mathrm{REF}}^3} - \frac{I_{\mathrm{IN}}^2}{8 I_{\mathrm{REF}}} + I_{\mathrm{REF}}}{\frac{I_{\mathrm{IN}}^6}{7680 I_{\mathrm{REF}}^5} + \frac{I_{\mathrm{IN}}^4}{160 I_{\mathrm{REF}}^3} + \frac{I_{\mathrm{IN}}^2}{8 I_{\mathrm{REF}}} + I_{\mathrm{REF}}} \right)^4. \tag{6.131}$$

Using the definition of the x variable, the previous relation becomes:

$$I_{\mathrm{OUT}} = I_{\mathrm{REF}} = \left(\frac{1 - \frac{x^2}{8} + \frac{x^4}{160} - \frac{x^6}{7680}}{1 + \frac{x^2}{8} + \frac{x^4}{160} + \frac{x^6}{7680}} \right)^4. \tag{6.132}$$

In conclusion, the analog function synthesizer shown in Fig. 6.13 approximates the Gaussian activation function:

$$I_{\mathrm{OUT}} = I_{\mathrm{REF}} a_{14}(x) \cong I_{\mathrm{REF}} f_3(x). \tag{6.133}$$

6.3.13 13th Analog Function Synthesizer for Generating the Gaussian Activation Function

The 13th analog function synthesizer, designed for generating the Gaussian activation function, uses $a_{15}(x)$ approximation function described in Chap. 2.

The block diagram of the 13th analog function synthesizer for generating the Gaussian activation function is presented in Fig. 6.14.

The expressions of the currents from the circuit are:

$$I_{OUT1} = \frac{I_{IN}^2}{8I_{REF}}, \tag{6.134}$$

$$I_{OUT2} = -\frac{24}{5}I_{OUT1} = -\frac{3I_{IN}^2}{5I_{REF}}, \tag{6.135}$$

$$I_{OUT3} = 8I_{OUT1} = \frac{I_{IN}^2}{I_{REF}}, \tag{6.136}$$

$$I_{OUT4} = \frac{I_{OUT3}^2}{8(5I_{REF}/6)} = \frac{3I_{IN}^4}{20I_{REF}^3}, \tag{6.137}$$

$$I_{OUT5} = I_{OUT4} = \frac{3I_{IN}^4}{20I_{REF}^3}, \tag{6.138}$$

$$I_{OUT6} = I_{OUT4} = \frac{3I_{IN}^4}{20I_{REF}^3}, \tag{6.139}$$

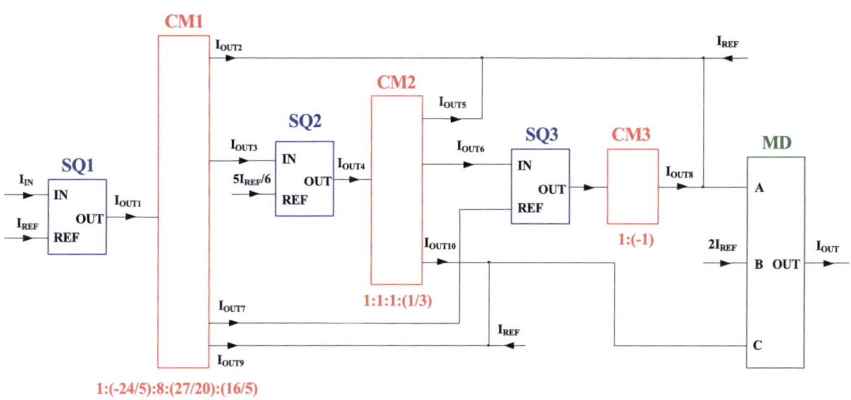

Fig. 6.14 Block diagram of the 13th analog function synthesizer for generating the Gaussian activation function

$$I_{OUT7} = \frac{27}{20} I_{OUT1} = \frac{27 I_{IN}^2}{160 I_{REF}}, \tag{6.140}$$

$$I_{OUT8} = -\frac{I_{OUT6}^2}{8 I_{OUT7}} = -\frac{I_{IN}^6}{60 I_{REF}^5}, \tag{6.141}$$

$$I_{OUT9} = \frac{16}{5} I_{OUT1} = \frac{2 I_{IN}^2}{5 I_{REF}} \tag{6.142}$$

and

$$I_{OUT10} = \frac{1}{3} I_{OUT4} = \frac{I_{IN}^4}{20 I_{REF}^3}. \tag{6.143}$$

So, the output current of the circuit from Fig. 6.14 can be expressed as follows:

$$I_{OUT} = I_{REF} \frac{-\frac{I_{IN}^6}{60 I_{REF}^5} + \frac{3 I_{IN}^4}{20 I_{REF}^3} - \frac{3 I_{IN}^2}{5 I_{REF}} + I_{REF}}{\frac{I_{IN}^4}{20 I_{REF}^3} + \frac{2 I_{IN}^2}{5 I_{REF}} + I_{REF}}. \tag{6.144}$$

Using the definition of the x variable, the previous relation becomes:

$$I_{OUT} = I_{REF} \frac{1 - \frac{3x^2}{5} + \frac{3x^4}{20} - \frac{x^6}{60}}{1 + \frac{2x^2}{5} + \frac{x^4}{20}}. \tag{6.145}$$

In conclusion, the analog function synthesizer shown in Fig. 6.14 approximates the Gaussian activation function:

$$I_{OUT} = I_{REF} a_{15}(x) \cong I_{REF} f_3(x). \tag{6.146}$$

6.3.14 14th Analog Function Synthesizer for Generating the Gaussian Activation Function

The 14th analog function synthesizer, designed for generating the Gaussian activation function, uses $a_{16}(x)$ approximation function described in Chap. 2.

The block diagram of the 14th analog function synthesizer for generating the Gaussian activation function is presented in Fig. 6.15.

The expressions of the currents from the circuit are:

$$I_{OUT1} = \frac{I_{IN}^2}{8 I_{REF}}, \tag{6.147}$$

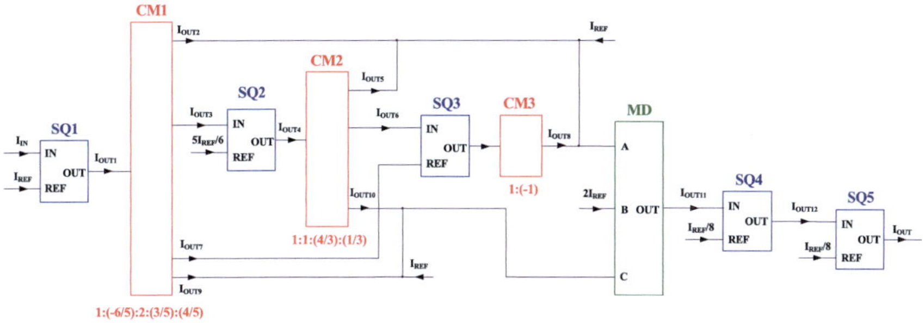

Fig. 6.15 Block diagram of the 14th analog function synthesizer for generating the Gaussian activation function

$$I_{\mathrm{OUT2}} = -\frac{6}{5} I_{\mathrm{OUT1}} = -\frac{3I_{\mathrm{IN}}^2}{20I_{\mathrm{REF}}}, \tag{6.148}$$

$$I_{\mathrm{OUT3}} = 2I_{\mathrm{OUT1}} = \frac{I_{\mathrm{IN}}^2}{4I_{\mathrm{REF}}}, \tag{6.149}$$

$$I_{\mathrm{OUT4}} = \frac{I_{\mathrm{OUT3}}^2}{8(5I_{\mathrm{REF}}/6)} = \frac{3I_{\mathrm{IN}}^4}{320I_{\mathrm{REF}}^3}, \tag{6.150}$$

$$I_{\mathrm{OUT5}} = I_{\mathrm{OUT4}} = \frac{3I_{\mathrm{IN}}^4}{320I_{\mathrm{REF}}^3}, \tag{6.151}$$

$$I_{\mathrm{OUT6}} = \frac{4}{3} I_{\mathrm{OUT4}} = \frac{I_{\mathrm{IN}}^4}{80I_{\mathrm{REF}}^3}, \tag{6.152}$$

$$I_{\mathrm{OUT7}} = \frac{3}{5} I_{\mathrm{OUT1}} = \frac{3I_{\mathrm{IN}}^2}{40I_{\mathrm{REF}}}, \tag{6.153}$$

$$I_{\mathrm{OUT8}} = -\frac{I_{\mathrm{OUT6}}^2}{8I_{\mathrm{OUT7}}} = -\frac{I_{\mathrm{IN}}^6}{3840I_{\mathrm{REF}}^5}, \tag{6.154}$$

$$I_{\mathrm{OUT9}} = \frac{4}{5} I_{\mathrm{OUT1}} = \frac{I_{\mathrm{IN}}^2}{10I_{\mathrm{REF}}}, \tag{6.155}$$

$$I_{\mathrm{OUT10}} = \frac{1}{3} I_{\mathrm{OUT4}} = \frac{I_{\mathrm{IN}}^4}{320I_{\mathrm{REF}}^3}, \tag{6.156}$$

$$I_{\mathrm{OUT11}} = I_{\mathrm{REF}} \frac{-\frac{I_{\mathrm{IN}}^6}{3840I_{\mathrm{REF}}^5} + \frac{3I_{\mathrm{IN}}^4}{320I_{\mathrm{REF}}^3} - \frac{3I_{\mathrm{IN}}^2}{20I_{\mathrm{REF}}} + I_{\mathrm{REF}}}{\frac{I_{\mathrm{IN}}^4}{320I_{\mathrm{REF}}^3} + \frac{I_{\mathrm{IN}}^2}{10I_{\mathrm{REF}}} + I_{\mathrm{REF}}} \tag{6.157}$$

and

$$I_{OUT12} = \frac{I_{OUT11}^2}{8(I_{REF}/8)}. \tag{6.158}$$

So, the output current of the circuit from Fig. 6.15 can be expressed as follows:

$$I_{OUT} = I_{REF}\left(\frac{-\frac{I_{IN}^6}{3840I_{REF}^5} + \frac{3I_{IN}^4}{320I_{REF}^3} - \frac{3I_{IN}^2}{20I_{REF}} + I_{REF}}{\frac{I_{IN}^4}{320I_{REF}^3} + \frac{I_{IN}^2}{10I_{REF}} + I_{REF}}\right)^4. \tag{6.159}$$

Using the definition of the x variable, the previous relation becomes:

$$I_{OUT} = I_{REF}\left(\frac{1 - \frac{3x^2}{20} + \frac{3x^4}{320} - \frac{x^6}{3840}}{1 + \frac{x^2}{10} + \frac{x^4}{320}}\right)^4. \tag{6.160}$$

In conclusion, the analog function synthesizer shown in Fig. 6.15 approximates the Gaussian activation function:

$$I_{OUT} = I_{REF}a_{16}(x) \cong I_{REF}f_3(x). \tag{6.161}$$

6.3.15 *15th Analog Function Synthesizer for Generating the Gaussian Activation Function*

The 15th analog function synthesizer, designed for generating the Gaussian activation function, uses $a_{17}(x)$ approximation function described in Chap. 2.

The block diagram of the 15th analog function synthesizer for generating the Gaussian activation function is presented in Fig. 6.16.

The expressions of the currents from the circuit are:

$$I_{OUT1} = \frac{I_{IN}^2}{8I_{REF}}, \tag{6.162}$$

$$I_{OUT2} = -\frac{2}{5}I_{OUT1} = -\frac{I_{IN}^2}{20I_{REF}}, \tag{6.163}$$

$$I_{OUT3} = I_{OUT1} = \frac{I_{IN}^2}{8I_{REF}}, \tag{6.164}$$

$$I_{OUT4} = \frac{I_{OUT3}^2}{8(5I_{REF}/16)} = \frac{I_{IN}^4}{160I_{REF}^3}, \tag{6.165}$$

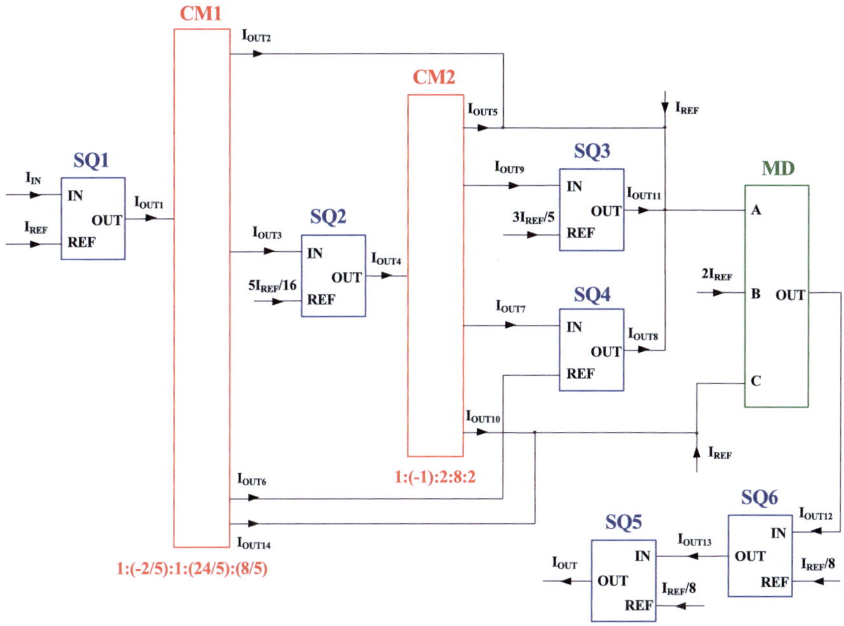

Fig. 6.16 Block diagram of the 15th analog function synthesizer for generating the Gaussian activation function

$$I_{\mathrm{OUT5}} = -I_{\mathrm{OUT4}} = -\frac{I_{\mathrm{IN}}^4}{160 I_{\mathrm{REF}}^3}, \tag{6.166}$$

$$I_{\mathrm{OUT6}} = \frac{24}{5} I_{\mathrm{OUT1}} = \frac{3 I_{\mathrm{IN}}^2}{5 I_{\mathrm{REF}}}, \tag{6.167}$$

$$I_{\mathrm{OUT7}} = 8 I_{\mathrm{OUT4}} = \frac{I_{\mathrm{IN}}^4}{20 I_{\mathrm{REF}}^3}, \tag{6.168}$$

$$I_{\mathrm{OUT8}} = \frac{I_{\mathrm{OUT7}}^2}{8 I_{\mathrm{OUT6}}} = \frac{I_{\mathrm{IN}}^6}{1920 I_{\mathrm{REF}}^5}, \tag{6.169}$$

$$I_{\mathrm{OUT9}} = 2 I_{\mathrm{OUT4}} = \frac{I_{\mathrm{IN}}^4}{80 I_{\mathrm{REF}}^3}, \tag{6.170}$$

$$I_{\mathrm{OUT10}} = 2 I_{\mathrm{OUT4}} = \frac{I_{\mathrm{IN}}^4}{80 I_{\mathrm{REF}}^3}, \tag{6.171}$$

$$I_{\mathrm{OUT11}} = \frac{I_{\mathrm{OUT9}}^2}{8(3 I_{\mathrm{REF}}/5)} = \frac{I_{\mathrm{IN}}^8}{30720 I_{\mathrm{REF}}^7}, \tag{6.172}$$

$$I_{OUT12} = I_{REF} \frac{I_{OUT2} + I_{OUT5} + I_{OUT8} + I_{OUT11} + I_{REF}}{I_{OUT10} + I_{OUT14} + I_{REF}}, \tag{6.173}$$

$$I_{OUT13} = \frac{I_{OUT12}^2}{8(I_{REF}/8)} \tag{6.174}$$

and

$$I_{OUT14} = \frac{8}{5} I_{OUT1} = \frac{I_{IN}^2}{5 I_{REF}}. \tag{6.175}$$

So, the output current of the circuit from Fig. 6.16 can be expressed as follows:

$$I_{OUT} = \frac{I_{OUT13}^2}{8(I_{REF}/8)}. \tag{6.176}$$

Using the definition of the x variable, the previous relation becomes:

$$I_{OUT} = I_{REF} \left(\frac{1 - \frac{x^2}{20} - \frac{x^4}{160} + \frac{x^6}{1920} + \frac{x^8}{30720}}{1 + \frac{x^2}{5} + \frac{x^4}{80}} \right)^4. \tag{6.177}$$

In conclusion, the analog function synthesizer shown in Fig. 6.16 approximates the Gaussian activation function:

$$I_{OUT} = I_{REF} a_{17}(x) \cong I_{REF} f_3(x). \tag{6.178}$$

6.3.16 16th Analog Function Synthesizer for Generating the Gaussian Activation Function

The 16th analog function synthesizer, designed for generating the Gaussian activation function, uses $a_{18}(x)$ approximation function described in Chap. 2.

The block diagram of the 16th analog function synthesizer for generating the Gaussian activation function is presented in Fig. 6.17.

The expressions of the currents from the circuit are:

$$I_{OUT1} = \frac{I_{IN}^2}{8 I_{REF}}, \tag{6.179}$$

$$I_{OUT2} = 4 I_{OUT1} = \frac{I_{IN}^2}{2 I_{REF}}, \tag{6.180}$$

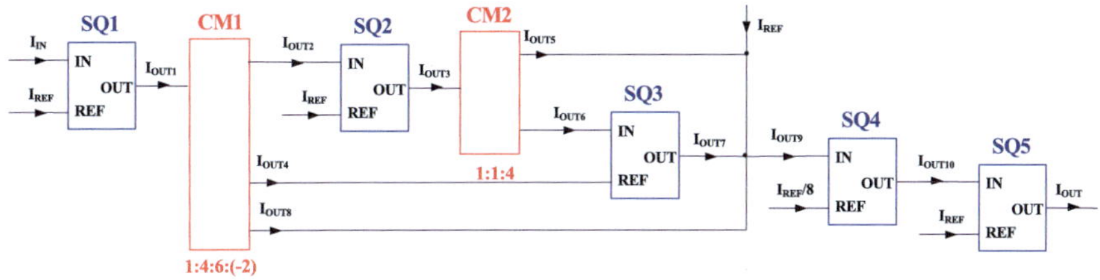

Fig. 6.17 Block diagram of the 16th analog function synthesizer for generating the Gaussian activation function

$$I_{\text{OUT3}} = \frac{I_{\text{OUT2}}^2}{8I_{\text{REF}}} = \frac{I_{\text{IN}}^4}{32I_{\text{REF}}^3}, \tag{6.181}$$

$$I_{\text{OUT4}} = 6I_{\text{OUT1}} = \frac{3I_{\text{IN}}^2}{4I_{\text{REF}}}, \tag{6.182}$$

$$I_{OUT5} = I_{OUT3} = \frac{I_{IN}^4}{32I_{REF}^3}, \tag{6.183}$$

$$I_{\text{OUT6}} = 4I_{\text{OUT3}} = \frac{I_{\text{IN}}^4}{8I_{\text{REF}}^3}, \tag{6.184}$$

$$I_{\text{OUT7}} = \frac{I_{\text{OUT6}}^2}{8I_{\text{OUT4}}} = \frac{I_{\text{IN}}^6}{384I_{\text{REF}}^5}, \tag{6.185}$$

$$I_{\text{OUT8}} = -2I_{\text{OUT1}} = -\frac{I_{\text{IN}}^2}{4I_{\text{REF}}}, \tag{6.186}$$

$$I_{\text{OUT9}} = I_{\text{REF}} - \frac{I_{\text{IN}}^2}{4I_{\text{REF}}} + \frac{I_{\text{IN}}^4}{32I_{\text{REF}}^3} + \frac{I_{\text{IN}}^6}{384I_{\text{REF}}^5} \tag{6.187}$$

and

$$I_{\text{OUT10}} = \frac{I_{\text{OUT9}}^2}{8(I_{\text{REF}}/8)} = \frac{\left(I_{\text{REF}} - \frac{I_{\text{IN}}^2}{4I_{\text{REF}}} + \frac{I_{\text{IN}}^4}{32I_{\text{REF}}^3} + \frac{I_{\text{IN}}^6}{384I_{\text{REF}}^5}\right)^2}{I_{\text{REF}}}. \tag{6.188}$$

So, the output current of the circuit from Fig. 6.17 can be expressed as follows:

$$I_{\text{OUT}} = \frac{I_{\text{OUT10}}^2}{8(I_{\text{REF}}/8)} = \frac{\left(I_{\text{REF}} - \frac{I_{\text{IN}}^2}{4I_{\text{REF}}} + \frac{I_{\text{IN}}^4}{32I_{\text{REF}}^3} + \frac{I_{\text{IN}}^6}{384I_{\text{REF}}^5}\right)^4}{I_{\text{REF}}^3}. \tag{6.189}$$

Using the definition of the x variable, the previous relation becomes:

$$I_{\mathrm{OUT}} = I_{\mathrm{REF}}\left(1 - \frac{x^2}{4} + \frac{x^4}{32} - \frac{x^6}{384}\right)^4. \quad (6.190)$$

In conclusion, the analog function synthesizer shown in Fig. 6.17 approximates the Gaussian activation function:

$$I_{\mathrm{OUT}} = I_{\mathrm{REF}} a_{18}(x) \cong I_{\mathrm{REF}} f_3(x). \quad (6.191)$$

6.4 Analog Function Synthesizers for Quadratic and Inverse Quadratic Activation Function

In order to implement in CMOS technology the quadratic activation function, the following simple circuit is used, based on a current-mode squaring circuit (SQ), analyzed in Chap. 4. The block diagram of the analog function synthesizer for generating the quadratic activation function is shown in Fig. 6.18.

The expression of the current I_{OUT1} is:

$$I_{\mathrm{OUT1}} = \frac{I_{\mathrm{IN}}^2}{8(I_{\mathrm{REF}}/8)} = \frac{I_{\mathrm{IN}}^2}{I_{\mathrm{REF}}}. \quad (6.192)$$

The output current of the circuit from Fig. 6.18 can be expressed as follows:

$$I_{\mathrm{OUT}} = I_{\mathrm{OUT1}} + a^2 I_{\mathrm{REF}} = I_{\mathrm{REF}}\left[\left(\frac{I_{\mathrm{IN}}}{I_{\mathrm{REF}}}\right)^2 + a^2\right]. \quad (6.193)$$

Using the definition of the x variable, the previous relation becomes:

$$I_{\mathrm{OUT}} = I_{\mathrm{REF}} f_4(x) \quad (6.194)$$

In conclusion, the analog function synthesizer shown in Fig. 6.18 generates the quadratic activation function.

In order to implement in CMOS technology the inverse quadratic activation function, the following simple circuit is used, based on two current-mode squaring circuits

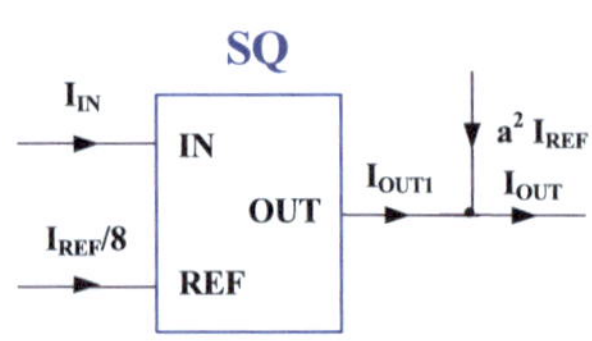

Fig. 6.18 Block diagram of the analog function synthesizer for generating the quadratic activation function

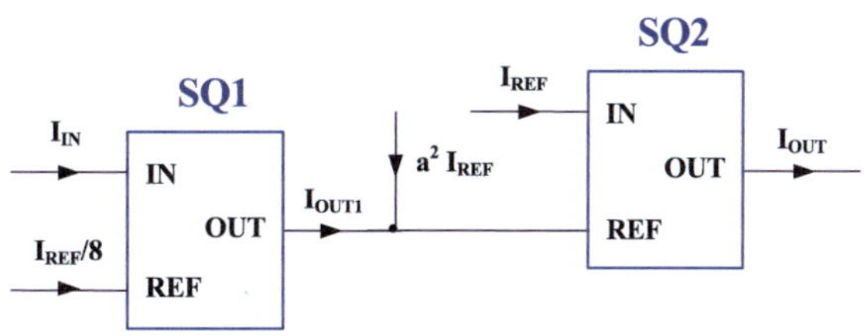

Fig. 6.19 Block diagram of the analog function synthesizer for generating the inverse quadratic activation function

(SQ), analyzed in Chap. 4. The block diagram of the analog function synthesizer for generating the inverse quadratic activation function is shown in Fig. 6.19.

The expression of the current I_{OUT1} is:

$$I_{OUT1} = \frac{I_{IN}^2}{8(I_{REF}/8)} = \frac{I_{IN}^2}{I_{REF}}. \tag{6.195}$$

The output current of the circuit from Fig. 6.19 can be expressed as follows:

$$I_{OUT} = \frac{I_{REF}^2}{8\left(a^2 I_{REF} + \frac{I_{IN}^2}{I_{REF}}\right)} = \frac{I_{REF}}{8} \frac{1}{\left(\frac{I_{IN}}{I_{REF}}\right)^2 + a^2}. \tag{6.196}$$

Using the definition of the x variable, the previous relation becomes:

$$I_{OUT} = \frac{I_{REF}}{8} f_5(x). \tag{6.197}$$

In conclusion, the analog function synthesizer shown in Fig. 6.19 generates the inverse quadratic activation function.

6.5 Analog Function Synthesizers for Multiquadratic Activation Function

In order to implement in CMOS technology the multiquadratic activation function, the approximation functions described in Chap. 2 and the CMOS squaring and multiplies/divider circuits from Chap. 4 are used.

6.5.1 First Analog Function Synthesizer for Generating the Multiquadratic Activation Function for $a = 1$

The first analog function synthesizer, designed for generating the multiquadratic activation function for $a = 1$, uses $a_{19}(x)$ approximation function described in Chap. 2.

The block diagram of the first analog function synthesizer for generating the multiquadratic activation function for $a = 1$ is presented in Fig. 6.20.

The expressions of the currents from the circuit are:

$$I_{\mathrm{OUT1}} = \frac{I_{\mathrm{IN}}^2}{8(I_{\mathrm{REF}}/8)} = \frac{I_{\mathrm{IN}}^2}{I_{\mathrm{REF}}}, \tag{6.198}$$

$$I_{\mathrm{OUT2}} = I_{\mathrm{OUT1}} = \frac{I_{\mathrm{IN}}^2}{I_{\mathrm{REF}}}, \tag{6.199}$$

$$I_{\mathrm{OUT3}} = I_{\mathrm{OUT1}} = \frac{I_{\mathrm{IN}}^2}{I_{\mathrm{REF}}}, \tag{6.200}$$

$$I_{\mathrm{OUT4}} = \frac{I_{\mathrm{OUT3}}^2}{8I_{\mathrm{REF}}} = \frac{I_{\mathrm{IN}}^4}{8I_{\mathrm{REF}}^3} \tag{6.201}$$

and

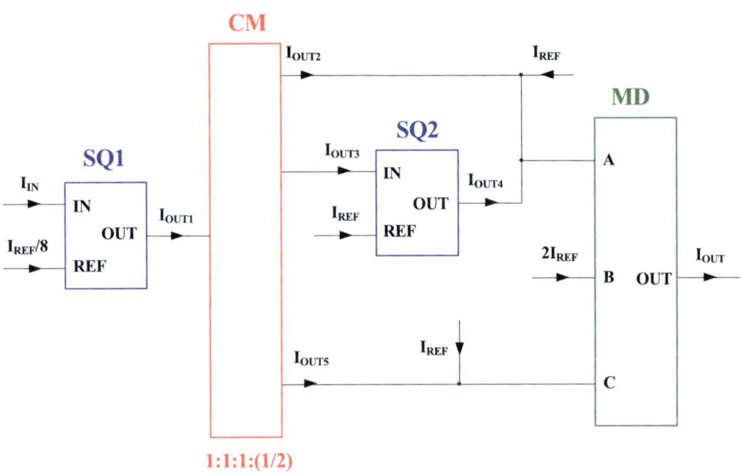

Fig. 6.20 Block diagram of the first analog function synthesizer for generating the multiquadratic activation function for $a = 1$

$$I_{OUT5} = \frac{1}{2} I_{OUT1} = \frac{I_{IN}^2}{2I_{REF}}. \tag{6.202}$$

The output current of the circuit from Fig. 6.20 can be expressed as follows:

$$I_{OUT} = I_{REF} \frac{\frac{I_{IN}^4}{8I_{REF}^3} + \frac{I_{IN}^2}{I_{REF}} + I_{REF}}{\frac{I_{IN}^2}{2I_{REF}} + I_{REF}}. \tag{6.203}$$

Using the definition of x variable, the previous relation becomes:

$$I_{OUT} = I_{REF} \frac{\frac{x^4}{8} + x^2 + 1}{\frac{x^2}{2} + 1}. \tag{6.204}$$

In conclusion, the analog function synthesizer shown in Fig. 6.20 approximates the multiquadratic activation function for $a = 1$:

$$I_{OUT} = I_{REF} a_{19}(x) \cong I_{REF} f_6(x). \tag{6.205}$$

6.5.2 Second Analog Function Synthesizer for Generating the Multiquadratic Activation Function for a = 1

The second analog function synthesizer, designed for generating the multiquadratic activation function for $a = 1$, uses $a_{20}(x)$ approximation function described in Chap. 2.

The block diagram of the second analog function synthesizer for generating the multiquadratic activation function for $a = 1$ is presented in Fig. 6.21.

The expressions of the currents from the circuit are:

$$I_{OUT1} = \frac{I_{IN}^2}{8(I_{REF}/8)} = \frac{I_{IN}^2}{I_{REF}}, \tag{6.206}$$

$$I_{OUT2} = \frac{5}{4} I_{OUT1} = \frac{5I_{IN}^2}{4I_{REF}}, \tag{6.207}$$

$$I_{OUT3} = I_{OUT1} = \frac{I_{IN}^2}{I_{REF}}, \tag{6.208}$$

$$I_{OUT4} = \frac{I_{OUT3}^2}{8(2I_{REF}/5)} = \frac{5I_{IN}^4}{16I_{REF}^3}, \tag{6.209}$$

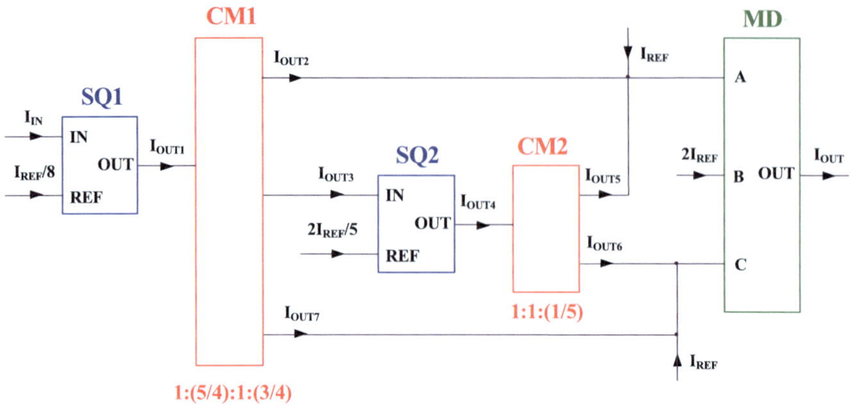

Fig. 6.21 Block diagram of the second analog function synthesizer for generating the multiquadratic activation function for $a = 1$

$$I_{\mathrm{OUT5}} = I_{\mathrm{OUT4}} = \frac{5I_{\mathrm{IN}}^4}{16I_{\mathrm{REF}}^3}, \tag{6.210}$$

$$I_{\mathrm{OUT6}} = \frac{1}{5}I_{\mathrm{OUT4}} = \frac{I_{\mathrm{IN}}^4}{16I_{\mathrm{REF}}^3} \tag{6.211}$$

and

$$I_{\mathrm{OUT7}} = \frac{3}{4}I_{\mathrm{OUT1}} = \frac{3I_{\mathrm{IN}}^2}{4I_{\mathrm{REF}}}. \tag{6.212}$$

So, the output current of the circuit from Fig. 6.21 can be expressed as follows:

$$I_{\mathrm{OUT}} = I_{\mathrm{REF}}\frac{I_{\mathrm{REF}} + \frac{5I_{\mathrm{IN}}^2}{4I_{\mathrm{REF}}} + \frac{5I_{\mathrm{IN}}^4}{16I_{\mathrm{REF}}^3}}{I_{\mathrm{REF}} + \frac{3I_{\mathrm{IN}}^2}{4I_{\mathrm{REF}}} + \frac{I_{\mathrm{IN}}^4}{16I_{\mathrm{REF}}^3}}. \tag{6.213}$$

Using the definition of the x variable, the previous relation becomes:

$$I_{\mathrm{OUT}} = I_{\mathrm{REF}}\frac{\frac{5x^4}{16} + \frac{5x^2}{4} + 1}{\frac{x^4}{16} + \frac{3x^2}{4} + 1}. \tag{6.214}$$

In conclusion, the analog function synthesizer shown in Fig. 6.21 approximates the multiquadratic activation function for $a = 1$:

$$I_{\mathrm{OUT}} = I_{\mathrm{REF}}a_{20}(x) \cong I_{\mathrm{REF}}f_6(x). \tag{6.215}$$

6.5.3 Third Analog Function Synthesizer for Generating the Multiquadratic Activation Function for a = 1

The third analog function synthesizer, designed for generating the multiquadratic activation function for $a = 1$, uses $a_{21}(x)$ approximation function described in Chap. 2.

The block diagram of the third analog function synthesizer for generating the multiquadratic activation function for $a = 1$ is presented in Fig. 6.22.

The expressions of the currents from the circuit are:

$$I_{\mathrm{OUT1}} = \frac{I_{\mathrm{IN}}^2}{8(I_{\mathrm{REF}}/4)} = \frac{I_{\mathrm{IN}}^2}{2I_{\mathrm{REF}}}, \tag{6.216}$$

$$I_{\mathrm{OUT2}} = 3I_{\mathrm{OUT1}} = \frac{3I_{\mathrm{IN}}^2}{2I_{\mathrm{REF}}}, \tag{6.217}$$

$$I_{\mathrm{OUT3}} = 4I_{\mathrm{OUT1}} = \frac{2I_{\mathrm{IN}}^2}{I_{\mathrm{REF}}}, \tag{6.218}$$

$$I_{\mathrm{OUT4}} = \frac{I_{\mathrm{OUT3}}^2}{8(8I_{\mathrm{REF}}/9)} = \frac{9I_{\mathrm{IN}}^4}{16I_{\mathrm{REF}}^3}, \tag{6.219}$$

$$I_{\mathrm{OUT5}} = I_{\mathrm{OUT4}} = \frac{9I_{\mathrm{IN}}^4}{16I_{\mathrm{REF}}^3}, \tag{6.220}$$

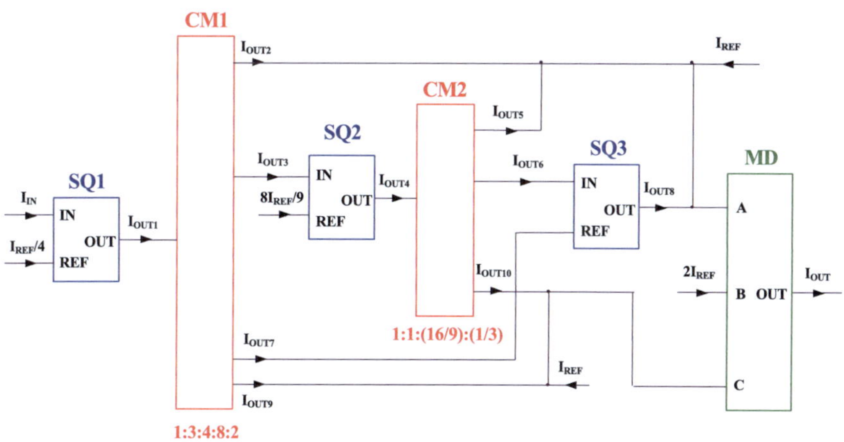

Fig. 6.22 Block diagram of the third analog function synthesizer for generating the multiquadratic activation function for $a = 1$

$$I_{\mathrm{OUT6}} = \frac{16}{9} I_{\mathrm{OUT4}} = \frac{I_{\mathrm{IN}}^4}{I_{\mathrm{REF}}^3}, \tag{6.221}$$

$$I_{\mathrm{OUT7}} = 8 I_{\mathrm{OUT1}} = \frac{4 I_{\mathrm{IN}}^2}{I_{\mathrm{REF}}}, \tag{6.222}$$

$$I_{\mathrm{OUT8}} = \frac{I_{\mathrm{OUT6}}^2}{8 I_{\mathrm{OUT7}}} = \frac{I_{\mathrm{IN}}^6}{32 I_{\mathrm{REF}}^5}, \tag{6.223}$$

$$I_{\mathrm{OUT9}} = 2 I_{\mathrm{OUT1}} = \frac{I_{\mathrm{IN}}^2}{I_{\mathrm{REF}}} \tag{6.224}$$

and

$$I_{\mathrm{OUT10}} = \frac{1}{3} I_{\mathrm{OUT4}} = \frac{3 I_{\mathrm{IN}}^4}{16 I_{\mathrm{REF}}^3}. \tag{6.225}$$

So, the output current of the circuit from Fig. 6.22 can be expressed as follows:

$$I_{\mathrm{OUT}} = I_{\mathrm{REF}} \frac{\frac{I_{\mathrm{IN}}^6}{32 I_{\mathrm{REF}}^5} + \frac{9 I_{\mathrm{IN}}^4}{16 I_{\mathrm{REF}}^3} + \frac{3 I_{\mathrm{IN}}^2}{2 I_{\mathrm{REF}}} + I_{\mathrm{REF}}}{\frac{3 I_{\mathrm{IN}}^4}{16 I_{\mathrm{REF}}^3} + \frac{I_{\mathrm{IN}}^2}{I_{\mathrm{REF}}} + I_{\mathrm{REF}}}. \tag{6.226}$$

Using the definition of the x variable, the previous relation becomes:

$$I_{\mathrm{OUT}} = I_{\mathrm{REF}} \frac{\frac{x^6}{32} + \frac{9x^4}{16} + \frac{3x^2}{2} + 1}{\frac{3x^4}{16} + x^2 + 1}. \tag{6.227}$$

In conclusion, the analog function synthesizer shown in Fig. 6.22 approximates the multiquadratic activation function for $a = 1$:

$$I_{\mathrm{OUT}} = I_{\mathrm{REF}} a_{21}(x) \cong I_{\mathrm{REF}} f_6(x). \tag{6.228}$$

6.5.4 *Fourth Analog Function Synthesizer for Generating the Multiquadratic Activation Function for a = 1*

The fourth analog function synthesizer, designed for generating the multiquadratic activation function for $a = 1$, uses $a_{22}(x)$ approximation function described in Chap. 2.

The block diagram of the fourth analog function synthesizer for generating the multiquadratic activation function for $a = 1$ is presented in Fig. 6.23.

The expressions of the currents from the circuit are:

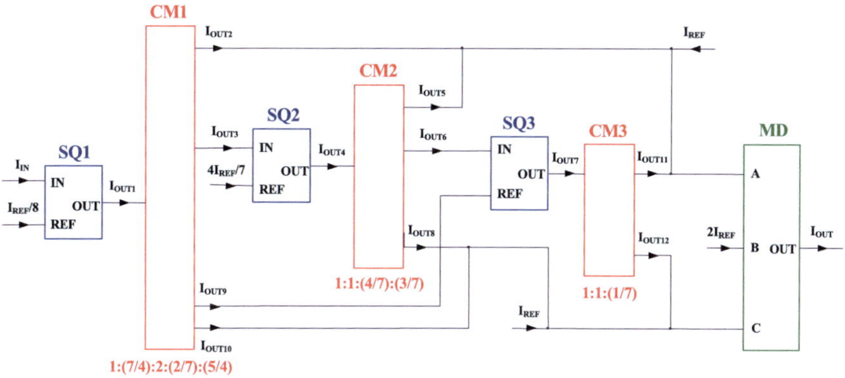

Fig. 6.23 Block diagram of the fourth analog function synthesizer for generating the multiquadratic activation function for $a = 1$

$$I_{\mathrm{OUT1}} = \frac{I_{\mathrm{IN}}^2}{8(I_{\mathrm{REF}}/8)} = \frac{I_{\mathrm{IN}}^2}{I_{\mathrm{REF}}}, \tag{6.229}$$

$$I_{\mathrm{OUT2}} = \frac{7}{4} I_{\mathrm{OUT1}} = \frac{7I_{\mathrm{IN}}^2}{4I_{\mathrm{REF}}}, \tag{6.230}$$

$$I_{\mathrm{OUT3}} = 2I_{\mathrm{OUT1}} = \frac{2I_{\mathrm{IN}}^2}{I_{\mathrm{REF}}}, \tag{6.231}$$

$$I_{\mathrm{OUT4}} = \frac{I_{\mathrm{OUT3}}^2}{8(4I_{\mathrm{REF}}/7)} = \frac{7I_{\mathrm{IN}}^4}{8I_{\mathrm{REF}}^3}, \tag{6.232}$$

$$I_{\mathrm{OUT5}} = I_{\mathrm{OUT4}} = \frac{7I_{\mathrm{IN}}^4}{8I_{\mathrm{REF}}^3}, \tag{6.233}$$

$$I_{\mathrm{OUT6}} = \frac{4}{7} I_{\mathrm{OUT4}} = \frac{I_{\mathrm{IN}}^4}{2I_{\mathrm{REF}}^3}, \tag{6.234}$$

$$I_{\mathrm{OUT7}} = \frac{I_{\mathrm{OUT6}}^2}{8I_{\mathrm{OUT9}}} = \frac{7I_{\mathrm{IN}}^6}{64I_{\mathrm{REF}}^5}, \tag{6.235}$$

$$I_{\mathrm{OUT8}} = \frac{3}{7} I_{\mathrm{OUT4}} = \frac{3I_{\mathrm{IN}}^4}{8I_{\mathrm{REF}}^3}, \tag{6.236}$$

$$I_{\mathrm{OUT9}} = \frac{2}{7} I_{\mathrm{OUT1}} = \frac{2}{7}\frac{I_{\mathrm{IN}}^2}{I_{\mathrm{REF}}}, \tag{6.237}$$

$$I_{\mathrm{OUT10}} = \frac{5}{4} I_{\mathrm{OUT1}} = \frac{5I_{\mathrm{IN}}^2}{4I_{\mathrm{REF}}}, \tag{6.238}$$

$$I_{OUT11} = I_{OUT7} = \frac{7I_{IN}^6}{64I_{REF}^5}, \tag{6.239}$$

$$I_{OUT12} = \frac{1}{7}I_{OUT7} = \frac{I_{IN}^6}{64I_{REF}^5}. \tag{6.240}$$

So, the output current of the circuit from Fig. 6.23 can be expressed as follows:

$$I_{OUT} = I_{REF}\frac{\frac{7I_{IN}^6}{64I_{REF}^5} + \frac{7I_{IN}^4}{8I_{REF}^3} + \frac{7I_{IN}^2}{4I_{REF}} + I_{REF}}{\frac{I_{IN}^6}{64I_{REF}^5} + \frac{3I_{IN}^4}{8I_{REF}^3} + \frac{5I_{IN}^2}{4I_{REF}} + I_{REF}}. \tag{6.241}$$

Using the definition of the x variable, the previous relation becomes:

$$I_{OUT} = I_{REF}\frac{\frac{7x^6}{64} + \frac{7x^4}{8} + \frac{7x^2}{4} + 1}{\frac{x^6}{64} + \frac{3x^4}{8} + \frac{5x^2}{4} + 1}. \tag{6.242}$$

In conclusion, the analog function synthesizer shown in Fig. 6.23 approximates the multiquadratic activation function for a = 1:

$$I_{OUT} = I_{REF}a_{22}(x) \cong I_{REF}f_6(x). \tag{6.243}$$

6.5.5 *Fifth Analog Function Synthesizer for Generating the Multiquadratic Activation Function for a = 1*

The fifth analog function synthesizer, designed for generating the multiquadratic activation function for $a = 1$, uses $a_{23}(x)$ approximation function described in Chap. 2.

The block diagram of the fifth analog function synthesizer for generating the multiquadratic activation function for $a = 1$ is presented in Fig. 6.24.

The expressions of the currents from the circuit are:

$$I_{OUT1} = \frac{I_{IN}^2}{8I_{REF}}, \tag{6.244}$$

$$I_{OUT2} = \frac{108}{7}I_{OUT1} = \frac{27I_{IN}^2}{14I_{REF}}, \tag{6.245}$$

$$I_{OUT3} = 8I_{OUT1} = \frac{I_{IN}^2}{I_{REF}}, \tag{6.246}$$

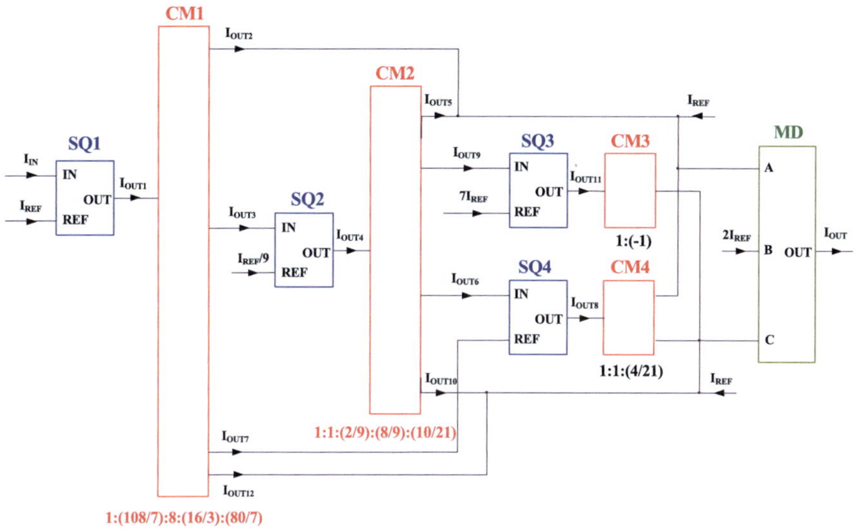

Fig. 6.24 Block diagram of the fifth analog function synthesizer for generating the multiquadratic activation function for $a = 1$

$$I_{\text{OUT4}} = \frac{I_{\text{OUT3}}^2}{8(I_{\text{REF}}/9)} = \frac{9I_{\text{IN}}^4}{8I_{\text{REF}}^3}, \tag{6.247}$$

$$I_{\text{OUT5}} = I_{\text{OUT4}} = \frac{9I_{\text{IN}}^4}{8I_{\text{REF}}^3}, \tag{6.248}$$

$$I_{\text{OUT6}} = \frac{8}{9}I_{\text{OUT4}} = \frac{I_{\text{IN}}^4}{I_{\text{REF}}^3}, \tag{6.249}$$

$$I_{\text{OUT7}} = \frac{16}{3}I_{\text{OUT1}} = \frac{2I_{\text{IN}}^2}{3I_{\text{REF}}}, \tag{6.250}$$

$$I_{\text{OUT8}} = \frac{I_{\text{OUT6}}^2}{8I_{\text{OUT7}}} = \frac{3I_{\text{IN}}^6}{16I_{\text{REF}}^5}, \tag{6.251}$$

$$I_{\text{OUT9}} = \frac{2}{9}I_{\text{OUT4}} = \frac{I_{\text{IN}}^4}{4I_{\text{REF}}^5}, \tag{6.252}$$

$$I_{\text{OUT10}} = \frac{10}{21}I_{\text{OUT4}} = \frac{15I_{\text{IN}}^2}{28I_{\text{REF}}}, \tag{6.253}$$

$$I_{\text{OUT11}} = \frac{I_{\text{OUT9}}^2}{8 \times 7I_{\text{REF}}} = \frac{I_{\text{IN}}^8}{896I_{\text{REF}}^7}, \tag{6.254}$$

and

$$I_{OUT12} = \frac{80}{7} I_{OUT1} = \frac{10 I_{IN}^2}{7 I_{REF}}. \tag{6.255}$$

So, the output current of the circuit from Fig. 6.24 can be expressed as follows:

$$I_{OUT} = I_{REF} \frac{\frac{3 I_{IN}^6}{16 I_{REF}^5} + \frac{9 I_{IN}^4}{8 I_{REF}^3} + \frac{27 I_{IN}^2}{14 I_{REF}} + I_{REF}}{-\frac{I_{IN}^8}{896 I_{REF}^7} + \frac{I_{IN}^6}{28 I_{REF}^5} + \frac{15 I_{IN}^4}{28 I_{REF}^3} + \frac{10 I_{IN}^2}{7 I_{REF}} + I_{REF}}. \tag{6.256}$$

Using the definition of the x variable, the previous relation becomes:

$$I_{OUT} = I_{REF} \frac{\frac{3x^6}{16} + \frac{9x^4}{8} + \frac{27x^2}{14} + 1}{-\frac{x^8}{896} + \frac{x^6}{28} + \frac{15x^4}{28} + \frac{10x^2}{7} + 1}. \tag{6.257}$$

In conclusion, the analog function synthesizer shown in Fig. 6.24 approximates the multiquadratic activation function for $a = 1$:

$$I_{OUT} = I_{REF} a_{23}(x) \cong I_{REF} f_6(x). \tag{6.258}$$

6.5.6 Sixth Analog Function Synthesizer for Generating the Multiquadratic Activation Function for a = 1

The sixth analog function synthesizer, designed for generating the multiquadratic activation function for $a = 1$, uses $a_{24}(x)$ approximation function described in Chap. 2.

The block diagram of the sixth analog function synthesizer for generating the multiquadratic activation function for $a = 1$ is presented in Fig. 6.25.

The expressions of the currents from the circuit are:

$$I_{OUT1} = \frac{I_{IN}^2}{8(I_{REF}/4)} = \frac{I_{IN}^2}{2 I_{REF}}, \tag{6.259}$$

$$I_{OUT2} = \frac{9}{2} I_{OUT1} = \frac{9 I_{IN}^2}{4 I_{REF}}, \tag{6.260}$$

$$I_{OUT3} = I_{OUT1} = \frac{I_{IN}^2}{2 I_{REF}}, \tag{6.261}$$

$$I_{OUT4} = \frac{I_{OUT3}^2}{8(4 I_{REF}/27)} = \frac{27 I_{IN}^4}{128 I_{REF}^3}, \tag{6.262}$$

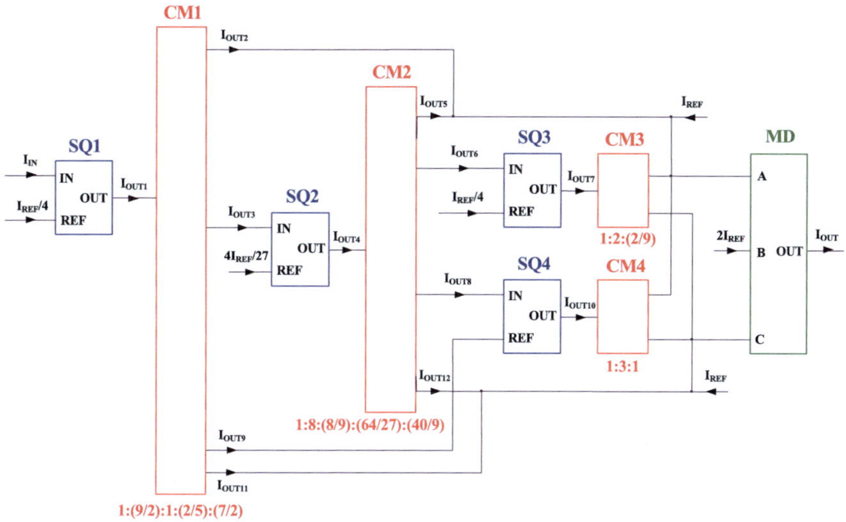

Fig. 6.25 Block diagram of the sixth analog function synthesizer for generating the multiquadratic activation function for $a = 1$

$$I_{\mathrm{OUT5}} = 8I_{\mathrm{OUT4}} = \frac{27I_{\mathrm{IN}}^4}{16I_{\mathrm{REF}}^3}, \tag{6.263}$$

$$I_{\mathrm{OUT6}} = \frac{8}{9}I_{\mathrm{OUT4}} = \frac{3I_{\mathrm{IN}}^4}{16I_{\mathrm{REF}}^3}, \tag{6.264}$$

$$I_{\mathrm{OUT7}} = \frac{I_{\mathrm{OUT6}}^2}{8(I_{\mathrm{REF}}/4)} = \frac{9I_{\mathrm{IN}}^8}{512I_{\mathrm{REF}}^7}, \tag{6.265}$$

$$I_{\mathrm{OUT8}} = \frac{64}{27}I_{\mathrm{OUT4}} = \frac{I_{\mathrm{IN}}^4}{2I_{\mathrm{REF}}^3}, \tag{6.266}$$

$$I_{\mathrm{OUT9}} = \frac{2}{5}I_{\mathrm{OUT1}} = \frac{I_{\mathrm{IN}}^2}{5I_{\mathrm{REF}}}, \tag{6.267}$$

$$I_{\mathrm{OUT10}} = \frac{I_{\mathrm{OUT8}}^2}{8I_{\mathrm{OUT9}}} = \frac{5I_{\mathrm{IN}}^6}{32I_{\mathrm{REF}}^5}, \tag{6.268}$$

$$I_{\mathrm{OUT11}} = \frac{7}{2}I_{\mathrm{OUT1}} = \frac{7I_{\mathrm{IN}}^2}{4I_{\mathrm{REF}}} \tag{6.269}$$

and

$$I_{\mathrm{OUT12}} = \frac{40}{9}I_{\mathrm{OUT4}} = \frac{15I_{\mathrm{IN}}^4}{16I_{\mathrm{REF}}^3}. \tag{6.270}$$

So, the output current of the circuit from Fig. 6.25 can be expressed as follows:

$$I_{OUT} = I_{REF} \frac{\frac{9I_{IN}^8}{256I_{REF}^7} + \frac{15I_{IN}^6}{32I_{REF}^5} + \frac{27I_{IN}^4}{16I_{REF}^3} + \frac{9I_{IN}^2}{4I_{REF}} + I_{REF}}{\frac{I_{IN}^8}{256I_{REF}^7} + \frac{5I_{IN}^6}{32I_{REF}^5} + \frac{15I_{IN}^4}{16I_{REF}^3} + \frac{7I_{IN}^2}{4I_{REF}} + I_{REF}}. \tag{6.271}$$

Using the definition of the x variable, the previous relation becomes:

$$I_{OUT} = I_{REF} \frac{\frac{9x^8}{256} + \frac{15x^6}{32} + \frac{27x^4}{16} + \frac{9x^2}{4} + 1}{\frac{x^8}{256} + \frac{5x^6}{32} + \frac{15x^4}{16} + \frac{7x^2}{4} + 1}. \tag{6.272}$$

In conclusion, the analog function synthesizer shown in Fig. 6.25 approximates the multiquadratic activation function for $a = 1$:

$$I_{OUT} = I_{REF} a_{24}(x) \cong I_{REF} f_6(x). \tag{6.273}$$

6.5.7 *First Analog Function Synthesizer for Generating the Multiquadratic Activation Function for a = 2*

The first analog function synthesizer, designed for generating the multiquadratic activation function for $a = 2$, uses $a_{25}(x)$ approximation function described in Chap. 2.

The block diagram of the first analog function synthesizer for generating the multiquadratic activation function for $a = 2$ is presented in Fig. 6.26.

The expressions of the currents from the circuit are:

$$I_{OUT1} = \frac{I_{IN}^2}{8I_{REF}}, \tag{6.274}$$

$$I_{OUT2} = 4I_{OUT1} = \frac{I_{IN}^2}{2I_{REF}}, \tag{6.275}$$

$$I_{OUT3} = 4I_{OUT1} = \frac{I_{IN}^2}{2I_{REF}}, \tag{6.276}$$

$$I_{OUT4} = \frac{I_{OUT3}^2}{8 \times 2I_{REF}} = \frac{I_{IN}^4}{64I_{REF}^3} \tag{6.277}$$

and

$$I_{OUT5} = I_{OUT1} = \frac{I_{IN}^2}{8I_{REF}}. \tag{6.278}$$

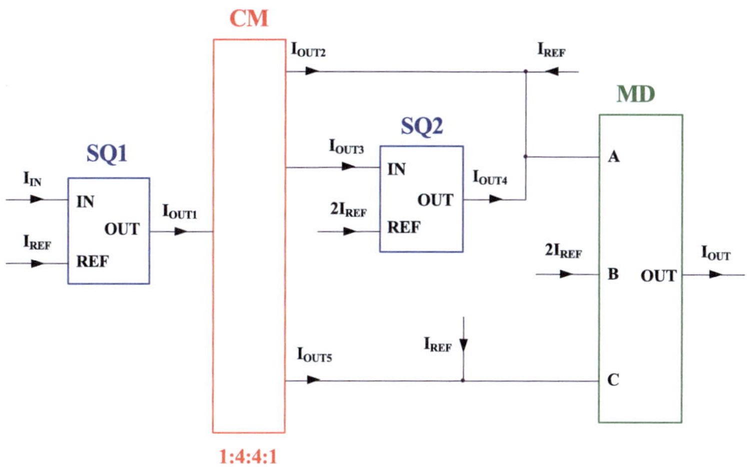

Fig. 6.26 Block diagram of the first analog function synthesizer for generating the multiquadratic activation function for $a = 2$

The output current of the circuit from Fig. 6.26 can be expressed as follows:

$$I_{OUT} = I_{REF}\frac{\frac{I_{IN}^4}{64I_{REF}^3} + \frac{I_{IN}^2}{2I_{REF}} + I_{REF}}{\frac{I_{IN}^2}{8I_{REF}} + I_{REF}}. \tag{6.279}$$

Using the definition of x variable, the previous relation becomes:

$$I_{OUT} = I_{REF}\frac{\frac{x^4}{64} + \frac{x^2}{2} + 2}{\frac{x^2}{8} + 1}. \tag{6.280}$$

In conclusion, the analog function synthesizer shown in Fig. 6.26 approximates the multiquadratic activation function for $a = 2$:

$$I_{OUT} = I_{REF}a_{25}(x) \cong I_{REF}f_6(x). \tag{6.281}$$

6.5.8 *Second Analog Function Synthesizer for Generating the Multiquadratic Activation Function for a = 2*

The second analog function synthesizer, designed for generating the multiquadratic activation function for $a = 2$, uses $a_{26}(x)$ approximation function described in Chap. 2.

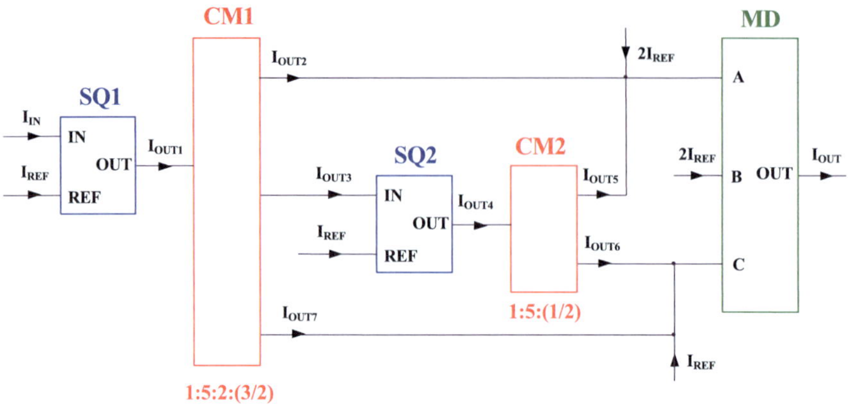

Fig. 6.27 Block diagram of the second analog function synthesizer for generating the multiquadratic activation function for $a = 2$

The block diagram of the second analog function synthesizer for generating the multiquadratic activation function for $a = 2$ is presented in Fig. 6.27.

The expressions of the currents from the circuit are:

$$I_{\mathrm{OUT1}} = \frac{I_{\mathrm{IN}}^2}{8I_{\mathrm{REF}}}, \tag{6.282}$$

$$I_{\mathrm{OUT2}} = 5I_{\mathrm{OUT1}} = \frac{5I_{\mathrm{IN}}^2}{8I_{\mathrm{REF}}}, \tag{6.283}$$

$$I_{\mathrm{OUT3}} = 2I_{\mathrm{OUT1}} = \frac{I_{\mathrm{IN}}^2}{4I_{\mathrm{REF}}}, \tag{6.284}$$

$$I_{\mathrm{OUT4}} = \frac{I_{\mathrm{OUT3}}^2}{8I_{\mathrm{REF}}} = \frac{I_{\mathrm{IN}}^4}{128I_{\mathrm{REF}}^3}, \tag{6.285}$$

$$I_{\mathrm{OUT5}} = 5I_{\mathrm{OUT4}} = \frac{5I_{\mathrm{IN}}^4}{128I_{\mathrm{REF}}^3}, \tag{6.286}$$

$$I_{\mathrm{OUT6}} = \frac{1}{2}I_{\mathrm{OUT4}} = \frac{I_{\mathrm{IN}}^4}{256I_{\mathrm{REF}}^3} \tag{6.287}$$

and

$$I_{\mathrm{OUT7}} = \frac{3}{2}I_{\mathrm{OUT1}} = \frac{3I_{\mathrm{IN}}^2}{16I_{\mathrm{REF}}}. \tag{6.288}$$

So, the output current of the circuit from Fig. 6.27 can be expressed as follows:

$$I_{OUT} = I_{REF} \frac{2I_{REF} + \frac{5I_{IN}^2}{8I_{REF}} + \frac{5I_{IN}^4}{128I_{REF}^3}}{I_{REF} + \frac{3I_{IN}^2}{16I_{REF}} + \frac{I_{IN}^4}{256I_{REF}^3}}. \quad (6.289)$$

Using the definition of the x variable, the previous relation becomes:

$$I_{OUT} = I_{REF} \frac{\frac{5x^4}{128} + \frac{5x^2}{8} + 2}{\frac{x^4}{256} + \frac{3x^2}{16} + 1}. \quad (6.290)$$

In conclusion, the analog function synthesizer shown in Fig. 6.27 approximates the multiquadratic activation function for $a = 2$:

$$I_{OUT} = I_{REF} a_{26}(x) \cong I_{REF} f_6(x). \quad (6.291)$$

6.5.9 *Third Analog Function Synthesizer for Generating the Multiquadratic Activation Function for a = 2*

The third analog function synthesizer, designed for generating the multiquadratic activation function for $a = 2$, uses $a_{27}(x)$ approximation function described in Chap. 2.

The block diagram of the third analog function synthesizer for generating the multiquadratic activation function for $a = 2$ is presented in Fig. 6.28.

The expressions of the currents from the circuit are:

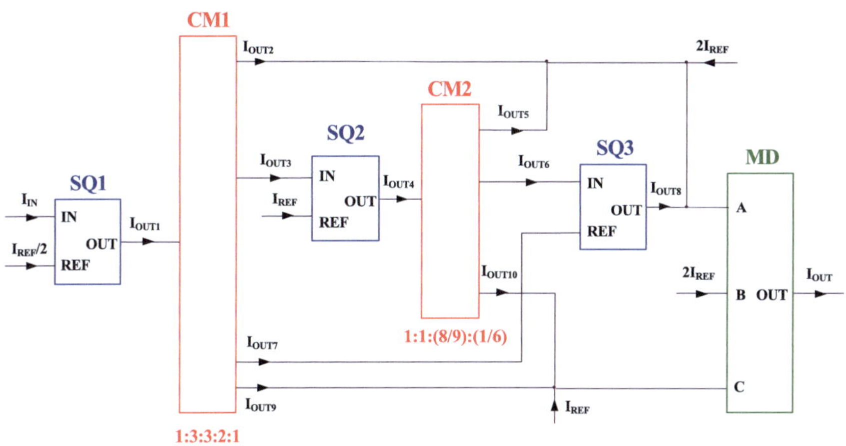

Fig. 6.28 Block diagram of the third analog function synthesizer for generating the multiquadratic activation function for $a = 2$

$$I_{\mathrm{OUT1}} = \frac{I_{\mathrm{IN}}^2}{8(I_{\mathrm{REF}}/2)} = \frac{I_{\mathrm{IN}}^2}{4I_{\mathrm{REF}}}, \tag{6.292}$$

$$I_{\mathrm{OUT2}} = 3I_{\mathrm{OUT1}} = \frac{3I_{\mathrm{IN}}^2}{4I_{\mathrm{REF}}}, \tag{6.293}$$

$$I_{\mathrm{OUT3}} = 3I_{\mathrm{OUT1}} = \frac{3I_{\mathrm{IN}}^2}{4I_{\mathrm{REF}}}, \tag{6.294}$$

$$I_{\mathrm{OUT4}} = \frac{I_{\mathrm{OUT3}}^2}{8I_{\mathrm{REF}}} = \frac{9I_{\mathrm{IN}}^4}{128I_{\mathrm{REF}}^3}, \tag{6.295}$$

$$I_{\mathrm{OUT5}} = I_{\mathrm{OUT4}} = \frac{9I_{\mathrm{IN}}^4}{128I_{\mathrm{REF}}^3}, \tag{6.296}$$

$$I_{\mathrm{OUT6}} = \frac{8}{9}I_{\mathrm{OUT4}} = \frac{I_{\mathrm{IN}}^4}{16I_{\mathrm{REF}}^3}, \tag{6.297}$$

$$I_{\mathrm{OUT7}} = 2I_{\mathrm{OUT1}} = \frac{I_{\mathrm{IN}}^2}{2I_{\mathrm{REF}}}, \tag{6.298}$$

$$I_{\mathrm{OUT8}} = \frac{I_{\mathrm{OUT6}}^2}{8I_{\mathrm{OUT7}}} = \frac{I_{\mathrm{IN}}^6}{1024I_{\mathrm{REF}}^5}, \tag{6.299}$$

$$I_{\mathrm{OUT9}} = I_{\mathrm{OUT1}} = \frac{I_{\mathrm{IN}}^2}{4I_{\mathrm{REF}}} \tag{6.300}$$

and

$$I_{\mathrm{OUT10}} = \frac{1}{6}I_{\mathrm{OUT4}} = \frac{3I_{\mathrm{IN}}^4}{256I_{\mathrm{REF}}^3}. \tag{6.301}$$

So, the output current of the circuit from Fig. 6.28 can be expressed as follows:

$$I_{\mathrm{OUT}} = I_{\mathrm{REF}} \frac{\frac{I_{\mathrm{IN}}^6}{1024I_{\mathrm{REF}}^5} + \frac{9I_{\mathrm{IN}}^4}{128I_{\mathrm{REF}}^3} + \frac{3I_{\mathrm{IN}}^2}{4I_{\mathrm{REF}}} + 2I_{\mathrm{REF}}}{\frac{3I_{\mathrm{IN}}^4}{256I_{\mathrm{REF}}^3} + \frac{I_{\mathrm{IN}}^2}{4I_{\mathrm{REF}}} + I_{\mathrm{REF}}}. \tag{6.302}$$

Using the definition of the x variable, the previous relation becomes:

$$I_{\mathrm{OUT}} = I_{\mathrm{REF}} \frac{\frac{x^6}{1024} + \frac{9x^4}{128} + \frac{3x^2}{4} + 2}{\frac{3x^4}{256} + \frac{x^2}{4} + 1}. \tag{6.303}$$

In conclusion, the analog function synthesizer shown in Fig. 6.28 approximates the multiquadratic activation function for $a = 2$:

$$I_{OUT} = I_{REF} a_{27}(x) \cong I_{REF} f_6(x). \tag{6.304}$$

6.5.10 Fourth Analog Function Synthesizer for Generating the Multiquadratic Activation Function for a = 2

The fourth analog function synthesizer, designed for generating the multiquadratic activation function for $a = 2$, uses $a_{28}(x)$ approximation function described in Chap. 2.

The block diagram of the fourth analog function synthesizer for generating the multiquadratic activation function for $a = 2$ is presented in Fig. 6.29.

The expressions of the currents from the circuit are:

$$I_{OUT1} = \frac{I_{IN}^2}{8(I_{REF}/8)} = \frac{I_{IN}^2}{I_{REF}}, \tag{6.305}$$

$$I_{OUT2} = \frac{7}{8} I_{OUT1} = \frac{7I_{IN}^2}{8I_{REF}}, \tag{6.306}$$

$$I_{OUT3} = I_{OUT1} = \frac{I_{IN}^2}{I_{REF}}, \tag{6.307}$$

$$I_{OUT4} = \frac{I_{OUT3}^2}{8(8I_{REF}/7)} = \frac{7I_{IN}^4}{64I_{REF}^3}, \tag{6.308}$$

$$I_{OUT5} = I_{OUT4} = \frac{7I_{IN}^4}{64I_{REF}^3}, \tag{6.309}$$

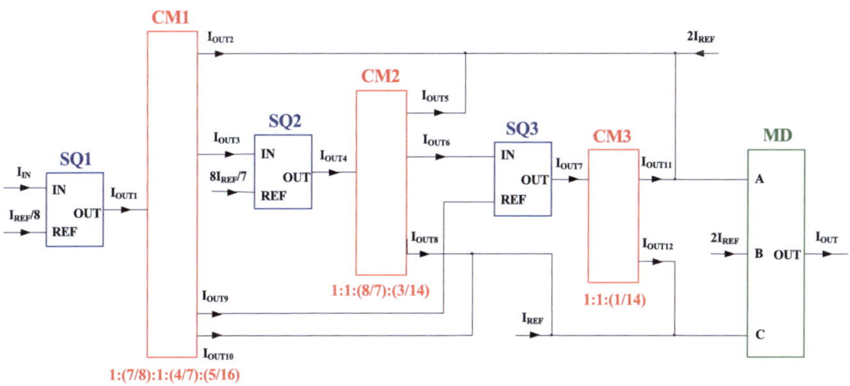

Fig. 6.29 Block diagram of the fourth analog function synthesizer for generating the multiquadratic activation function for $a = 2$

$$I_{OUT6} = \frac{8}{7} I_{OUT4} = \frac{I_{IN}^4}{8 I_{REF}^3}, \tag{6.310}$$

$$I_{OUT7} = \frac{I_{OUT6}^2}{8 I_{OUT9}} = \frac{7 I_{IN}^6}{2048 I_{REF}^5}, \tag{6.311}$$

$$I_{OUT8} = \frac{3}{14} I_{OUT4} = \frac{3 I_{IN}^4}{128 I_{REF}^3}, \tag{6.312}$$

$$I_{OUT9} = \frac{4}{7} I_{OUT1} = \frac{4 I_{IN}^2}{7 I_{REF}}, \tag{6.313}$$

$$I_{OUT10} = \frac{5}{16} I_{OUT1} = \frac{5 I_{IN}^2}{16 I_{REF}}, \tag{6.314}$$

$$I_{OUT11} = I_{OUT7} = \frac{7 I_{IN}^6}{2048 I_{REF}^5} \tag{6.315}$$

and

$$I_{OUT12} = \frac{1}{14} I_{OUT7} = \frac{I_{IN}^6}{4096 I_{REF}^5}. \tag{6.316}$$

So, the output current of the circuit from Fig. 6.29 can be expressed as follows:

$$I_{OUT} = I_{REF} \frac{\frac{7 I_{IN}^6}{2048 I_{REF}^5} + \frac{7 I_{IN}^4}{64 I_{REF}^3} + \frac{7 I_{IN}^2}{8 I_{REF}} + 2 I_{REF}}{\frac{I_{IN}^6}{4096 I_{REF}^5} + \frac{3 I_{IN}^4}{128 I_{REF}^3} + \frac{5 I_{IN}^2}{16 I_{REF}} + I_{REF}}. \tag{6.317}$$

Using the definition of the x variable, the previous relation becomes:

$$I_{OUT} = I_{REF} \frac{\frac{7x^6}{2048} + \frac{7x^4}{64} + \frac{7x^2}{8} + 2}{\frac{x^6}{4096} + \frac{3x^4}{128} + \frac{5x^2}{16} + 1}. \tag{6.318}$$

In conclusion, the analog function synthesizer shown in Fig. 6.29 approximates the multiquadratic activation function for $a = 2$:

$$I_{OUT} = I_{REF} a_{28}(x) \cong I_{REF} f_6(x). \tag{6.319}$$

6.5.11 *First Analog Function Synthesizer for Generating the Multiquadratic Activation Function for a = 3*

The first analog function synthesizer, designed for generating the multiquadratic activation function for $a = 3$, uses $a_{29}(x)$ approximation function described in Chap. 2.

The block diagram of the first analog function synthesizer for generating the multiquadratic activation function for $a = 3$ is presented in Fig. 6.30.

The expressions of the currents from the circuit are:

$$I_{\mathrm{OUT1}} = \frac{I_{\mathrm{IN}}^2}{8I_{\mathrm{REF}}}, \tag{6.320}$$

$$I_{\mathrm{OUT2}} = \frac{8}{3}I_{\mathrm{OUT1}} = \frac{I_{\mathrm{IN}}^2}{3I_{\mathrm{REF}}}, \tag{6.321}$$

$$I_{\mathrm{OUT3}} = 2I_{\mathrm{OUT1}} = \frac{I_{\mathrm{IN}}^2}{4I_{\mathrm{REF}}}, \tag{6.322}$$

$$I_{\mathrm{OUT4}} = \frac{I_{\mathrm{OUT3}}^2}{8(27I_{\mathrm{REF}}/16)} = \frac{I_{\mathrm{IN}}^4}{216I_{\mathrm{REF}}^3} \tag{6.323}$$

and

$$I_{\mathrm{OUT5}} = \frac{4}{9}I_{\mathrm{OUT1}} = \frac{I_{\mathrm{IN}}^2}{18I_{\mathrm{REF}}}. \tag{6.324}$$

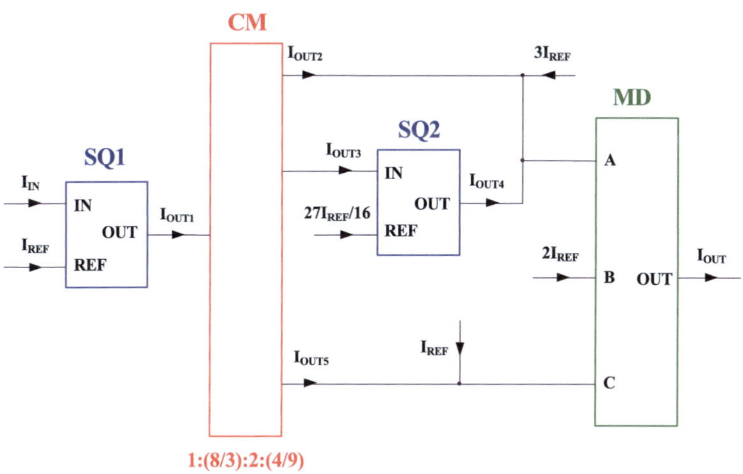

Fig. 6.30 Block diagram of the first analog function synthesizer for generating the multiquadratic activation function for $a = 3$

The output current of the circuit from Fig. 6.30 can be expressed as follows:

$$I_{\mathrm{OUT1}} = I_{\mathrm{REF}} \frac{\frac{I_{\mathrm{IN}}^4}{216 I_{\mathrm{REF}}^3} + \frac{I_{\mathrm{IN}}^2}{3 I_{\mathrm{REF}}} + 3 I_{\mathrm{REF}}}{\frac{I_{\mathrm{IN}}^2}{18 I_{\mathrm{REF}}} + I_{\mathrm{REF}}}. \tag{6.325}$$

Using the definition of x variable, the previous relation becomes:

$$I_{\mathrm{OUT}} = I_{\mathrm{REF}} \frac{\frac{x^4}{216} + \frac{x^2}{3} + 3}{\frac{x^2}{18} + 1}. \tag{6.326}$$

In conclusion, the analog function synthesizer shown in Fig. 6.30 approximates the multiquadratic activation function for $a = 3$:

$$I_{\mathrm{OUT}} = I_{\mathrm{REF}} a_{29}(x) \cong I_{\mathrm{REF}} f_6(x). \tag{6.327}$$

6.5.12 Second Analog Function Synthesizer for Generating the Multiquadratic Activation Function for a = 3

The second analog function synthesizer, designed for generating the multiquadratic activation function for $a = 3$, uses $a_{30}(x)$ approximation function described in Chap. 2.

The block diagram of the second analog function synthesizer for generating the multiquadratic activation function for $a = 3$ is presented in Fig. 6.31.

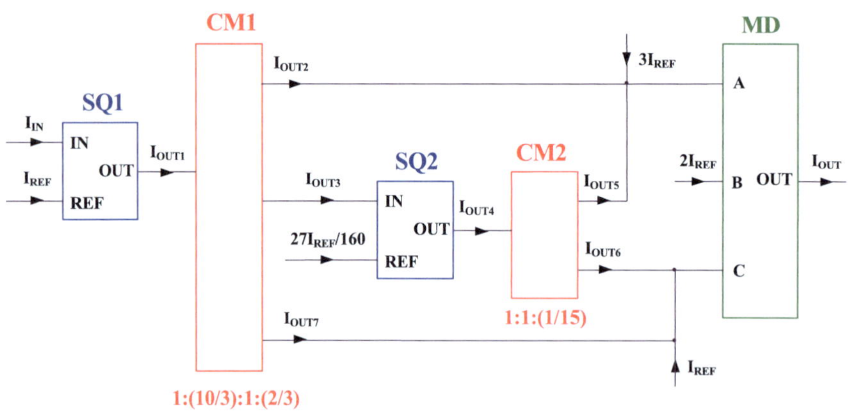

Fig. 6.31 Block diagram of the second analog function synthesizer for generating the multiquadratic activation function for $a = 3$

The expressions of the currents from the circuit are:

$$I_{OUT1} = \frac{I_{IN}^2}{8I_{REF}}, \tag{6.328}$$

$$I_{OUT2} = \frac{10}{3} I_{OUT1} = \frac{5I_{IN}^2}{12I_{REF}}, \tag{6.329}$$

$$I_{OUT3} = I_{OUT1} = \frac{I_{IN}^2}{8I_{REF}}, \tag{6.330}$$

$$I_{OUT4} = \frac{I_{OUT3}^2}{8(27I_{REF}/160)} = \frac{5I_{IN}^4}{432I_{REF}^3}, \tag{6.331}$$

$$I_{OUT5} = I_{OUT4} = \frac{5I_{IN}^4}{432I_{REF}^3}, \tag{6.332}$$

$$I_{OUT6} = \frac{1}{15} I_{OUT4} = \frac{I_{REF}^4}{1296I_{REF}^3} \tag{6.333}$$

and

$$I_{OUT7} = \frac{2}{3} I_{OUT1} = \frac{I_{IN}^2}{12I_{REF}}. \tag{6.334}$$

So, the output current of the circuit from Fig. 6.31 can be expressed as follows:

$$I_{OUT} = I_{REF} \frac{3I_{REF} + \frac{5I_{IN}^2}{12I_{REF}} + \frac{5I_{IN}^4}{432I_{REF}^3}}{I_{REF} + \frac{I_{IN}^2}{12I_{REF}} + \frac{I_{IN}^4}{1296I_{REF}^3}}. \tag{6.335}$$

Using the definition of the x variable, the previous relation becomes:

$$I_{OUT} = I_{REF} \frac{\frac{5x^4}{432} + \frac{5x^2}{12} + 3}{\frac{x^4}{1296} + \frac{x^2}{12} + 1}. \tag{6.336}$$

In conclusion, the analog function synthesizer shown in Fig. 6.31 approximates the multiquadratic activation function for $a = 3$:

$$I_{OUT} = I_{REF} a_{30}(x) \cong I_{REF} f_6(x). \tag{6.337}$$

6.5.13 Third Analog Function Synthesizer for Generating the Multiquadratic Activation Function for a = 3

The third analog function synthesizer, designed for generating the multiquadratic activation function for $a = 3$, uses $a_{31}(x)$ approximation function described in Chap. 2.

The block diagram of the third analog function synthesizer for generating the multiquadratic activation function for $a = 3$ is presented in Fig. 6.32.

The expressions of the currents from the circuit are:

$$I_{\text{OUT1}} = \frac{I_{\text{IN}}^2}{8(I_{\text{REF}}/4)} = \frac{I_{\text{IN}}^2}{2I_{\text{REF}}}, \tag{6.338}$$

$$I_{\text{OUT2}} = I_{\text{OUT1}} = \frac{I_{\text{IN}}^2}{2I_{\text{REF}}}, \tag{6.339}$$

$$I_{\text{OUT3}} = I_{\text{OUT1}} = \frac{I_{\text{IN}}^2}{2I_{\text{REF}}}, \tag{6.340}$$

$$I_{\text{OUT4}} = \frac{I_{\text{OUT3}}^2}{8(3I_{\text{REF}}/2)} = \frac{I_{\text{IN}}^4}{48I_{\text{REF}}^3}, \tag{6.341}$$

$$I_{\text{OUT5}} = I_{\text{OUT4}} = \frac{I_{\text{IN}}^4}{48I_{\text{REF}}^3}, \tag{6.342}$$

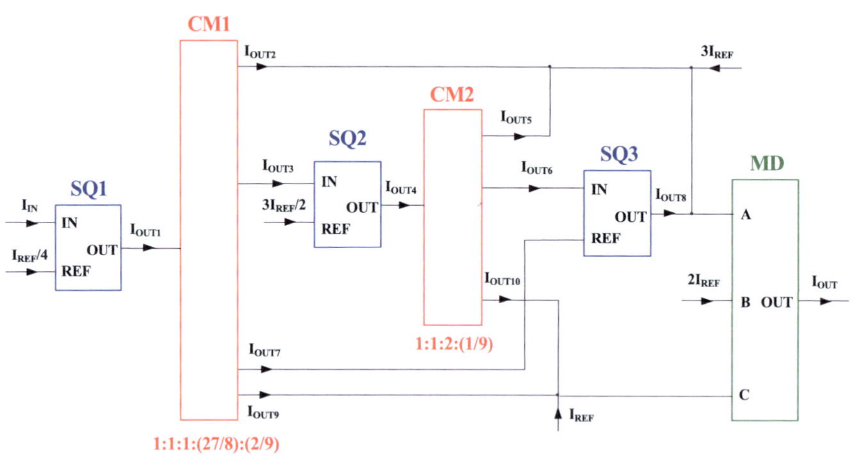

Fig. 6.32 Block diagram of the third analog function synthesizer for generating the multiquadratic activation function for $a = 3$

$$I_{OUT6} = 2I_{OUT4} = \frac{I_{IN}^4}{24I_{REF}^3}, \tag{6.343}$$

$$I_{OUT7} = \frac{27}{8}I_{OUT1} = \frac{27I_{IN}^2}{16I_{REF}}, \tag{6.344}$$

$$I_{OUT8} = \frac{I_{OUT6}^2}{8I_{OUT7}} = \frac{I_{IN}^6}{7776I_{REF}^5}, \tag{6.345}$$

$$I_{OUT9} = \frac{2}{9}I_{OUT1} = \frac{I_{IN}^2}{9I_{REF}} \tag{6.346}$$

and

$$I_{OUT10} = \frac{1}{9}I_{OUT4} = \frac{I_{IN}^4}{432I_{REF}^3}. \tag{6.347}$$

So, the output current of the circuit from Fig. 6.32 can be expressed as follows:

$$I_{OUT} = I_{REF}\frac{\frac{I_{IN}^6}{7776I_{REF}^5} + \frac{I_{IN}^4}{48I_{REF}^3} + \frac{I_{IN}^2}{2I_{REF}} + 3I_{REF}}{\frac{I_{IN}^4}{432I_{REF}^3} + \frac{I_{IN}^2}{9I_{REF}} + I_{REF}}. \tag{6.348}$$

Using the definition of the x variable, the previous relation becomes:

$$I_{OUT} = I_{REF}\frac{\frac{x^6}{7776} + \frac{x^4}{48} + \frac{x^2}{2} + 3}{\frac{x^4}{432} + \frac{x^2}{9} + 1}. \tag{6.349}$$

In conclusion, the analog function synthesizer shown in Fig. 6.32 approximates the multiquadratic activation function for $a = 3$:

$$I_{OUT} = I_{REF}a_{31}(x) \cong I_{REF}f_6(x). \tag{6.350}$$

6.5.14 *Fourth Analog Function Synthesizer for Generating the Multiquadratic Activation Function for a = 3*

The fourth analog function synthesizer, designed for generating the multiquadratic activation function for $a = 3$, uses $a_{32}(x)$ approximation function described in Chap. 2.

The block diagram of the fourth analog function synthesizer for generating the multiquadratic activation function for $a = 3$ is presented in Fig. 6.33.

The expressions of the currents from the circuit are:

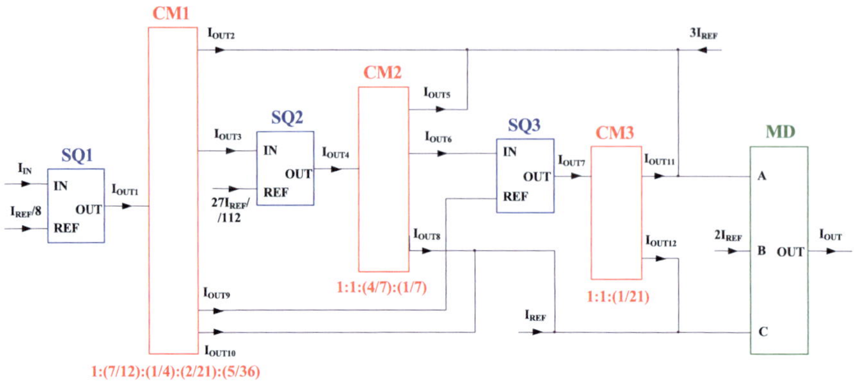

Fig. 6.33 Block diagram of the fourth analog function synthesizer for generating the multiquadratic activation function for $a = 3$

$$I_{\mathrm{OUT1}} = \frac{I_{\mathrm{IN}}^2}{8(I_{\mathrm{REF}}/8)} = \frac{I_{\mathrm{IN}}^2}{I_{\mathrm{REF}}}, \tag{6.351}$$

$$I_{\mathrm{OUT2}} = \frac{7}{12} I_{\mathrm{OUT1}} = \frac{7I_{\mathrm{IN}}^2}{12I_{\mathrm{REF}}}, \tag{6.352}$$

$$I_{\mathrm{OUT3}} = \frac{1}{4} I_{\mathrm{OUT1}} = \frac{I_{\mathrm{IN}}^2}{4I_{\mathrm{REF}}}, \tag{6.353}$$

$$I_{\mathrm{OUT4}} = \frac{I_{\mathrm{OUT3}}^2}{8(27I_{\mathrm{REF}}/112)} = \frac{7I_{\mathrm{IN}}^4}{216I_{\mathrm{REF}}^3}, \tag{6.354}$$

$$I_{\mathrm{OUT5}} = I_{\mathrm{OUT4}} = \frac{7I_{\mathrm{IN}}^4}{216I_{\mathrm{REF}}^3}, \tag{6.355}$$

$$I_{\mathrm{OUT6}} = \frac{4}{7} I_{\mathrm{OUT4}} = \frac{I_{\mathrm{IN}}^4}{54I_{\mathrm{REF}}^3}, \tag{6.356}$$

$$I_{\mathrm{OUT7}} = \frac{I_{\mathrm{OUT6}}^2}{8I_{\mathrm{OUT9}}} = \frac{7I_{\mathrm{IN}}^6}{15552I_{\mathrm{REF}}^5}, \tag{6.357}$$

$$I_{\mathrm{OUT8}} = \frac{1}{7} I_{\mathrm{OUT4}} = \frac{I_{\mathrm{IN}}^4}{216I_{\mathrm{REF}}^3}, \tag{6.358}$$

$$I_{\mathrm{OUT9}} = \frac{2}{21} I_{\mathrm{OUT1}} = \frac{2I_{\mathrm{IN}}^2}{21I_{\mathrm{REF}}}, \tag{6.359}$$

$$I_{\mathrm{OUT10}} = \frac{5}{36} I_{\mathrm{OUT1}} = \frac{5I_{\mathrm{IN}}^2}{36I_{\mathrm{REF}}}, \tag{6.360}$$

$$I_{\mathrm{OUT11}} = I_{\mathrm{OUT7}} = \frac{7I_{\mathrm{IN}}^6}{15552I_{\mathrm{REF}}^5} \tag{6.361}$$

and

$$I_{\mathrm{OUT12}} = \frac{1}{21}I_{\mathrm{OUT7}} = \frac{I_{\mathrm{IN}}^6}{46656I_{\mathrm{REF}}^5}. \tag{6.362}$$

So, the output current of the circuit from Fig. 6.33 can be expressed as follows:

$$I_{\mathrm{OUT}} = I_{\mathrm{REF}}\frac{\frac{7I_{\mathrm{IN}}^6}{15552I_{\mathrm{REF}}^5} + \frac{7I_{\mathrm{IN}}^4}{216I_{\mathrm{REF}}^3} + \frac{7I_{\mathrm{IN}}^2}{12I_{\mathrm{REF}}} + 3I_{\mathrm{REF}}}{\frac{I_{\mathrm{IN}}^6}{46656I_{\mathrm{REF}}^5} + \frac{I_{\mathrm{IN}}^4}{216I_{\mathrm{REF}}^3} + \frac{5I_{\mathrm{IN}}^2}{36I_{\mathrm{REF}}} + I_{\mathrm{REF}}}. \tag{6.363}$$

Using the definition of the x variable, the previous relation becomes:

$$I_{\mathrm{OUT}} = I_{\mathrm{REF}}\frac{\frac{7x^6}{15552} + \frac{7x^4}{216} + \frac{7x^2}{12} + 3}{\frac{x^6}{46656} + \frac{x^4}{216} + \frac{5x^2}{36} + 1}. \tag{6.364}$$

In conclusion, the analog function synthesizer shown in Fig. 6.33 approximates the multiquadratic activation function for $a = 3$:

$$I_{\mathrm{OUT}} = I_{\mathrm{REF}}a_{32}(x) \cong I_{\mathrm{REF}}f_6(x). \tag{6.365}$$

6.6 Analog Function Synthesizers for Inverse Multiquadratic Activation Function

In order to implement in CMOS technology the inverse multiquadratic activation function, the approximation functions described in Chap. 2 and the CMOS squaring and multiplies/divider circuits from Chap. 4 are used.

6.6.1 First Analog Function Synthesizer for Generating the Inverse Multiquadratic Activation Function for a = 1

The first analog function synthesizer, designed for generating the inverse multiquadratic activation function for $a = 1$, uses $a_{33}(x)$ approximation function described in Chap. 2.

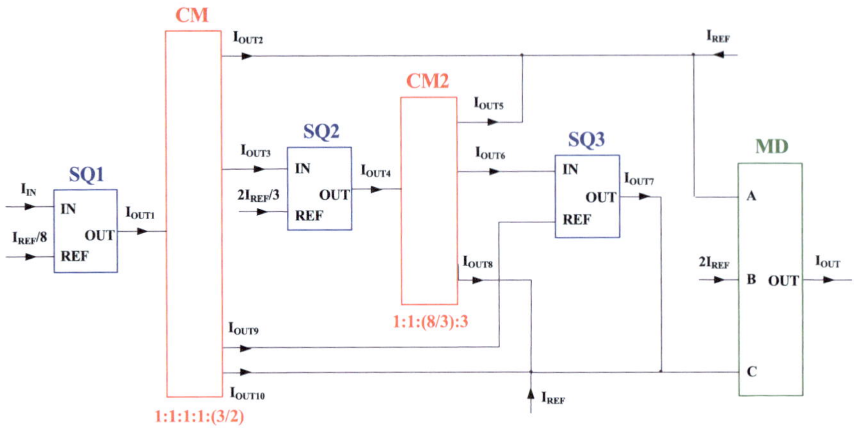

Fig. 6.34 Block diagram of the first analog function synthesizer for generating the inverse multiquadratic activation function for $a = 1$

The block diagram of the first analog function synthesizer for generating the inverse multiquadratic activation function for $a = 1$ is presented in Fig. 6.34.

The expressions of the currents from the circuit are:

$$I_{\mathrm{OUT1}} = \frac{I_{\mathrm{IN}}^2}{8(I_{\mathrm{REF}}/8)} = \frac{I_{\mathrm{IN}}^2}{I_{\mathrm{REF}}}, \tag{6.366}$$

$$I_{\mathrm{OUT2}} = I_{\mathrm{OUT1}} = \frac{I_{\mathrm{IN}}^2}{I_{\mathrm{REF}}}, \tag{6.367}$$

$$I_{\mathrm{OUT3}} = I_{\mathrm{OUT1}} = \frac{I_{\mathrm{IN}}^2}{I_{\mathrm{REF}}}, \tag{6.368}$$

$$I_{\mathrm{OUT4}} = \frac{I_{\mathrm{OUT3}}^2}{8(2I_{\mathrm{REF}}/3)} = \frac{3I_{\mathrm{IN}}^4}{16I_{\mathrm{REF}}^3}, \tag{6.369}$$

$$I_{\mathrm{OUT5}} = I_{\mathrm{OUT4}} = \frac{3I_{\mathrm{IN}}^4}{16I_{\mathrm{REF}}^3}, \tag{6.370}$$

$$I_{\mathrm{OUT6}} = \frac{8}{3}I_{\mathrm{OUT4}} = \frac{I_{\mathrm{IN}}^4}{2I_{\mathrm{REF}}^3}, \tag{6.371}$$

$$I_{\mathrm{OUT7}} = \frac{I_{\mathrm{OUT6}}^2}{8I_{\mathrm{OUT9}}} = \frac{I_{\mathrm{IN}}^6}{32I_{\mathrm{REF}}^5}, \tag{6.372}$$

$$I_{\mathrm{OUT8}} = 3I_{\mathrm{OUT4}} = \frac{9I_{\mathrm{IN}}^4}{16I_{\mathrm{REF}}^3}, \tag{6.373}$$

$$I_{OUT9} = I_{OUT1} = \frac{I_{IN}^2}{I_{REF}} \tag{6.374}$$

and

$$I_{OUT10} = \frac{3}{2} I_{OUT1} = \frac{3I_{IN}^2}{2I_{REF}}. \tag{6.375}$$

So, the output current of the circuit from Fig. 6.34 can be expressed as follows:

$$I_{OUT} = I_{REF} \frac{\frac{3I_{IN}^4}{16I_{REF}^3} + \frac{I_{IN}^2}{I_{REF}} + I_{REF}}{\frac{I_{IN}^6}{32I_{REF}^5} + \frac{9I_{IN}^4}{16I_{REF}^3} + \frac{3I_{IN}^2}{2I_{REF}} + I_{REF}}. \tag{6.376}$$

Using the definition of the x variable, the previous relation becomes:

$$I_{OUT} = I_{REF} \frac{\frac{3x^4}{16} + x^2 + 1}{\frac{x^6}{32} + \frac{9x^4}{16} + \frac{3x^2}{2} + 1}. \tag{6.377}$$

In conclusion, the analog function synthesizer shown in Fig. 6.34 approximates the inverse multiquadratic activation function for $a = 1$:

$$I_{OUT} = I_{REF} a_{33}(x) \cong I_{REF} f_7(x). \tag{6.378}$$

6.6.2 *Second Analog Function Synthesizer for Generating the Inverse Multiquadratic Activation Function for a = 1*

The second analog function synthesizer, designed for generating the inverse multiquadratic activation function for $a = 1$, uses $a_{34}(x)$ approximation function described in Chap. 2.

The block diagram of the second analog function synthesizer for generating the inverse multiquadratic activation function for $a = 1$ is presented in Fig. 6.35.

The expressions of the currents from the circuit are:

$$I_{OUT1} = \frac{I_{IN}^2}{8(I_{REF}/8)} = \frac{I_{IN}^2}{I_{REF}}, \tag{6.379}$$

$$I_{OUT2} = \frac{5}{4} I_{OUT1} = \frac{5I_{IN}^2}{4I_{REF}}, \tag{6.380}$$

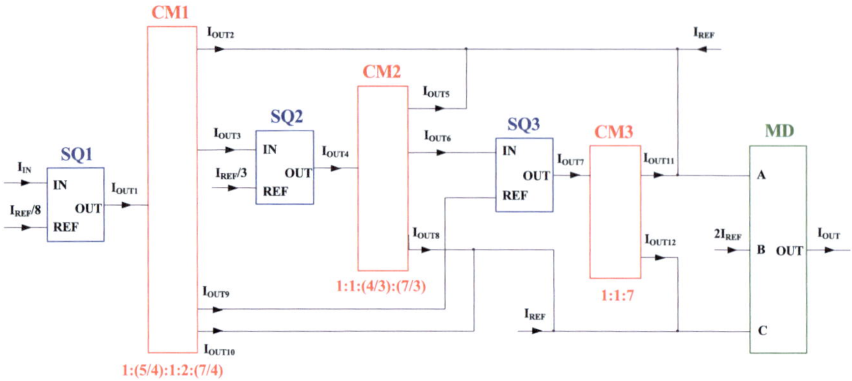

Fig. 6.35 Block diagram of the second analog function synthesizer for generating the inverse multiquadratic activation function for $a = 1$

$$I_{\text{OUT3}} = I_{\text{OUT1}} = \frac{I_{\text{IN}}^2}{I_{\text{REF}}}, \tag{6.381}$$

$$I_{\text{OUT4}} = \frac{I_{\text{OUT3}}^2}{8(I_{\text{REF}}/3)} = \frac{3I_{\text{IN}}^4}{8I_{\text{REF}}^3}, \tag{6.382}$$

$$I_{\text{OUT5}} = I_{\text{OUT4}} = \frac{3I_{\text{IN}}^4}{8I_{\text{REF}}^3} \tag{6.383}$$

$$I_{\text{OUT6}} = \frac{4}{3} I_{\text{OUT4}} = \frac{I_{\text{IN}}^4}{2I_{\text{REF}}^3}, \tag{6.384}$$

$$I_{\text{OUT7}} = \frac{I_{\text{OUT6}}^2}{8I_{\text{OUT9}}} = \frac{I_{\text{IN}}^6}{64I_{\text{REF}}^5}, \tag{6.385}$$

$$I_{\text{OUT8}} = \frac{7}{3} I_{\text{OUT4}} = \frac{7I_{\text{IN}}^4}{8I_{\text{REF}}^3}, \tag{6.386}$$

$$I_{\text{OUT9}} = 2I_{\text{OUT1}} = \frac{2I_{\text{IN}}^2}{I_{\text{REF}}}, \tag{6.387}$$

$$I_{\text{OUT10}} = \frac{7}{4} I_{\text{OUT1}} = \frac{7I_{\text{IN}}^2}{4I_{\text{REF}}}, \tag{6.388}$$

$$I_{\text{OUT11}} = I_{\text{OUT7}} = \frac{I_{\text{IN}}^6}{64I_{\text{REF}}^5} \tag{6.389}$$

and

$$I_{OUT12} = 7I_{OUT7} = \frac{7I_{IN}^6}{64I_{REF}^5}. \tag{6.390}$$

So, the output current of the circuit from Fig. 6.35 can be expressed as follows:

$$I_{OUT} = I_{REF}\frac{\frac{I_{IN}^6}{64I_{REF}^5} + \frac{3I_{IN}^4}{8I_{REF}^3} + \frac{5I_{IN}^2}{4I_{REF}} + I_{REF}}{\frac{7I_{IN}^6}{64I_{REF}^5} + \frac{7I_{IN}^4}{8I_{REF}^3} + \frac{7I_{IN}^2}{4I_{REF}} + I_{REF}}. \tag{6.391}$$

Using the definition of the x variable, the previous relation becomes:

$$I_{OUT} = I_{REF}\frac{\frac{x^6}{64} + \frac{3x^4}{8} + \frac{5x^2}{4} + 1}{\frac{7x^6}{64} + \frac{7x^4}{8} + \frac{7x^2}{4} + 1}. \tag{6.392}$$

In conclusion, the analog function synthesizer shown in Fig. 6.35 approximates the inverse multiquadratic activation function for $a = 1$:

$$I_{OUT} = I_{REF}a_{34}(x) \cong I_{REF}f_7(x). \tag{6.393}$$

6.6.3 *First Analog Function Synthesizer for Generating the Inverse Multiquadratic Activation Function for a = 2*

The first analog function synthesizer, designed for generating the inverse multiquadratic activation function for $a = 2$, uses $a_{35}(x)$ approximation function described in Chap. 2.

The block diagram of the first analog function synthesizer for generating the inverse multiquadratic activation function for $a = 2$ is presented in Fig. 6.36.

The expressions of the currents from the circuit are:

$$I_{OUT1} = \frac{I_{IN}^2}{8I_{REF}}, \tag{6.394}$$

$$I_{OUT2} = \frac{3}{4}I_{OUT1} = \frac{3I_{IN}^2}{32I_{REF}}, \tag{6.395}$$

$$I_{OUT3} = I_{OUT1} = \frac{I_{IN}^2}{8I_{REF}}, \tag{6.396}$$

$$I_{OUT4} = \frac{I_{OUT3}^2}{8I_{REF}} = \frac{I_{IN}^4}{512I_{REF}^3}, \tag{6.397}$$

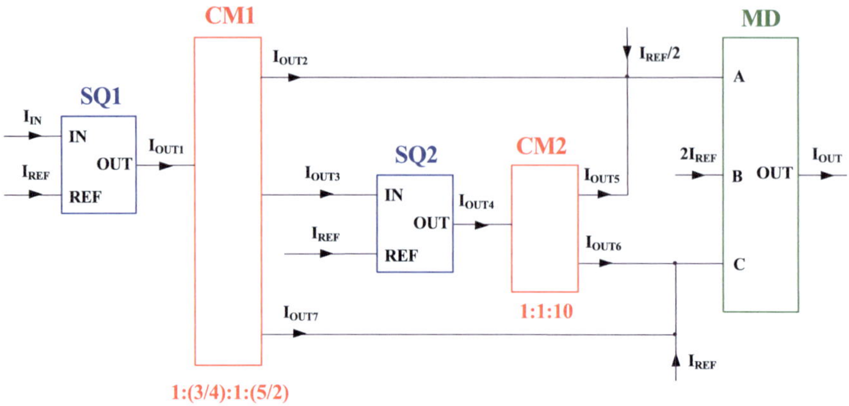

Fig. 6.36 Block diagram of the first analog function synthesizer for generating the inverse multiquadratic activation function for $a = 2$

$$I_{\mathrm{OUT5}} = I_{\mathrm{OUT4}} = \frac{I_{\mathrm{IN}}^4}{512 I_{\mathrm{REF}}^3}, \tag{6.398}$$

$$I_{\mathrm{OUT6}} = 10 I_{\mathrm{OUT4}} = \frac{5 I_{\mathrm{IN}}^4}{256 I_{\mathrm{REF}}^3} \tag{6.399}$$

and

$$I_{\mathrm{OUT7}} = \frac{5}{2} I_{\mathrm{OUT1}} = \frac{5 I_{\mathrm{IN}}^2}{16 I_{\mathrm{REF}}}. \tag{6.400}$$

So, the output current of the circuit from Fig. 6.36 can be expressed as follows:

$$I_{\mathrm{OUT}} = I_{\mathrm{REF}} \frac{\frac{I_{\mathrm{REF}}}{2} + \frac{3 I_{\mathrm{IN}}^2}{32 I_{\mathrm{REF}}} + \frac{I_{\mathrm{IN}}^4}{512 I_{\mathrm{REF}}^3}}{I_{\mathrm{REF}} + \frac{5 I_{\mathrm{IN}}^2}{16 I_{\mathrm{REF}}} + \frac{5 I_{\mathrm{IN}}^4}{256 I_{\mathrm{REF}}^3}}. \tag{6.401}$$

Using the definition of the x variable, the previous relation becomes:

$$I_{\mathrm{OUT}} = I_{\mathrm{REF}} \frac{\frac{x^4}{512} + \frac{3x^2}{32} + \frac{1}{2}}{\frac{5x^4}{256} + \frac{5x^2}{16} + 1}. \tag{6.402}$$

In conclusion, the analog function synthesizer shown in Fig. 6.36 approximates the multiquadratic activation function for $a = 2$:

$$I_{\mathrm{OUT}} = I_{\mathrm{REF}} a_{35}(x) \cong I_{\mathrm{REF}} f_7(x). \tag{6.403}$$

6.6.4 Second Analog Function Synthesizer for Generating the Inverse Multiquadratic Activation Function for a = 2

The second analog function synthesizer, designed for generating the inverse multiquadratic activation function for $a = 2$, uses $a_{36}(x)$ approximation function described in Chap. 2.

The block diagram of the second analog function synthesizer for generating the inverse multiquadratic activation function for $a = 2$ is presented in Fig. 6.37.

The expressions of the currents from the circuit are:

$$I_{\mathrm{OUT1}} = \frac{I_{\mathrm{IN}}^2}{8I_{\mathrm{REF}}}, \tag{6.404}$$

$$I_{\mathrm{OUT2}} = I_{\mathrm{OUT1}} = \frac{I_{\mathrm{IN}}^2}{8I_{\mathrm{REF}}}, \tag{6.405}$$

$$I_{\mathrm{OUT3}} = I_{\mathrm{OUT1}} = \frac{I_{\mathrm{IN}}^2}{8I_{\mathrm{REF}}}, \tag{6.406}$$

$$I_{\mathrm{OUT4}} = \frac{I_{\mathrm{OUT3}}^2}{8I_{\mathrm{REF}}} = \frac{I_{\mathrm{IN}}^4}{512I_{\mathrm{REF}}^3}, \tag{6.407}$$

$$I_{\mathrm{OUT5}} = 3I_{\mathrm{OUT4}} = \frac{3I_{\mathrm{IN}}^4}{512I_{\mathrm{REF}}^3}, \tag{6.408}$$

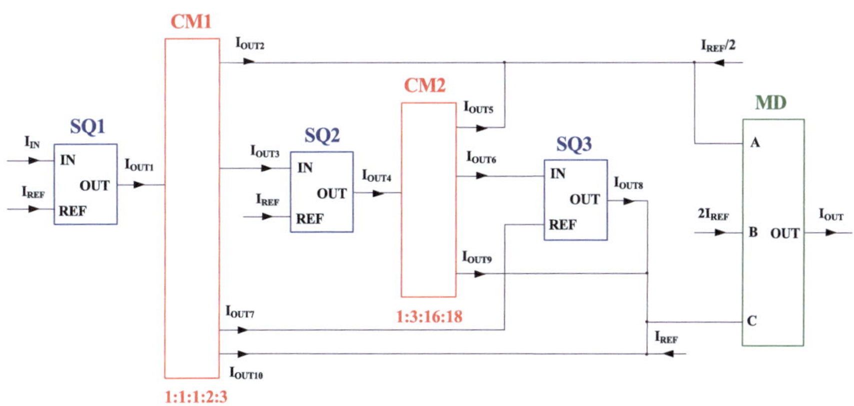

Fig. 6.37 Block diagram of the second analog function synthesizer for generating the inverse multiquadratic activation function for $a = 2$

$$I_{\mathrm{OUT6}} = 16 I_{\mathrm{OUT4}} = \frac{I_{\mathrm{IN}}^4}{32 I_{\mathrm{REF}}^3}, \tag{6.409}$$

$$I_{\mathrm{OUT7}} = 2 I_{\mathrm{OUT1}} = \frac{I_{\mathrm{IN}}^2}{4 I_{\mathrm{REF}}}, \tag{6.410}$$

$$I_{\mathrm{OUT8}} = \frac{I_{\mathrm{OUT6}}^2}{8 I_{\mathrm{OUT7}}} = \frac{I_{\mathrm{IN}}^6}{2048 I_{\mathrm{REF}}^5}, \tag{6.411}$$

$$I_{\mathrm{OUT9}} = 18 I_{\mathrm{OUT4}} = \frac{9 I_{\mathrm{IN}}^4}{256 I_{\mathrm{REF}}^3} \tag{6.412}$$

and

$$I_{\mathrm{OUT10}} = 3 I_{\mathrm{OUT1}} = \frac{3 I_{\mathrm{IN}}^2}{8 I_{\mathrm{REF}}}. \tag{6.413}$$

So, the output current of the circuit from Fig. 6.37 can be expressed as follows:

$$I_{\mathrm{OUT}} = I_{\mathrm{REF}} \frac{\frac{I_{\mathrm{REF}}}{2} + \frac{I_{\mathrm{IN}}^2}{8 I_{\mathrm{REF}}} + \frac{3 I_{\mathrm{IN}}^4}{512 I_{\mathrm{REF}}^3}}{I_{\mathrm{REF}} + \frac{3 I_{\mathrm{IN}}^2}{8 I_{\mathrm{REF}}} + \frac{9 I_{\mathrm{IN}}^4}{256 I_{\mathrm{REF}}^3} + \frac{I_{\mathrm{IN}}^6}{2048 I_{\mathrm{REF}}^5}}. \tag{6.414}$$

Using the definition of the x variable, the previous relation becomes:

$$I_{\mathrm{OUT}} = I_{\mathrm{REF}} \frac{\frac{3x^4}{512} + \frac{x^2}{8} + \frac{1}{2}}{\frac{x^6}{2048} + \frac{9x^4}{256} + \frac{3x^2}{8} + 1}. \tag{6.415}$$

In conclusion, the analog function synthesizer shown in Fig. 6.37 approximates the multiquadratic activation function for $a = 2$:

$$I_{\mathrm{OUT}} = I_{\mathrm{REF}} a_{36}(x) \cong I_{\mathrm{REF}} f_7(x). \tag{6.416}$$

6.6.5 *Third Analog Function Synthesizer for Generating the Inverse Multiquadratic Activation Function for a = 2*

The third analog function synthesizer, designed for generating the inverse multiquadratic activation function for $a = 2$, uses $a_{37}(x)$ approximation function described in Chap. 2.

The block diagram of the third analog function synthesizer for generating the inverse multiquadratic activation function for $a = 2$ is presented in Fig. 6.38.

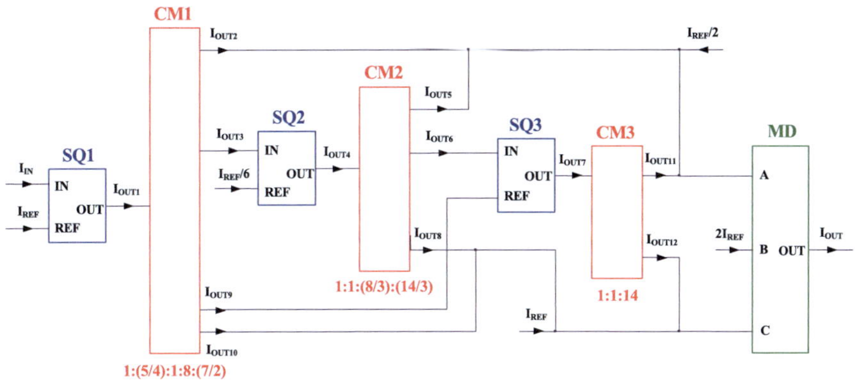

Fig. 6.38 Block diagram of the third analog function synthesizer for generating the inverse multiquadratic activation function for $a = 2$

The expressions of the currents from the circuit are:

$$I_{\mathrm{OUT1}} = \frac{I_{\mathrm{IN}}^2}{8I_{\mathrm{REF}}}, \tag{6.417}$$

$$I_{\mathrm{OUT2}} = \frac{5}{4}I_{\mathrm{OUT1}} = \frac{5I_{\mathrm{IN}}^2}{32I_{\mathrm{REF}}}, \tag{6.418}$$

$$I_{\mathrm{OUT3}} = I_{\mathrm{OUT1}} = \frac{I_{\mathrm{IN}}^2}{8I_{\mathrm{REF}}}, \tag{6.419}$$

$$I_{\mathrm{OUT4}} = \frac{I_{\mathrm{OUT3}}^2}{8(I_{\mathrm{REF}}/6)} = \frac{3I_{\mathrm{IN}}^4}{256I_{\mathrm{REF}}^3}, \tag{6.420}$$

$$I_{\mathrm{OUT5}} = I_{\mathrm{OUT4}} = \frac{3I_{\mathrm{IN}}^4}{256I_{\mathrm{REF}}^3}, \tag{6.421}$$

$$I_{\mathrm{OUT6}} = \frac{8}{3}I_{\mathrm{OUT4}} = \frac{I_{\mathrm{IN}}^4}{32I_{\mathrm{REF}}^3}, \tag{6.422}$$

$$I_{\mathrm{OUT7}} = \frac{I_{\mathrm{OUT6}}^2}{8I_{\mathrm{OUT9}}} = \frac{I_{\mathrm{IN}}^6}{8192I_{\mathrm{REF}}^5}, \tag{6.423}$$

$$I_{\mathrm{OUT8}} = \frac{14}{3}I_{\mathrm{OUT4}} = \frac{7I_{\mathrm{IN}}^4}{128I_{\mathrm{REF}}^3}, \tag{6.424}$$

$$I_{\mathrm{OUT9}} = 8I_{\mathrm{OUT1}} = \frac{I_{\mathrm{IN}}^2}{I_{\mathrm{REF}}}, \tag{6.425}$$

$$I_{OUT10} = \frac{7}{2} I_{OUT1} = \frac{7I_{IN}^2}{16I_{REF}}, \tag{6.426}$$

$$I_{OUT11} = I_{OUT7} = \frac{I_{IN}^6}{8192 I_{REF}^5} \tag{6.427}$$

and

$$I_{OUT12} = 14 I_{OUT7} = \frac{7I_{IN}^6}{4096 I_{REF}^5}. \tag{6.428}$$

So, the output current of the circuit from Fig. 6.38 can be expressed as follows:

$$I_{OUT} = I_{REF} \frac{\frac{I_{IN}^6}{8192 I_{REF}^5} + \frac{3I_{IN}^4}{256 I_{REF}^3} + \frac{5I_{IN}^2}{32 I_{REF}} + \frac{I_{REF}}{2}}{\frac{7I_{IN}^6}{4096 I_{REF}^5} + \frac{7I_{IN}^4}{128 I_{REF}^3} + \frac{7I_{IN}^2}{16 I_{REF}} + I_{REF}}. \tag{6.429}$$

Using the definition of the x variable, the previous relation becomes:

$$I_{OUT} = I_{REF} \frac{\frac{x^6}{8192} + \frac{3x^4}{256} + \frac{5x^2}{32} + \frac{1}{2}}{\frac{7x^6}{4096} + \frac{7x^4}{128} + \frac{7x^2}{16} + 1}. \tag{6.430}$$

In conclusion, the analog function synthesizer shown in Fig. 6.38 approximates the inverse multiquadratic activation function for $a = 2$:

$$I_{OUT} = I_{REF} a_{37}(x) \cong I_{REF} f_7(x). \tag{6.431}$$

6.6.6 *First Analog Function Synthesizer for Generating the Inverse Multiquadratic Activation Function for a = 3*

The first analog function synthesizer, designed for generating the inverse multiquadratic activation function for $a = 3$, uses $a_{38}(x)$ approximation function described in Chap. 2.

The block diagram of the first analog function synthesizer for generating the inverse multiquadratic activation function for $a = 3$ is presented in Fig. 6.39.

The expressions of the currents from the circuit are:

$$I_{OUT1} = \frac{I_{IN}^2}{8I_{REF}}, \tag{6.432}$$

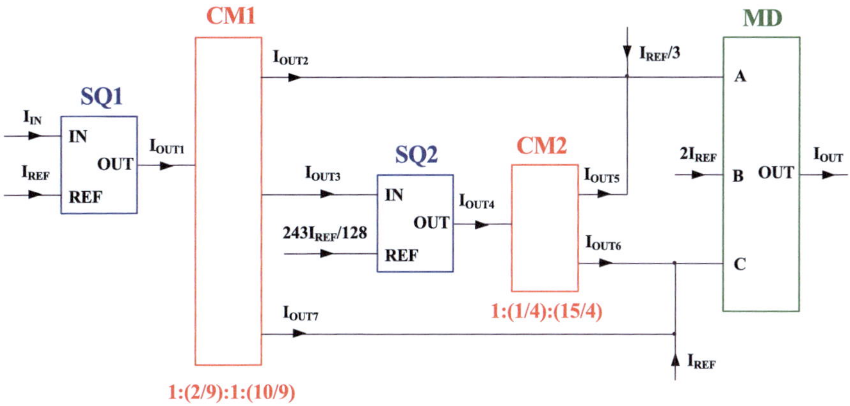

Fig. 6.39 Block diagram of the first analog function synthesizer for generating the inverse multiquadratic activation function for $a = 3$

$$I_{OUT2} = \frac{2}{9} I_{OUT1} = \frac{I_{IN}^2}{36 I_{REF}}, \tag{6.433}$$

$$I_{OUT3} = I_{OUT1} = \frac{I_{IN}^2}{8 I_{REF}}, \tag{6.434}$$

$$I_{OUT4} = \frac{I_{OUT3}^2}{8(243 I_{REF}/128)} = \frac{I_{IN}^4}{972 I_{REF}^3}, \tag{6.435}$$

$$I_{OUT5} = \frac{1}{4} I_{OUT4} = \frac{I_{IN}^4}{3888 I_{REF}^3}, \tag{6.436}$$

$$I_{OUT6} = \frac{15}{4} I_{OUT4} = \frac{5 I_{IN}^4}{1296 I_{REF}^3} \tag{6.437}$$

and

$$I_{OUT7} = \frac{10}{9} I_{OUT1} = \frac{5 I_{IN}^2}{36 I_{REF}}. \tag{6.438}$$

So, the output current of the circuit from Fig. 6.39 can be expressed as follows:

$$I_{OUT} = I_{REF} \frac{\frac{I_{REF}}{3} + \frac{I_{IN}^2}{36 I_{REF}} + \frac{I_{IN}^4}{3888 I_{REF}^3}}{I_{REF} + \frac{5 I_{IN}^2}{36 I_{REF}} + \frac{5 I_{IN}^4}{1296 I_{REF}^3}}. \tag{6.439}$$

Using the definition of the x variable, the previous relation becomes:

$$I_{OUT} = I_{REF} \frac{\frac{x^4}{3888} + \frac{x^2}{36} + \frac{1}{3}}{\frac{5x^4}{1296} + \frac{5x^2}{36} + 1}. \tag{6.440}$$

In conclusion, the analog function synthesizer shown in Fig. 6.39 approximates the inverse multiquadratic activation function for $a = 3$:

$$I_{OUT} = I_{REF} a_{38}(x) \cong I_{REF} f_7(x). \tag{6.441}$$

6.6.7 Second Analog Function Synthesizer for Generating the Inverse Multiquadratic Activation Function for a = 3

The second analog function synthesizer, designed for generating the inverse multiquadratic activation function for $a = 3$, uses $a_{39}(x)$ approximation function described in Chap. 2.

The block diagram of the second analog function synthesizer for generating the inverse multiquadratic activation function for $a = 3$ is presented in Fig. 6.40.

The expressions of the currents from the circuit are:

$$I_{OUT1} = \frac{I_{IN}^2}{8I_{REF}}, \tag{6.442}$$

$$I_{OUT2} = \frac{8}{27} I_{OUT1} = \frac{I_{IN}^2}{27I_{REF}}, \tag{6.443}$$

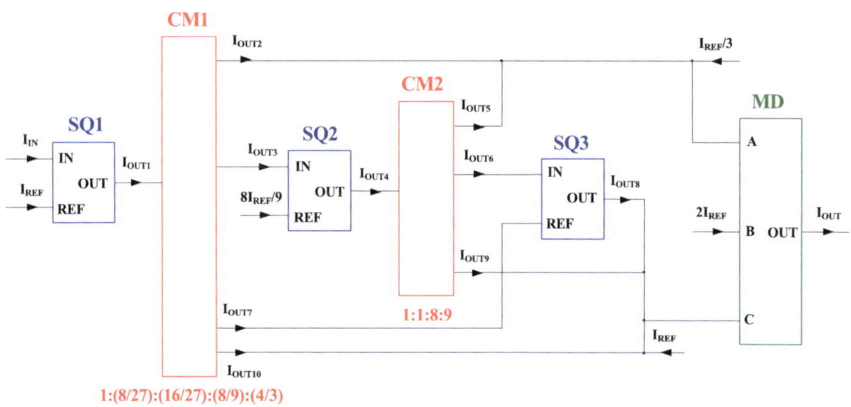

Fig. 6.40 Block diagram of the second analog function synthesizer for generating the inverse multiquadratic activation function for $a = 3$

$$I_{\mathrm{OUT3}} = \frac{16}{27} I_{\mathrm{OUT1}} = \frac{2I_{\mathrm{IN}}^2}{27 I_{\mathrm{REF}}}, \tag{6.444}$$

$$I_{\mathrm{OUT4}} = \frac{I_{\mathrm{OUT3}}^2}{8(8I_{\mathrm{REF}}/9)} = \frac{I_{\mathrm{IN}}^4}{1296 I_{\mathrm{REF}}^3}, \tag{6.445}$$

$$I_{\mathrm{OUT5}} = I_{\mathrm{OUT4}} = \frac{I_{\mathrm{IN}}^4}{1296 I_{\mathrm{REF}}^3}, \tag{6.446}$$

$$I_{\mathrm{OUT6}} = 8 I_{\mathrm{OUT4}} = \frac{I_{\mathrm{IN}}^4}{162 I_{\mathrm{REF}}^3}, \tag{6.447}$$

$$I_{\mathrm{OUT7}} = \frac{8}{9} I_{\mathrm{OUT1}} = \frac{I_{\mathrm{IN}}^2}{9 I_{\mathrm{REF}}}, \tag{6.448}$$

$$I_{\mathrm{OUT8}} = \frac{I_{\mathrm{OUT6}}^2}{8 I_{\mathrm{OUT7}}} = \frac{I_{\mathrm{IN}}^6}{23328 I_{\mathrm{REF}}^5}, \tag{6.449}$$

$$I_{\mathrm{OUT9}} = 9 I_{\mathrm{OUT4}} = \frac{I_{\mathrm{IN}}^4}{144 I_{\mathrm{REF}}^3} \tag{6.450}$$

and

$$I_{\mathrm{OUT10}} = \frac{4}{3} I_{\mathrm{OUT1}} = \frac{I_{\mathrm{IN}}^2}{6 I_{\mathrm{REF}}}. \tag{6.451}$$

So, the output current of the circuit from Fig. 6.40 can be expressed as follows:

$$I_{\mathrm{OUT}} = I_{\mathrm{REF}} \frac{\frac{I_{\mathrm{REF}}}{3} + \frac{I_{\mathrm{IN}}^2}{27 I_{\mathrm{REF}}} + \frac{I_{\mathrm{IN}}^4}{1296 I_{\mathrm{REF}}^3}}{I_{\mathrm{REF}} + \frac{I_{IN}^2}{6 I_{\mathrm{REF}}} + \frac{I_{\mathrm{IN}}^4}{144 I_{\mathrm{REF}}^3} + \frac{I_{\mathrm{IN}}^6}{23328 I_{\mathrm{REF}}^5}}. \tag{6.452}$$

Using the definition of the x variable, the previous relation becomes:

$$I_{\mathrm{OUT}} = I_{\mathrm{REF}} \frac{\frac{x^4}{1296} + \frac{x^2}{27} + \frac{1}{3}}{\frac{x^6}{23328} + \frac{x^4}{144} + \frac{x^2}{6} + 1}. \tag{6.453}$$

In conclusion, the analog function synthesizer shown in Fig. 6.40 approximates the multiquadratic activation function for $a = 3$:

$$I_{\mathrm{OUT}} = I_{\mathrm{REF}} a_{39}(x) \cong I_{\mathrm{REF}} f_7(x). \tag{6.454}$$

References

Popa C (2012) High-accuracy function synthesizer circuit with applications in signal processing. EURASIP J Adv Signal Process 2012:129. https://doi.org/10.1186/1687-6180-2012-129

Popa CR (2023) High-accuracy Gaussian function generator for neural networks. Electronics 12(1):24. https://doi.org/10.3390/electronics12010024

Chapter 7
Analysis and Design of Analog Function Synthesizers for Artificial Neural Networks Applications

7.1 Introduction

This chapter presents a multitude of analog function synthesizers (Popa 2012, 2023), developed for implementing in CMOS technology important functions, having a multitude of applications in analog signal processing and artificial neural networks, analyzed in Chap. 3 (exponential function, exponential ramp function, hyperbolic sinusoidal function, hyperbolic cosinusoidal function and, also, any mathematical function with continuous behavior, useful for previous mentioned applications). The superior-order approximation functions described in Chap. 3 represent the basis for elaborating the block diagrams from this chapter, while the concrete implementations in CMOS technology based on reasonable hardware resources use the fundamental CMOS computational circuits (current-mode squaring and multiplier/divider circuits), analyzed in details in Chap. 4.

7.2 Analog Function Synthesizers for the Exponential Function

In order to implement in CMOS technology the exponential function, the approximation functions described in Chap. 3 and the CMOS squaring and multiplies/divider circuits from Chap. 4 are used.

C. R. Popa, *Analog Current-Mode Computational Circuits for Artificial Neural Networks*, Analog Circuits and Signal Processing,
https://doi.org/10.1007/978-3-032-03989-7_7

7.2.1 *First Analog Function Synthesizer for Generating the Exponential Function*

The first analog function synthesizer, designed for generating the exponential function, uses $a_1(x)$ approximation function described in Chap. 3. The utilization of the current-mode squaring circuits (SQ) and multiplier/divider circuits (MD) analyzed in Chap. 4 allows to obtain CMOS implementations of the analog function synthesizer, having very good accuracy and reasonable complexity.

The block diagram of the first analog function synthesizer for generating the exponential function is presented in Fig. 7.1.

The expression of the I_{OUT1} current is:

$$I_{OUT1} = I_{REF}\frac{I_{REF} + \frac{I_{IN}}{4}}{I_{REF} - \frac{I_{IN}}{4}} \tag{7.1}$$

and

$$I_{OUT} = \frac{I_{OUT1}^2}{8(I_{REF}/8)} = I_{REF}\left(\frac{I_{REF} + \frac{I_{IN}}{4}}{I_{REF} - \frac{I_{IN}}{4}}\right)^2. \tag{7.2}$$

Using the definition of x variable, the previous relation becomes:

$$I_{OUT} = I_{REF}\left(\frac{1 + \frac{x}{4}}{1 - \frac{x}{4}}\right)^2. \tag{7.3}$$

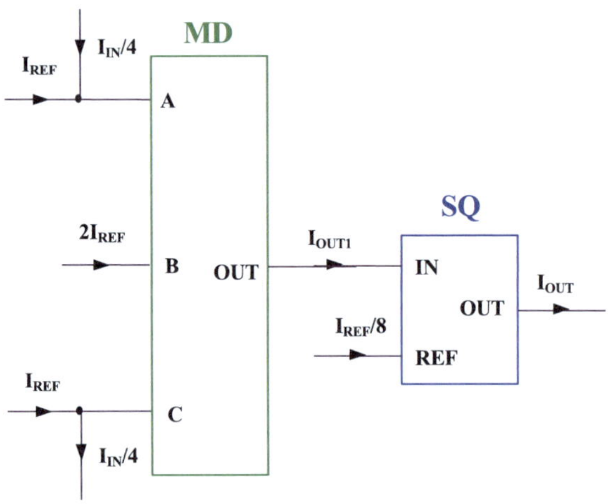

Fig. 7.1 Block diagram of the first analog function synthesizer for generating the exponential function

In conclusion, the analog function synthesizer shown in Fig. 7.1 approximates the exponential function:

$$I_{\mathrm{OUT}} = I_{\mathrm{REF}} a_1(x) \cong I_{\mathrm{REF}} f_1(x). \tag{7.4}$$

7.2.2 *Second Analog Function Synthesizer for Generating the Exponential Function*

The second analog function synthesizer, designed for generating the exponential function, uses $a_2(x)$ approximation function described in Chap. 3.

The block diagram of the second analog function synthesizer for generating the exponential function is presented in Fig. 7.2.

The expressions of the currents from the circuit are:

$$I_{\mathrm{OUT1}} = I_{\mathrm{REF}} \frac{I_{\mathrm{REF}} + \frac{I_{\mathrm{IN}}}{8}}{I_{\mathrm{REF}} - \frac{I_{\mathrm{IN}}}{8}}, \tag{7.5}$$

$$I_{\mathrm{OUT2}} = \frac{I_{\mathrm{OUT1}}^2}{8(I_{\mathrm{REF}}/8)} = I_{\mathrm{REF}} \left(\frac{I_{\mathrm{REF}} + \frac{I_{\mathrm{IN}}}{8}}{I_{\mathrm{REF}} - \frac{I_{\mathrm{IN}}}{8}} \right)^2 \tag{7.6}$$

and

$$I_{\mathrm{OUT}} = \frac{I_{\mathrm{OUT1}}^2}{8(I_{\mathrm{REF}}/8)} = I_{\mathrm{REF}} \left(\frac{I_{\mathrm{REF}} + \frac{I_{\mathrm{IN}}}{8}}{I_{\mathrm{REF}} - \frac{I_{\mathrm{IN}}}{8}} \right)^4. \tag{7.7}$$

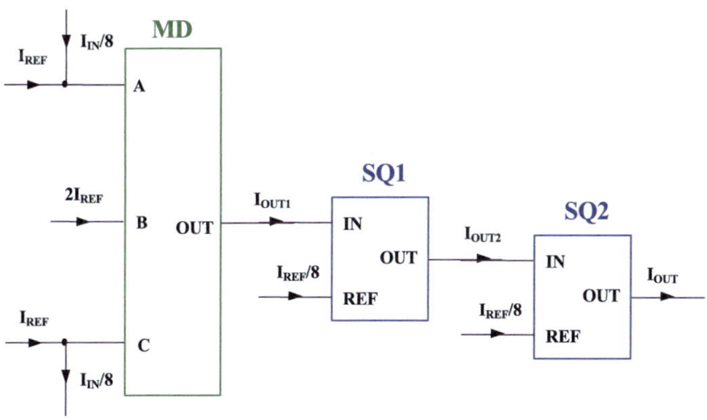

Fig. 7.2 Block diagram of the second analog function synthesizer for generating the exponential function

Using the definition of x variable, the previous relation becomes:

$$I_{\rm OUT} = I_{\rm REF}\left(\frac{1+\frac{x}{8}}{1-\frac{x}{8}}\right)^4. \tag{7.8}$$

In conclusion, the analog function synthesizer shown in Fig. 7.2 approximates the exponential function:

$$I_{\rm OUT} = I_{\rm REF}a_2(x) \cong I_{\rm REF}f_1(x). \tag{7.9}$$

7.2.3 Third Analog Function Synthesizer for Generating the Exponential Function

The third analog function synthesizer, designed for generating the exponential function, uses $a_3(x)$ approximation function described in Chap. 3.

The block diagram of the third analog function synthesizer for generating the exponential function is presented in Fig. 7.3.

The expressions of the currents from the circuit are:

$$I_{\rm OUT1} = \frac{I_{\rm IN}^2}{8I_{\rm REF}}, \tag{7.10}$$

$$I_{\rm OUT2} = \frac{2}{3}I_{\rm OUT1} = \frac{I_{\rm IN}^2}{12I_{\rm REF}} \tag{7.11}$$

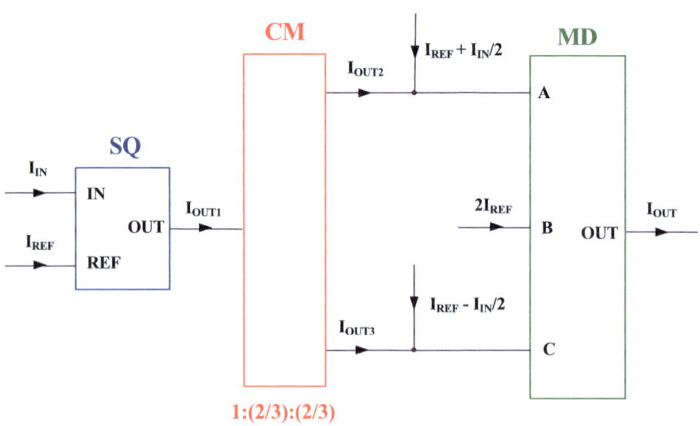

Fig. 7.3 Block diagram of the third analog function synthesizer for generating the exponential function

and

$$I_{OUT3} = \frac{2}{3} I_{OUT1} = \frac{I_{IN}^2}{12 I_{REF}}. \tag{7.12}$$

The output current of the circuit from Fig. 7.3 can be expressed as follows:

$$I_{OUT} = I_{REF} \frac{I_{REF} + \frac{I_{IN}}{2} + \frac{I_{IN}^2}{12 I_{REF}}}{I_{REF} - \frac{I_{IN}}{2} + \frac{I_{IN}^2}{12 I_{REF}}}. \tag{7.13}$$

Using the definition of x variable, the previous relation becomes:

$$I_{OUT} = I_{REF} \frac{x^2 + 6x + 12}{x^2 - 6x + 12}. \tag{7.14}$$

In conclusion, the analog function synthesizer shown in Fig. 7.3 approximates the exponential function:

$$I_{OUT} = I_{REF} a_3(x) \cong I_{REF} f_1(x). \tag{7.15}$$

7.2.4 *Fourth Analog Function Synthesizer for Generating the Exponential Function*

The fourth analog function synthesizer, designed for generating the exponential function, uses $a_4(x)$ approximation function described in Chap. 3.

The block diagram of the fourth analog function synthesizer for generating the exponential function is presented in Fig. 7.4.

The expressions of the currents from the circuit are:

$$I_{OUT1} = \frac{I_{IN}^2}{8 I_{REF}}, \tag{7.16}$$

$$I_{OUT2} = \frac{1}{6} I_{OUT1} = \frac{I_{IN}^2}{48 I_{REF}}, \tag{7.17}$$

$$I_{OUT3} = \frac{1}{6} I_{OUT1} = \frac{I_{IN}^2}{48 I_{REF}} \tag{7.18}$$

and

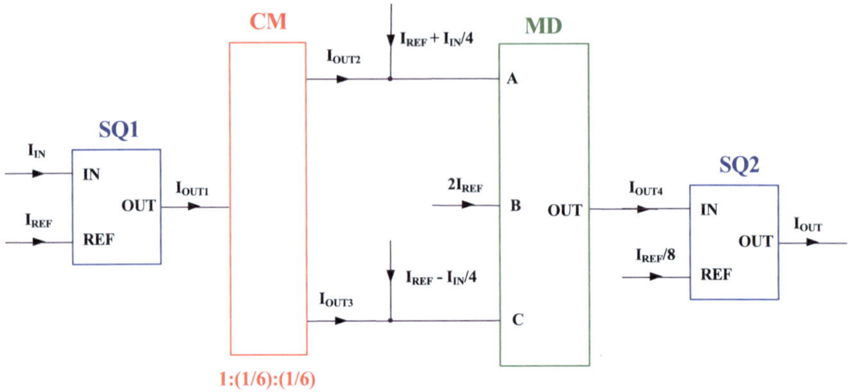

Fig. 7.4 Block diagram of the fourth analog function synthesizer for generating the exponential function

$$I_{OUT4} = I_{REF}\frac{I_{REF} + \frac{I_{IN}}{4} + \frac{I_{IN}^2}{48I_{REF}}}{I_{REF} - \frac{I_{IN}}{4} + \frac{I_{IN}^2}{48I_{REF}}}. \tag{7.19}$$

The output current of the circuit from Fig. 7.4 can be expressed as follows:

$$I_{OUT} = \frac{I_{OUT4}^2}{8(I_{REF}/8)} = I_{REF}\left(\frac{I_{REF} + \frac{I_{IN}}{4} + \frac{I_{IN}^2}{48I_{REF}}}{I_{REF} - \frac{I_{IN}}{4} + \frac{I_{IN}^2}{48I_{REF}}}\right)^2. \tag{7.20}$$

Using the definition of x variable, the previous relation becomes:

$$I_{OUT} = I_{REF}\left(\frac{x^2 + 12x + 48}{x^2 - 12x + 48}\right)^2. \tag{7.21}$$

In conclusion, the analog function synthesizer shown in Fig. 7.4 approximates the exponential function:

$$I_{OUT} = I_{REF}a_4(x) \cong I_{REF}f_1(x). \tag{7.22}$$

7.2.5 *Fifth Analog Function Synthesizer for Generating the Exponential Function*

The fifth analog function synthesizer, designed for generating the exponential function, uses $a_5(x)$ approximation function described in Chap. 3.

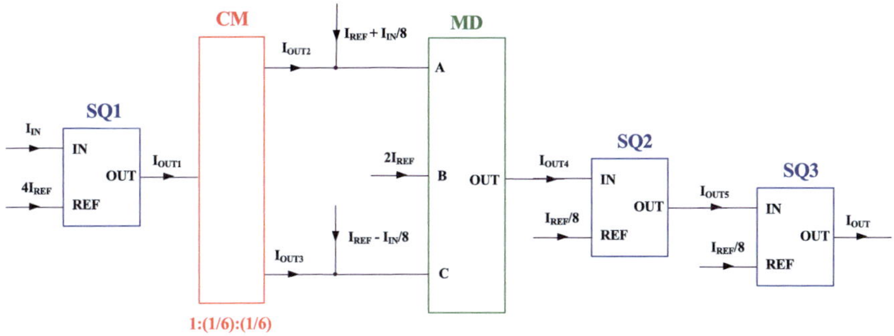

Fig. 7.5 Block diagram of the fifth analog function synthesizer for generating the exponential function

The block diagram of the fifth analog function synthesizer for generating the exponential function is presented in Fig. 7.5.

The expressions of the currents from the circuit are:

$$I_{\mathrm{OUT1}} = \frac{I_{\mathrm{IN}}^2}{32 I_{\mathrm{REF}}}, \tag{7.23}$$

$$I_{\mathrm{OUT2}} = \frac{1}{6} I_{\mathrm{OUT1}} = \frac{I_{\mathrm{IN}}^2}{192 I_{\mathrm{REF}}}, \tag{7.24}$$

$$I_{\mathrm{OUT3}} = \frac{1}{6} I_{\mathrm{OUT1}} = \frac{I_{\mathrm{IN}}^2}{192 I_{\mathrm{REF}}}, \tag{7.25}$$

$$I_{\mathrm{OUT4}} = I_{\mathrm{REF}} \frac{I_{\mathrm{REF}} + \frac{I_{\mathrm{IN}}}{8} + \frac{I_{\mathrm{IN}}^2}{192 I_{\mathrm{REF}}}}{I_{\mathrm{REF}} - \frac{I_{\mathrm{IN}}}{8} + \frac{I_{\mathrm{IN}}^2}{192 I_{\mathrm{REF}}}} \tag{7.26}$$

and

$$I_{\mathrm{OUT5}} = \frac{I_{\mathrm{OUT4}}^2}{8(I_{\mathrm{REF}}/8)} = I_{\mathrm{REF}} \left(\frac{I_{\mathrm{REF}} + \frac{I_{\mathrm{IN}}}{8} + \frac{I_{\mathrm{IN}}^2}{192 I_{\mathrm{REF}}}}{I_{\mathrm{REF}} - \frac{I_{\mathrm{IN}}}{8} + \frac{I_{\mathrm{IN}}^2}{192 I_{\mathrm{REF}}}} \right)^2 . \tag{7.27}$$

The output current of the circuit from Fig. 7.5 can be expressed as follows:

$$I_{\mathrm{OUT}} = \frac{I_{\mathrm{OUT5}}^2}{8(I_{\mathrm{REF}}/8)} = I_{\mathrm{REF}} \left(\frac{I_{\mathrm{REF}} + \frac{I_{\mathrm{IN}}}{8} + \frac{I_{\mathrm{IN}}^2}{192 I_{\mathrm{REF}}}}{I_{\mathrm{REF}} - \frac{I_{\mathrm{IN}}}{8} + \frac{I_{\mathrm{IN}}^2}{192 I_{\mathrm{REF}}}} \right)^4 . \tag{7.28}$$

Using the definition of x variable, the previous relation becomes:

$$I_{\mathrm{OUT}} = I_{\mathrm{REF}} \left(\frac{x^2 + 24x + 192}{x^2 - 24x + 192} \right)^4. \tag{7.29}$$

In conclusion, the analog function synthesizer shown in Fig. 7.5 approximates the exponential function:

$$I_{\mathrm{OUT}} = I_{\mathrm{REF}} a_5(x) \cong I_{\mathrm{REF}} f_1(x). \tag{7.30}$$

7.2.6 *Sixth Analog Function Synthesizer for Generating the Exponential Function*

The sixth analog function synthesizer, designed for generating the exponential function, uses $a_6(x)$ approximation function described in Chap. 3.

The block diagram of the sixth analog function synthesizer for generating the exponential function is presented in Fig. 7.6.

The expressions of the currents from the circuit are:

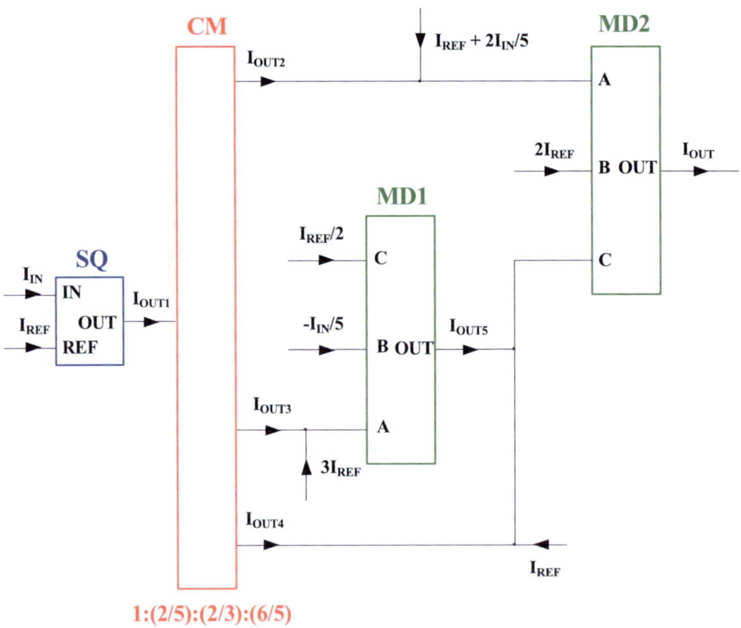

Fig. 7.6 Block diagram of the sixth analog function synthesizer for generating the exponential function

$$I_{OUT1} = \frac{I_{IN}^2}{8I_{REF}}, \tag{7.31}$$

$$I_{OUT2} = \frac{2}{5}I_{OUT1} = \frac{I_{IN}^2}{20I_{REF}}, \tag{7.32}$$

$$I_{OUT3} = \frac{2}{3}I_{OUT1} = \frac{I_{IN}^2}{12I_{REF}}, \tag{7.33}$$

$$I_{OUT4} = \frac{6}{5}I_{OUT1} = \frac{3I_{IN}^2}{20I_{REF}} \tag{7.34}$$

and

$$I_{OUT5} = -\frac{I_{IN}}{5I_{REF}}\left(\frac{I_{IN}^2}{12I_{REF}} + 3I_{REF}\right). \tag{7.35}$$

The output current of the circuit from Fig. 7.6 can be expressed as follows:

$$I_{OUT} = I_{REF}\frac{\frac{I_{IN}^2}{20I_{REF}} + \frac{2I_{IN}}{5} + I_{REF}}{-\frac{I_{IN}}{5I_{REF}}\left(\frac{I_{IN}^2}{12I_{REF}} + 3I_{REF}\right) + \frac{3I_{IN}^2}{20I_{REF}} + I_{REF}}. \tag{7.36}$$

Using the definition of the x variable, the previous relation becomes:

$$I_{OUT} = I_{REF}\frac{\frac{x^2}{20} + \frac{2x}{5} + 1}{-\frac{x^3}{60} + \frac{3x^2}{20} - \frac{3x}{5} + 1}. \tag{7.37}$$

In conclusion, the analog function synthesizer shown in Fig. 7.6 approximates the exponential function:

$$I_{OUT} = I_{REF}a_6(x) \cong I_{REF}f_1(x). \tag{7.38}$$

7.2.7 *Seventh Analog Function Synthesizer for Generating the Exponential Function*

The seventh analog function synthesizer, designed for generating the exponential function, uses $a_7(x)$ approximation function described in Chap. 3.

The block diagram of the seventh analog function synthesizer for generating the exponential function is presented in Fig. 7.7.

The expressions of the currents from the circuit are:

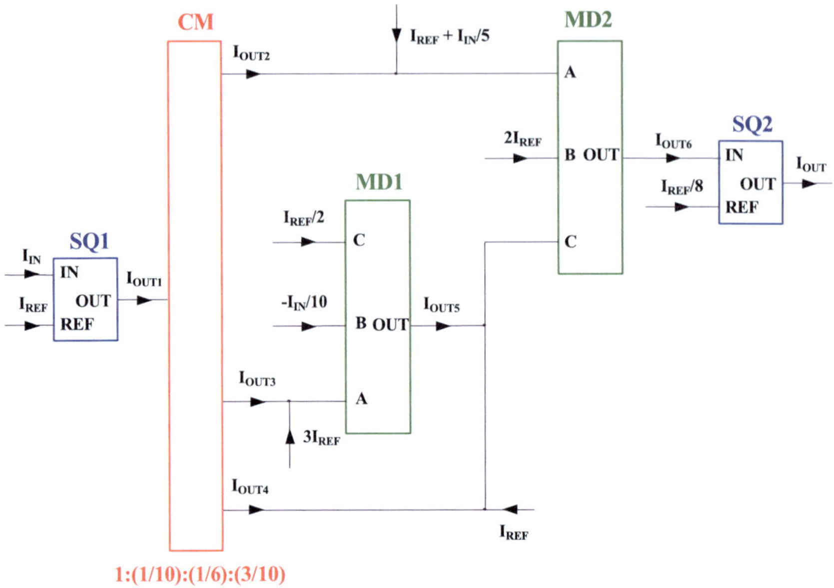

Fig. 7.7 Block diagram of the seventh analog function synthesizer for generating the exponential function

$$I_{OUT1} = \frac{I_{IN}^2}{8I_{REF}}, \tag{7.39}$$

$$I_{OUT2} = \frac{1}{10} I_{OUT1} = \frac{I_{IN}^2}{80I_{REF}}, \tag{7.40}$$

$$I_{OUT3} = \frac{1}{6} I_{OUT1} = \frac{I_{IN}^2}{48I_{REF}}, \tag{7.41}$$

$$I_{OUT4} = \frac{3}{10} I_{OUT1} = \frac{3I_{IN}^2}{80I_{REF}}, \tag{7.42}$$

$$I_{OUT5} = -\frac{I_{IN}}{10I_{REF}} \left(\frac{I_{IN}^2}{48I_{REF}} + 3I_{REF} \right) \tag{7.43}$$

and

$$I_{OUT6} = I_{REF} \frac{\frac{I_{IN}^2}{80I_{REF}} + \frac{I_{IN}}{5} + I_{REF}}{-\frac{I_{IN}}{10I_{REF}} \left(\frac{I_{IN}^2}{48I_{REF}} + 3I_{REF} \right) + \frac{3I_{IN}^2}{80I_{REF}} + I_{REF}}. \tag{7.44}$$

The output current of the circuit from Fig. 7.7 can be expressed as follows:

$$I_{\mathrm{OUT}} = \frac{I_{\mathrm{OUT6}}^2}{8(I_{\mathrm{REF}}/8)}. \tag{7.45}$$

Using the definition of the x variable, the previous relation becomes:

$$I_{\mathrm{OUT}} = I_{\mathrm{REF}}\left(\frac{\frac{x^2}{80} + \frac{x}{5} + 1}{-\frac{x^3}{480} + \frac{3x^2}{80} - \frac{3x}{10} + 1}\right)^2. \tag{7.46}$$

In conclusion, the analog function synthesizer shown in Fig. 7.7 approximates the exponential function:

$$I_{\mathrm{OUT}} = I_{\mathrm{REF}} a_7(x) \cong I_{\mathrm{REF}} f_1(x). \tag{7.47}$$

7.2.8 *Eight Analog Function Synthesizer for Generating the Exponential Function*

The eighth analog function synthesizer, designed for generating the exponential function, uses $a_8(x)$ approximation function described in Chap. 3.

The block diagram of the eighth analog function synthesizer for generating the exponential function is presented in Fig. 7.8.

The expressions of the currents from the circuit are:

$$I_{\mathrm{OUT1}} = \frac{I_{\mathrm{IN}}^2}{8I_{\mathrm{REF}}}, \tag{7.48}$$

$$I_{\mathrm{OUT2}} = \frac{4}{5} I_{\mathrm{OUT1}} = \frac{I_{\mathrm{IN}}^2}{10I_{\mathrm{REF}}}, \tag{7.49}$$

$$I_{\mathrm{OUT3}} = \frac{2}{15} I_{\mathrm{OUT1}} = \frac{I_{\mathrm{IN}}^2}{60I_{\mathrm{REF}}}, \tag{7.50}$$

$$I_{\mathrm{OUT4}} = \frac{I_{\mathrm{IN}}}{2I_{\mathrm{REF}}}\left(\frac{I_{\mathrm{IN}}^2}{60I_{\mathrm{REF}}} + I_{\mathrm{REF}}\right), \tag{7.51}$$

$$I_{\mathrm{OUT5}} = \frac{2}{15} I_{\mathrm{OUT1}} = \frac{I_{\mathrm{IN}}^2}{60I_{\mathrm{REF}}}, \tag{7.52}$$

$$I_{\mathrm{OUT6}} = -\frac{I_{\mathrm{IN}}}{2I_{\mathrm{REF}}}\left(\frac{I_{\mathrm{IN}}^2}{60I_{\mathrm{REF}}} + I_{\mathrm{REF}}\right) \tag{7.53}$$

and

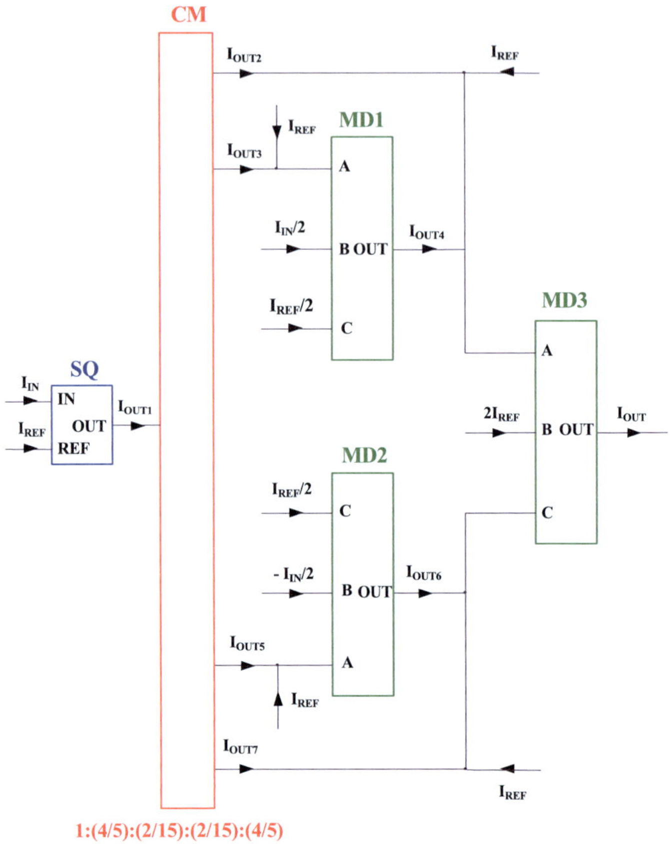

Fig. 7.8 Block diagram of the eighth analog function synthesizer for generating the exponential function

$$I_{\mathrm{OUT7}} = \frac{4}{5} I_{\mathrm{OUT1}} = \frac{I_{\mathrm{IN}}^2}{10 I_{\mathrm{REF}}}. \tag{7.54}$$

The output current of the circuit from Fig. 7.8 can be expressed as follows:

$$I_{\mathrm{OUT}} = I_{\mathrm{REF}} \frac{\frac{I_{\mathrm{IN}}^3}{120 I_{\mathrm{REF}}^2} + \frac{I_{\mathrm{IN}}^2}{10 I_{\mathrm{REF}}} + \frac{I_{\mathrm{IN}}}{2} + I_{\mathrm{REF}}}{-\frac{I_{\mathrm{IN}}^3}{120 I_{\mathrm{REF}}^2} + \frac{I_{\mathrm{IN}}^2}{10 I_{\mathrm{REF}}} - \frac{I_{\mathrm{IN}}}{2} + I_{\mathrm{REF}}}. \tag{7.55}$$

Using the definition of the x variable, the previous relation becomes:

$$I_{\mathrm{OUT}} = I_{\mathrm{REF}} \frac{\frac{x^3}{120} + \frac{x^2}{10} + \frac{x}{2} + 1}{-\frac{x^3}{120} + \frac{x^2}{10} - \frac{x}{2} + 1}. \tag{7.56}$$

In conclusion, the analog function synthesizer shown in Fig. 7.8 approximates the exponential function:

$$I_{OUT} = I_{REF} a_8(x) \cong I_{REF} f_1(x). \tag{7.57}$$

7.2.9 Ninth Analog Function Synthesizer for Generating the Exponential Function

The ninth analog function synthesizer, designed for generating the exponential function, uses $a_9(x)$ approximation function described in Chap. 3.

The block diagram of the ninth analog function synthesizer for generating the exponential function is presented in Fig. 7.9.

The expressions of the currents from the circuit are:

$$I_{OUT1} = \frac{I_{IN}^2}{40 I_{REF}}, \tag{7.58}$$

$$I_{OUT2} = I_{OUT1} = \frac{I_{IN}^2}{40 I_{REF}}, \tag{7.59}$$

$$I_{OUT3} = \frac{1}{6} I_{OUT1} = \frac{I_{IN}^2}{240 I_{REF}}, \tag{7.60}$$

$$I_{OUT4} = \frac{I_{IN}}{4 I_{REF}} \left(\frac{I_{IN}^2}{240 I_{REF}} + I_{REF} \right), \tag{7.61}$$

$$I_{OUT5} = \frac{1}{6} I_{OUT1} = \frac{I_{IN}^2}{240 I_{REF}}, \tag{7.62}$$

$$I_{OUT6} = -\frac{I_{IN}}{4 I_{REF}} \left(\frac{I_{IN}^2}{240 I_{REF}} + I_{REF} \right), \tag{7.63}$$

$$I_{OUT7} = I_{OUT1} = \frac{I_{IN}^2}{40 I_{REF}} \tag{7.64}$$

and

$$I_{OUT8} = I_{REF} \frac{\frac{I_{IN}^3}{960 I_{REF}^2} + \frac{I_{IN}^2}{40 I_{REF}} + \frac{I_{IN}}{4} + I_{REF}}{-\frac{I_{IN}^3}{960 I_{REF}^2} + \frac{I_{IN}^2}{40 I_{REF}} - \frac{I_{IN}}{4} + I_{REF}}. \tag{7.65}$$

The output current of the circuit from Fig. 7.9 can be expressed as follows:

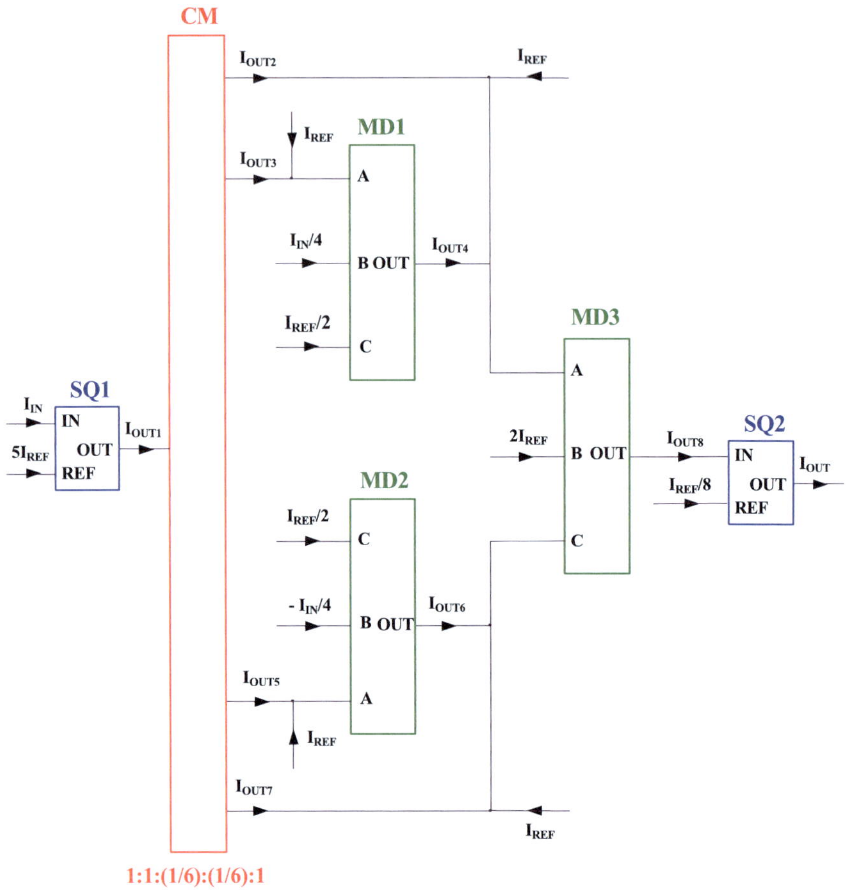

Fig. 7.9 Block diagram of the ninth analog function synthesizer for generating the exponential function

$$I_{\mathrm{OUT}} = \frac{I_{\mathrm{OUT8}}^2}{8(I_{\mathrm{REF}}/8)}. \tag{7.66}$$

Using the definition of the x variable, the previous relation becomes:

$$I_{\mathrm{OUT}} = I_{\mathrm{REF}} \left(\frac{\frac{x^3}{960} + \frac{x^2}{40} + \frac{x}{4} + 1}{-\frac{x^3}{960} + \frac{x^2}{40} - \frac{x}{4} + 1} \right)^2. \tag{7.67}$$

In conclusion, the analog function synthesizer shown in Fig. 7.9 approximates the exponential function:

$$I_{\mathrm{OUT}} = I_{\mathrm{REF}} a_9(x) \cong I_{\mathrm{REF}} f_1(x). \tag{7.68}$$

7.2.10 Tenth Analog Function Synthesizer for Generating the Exponential Function

The tenth analog function synthesizer, designed for generating the exponential function, uses $a_{10}(x)$ approximation function described in Chap. 3.

The block diagram of the tenth analog function synthesizer for generating the exponential function is presented in Fig. 7.10.

The expressions of the currents from the circuit are:

$$I_{\text{OUT1}} = \frac{I_{\text{IN}}^2}{8I_{\text{REF}}}, \tag{7.69}$$

$$I_{\text{OUT2}} = \frac{4}{7}I_{\text{OUT1}} = \frac{I_{\text{IN}}^2}{14I_{\text{REF}}}, \tag{7.70}$$

$$I_{\text{OUT3}} = 2I_{\text{OUT1}} = \frac{I_{\text{IN}}^2}{4I_{\text{REF}}}, \tag{7.71}$$

$$I_{\text{OUT4}} = \frac{I_{\text{OUT3}}^2}{8(105I_{\text{IN}}/64)} = \frac{I_{\text{IN}}^3}{210I_{\text{REF}}^2}, \tag{7.72}$$

$$I_{\text{OUT5}} = I_{\text{OUT4}} = \frac{I_{\text{IN}}^3}{210I_{\text{REF}}^2}, \tag{7.73}$$

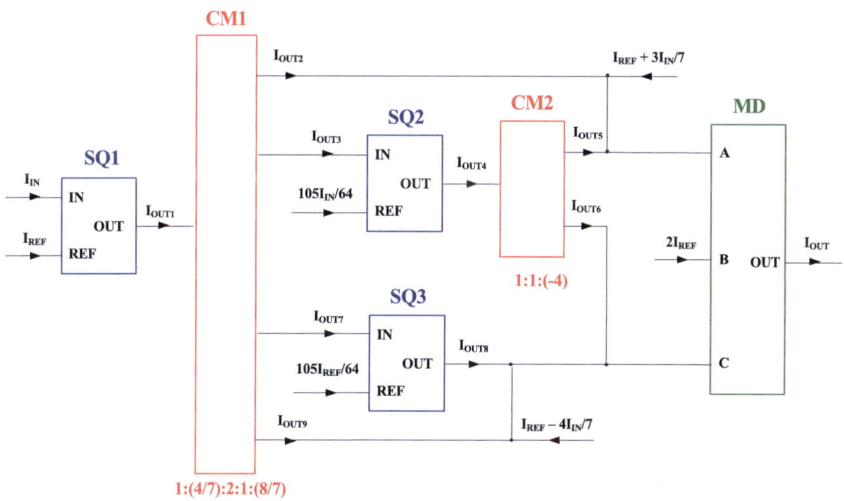

Fig. 7.10 Block diagram of the tenth analog function synthesizer for generating the exponential function

$$I_{\text{OUT6}} = -4I_{\text{OUT4}} = -\frac{2I_{\text{IN}}^3}{105I_{\text{REF}}^2}, \tag{7.74}$$

$$I_{\text{OUT7}} = I_{\text{OUT1}} = \frac{I_{\text{IN}}^2}{8I_{\text{REF}}}, \tag{7.75}$$

$$I_{\text{OUT8}} = \frac{I_{\text{OUT7}}^2}{8(105I_{\text{REF}}/64)} = \frac{I_{\text{IN}}^4}{840I_{\text{REF}}^3} \tag{7.76}$$

and

$$I_{\text{OUT9}} = \frac{8}{7}I_{\text{OUT1}} = \frac{I_{\text{IN}}^2}{7I_{\text{REF}}}. \tag{7.77}$$

So, the output current of the circuit from Fig. 7.10 can be expressed as follows:

$$I_{\text{OUT}} = I_{\text{REF}}\frac{\frac{I_{\text{IN}}^3}{210I_{\text{REF}}^2} + \frac{I_{\text{IN}}^2}{14I_{\text{REF}}} + \frac{3I_{\text{IN}}}{7} + I_{\text{REF}}}{\frac{I_{\text{IN}}^4}{840I_{\text{REF}}^3} - \frac{2I_{\text{IN}}^3}{105I_{\text{REF}}^2} + \frac{I_{\text{IN}}^2}{7I_{\text{REF}}} - \frac{4I_{\text{IN}}}{7} + I_{\text{REF}}}. \tag{7.78}$$

Using the definition of the x variable, the previous relation becomes:

$$I_{\text{OUT}} = I_{\text{REF}}\frac{\frac{x^3}{210} + \frac{x^2}{14} + \frac{3x}{7} + 1}{\frac{x^4}{840} - \frac{2x^3}{105} + \frac{x^2}{7} - \frac{4x}{7} + 1}. \tag{7.79}$$

In conclusion, the analog function synthesizer shown in Fig. 7.10 approximates the exponential function:

$$I_{\text{OUT}} = I_{\text{REF}}a_{10}(x) \cong I_{\text{REF}}f_1(x). \tag{7.80}$$

7.2.11 *Eleventh Analog Function Synthesizer for Generating the Exponential Function*

The 11th analog function synthesizer, designed for generating the exponential function, uses $a_{11}(x)$ approximation function described in Chap. 3.

The block diagram of the 11th analog function synthesizer for generating the exponential function is presented in Fig. 7.11.

The expressions of the currents from the circuit are:

$$I_{\text{OUT1}} = \frac{I_{\text{IN}}^2}{8I_{\text{REF}}}, \tag{7.81}$$

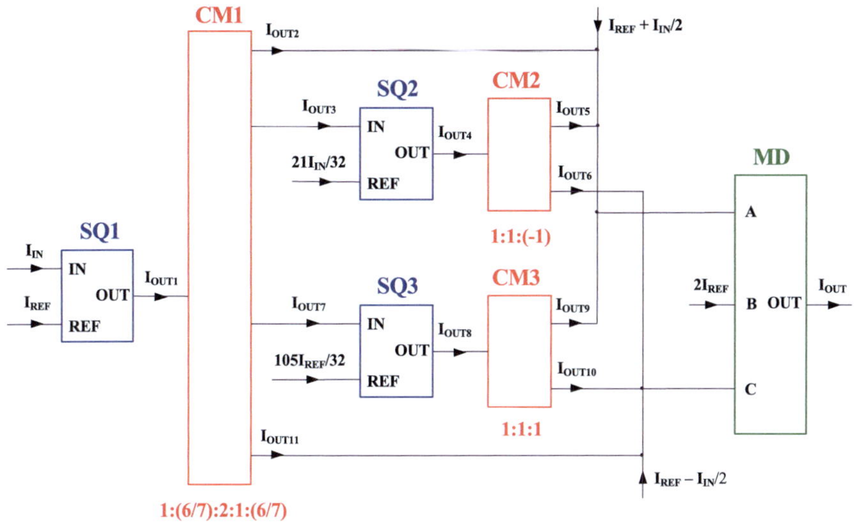

Fig. 7.11 Block diagram of the 11th analog function synthesizer for generating the exponential function

$$I_{\text{OUT2}} = \frac{6}{7} I_{\text{OUT1}} = \frac{3I_{\text{IN}}^2}{28I_{\text{REF}}}, \tag{7.82}$$

$$I_{\text{OUT3}} = 2I_{\text{OUT1}} = \frac{I_{\text{IN}}^2}{4I_{\text{REF}}}, \tag{7.83}$$

$$I_{\text{OUT4}} = \frac{I_{\text{OUT3}}^2}{8(211_{\text{IN}}/32)} = \frac{I_{\text{IN}}^3}{84I_{\text{REF}}^2}, \tag{7.84}$$

$$I_{\text{OUT5}} = I_{\text{OUT4}} = \frac{I_{\text{IN}}^3}{84I_{\text{REF}}^2}, \tag{7.85}$$

$$I_{\text{OUT6}} = -I_{\text{OUT4}} = -\frac{I_{\text{IN}}^3}{84I_{\text{REF}}^2}, \tag{7.86}$$

$$I_{\text{OUT7}} = I_{\text{OUT1}} = \frac{I_{\text{IN}}^2}{8I_{\text{REF}}}, \tag{7.87}$$

$$I_{\text{OUT8}} = \frac{I_{\text{OUT7}}^2}{8(1051_{\text{REF}}/32)} = \frac{I_{\text{IN}}^4}{1680I_{\text{REF}}^3}, \tag{7.88}$$

$$I_{\text{OUT9}} = I_{\text{OUT8}} = \frac{I_{\text{IN}}^4}{1680I_{\text{REF}}^3}, \tag{7.89}$$

$$I_{OUT10} = I_{OUT8} = \frac{I_{IN}^4}{1680 I_{REF}^3} \tag{7.90}$$

and

$$I_{OUT11} = \frac{6}{7} I_{OUT1} = \frac{3 I_{IN}^2}{28 I_{REF}}. \tag{7.91}$$

So, the output current of the circuit from Fig. 7.11 can be expressed as follows:

$$I_{OUT} = I_{REF} \frac{I_{REF} + \frac{I_{IN}}{2} + \frac{31_{IN}^2}{28 I_{REF}} + \frac{I_{IN}^3}{84 I_{REF}^2} + \frac{I_{IN}^4}{1680 I_{REF}^3}}{I_{REF} - \frac{I_{IN}}{2} + \frac{31_{IN}^2}{28 I_{REF}} - \frac{I_{IN}^3}{84 I_{REF}^2} + \frac{I_{IN}^4}{1680 I_{REF}^3}}. \tag{7.92}$$

Using the definition of the x variable, the previous relation becomes:

$$I_{OUT} = I_{REF} \frac{x^4 + 20x^3 + 180x^2 + 840x + 1680}{x^4 - 20x^3 + 180x^2 - 840x + 1680}. \tag{7.93}$$

In conclusion, the analog function synthesizer shown in Fig. 7.11 approximates the exponential function:

$$I_{OUT} = I_{REF} a_{11}(x) \cong I_{REF} f_1(x). \tag{7.94}$$

7.2.12 *Twelfth Analog Function Synthesizer for Generating the Exponential Function*

The 12th analog function synthesizer, designed for generating the exponential function, uses $a_{12}(x)$ approximation function described in Chap. 3.

The block diagram of the 12th analog function synthesizer for generating the exponential function is presented in Fig. 7.12.

The expressions of the currents from the circuit are:

$$I_{OUT1} = \frac{(64 I_{REF})^2}{8\left(I_{REF} - \frac{I_{IN}}{16}\right)}, \tag{7.95}$$

$$I_{OUT2} = \frac{(I_{IN}/2)^2}{8(2 I_{REF})} = \frac{I_{IN}^2}{64 I_{REF}}, \tag{7.96}$$

$$I_{OUT3} = -\frac{1}{3} I_{OUT1} + \frac{I_{IN}^2}{64 I_{REF}} + \frac{29 I_{REF}}{3} + \frac{5 I_{IN}}{12}, \tag{7.97}$$

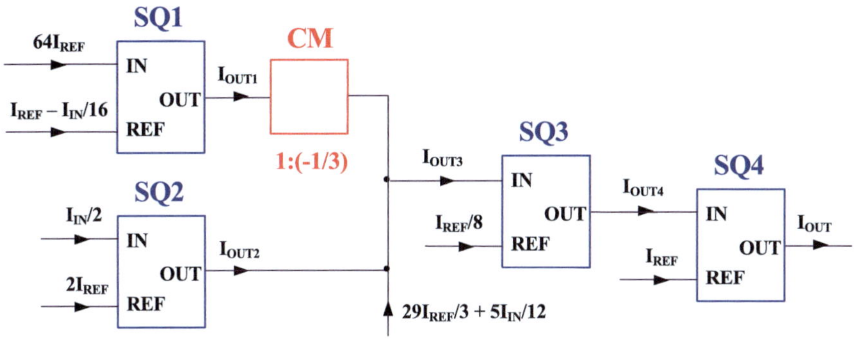

Fig. 7.12 Block diagram of the 12th analog function synthesizer for generating the exponential function

equivalent with:

$$I_{\text{OUT3}} = -\frac{512}{3}\frac{I_{\text{REF}}^2}{I_{\text{REF}} - \frac{I_{\text{IN}}}{6}} + \frac{I_{\text{IN}}^2}{64 I_{\text{REF}}} + \frac{29 I_{\text{REF}}}{3} + \frac{5 I_{\text{IN}}}{12}, \tag{7.98}$$

$$I_{\text{OUT4}} = \frac{I_{\text{OUT3}}^2}{8(I_{\text{REF}}/8)}. \tag{7.99}$$

The output current of the circuit from Fig. 7.12 can be expressed as follows:

$$I_{\text{OUT}} = \frac{I_{\text{OUT4}}^2}{8(I_{\text{REF}}/8)}. \tag{7.100}$$

Using the definition of the x variable, the previous relation becomes:

$$I_{\text{OUT}} = I_{\text{REF}}\left(\frac{512}{3}\frac{1}{1 - \frac{x}{16}} - \frac{x^2}{64} - \frac{5x}{12} - \frac{29}{3}\right)^4. \tag{7.101}$$

In conclusion, the analog function synthesizer shown in Fig. 7.12 approximates the exponential function:

$$I_{\text{OUT}} = I_{\text{REF}} a_{12}(x) \cong I_{\text{REF}} f_1(x). \tag{7.102}$$

7.2.13 Thirteenth Analog Function Synthesizer for Generating the Exponential Function

The 13th analog function synthesizer, designed for generating the exponential function, uses $a_{13}(x)$ approximation function described in Chap. 3.

The block diagram of the 13th analog function synthesizer for generating the exponential function is presented in Fig. 7.13.

The expressions of the currents from the circuit are:

$$I_{\mathrm{OUT1}} = \frac{I_{\mathrm{IN}}^2}{8I_{\mathrm{REF}}}, \tag{7.103}$$

$$I_{\mathrm{OUT2}} = \frac{262}{45}I_{\mathrm{OUT1}} = \frac{131I_{\mathrm{IN}}^2}{180I_{\mathrm{REF}}}, \tag{7.104}$$

$$I_{\mathrm{OUT3}} = \frac{4}{15}I_{\mathrm{OUT1}} = \frac{I_{\mathrm{IN}}^2}{30I_{\mathrm{REF}}}, \tag{7.105}$$

$$I_{\mathrm{OUT4}} = I_{\mathrm{REF}}\frac{\frac{131I_{\mathrm{IN}}^2}{180I_{\mathrm{REF}}} - \frac{7I_{\mathrm{IN}}}{6} + I_{\mathrm{REF}}}{\frac{I_{\mathrm{IN}}^2}{30I_{\mathrm{REF}}} - \frac{I_{IN}}{3} + I_{\mathrm{REF}}} \tag{7.106}$$

and

$$I_{\mathrm{OUT5}} = \frac{I_{\mathrm{IN}}^2}{8(3I_{\mathrm{REF}}/2)} = \frac{I_{\mathrm{IN}}^2}{12I_{\mathrm{REF}}}. \tag{7.107}$$

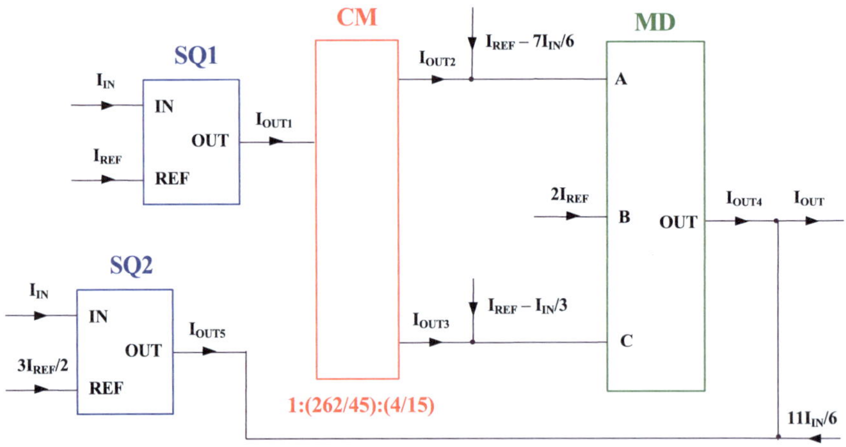

Fig. 7.13 Block diagram of the 13th analog function synthesizer for generating the exponential function

The output current of the circuit from Fig. 7.13 can be expressed as follows:

$$I_{\mathrm{OUT}} = I_{\mathrm{REF}} \frac{\frac{131 I_{\mathrm{IN}}^2}{180 I_{\mathrm{REF}}} - \frac{7 I_{\mathrm{IN}}}{6} + I_{\mathrm{REF}}}{\frac{I_{\mathrm{IN}}^2}{30 I_{\mathrm{REF}}} - \frac{I_{\mathrm{IN}}}{3} + I_{\mathrm{REF}}} + \frac{I_{\mathrm{IN}}^2}{12 I_{\mathrm{REF}}} + \frac{11 I_{\mathrm{IN}}}{6}. \tag{7.108}$$

Using the definition of x variable, the previous relation becomes:

$$I_{\mathrm{OUT}} = I_{\mathrm{REF}} \left(\frac{131x^2 - 210x + 180}{6x^2 - 60x + 180} + \frac{x^2 + 22x}{12} \right). \tag{7.109}$$

In conclusion, the analog function synthesizer shown in Fig. 7.13 approximates the exponential function:

$$I_{\mathrm{OUT}} = I_{\mathrm{REF}} a_{13}(x) \cong I_{\mathrm{REF}} f_1(x). \tag{7.110}$$

7.2.14 *Fourteenth Analog Function Synthesizer for Generating the Exponential Function*

The 14th analog function synthesizer, designed for generating the exponential function, uses $a_{14}(x)$ approximation function described in Chap. 3.

The block diagram of the 14th analog function synthesizer for generating the exponential function is presented in Fig. 7.14.

The expressions of the currents from the circuit are:

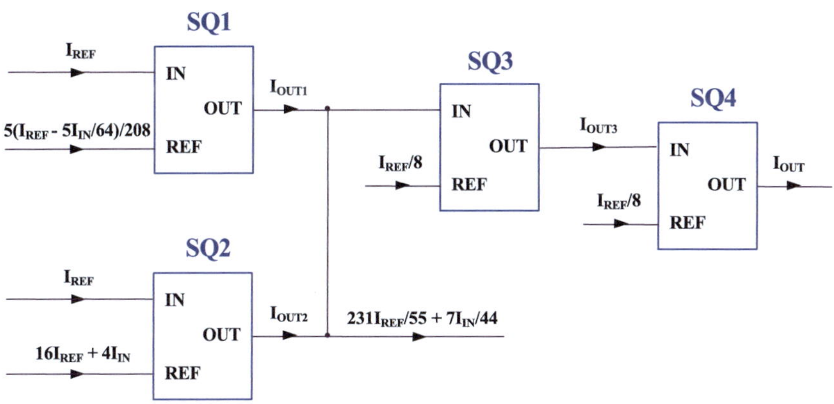

Fig. 7.14 Block diagram of the 14th analog function synthesizer for generating the exponential function

$$I_{OUT1} = \frac{26 I_{REF}^2}{5\left(I_{REF} - \frac{5 I_{IN}}{64}\right)}, \tag{7.111}$$

$$I_{OUT2} = \frac{I_{REF}^2}{8(16 I_{REF} + 4 I_{IN})} \tag{7.112}$$

and

$$I_{OUT3} = \frac{1}{I_{REF}}\left[\frac{26 I_{REF}^2}{5\left(I_{REF} - \frac{5 I_{IN}}{64}\right)} + \frac{I_{REF}^2}{8(16 I_{REF} + 4 I_{IN})} - \frac{231 I_{REF}}{55} - \frac{7 I_{IN}}{44}\right]^2. \tag{7.113}$$

The output current of the circuit from Fig. 7.14 can be expressed as follows:

$$I_{OUT} = \frac{I_{OUT3}^3}{8(I_{REF}/8)}. \tag{7.114}$$

Using the definition of x variable, the previous relation becomes:

$$I_{OUT} = I_{REF}\left(\frac{1664}{5}\frac{1}{64 - 5x} + \frac{1}{32}\frac{1}{x + 4} - \frac{35x + 924}{220}\right)^4. \tag{7.115}$$

In conclusion, the analog function synthesizer shown in Fig. 7.14 approximates the exponential function:

$$I_{OUT} = I_{REF} a_{14}(x) \cong I_{REF} f_1(x). \tag{7.116}$$

7.3 Analog Function Synthesizers for the Exponential Ramp Function

In order to implement in CMOS technology the exponential ramp activation function, the approximation functions described in Chap. 3 and the CMOS squaring and multiplies/divider circuits from Chap. 4 are used.

7.3.1 First Analog Function Synthesizer for Generating the Exponential Ramp Function (for a = 1)

The first analog function synthesizer, designed for generating the exponential ramp function (for $a = 1$), uses $a_{15}(x)$ approximation function described in Chap. 3.

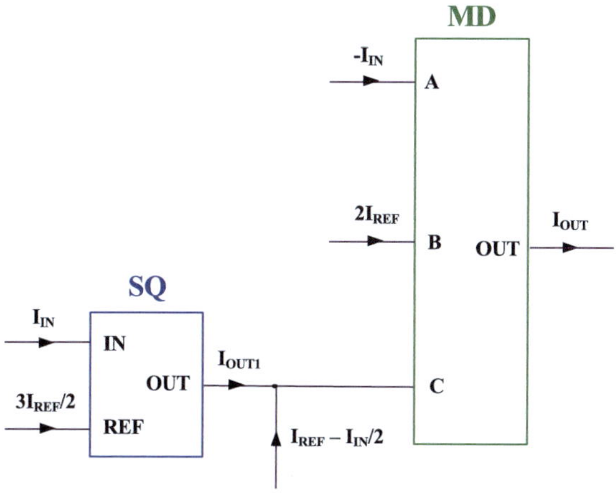

Fig. 7.15 Block diagram of the first analog function synthesizer for generating the exponential ramp function (for $a = 1$)

The block diagram of the first analog function synthesizer for generating the exponential ramp function (for $a = 1$) is presented in Fig. 7.15.

The expression of I_{OUT1} current is:

$$I_{OUT1} = \frac{I_{IN}^2}{8(3I_{REF}/2)} = \frac{I_{IN}^2}{12I_{REF}}. \tag{7.117}$$

The output current of the circuit from Fig. 7.15 can be expressed as follows:

$$I_{OUT} = I_{REF}\frac{-I_{IN}}{\frac{I_{IN}^2}{12I_{REF}} - \frac{I_{IN}}{2} + I_{REF}}. \tag{7.118}$$

Using the definition of x variable, the previous relation becomes:

$$I_{OUT} = -I_{REF}\frac{x}{\frac{x^2}{12} - \frac{x}{2} + 1}. \tag{7.119}$$

In conclusion, the analog function synthesizer shown in Fig. 7.15 approximates the exponential ramp function (for $a = 1$):

$$I_{OUT} = I_{REF}a_{15}(x) \cong I_{REF}f_3(x). \tag{7.120}$$

7.3.2 Second Analog Function Synthesizer for Generating the Exponential Ramp Function (for a = 1)

The second analog function synthesizer, designed for generating the exponential ramp function (for $a = 1$), uses $a_{16}(x)$ approximation function described in Chap. 3.

The block diagram of the second analog function synthesizer for generating the exponential ramp function (for $a = 1$) is presented in Fig. 7.16.

The expressions of the currents from the circuit are:

$$I_{OUT1} = \frac{I_{IN}^2}{8I_{REF}}, \tag{7.121}$$

$$I_{OUT2} = \frac{2}{15} I_{OUT1} = \frac{I_{IN}^2}{60I_{REF}}, \tag{7.122}$$

$$I_{OUT3} = -\frac{4}{5} I_{OUT1} = -\frac{I_{IN}^2}{10I_{REF}}, \tag{7.123}$$

$$I_{OUT4} = \frac{2}{5} I_{OUT1} = \frac{I_{IN}^2}{20I_{REF}} \tag{7.124}$$

and

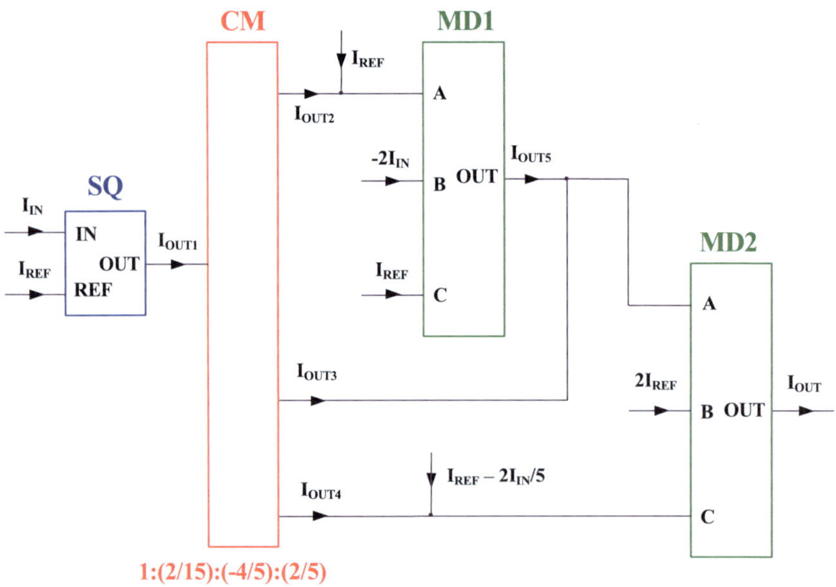

Fig. 7.16 Block diagram of the second analog function synthesizer for generating the exponential ramp function (for $a = 1$)

$$I_{\mathrm{OUT5}} = -\frac{I_{\mathrm{IN}}}{I_{\mathrm{REF}}}\left(\frac{I_{\mathrm{IN}}^2}{60 I_{\mathrm{REF}}} + I_{\mathrm{REF}}\right). \tag{7.125}$$

So, the output current of the circuit from Fig. 7.16 can be expressed as follows:

$$I_{\mathrm{OUT}} = I_{\mathrm{REF}} \frac{-\frac{I_{\mathrm{IN}}^3}{60 I_{\mathrm{REF}}^2} - \frac{I_{\mathrm{IN}}^2}{10 I_{\mathrm{REF}}} - I_{\mathrm{IN}}}{\frac{I_{\mathrm{IN}}^2}{20 I_{\mathrm{REF}}} - \frac{2 I_{\mathrm{IN}}}{5} + I_{\mathrm{REF}}}. \tag{7.126}$$

Using the definition of the x variable, the previous relation becomes:

$$I_{\mathrm{OUT}} = I_{\mathrm{REF}} \frac{-\frac{x^3}{60} - \frac{x^2}{10} - x}{\frac{x^2}{20} - \frac{2x}{5} + 1}. \tag{7.127}$$

In conclusion, the analog function synthesizer shown in Fig. 7.16 approximates the exponential ramp function (for $a = 1$):

$$I_{\mathrm{OUT}} = I_{\mathrm{REF}} a_{16}(x) \cong I_{\mathrm{REF}} f_3(x). \tag{7.128}$$

7.3.3 Third Analog Function Synthesizer for Generating the Exponential Ramp Function (for a = 1)

The third analog function synthesizer, designed for generating the exponential ramp function (for $a = 1$), uses $a_{17}(x)$ approximation function described in Chap. 3.

The block diagram of the second analog function synthesizer for generating the exponential ramp function (for $a = 1$) is presented in Fig. 7.17.

The expressions of the currents from the circuit are:

$$I_{\mathrm{OUT1}} = \frac{I_{\mathrm{IN}}^2}{8 I_{\mathrm{REF}}}, \tag{7.129}$$

$$I_{\mathrm{OUT2}} = \frac{2}{15} I_{\mathrm{OUT1}} = \frac{I_{\mathrm{IN}}^2}{60 I_{\mathrm{REF}}}, \tag{7.130}$$

$$I_{\mathrm{OUT3}} = -\frac{I_{\mathrm{IN}}}{I_{\mathrm{REF}}}\left(\frac{I_{\mathrm{IN}}^2}{60 I_{\mathrm{REF}}} + I_{\mathrm{REF}}\right), \tag{7.131}$$

$$I_{\mathrm{OUT4}} = \frac{2}{15} I_{\mathrm{OUT1}} = \frac{I_{\mathrm{IN}}^2}{60 I_{\mathrm{REF}}}, \tag{7.132}$$

$$I_{\mathrm{OUT5}} = -\frac{I_{\mathrm{IN}}}{2 I_{\mathrm{REF}}}\left(\frac{I_{\mathrm{IN}}^2}{60 I_{\mathrm{REF}}} + I_{\mathrm{REF}}\right) \tag{7.133}$$

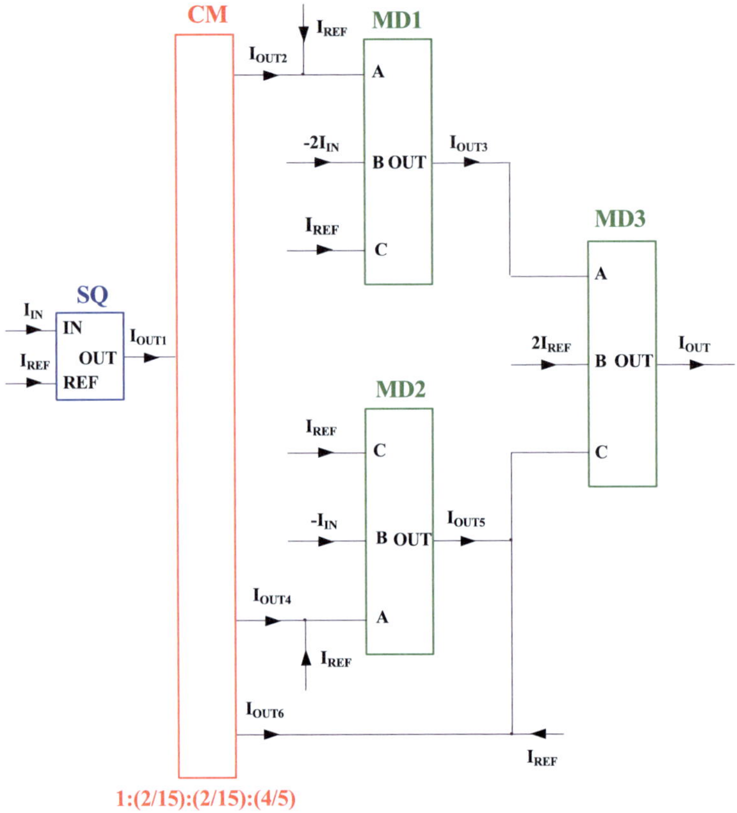

Fig. 7.17 Block diagram of the third analog function synthesizer for generating the exponential ramp function (for $a = 1$)

and

$$I_{\mathrm{OUT6}} = \frac{4}{5} I_{\mathrm{OUT1}} = \frac{I_{\mathrm{IN}}^2}{10 I_{\mathrm{REF}}}. \tag{7.134}$$

The output current of the circuit from Fig. 7.17 can be expressed as follows:

$$I_{\mathrm{OUT}} = I_{\mathrm{REF}} \frac{-\frac{I_{\mathrm{IN}}^3}{60 I_{\mathrm{REF}}^2} - I_{\mathrm{IN}}}{-\frac{I_{\mathrm{IN}}^3}{120 I_{\mathrm{REF}}^2} + \frac{I_{\mathrm{IN}}^2}{10 I_{\mathrm{REF}}} - \frac{I_{\mathrm{IN}}}{2} + I_{\mathrm{REF}}}. \tag{7.135}$$

Using the definition of the x variable, the previous relation becomes:

$$I_{\text{OUT}} = I_{\text{REF}} \frac{-\frac{x^3}{60} - x}{-\frac{x^3}{120} + \frac{x^2}{10} - \frac{x}{2} + 1}. \tag{7.136}$$

In conclusion, the analog function synthesizer shown in Fig. 7.17 approximates the exponential ramp function (for $a = 1$):

$$I_{\text{OUT}} = I_{\text{REF}} a_{17}(x) \cong I_{\text{REF}} f_3(x). \tag{7.137}$$

7.3.4 *Fourth Analog Function Synthesizer for Generating the Exponential Ramp Function (for a = 1)*

The fourth analog function synthesizer, designed for generating the exponential ramp function (for $a = 1$), uses $a_{18}(x)$ approximation function described in Chap. 3.

The block diagram of the fourth analog function synthesizer for generating the exponential ramp function (for $a = 1$) is presented in Fig. 7.18.

The expressions of the currents from the circuit are:

$$I_{\text{OUT1}} = \frac{I_{\text{IN}}^2}{8I_{\text{REF}}}, \tag{7.138}$$

$$I_{\text{OUT2}} = \frac{6}{7} I_{\text{OUT1}} = \frac{3I_{\text{IN}}^2}{28I_{\text{REF}}}, \tag{7.139}$$

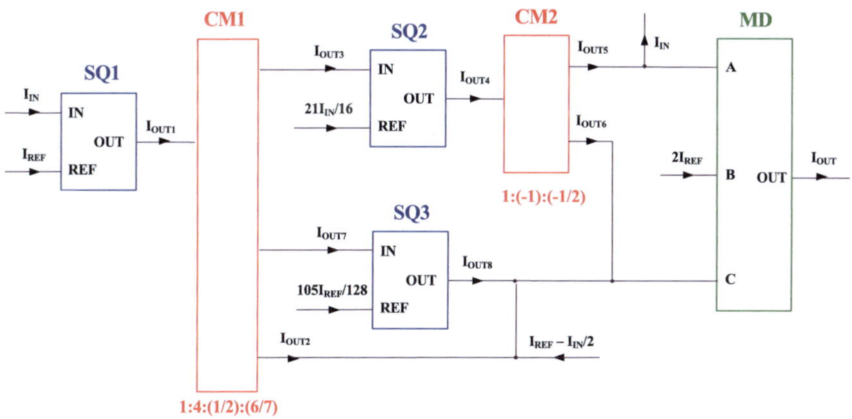

Fig. 7.18 Block diagram of the fourth analog function synthesizer for generating the exponential ramp function (for $a = 1$)

$$I_{\text{OUT3}} = 4I_{\text{OUT1}} = \frac{I_{\text{IN}}^2}{2I_{\text{REF}}}, \tag{7.140}$$

$$I_{\text{OUT4}} = \frac{I_{\text{OUT3}}^2}{8(21I_{\text{IN}}/16)} = \frac{I_{\text{IN}}^3}{42I_{\text{REF}}^2}, \tag{7.141}$$

$$I_{\text{OUT5}} = -I_{\text{OUT4}} = -\frac{I_{\text{IN}}^3}{42I_{\text{REF}}^2}, \tag{7.142}$$

$$I_{\text{OUT6}} = -\frac{1}{2}I_{\text{OUT4}} = -\frac{I_{\text{IN}}^3}{84I_{\text{REF}}^2}, \tag{7.143}$$

$$I_{\text{OUT7}} = \frac{I_{\text{OUT1}}}{2} = \frac{I_{\text{IN}}^2}{16I_{\text{REF}}} \tag{7.144}$$

and

$$I_{\text{OUT8}} = \frac{I_{\text{OUT7}}^2}{8(105I_{\text{REF}}/128)} = \frac{I_{\text{IN}}^4}{1680I_{\text{REF}}^3}. \tag{7.145}$$

So, the output current of the circuit from Fig. 7.18 can be expressed as follows:

$$I_{\text{OUT}} = I_{\text{REF}} \frac{-\frac{I_{\text{IN}}^3}{42I_{\text{REF}}^2} - I_{\text{IN}}}{\frac{I_{\text{IN}}^4}{1680I_{\text{REF}}^3} - \frac{I_{\text{IN}}^3}{84I_{\text{REF}}^2} + \frac{3I_{\text{IN}}^2}{28I_{\text{REF}}} - \frac{I_{\text{IN}}}{2} + I_{\text{REF}}}. \tag{7.146}$$

Using the definition of the x variable, the previous relation becomes:

$$I_{\text{OUT}} = I_{\text{REF}} \frac{-\frac{x^3}{42} - x}{\frac{x^4}{1680} - \frac{x^3}{84} + \frac{3x^2}{28} - \frac{x}{2} + 1}. \tag{7.147}$$

In conclusion, the analog function synthesizer shown in Fig. 7.18 approximates the exponential ramp function (for $a = 1$):

$$I_{\text{OUT}} = I_{\text{REF}} a_{18}(x) \cong I_{\text{REF}} f_3(x). \tag{7.148}$$

7.3.5 *First Analog Function Synthesizer for Generating the Exponential Ramp Function (for a = 2)*

The first analog function synthesizer, designed for generating the exponential ramp function (for $a = 2$), uses $a_{19}(x)$ approximation function described in Chap. 3.

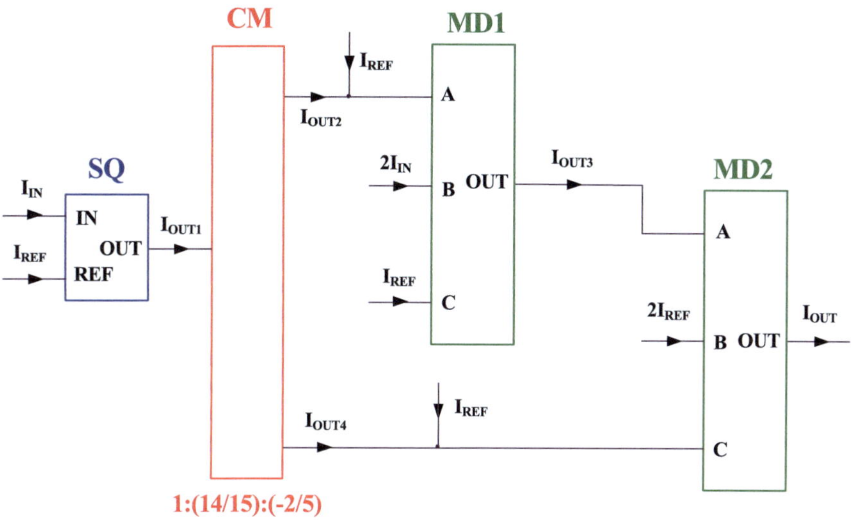

Fig. 7.19 Block diagram of the first analog function synthesizer for generating the exponential ramp function (for $a = 2$)

The block diagram of the first analog function synthesizer for generating the exponential ramp function (for $a = 2$) is presented in Fig. 7.19.

The expressions of the currents from the circuit are:

$$I_{\mathrm{OUT1}} = \frac{I_{\mathrm{IN}}^2}{8I_{\mathrm{REF}}}, \tag{7.149}$$

$$I_{\mathrm{OUT2}} = \frac{6}{7}I_{\mathrm{OUT1}} = \frac{3I_{\mathrm{IN}}^2}{28I_{\mathrm{REF}}}, \tag{7.150}$$

$$I_{\mathrm{OUT3}} = 4I_{\mathrm{OUT1}} = \frac{I_{\mathrm{IN}}^2}{2I_{\mathrm{REF}}}, \tag{7.151}$$

$$I_{\mathrm{OUT4}} = \frac{I_{\mathrm{OUT3}}^2}{8(21I_{\mathrm{IN}}/16)} = \frac{I_{\mathrm{IN}}^3}{42I_{\mathrm{REF}}^2}, \tag{7.152}$$

$$I_{OUT5} = -I_{OUT4} = -\frac{I_{IN}^3}{42I_{REF}^2}, \tag{7.153}$$

$$I_{\mathrm{OUT6}} = -\frac{1}{2}I_{\mathrm{OUT4}} = -\frac{I_{\mathrm{IN}}^3}{84I_{\mathrm{REF}}^2}, \tag{7.154}$$

$$I_{\mathrm{OUT7}} = \frac{I_{\mathrm{OUT1}}}{2} = \frac{I_{\mathrm{IN}}^2}{16I_{\mathrm{REF}}} \tag{7.155}$$

and

$$I_{\mathrm{OUT8}} = \frac{I_{\mathrm{OUT7}}^2}{8(105I_{\mathrm{REF}}/128)} = \frac{I_{\mathrm{IN}}^4}{1680I_{\mathrm{REF}}^3}. \tag{7.156}$$

So, the output current of the circuit from Fig. 7.19 can be expressed as follows:

$$I_{\mathrm{OUT}} = I_{\mathrm{REF}} \frac{-\frac{I_{\mathrm{IN}}^3}{42I_{\mathrm{REF}}^2} - I_{\mathrm{IN}}}{\frac{I_{\mathrm{IN}}^4}{1680I_{\mathrm{REF}}^3} - \frac{I_{\mathrm{IN}}^3}{84I_{\mathrm{REF}}^2} + \frac{3I_{\mathrm{IN}}^2}{28I_{\mathrm{REF}}} - \frac{I_{\mathrm{IN}}}{2} + I_{\mathrm{REF}}}. \tag{7.157}$$

Using the definition of the x variable, the previous relation becomes:

$$I_{\mathrm{OUT}} = I_{\mathrm{REF}} \frac{-\frac{x^3}{42} - x}{\frac{x^4}{1680} - \frac{x^3}{84} + \frac{3x^2}{28} - \frac{x}{2} + 1}. \tag{7.158}$$

In conclusion, the analog function synthesizer shown in Fig. 7.19 approximates the exponential ramp function (for $a = 2$):

$$I_{\mathrm{OUT}} = I_{\mathrm{REF}} a_{19}(x) \cong I_{\mathrm{REF}} f_4(x). \tag{7.159}$$

7.3.6 First Analog Function Synthesizer for Generating the Exponential Ramp Function (for a = 1/2)

The first analog function synthesizer, designed for generating the exponential ramp function (for $a = 1/2$), uses $a_{20}(x)$ approximation function described in Chap. 3.

The block diagram of the first analog function synthesizer for generating the exponential ramp function (for $a = 1/2$) is presented in Fig. 7.20.

The expression of I_{OUT1} current is:

$$I_{OUT1} = \frac{I_{IN}^2}{8(3I_{REF})} = \frac{I_{IN}^2}{24I_{REF}}. \tag{7.160}$$

The output current of the circuit from Fig. 7.20 can be expressed as follows:

$$I_{\mathrm{OUT}} = I_{\mathrm{REF}} \frac{-I_{\mathrm{IN}}}{\frac{I_{\mathrm{IN}}^2}{24I_{\mathrm{REF}}} - \frac{I_{\mathrm{IN}}}{2} + 2I_{\mathrm{REF}}}. \tag{7.161}$$

Using the definition of x variable, the previous relation becomes:

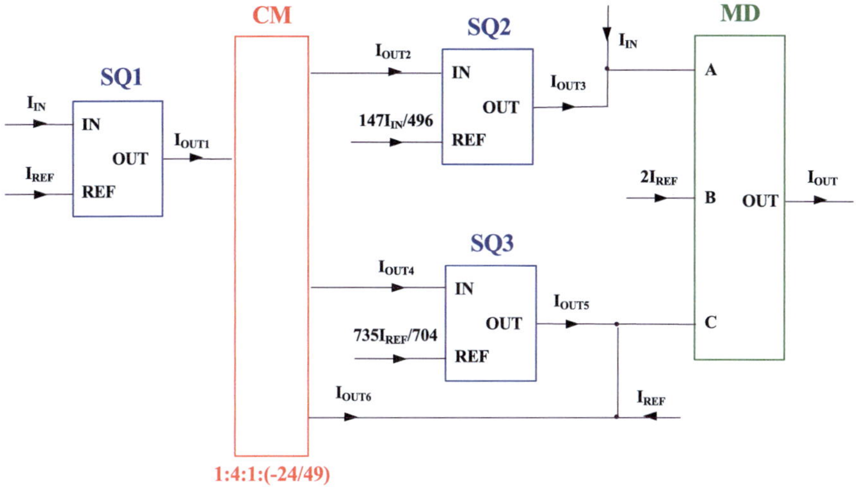

Fig. 7.20 Block diagram of the first analog function synthesizer for generating the exponential ramp function (for $a = 1/2$)

$$I_{\mathrm{OUT}} = -I_{\mathrm{REF}} \frac{x}{\frac{x^2}{24} - \frac{x}{2} + 2}. \tag{7.162}$$

In conclusion, the analog function synthesizer shown in Fig. 7.20 approximates the exponential ramp function (for $a = 1/2$):

$$I_{\mathrm{OUT}} = I_{\mathrm{REF}} a_{20}(x) \cong I_{\mathrm{REF}} f_5(x). \tag{7.163}$$

7.3.7 *Second Analog Function Synthesizer for Generating the Exponential Ramp Function (for a = 1/2)*

The second analog function synthesizer, designed for generating the exponential ramp function (for $a = 1/2$), uses $a_{21}(x)$ approximation function described in Chap. 3.

The block diagram of the second analog function synthesizer for generating the exponential ramp function (for $a = 1/2$) is presented in Fig. 7.21.

The expressions of the currents from the circuit are:

$$I_{\mathrm{OUT1}} = \frac{I_{\mathrm{IN}}^2}{32 I_{\mathrm{REF}}}, \tag{7.164}$$

$$I_{\mathrm{OUT2}} = \frac{2}{15} I_{\mathrm{OUT1}} = \frac{I_{\mathrm{IN}}^2}{240 I_{\mathrm{REF}}}, \tag{7.165}$$

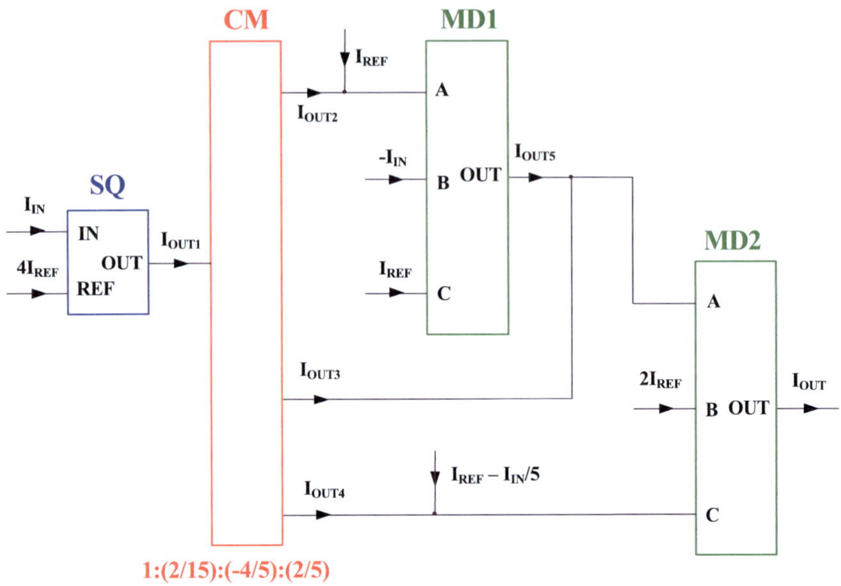

Fig. 7.21 Block diagram of the second analog function synthesizer for generating the exponential ramp function (for $a = 1/2$)

$$I_{\mathrm{OUT3}} = -\frac{4}{5} I_{\mathrm{OUT1}} = -\frac{I_{\mathrm{IN}}^2}{40 I_{\mathrm{REF}}}, \tag{7.166}$$

$$I_{\mathrm{OUT4}} = \frac{2}{5} I_{\mathrm{OUT1}} = \frac{I_{\mathrm{IN}}^2}{80 I_{\mathrm{REF}}} \tag{7.167}$$

and

$$I_{\mathrm{OUT5}} = -\frac{I_{\mathrm{IN}}}{2 I_{\mathrm{REF}}} \left(\frac{I_{\mathrm{IN}}^2}{240 I_{\mathrm{REF}}} + I_{\mathrm{REF}} \right). \tag{7.168}$$

So, the output current of the circuit from Fig. 7.21 can be expressed as follows:

$$I_{\mathrm{OUT}} = I_{\mathrm{REF}} \frac{-\frac{I_{\mathrm{IN}}^3}{480 I_{\mathrm{REF}}^2} - \frac{I_{\mathrm{IN}}^2}{40 I_{\mathrm{REF}}} - \frac{I_{\mathrm{IN}}}{2}}{\frac{I_{\mathrm{IN}}^2}{80 I_{\mathrm{REF}}} - \frac{I_{\mathrm{IN}}}{5} + I_{\mathrm{REF}}}. \tag{7.169}$$

Using the definition of the x variable, the previous relation becomes:

$$I_{\mathrm{OUT}} = I_{\mathrm{REF}} \frac{-\frac{x^3}{480} - \frac{x^2}{40} - \frac{x}{2}}{\frac{x^2}{80} - \frac{x}{5} + 1}. \tag{7.170}$$

In conclusion, the analog function synthesizer shown in Fig. 7.21 approximates the exponential ramp function (for $a = 1/2$):

$$I_{OUT} = I_{REF} a_{21}(x) \cong I_{REF} f_5(x). \tag{7.171}$$

7.3.8 Third Analog Function Synthesizer for Generating the Exponential Ramp Function (for a = 1/2)

The third analog function synthesizer, designed for generating the exponential ramp function (for $a = 1/2$), uses $a_{22}(x)$ approximation function described in Chap. 3.

The block diagram of the third analog function synthesizer for generating the exponential ramp function (for $a = 1/2$) is presented in Fig. 7.22.

The expressions of the currents from the circuit are:

$$I_{OUT1} = \frac{I_{IN}^2}{32 I_{REF}}, \tag{7.172}$$

$$I_{OUT2} = \frac{2}{15} I_{OUT1} = \frac{I_{IN}^2}{240 I_{REF}}, \tag{7.173}$$

$$I_{OUT3} = -\frac{I_{IN}}{2 I_{REF}}, \tag{7.174}$$

$$I_{OUT4} = \frac{2}{15} I_{OUT1} = \frac{I_{IN}^2}{240 I_{REF}}, \tag{7.175}$$

$$I_{OUT5} = -\frac{I_{IN}}{4 I_{REF}} \left(\frac{I_{IN}^2}{240 I_{REF}} + I_{REF} \right) \tag{7.176}$$

and

$$I_{OUT6} = \frac{4}{5} I_{OUT1} = \frac{I_{IN}^2}{40 I_{REF}}. \tag{7.177}$$

The output current of the circuit from Fig. 7.22 can be expressed as follows:

$$I_{OUT} = I_{REF} \frac{-\frac{I_{IN}^3}{480 I_{REF}^2} - \frac{I_{IN}}{2}}{-\frac{I_{IN}^3}{960 I_{REF}^2} + \frac{I_{IN}^2}{40 I_{REF}} - \frac{I_{IN}}{4} + I_{REF}}. \tag{7.178}$$

Using the definition of the x variable, the previous relation becomes:

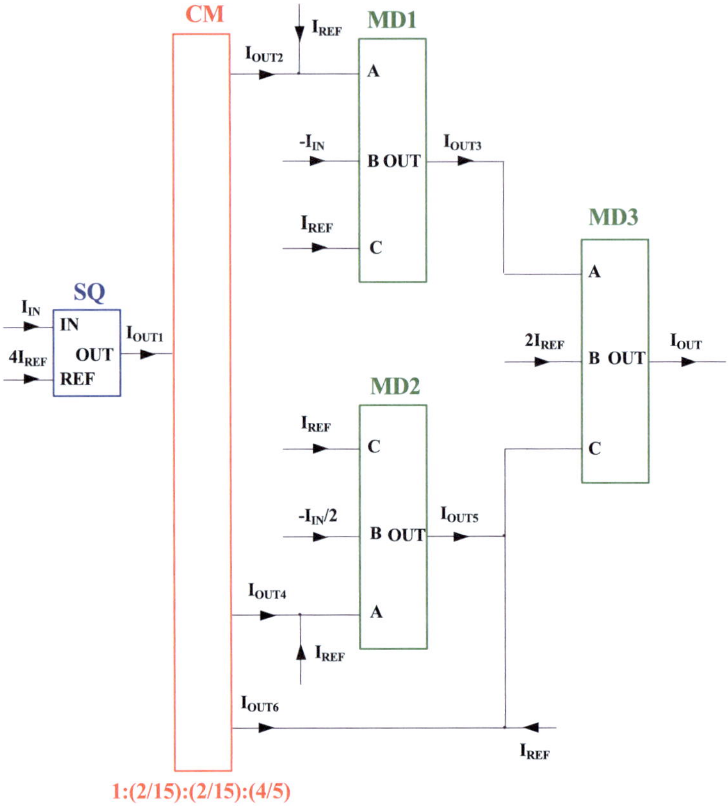

Fig. 7.22 Block diagram of the third analog function synthesizer for generating the exponential ramp function (for $a = 1/2$)

$$I_{\mathrm{OUT}} = I_{\mathrm{REF}} \frac{-\frac{x^3}{480} - \frac{x}{2}}{-\frac{x^3}{960} + \frac{x^2}{40} - \frac{x}{4} + 1}. \tag{7.179}$$

In conclusion, the analog function synthesizer shown in Fig. 7.22 approximates the exponential ramp function (for $a = 1/2$):

$$I_{\mathrm{OUT}} = I_{\mathrm{REF}} a_{22}(x) \cong I_{\mathrm{REF}} f_5(x). \tag{7.180}$$

7.4 Analog Function Synthesizers for the Hyperbolic Sinusoidal Function

In order to implement in CMOS technology the hyperbolic sinusoidal function, the approximation functions described in Chap. 3 and the CMOS squaring and multiplies/divider circuits from Chap. 4 are used.

7.4.1 *First Analog Function Synthesizer for Generating the Hyperbolic Sinusoidal Function*

The first analog function synthesizer, designed for generating the hyperbolic sinusoidal function, uses $a_{23}(x)$ approximation function described in Chap. 3.

The block diagram of the first analog function synthesizer for generating the hyperbolic sinusoidal function is presented in Fig. 7.23.

The expressions of the currents from the circuit are:

$$I_{OUT1} = \frac{I_{IN}^2}{8I_{REF}}, \tag{7.181}$$

$$I_{OUT2} = \frac{14}{15}I_{OUT1} = \frac{7I_{IN}^2}{60I_{REF}}, \tag{7.182}$$

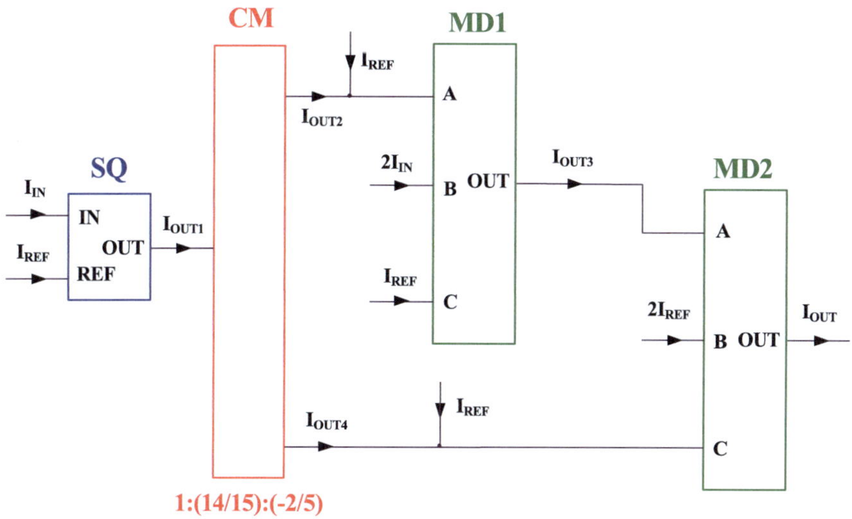

Fig. 7.23 Block diagram of the first analog function synthesizer for generating the hyperbolic sinusoidal function

$$I_{OUT3} = \frac{I_{IN}}{I_{REF}}\left(\frac{7I_{IN}^2}{60I_{REF}} + I_{REF}\right) \tag{7.183}$$

and

$$I_{OUT4} = -\frac{2}{5}I_{OUT1} = -\frac{I_{IN}^2}{20I_{REF}}. \tag{7.184}$$

So, the output current of the circuit from Fig. 7.23 can be expressed as follows:

$$I_{OUT} = I_{REF}\frac{\frac{7I_{IN}^3}{60I_{REF}^2} + I_{IN}}{I_{REF} - \frac{I_{IN}^2}{20I_{REF}}}. \tag{7.185}$$

Using the definition of the x variable, the previous relation becomes:

$$I_{OUT} = I_{REF}\frac{\frac{7x^3}{60} + x}{1 - \frac{x^2}{20}}. \tag{7.186}$$

In conclusion, the analog function synthesizer shown in Fig. 7.23 approximates the hyperbolic sinusoidal function:

$$I_{OUT} = I_{REF}a_{23}(x) \cong I_{REF}f_6(x). \tag{7.187}$$

7.4.2 Second Analog Function Synthesizer for Generating the Hyperbolic Sinusoidal Function

The second analog function synthesizer, designed for generating the hyperbolic sinusoidal function, uses $a_{24}(x)$ approximation function described in Chap. 3.

The block diagram of the second analog function synthesizer for generating the hyperbolic sinusoidal function is presented in Fig. 7.24.

The expressions of the currents from the circuit are:

$$I_{OUT1} = \frac{I_{IN}^2}{8I_{REF}}, \tag{7.188}$$

$$I_{OUT2} = 4I_{OUT1} = \frac{I_{IN}^2}{2I_{REF}}, \tag{7.189}$$

$$I_{OUT3} = \frac{I_{OUT2}^2}{8(147I_{IN}/496)} = \frac{31I_{IN}^3}{294I_{REF}^2}, \tag{7.190}$$

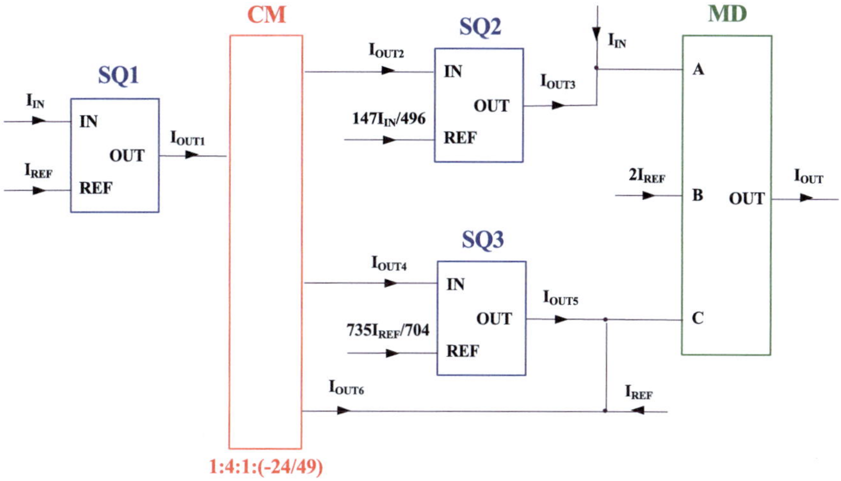

Fig. 7.24 Block diagram of the second analog function synthesizer for generating the hyperbolic sinusoidal function

$$I_{\mathrm{OUT4}} = I_{\mathrm{OUT1}} = \frac{I_{\mathrm{IN}}^2}{8I_{\mathrm{REF}}}, \tag{7.191}$$

$$I_{\mathrm{OUT5}} = \frac{I_{\mathrm{OUT4}}^2}{8(735I_{\mathrm{REF}}/704)} = \frac{11I_{\mathrm{IN}}^4}{5880I_{\mathrm{REF}}^3} \tag{7.192}$$

and

$$I_{\mathrm{OUT6}} = -\frac{24}{49}I_{\mathrm{OUT1}} = -\frac{3}{49}\frac{I_{\mathrm{IN}}^2}{I_{\mathrm{REF}}}. \tag{7.193}$$

So, the output current of the circuit from Fig. 7.24 can be expressed as follows:

$$I_{\mathrm{OUT}} = I_{\mathrm{REF}}\frac{\frac{31I_{\mathrm{IN}}^3}{294I_{\mathrm{REF}}^2} + I_{\mathrm{IN}}}{\frac{11I_{\mathrm{IN}}^4}{5880I_{\mathrm{REF}}^3} - \frac{3I_{\mathrm{IN}}^2}{49I_{\mathrm{REF}}} + I_{\mathrm{REF}}}. \tag{7.194}$$

Using the definition of the x variable, the previous relation becomes:

$$I_{\mathrm{OUT}} = I_{\mathrm{REF}}\frac{\frac{31x^3}{294} + x}{1 - \frac{3x^2}{49} + \frac{11x^4}{5880}}. \tag{7.195}$$

In conclusion, the analog function synthesizer shown in Fig. 7.24 approximates the hyperbolic sinusoidal function:

$$I_{\text{OUT}} = I_{\text{REF}} a_{24}(x) \cong I_{\text{REF}} f_6(x). \tag{7.196}$$

7.4.3 Third Analog Function Synthesizer for Generating the Hyperbolic Sinusoidal Function

The third analog function synthesizer, designed for generating the hyperbolic sinusoidal function, uses $a_{25}(x)$ approximation function described in Chap. 3.

The block diagram of the third analog function synthesizer for generating the hyperbolic sinusoidal function is presented in Fig. 7.25.

The expressions of the currents from the circuit are:

$$I_{\text{OUT1}} = \frac{I_{\text{IN}}^2}{8(21\,I_{\text{REF}}/4)} = \frac{I_{\text{IN}}^2}{42 I_{\text{REF}}}, \tag{7.197}$$

$$I_{\text{OUT2}} = \frac{(8 I_{\text{REF}})^2}{8(I_{\text{OUT1}} - I_{\text{REF}})} = \frac{8 I_{\text{REF}}^2}{\frac{I_{\text{IN}}^2}{42 I_{\text{REF}}} - I_{\text{REF}}}, \tag{7.198}$$

$$I_{\text{OUT3}} = \frac{147}{80} I_{\text{OUT2}} = \frac{147}{10} \frac{I_{\text{REF}}^2}{\frac{I_{\text{IN}}^2}{42 I_{\text{REF}}} - I_{\text{REF}}} \tag{7.199}$$

and

$$I_{\text{OUT4}} = \frac{I_{\text{IN}}^2}{8(15 I_{\text{REF}}/22)} = \frac{11 I_{\text{IN}}^2}{60 I_{\text{REF}}}. \tag{7.200}$$

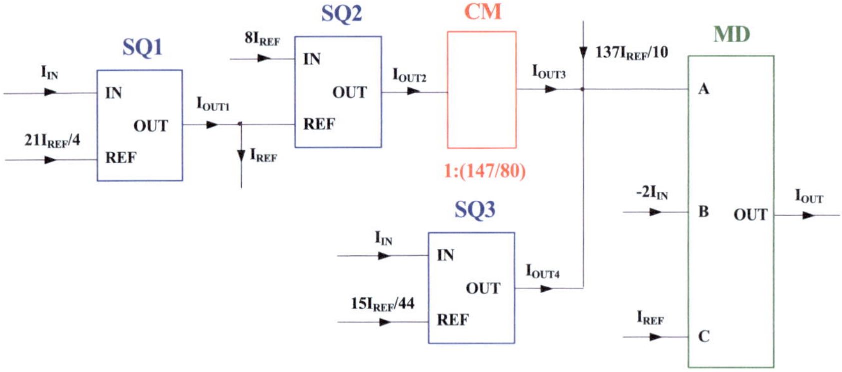

Fig. 7.25 Block diagram of the third analog function synthesizer for generating the hyperbolic sinusoidal function

So, the output current of the circuit from Fig. 7.25 can be expressed as follows:

$$I_{OUT} = -\frac{I_{IN}}{I_{REF}}\left(I_{OUT3} + I_{OUT4} + \frac{137}{10}I_{REF}\right). \tag{7.201}$$

Using the definition of the x variable, the previous relation becomes:

$$I_{OUT} = I_{REF}\frac{x}{10}\left(\frac{6174}{42 - x^2} - \frac{822 + 11x^2}{6}\right). \tag{7.202}$$

In conclusion, the analog function synthesizer shown in Fig. 7.25 approximates the hyperbolic sinusoidal function:

$$I_{OUT} = I_{REF}a_{25}(x) \cong I_{REF}f_6(x). \tag{7.203}$$

7.4.4 Fourth Analog Function Synthesizer for Generating the Hyperbolic Sinusoidal Function

The fourth analog function synthesizer, designed for generating the hyperbolic sinusoidal function, uses $a_{26}(x)$ approximation function described in Chap. 3.

The block diagram of the fourth analog function synthesizer for generating the hyperbolic sinusoidal function is presented in Fig. 7.26.

The expressions of the currents from the circuit are:

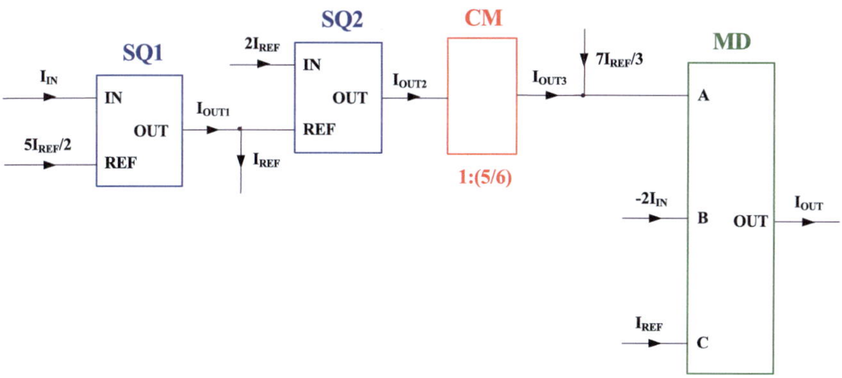

Fig. 7.26 Block diagram of the fourth analog function synthesizer for generating the hyperbolic sinusoidal function

$$I_{OUT1} = \frac{I_{IN}^2}{8(5I_{REF}/2)} = \frac{I_{IN}^2}{20I_{REF}}, \tag{7.204}$$

$$I_{OUT2} = \frac{4I_{REF}^2}{\frac{I_{IN}^2}{20I_{REF}} - I_{REF}} \tag{7.205}$$

and

$$I_{OUT3} = \frac{5}{6}I_{OUT2} = \frac{10}{3}\frac{I_{REF}^2}{\frac{I_{IN}^2}{20I_{REF}} - I_{REF}}. \tag{7.206}$$

So, the output current of the circuit from Fig. 7.26 can be expressed as follows:

$$I_{OUT} = -\frac{I_{IN}}{I_{REF}}\left(\frac{10}{3}\frac{I_{REF}^2}{\frac{I_{IN}^2}{20I_{REF}} - I_{REF}} + \frac{7I_{REF}}{3}\right). \tag{7.207}$$

Using the definition of the x variable, the previous relation becomes:

$$I_{OUT} = I_{REF}x\left(\frac{10}{3}\frac{1}{1 - \frac{x^2}{20}} - \frac{7}{3}\right). \tag{7.208}$$

In conclusion, the analog function synthesizer shown in Fig. 7.26 approximates the hyperbolic sinusoidal function:

$$I_{OUT} = I_{REF}a_{26}(x) \cong I_{REF}f_6(x). \tag{7.209}$$

7.5 Analog Function Synthesizers for the Hyperbolic Cosinusoidal Function

In order to implement in CMOS technology the hyperbolic cosinusoidal function, the approximation functions described in Chap. 3 and the CMOS squaring and multiplies/divider circuits from Chap. 4 are used.

7.5.1 First Analog Function Synthesizer for Generating the Hyperbolic Cosinusoidal Function

The first analog function synthesizer, designed for generating the hyperbolic cosinusoidal function, uses $a_{27}(x)$ approximation function described in Chap. 3.

The block diagram of the first analog function synthesizer for generating the hyperbolic cosinusoidal function is presented in Fig. 7.27.

The expressions of the currents from the circuit are:

$$I_{\mathrm{OUT1}} = \frac{I_{\mathrm{IN}}^2}{8(I_{\mathrm{REF}}/8)} = \frac{I_{\mathrm{IN}}^2}{I_{\mathrm{REF}}}, \tag{7.210}$$

$$I_{\mathrm{OUT2}} = I_{\mathrm{OUT1}} = \frac{I_{\mathrm{IN}}^2}{I_{\mathrm{REF}}}, \tag{7.211}$$

$$I_{\mathrm{OUT3}} = \frac{I_{\mathrm{OUT2}}^2}{8(75I_{\mathrm{IN}}/244)} = \frac{61I_{\mathrm{IN}}^3}{150I_{\mathrm{REF}}^2}, \tag{7.212}$$

$$I_{\mathrm{OUT4}} = \frac{1}{4}I_{\mathrm{OUT1}} = \frac{I_{\mathrm{IN}}^2}{4I_{\mathrm{REF}}}, \tag{7.213}$$

$$I_{\mathrm{OUT5}} = \frac{I_{\mathrm{OUT4}}^2}{8(25I_{\mathrm{REF}}/16)} = \frac{I_{\mathrm{IN}}^4}{200I_{\mathrm{REF}}^3} \tag{7.214}$$

and

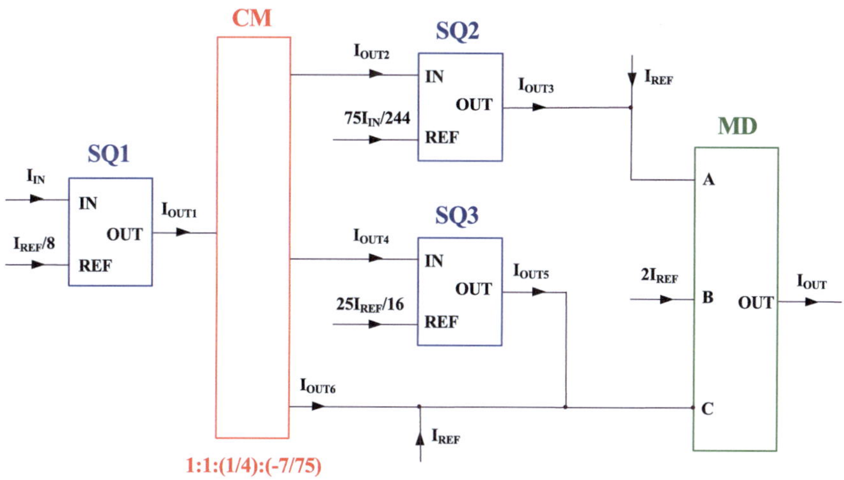

Fig. 7.27 Block diagram of the first analog function synthesizer for generating the hyperbolic cosinusoidal function

$$I_{\mathrm{OUT6}} = -\frac{7}{75} I_{\mathrm{OUT1}} = -\frac{7 I_{\mathrm{IN}}^2}{75 I_{\mathrm{REF}}}. \tag{7.215}$$

So, the output current of the circuit from Fig. 7.27 can be expressed as follows:

$$I_{\mathrm{OUT}} = I_{\mathrm{REF}} \frac{\frac{61 I_{\mathrm{IN}}^3}{150 I_{\mathrm{REF}}^2} + I_{\mathrm{REF}}}{I_{\mathrm{REF}} - \frac{7 I_{\mathrm{IN}}^2}{75 I_{\mathrm{REF}}} + \frac{I_{\mathrm{IN}}^4}{200 I_{\mathrm{REF}}^3}}. \tag{7.216}$$

Using the definition of the x variable, the previous relation becomes:

$$I_{\mathrm{OUT}} = I_{\mathrm{REF}} \frac{\frac{61x^3}{150} + 1}{1 - \frac{7x^2}{75} + \frac{x^4}{200}}. \tag{7.217}$$

In conclusion, the analog function synthesizer shown in Fig. 7.27 approximates the hyperbolic cosinusoidal function:

$$I_{\mathrm{OUT}} = I_{\mathrm{REF}} a_{27}(x) \cong I_{\mathrm{REF}} f_7(x). \tag{7.218}$$

7.5.2 *Second Analog Function Synthesizer for Generating the Hyperbolic Cosinusoidal Function*

The second analog function synthesizer, designed for generating the hyperbolic cosinusoidal function, uses $a_{28}(x)$ approximation function described in Chap. 3.

The block diagram of the second analog function synthesizer for generating the hyperbolic cosinusoidal function is presented in Fig. 7.28.

The expressions of the currents from the circuit are:

$$I_{\mathrm{OUT1}} = \frac{I_{\mathrm{IN}}^2}{8(I_{\mathrm{REF}}/4)} = \frac{I_{\mathrm{IN}}^2}{2 I_{\mathrm{REF}}}, \tag{7.219}$$

$$I_{\mathrm{OUT2}} = \frac{14}{15} I_{\mathrm{OUT1}} = \frac{7 I_{\mathrm{IN}}^2}{15 I_{\mathrm{REF}}}, \tag{7.220}$$

$$I_{\mathrm{OUT3}} = I_{\mathrm{OUT1}} = \frac{I_{\mathrm{IN}}^2}{2 I_{\mathrm{REF}}}, \tag{7.221}$$

$$I_{\mathrm{OUT4}} = \frac{I_{\mathrm{OUT3}}^2}{8(5 I_{\mathrm{REF}}/4)} = \frac{I_{\mathrm{IN}}^4}{40 I_{\mathrm{REF}}^3} \tag{7.222}$$

and

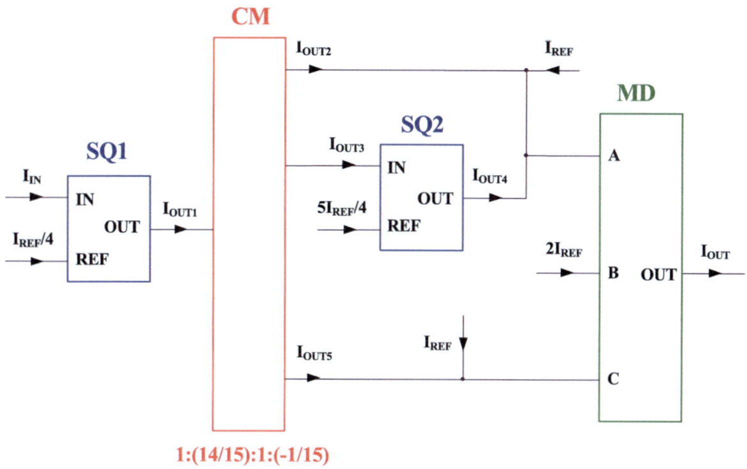

Fig. 7.28 Block diagram of the second analog function synthesizer for generating the hyperbolic cosinusoidal function

$$I_{OUT5} = -\frac{1}{15} I_{OUT1} = -\frac{I_{IN}^2}{30 I_{REF}}. \tag{7.223}$$

The output current of the circuit from Fig. 7.28 can be expressed as follows:

$$I_{OUT} = I_{REF} \frac{\frac{I_{IN}^4}{40 I_{REF}^3} + \frac{7 I_{IN}^2}{15 I_{REF}} + I_{REF}}{I_{REF} - \frac{I_{IN}^2}{30 I_{REF}}}. \tag{7.224}$$

Using the definition of x variable, the previous relation becomes:

$$I_{OUT} = I_{REF} \frac{\frac{x^4}{40} + \frac{7x^3}{15} + 1}{1 - \frac{x^2}{30}}. \tag{7.225}$$

In conclusion, the analog function synthesizer shown in Fig. 7.28 approximates the hyperbolic cosinusoidal function:

$$I_{OUT} = I_{REF} a_{28}(x) \cong I_{REF} f_7(x). \tag{7.226}$$

7.6 Analog Function Synthesizers for Continuous Mathematical Functions

In order to implement in CMOS technology a continuous mathematical function $f(x)$, the approximation functions described in Chap. 3 and the CMOS squaring and multiplies/divider circuits from Chap. 4 are used.

7.6.1 *First Analog Function Synthesizer for Generating the Continuous Mathematical Functions*

The first analog function synthesizer, designed for generating continuous mathematical functions $f(x)$, uses $a_{29}(x)$ approximation function described in Chap. 2.

The block diagram of the first analog function synthesizer for generating continuous mathematical functions $f(x)$ is presented in Fig. 7.29.

The expression of the I_{OUT1} current is:

$$I_{\mathrm{OUT1}} = \frac{c^3 I_{\mathrm{REF}}^2}{d^2\left(I_{\mathrm{REF}} - \frac{dI_{\mathrm{IN}}}{c}\right)}. \tag{7.227}$$

So, the output current of the circuit from Fig. 7.29 can be expressed as follows:

$$I_{\mathrm{OUT}} = \frac{c^3 I_{\mathrm{REF}}^2}{d^2\left(I_{\mathrm{REF}} - \frac{dI_{\mathrm{IN}}}{c}\right)} + \left(b - \frac{c^2}{d}\right)I_{\mathrm{IN}} + \left(a - \frac{c^3}{d^2}\right)I_{\mathrm{REF}}. \tag{7.228}$$

Using the definition of the x variable, the previous relation becomes:

$$I_{\mathrm{OUT}} = I_{\mathrm{REF}}\left[\frac{c^3}{d^2}\frac{1}{1 - \frac{dx}{c}} + \left(b - \frac{c^2}{d}\right)x + \left(a - \frac{c^3}{d^2}\right)\right]. \tag{7.229}$$

In conclusion, the analog function synthesizer shown in Fig. 7.29 approximates continuous mathematical functions $f(x)$:

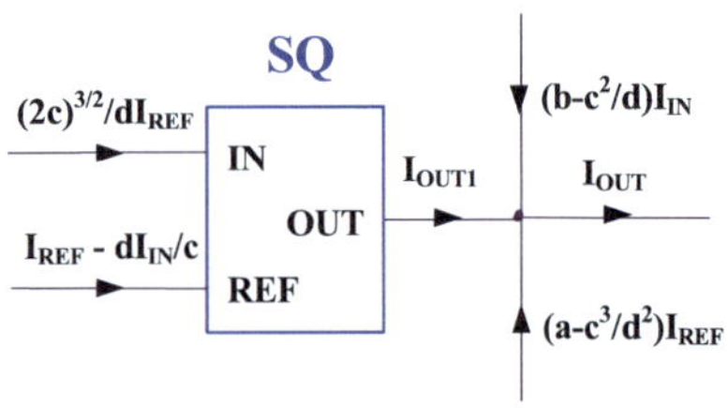

Fig. 7.29 Block diagram of the first analog function synthesizer for generating continuous mathematical functions $f(x)$

$$I_{OUT} = I_{REF} a_{29}(x) \cong I_{REF} f(x). \tag{7.230}$$

7.6.2 Second Analog Function Synthesizer for Generating the Continuous Mathematical Functions

The second analog function synthesizer, designed for generating continuous mathematical functions $f(x)$, uses $a_{30}(x)$ approximation function described in Chap. 3.

The block diagram of the second analog function synthesizer for generating continuous mathematical functions $f(x)$ is presented in Fig. 7.30.

The expressions of the currents from the circuit are:

$$I_{OUT1} = \frac{d^4 I_{REF}^2}{e^3 \left(I_{REF} - \frac{e I_{IN}}{d} \right)} \tag{7.231}$$

and

$$I_{OUT2} = \left(c - \frac{d^2}{e} \right) \frac{I_{IN}^2}{I_{REF}}. \tag{7.232}$$

So, the output current of the circuit from Fig. 7.30 can be expressed as follows:

$$I_{OUT} = I_{OUT1} + I_{OUT2} + \left(b - \frac{d^3}{e^2} \right) I_{IN} + \left(a - \frac{d^4}{e^3} \right) I_{REF}. \tag{7.233}$$

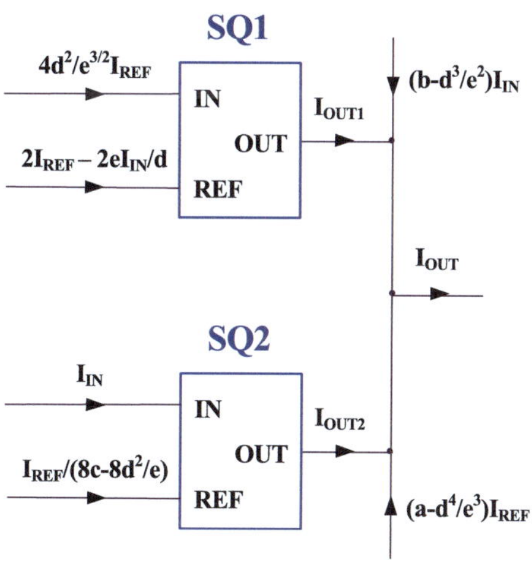

Fig. 7.30 Block diagram of the second analog function synthesizer for generating continuous mathematical functions $f(x)$

Using the definition of the x variable, the previous relation becomes:

$$I_{\mathrm{OUT}} = I_{\mathrm{REF}}\left[\frac{d^4}{e^3}\frac{1}{1-\frac{ex}{d}} + \left(c - \frac{d^2}{e}\right)x^2 + \left(b - \frac{d^3}{e^2}\right)x + \left(a - \frac{d^4}{e^3}\right)\right]. \quad (7.234)$$

In conclusion, the analog function synthesizer shown in Fig. 7.30 approximates continuous mathematical functions $f(x)$:

$$I_{\mathrm{OUT}} = I_{\mathrm{REF}} a_{30}(x) \cong I_{\mathrm{REF}} f(x). \quad (7.235)$$

7.6.3 Third Analog Function Synthesizer for Generating the Continuous Mathematical Functions

The third analog function synthesizer, designed for generating continuous mathematical functions $f(x)$, uses $a_{31}(x)$ approximation function described in Chap. 2.

The block diagram of the third analog function synthesizer for generating continuous mathematical functions $f(x)$ is presented in Fig. 7.31.

The expressions of the currents from the circuit are:

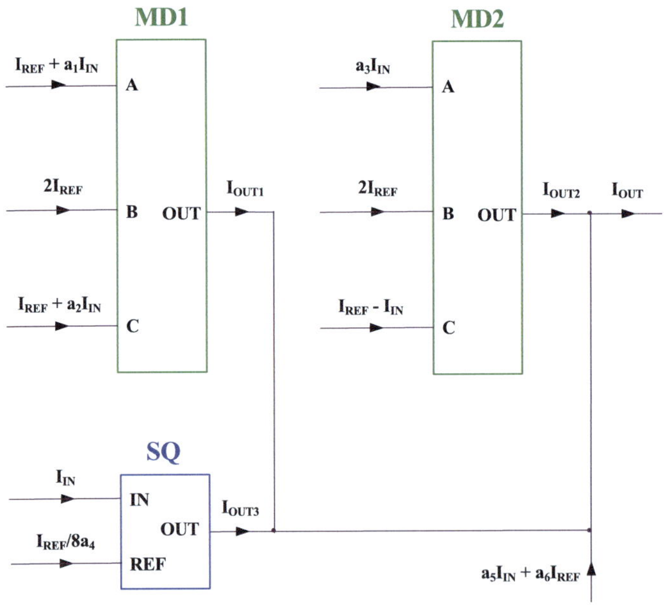

Fig. 7.31 Block diagram of the third analog function synthesizer for generating continuous mathematical functions $f(x)$

$$I_{OUT1} = I_{REF} \frac{I_{REF} + a_1 I_{IN}}{I_{REF} + a_2 I_{IN}}, \tag{7.236}$$

$$I_{OUT2} = \frac{a_3 I_{IN}}{I_{REF} - I_{IN}} \tag{7.237}$$

and

$$I_{OUT3} = \frac{a_4 I_{IN}^2}{I_{REF}}. \tag{7.238}$$

So, the output current of the circuit from Fig. 7.31 can be expressed as follows:

$$I_{OUT} = I_{OUT1} + I_{OUT2} + I_{OUT4} + a_5 I_{IN} + a_6 I_{REF}. \tag{7.239}$$

Using the definition of the x variable, the previous relation becomes:

$$I_{OUT} = I_{REF}\left(\frac{1 + a_1 x}{1 + a_2 x} + \frac{a_3 x}{1 - x} + a_4 x^2 + a_5 x + a_6\right). \tag{7.240}$$

In conclusion, the analog function synthesizer shown in Fig. 7.31 approximates continuous mathematical functions $f(x)$:

$$I_{OUT} = I_{REF} a_{31}(x) \cong I_{REF} f(x). \tag{7.241}$$

References

Popa C (2012) High-accuracy function synthesizer circuit with applications in signal processing. EURASIP J Adv Signal Process 2012:129. https://doi.org/10.1186/1687-6180-2012-129

Popa CR (2023) High-accuracy Gaussian function generator for neural networks. Electronics 12(1):24. https://doi.org/10.3390/electronics12010024

MIX
Papier aus verantwortungsvollen Quellen
Paper from responsible sources
FSC® C105338

If you have any concerns about our products,
you can contact us on
ProductSafety@springernature.com

In case Publisher is established outside the EU,
the EU authorized representative is:
Springer Nature Customer Service Center GmbH
Europaplatz 3, 69115 Heidelberg, Germany

Printed by Libri Plureos GmbH
in Hamburg, Germany